AVIAN BIOLOGY
Volume II

EDITED BY

DONALD S. FARNER
Department of Zoology
University of Washington
Seattle, Washington

JAMES R. KING
Department of Zoology
Washington State University
Pullman, Washington

TAXONOMIC EDITOR

KENNETH C. PARKES
Curator of Birds
Carnegie Museum
Pittsburgh, Pennsylvania

1972

ACADEMIC PRESS New York and London

ACADEMIC PRESS, INC.
111 Fifth Avenue, New York, New York 10003

United Kingdom Edition published by
ACADEMIC PRESS, INC. (LONDON) LTD.
24/28 Oval Road, London NW1

LIBRARY OF CONGRESS CATALOG CARD NUMBER: 79-178216

PRINTED IN THE UNITED STATES OF AMERICA

These volumes are dedicated to the memory of

A. J. "JOCK" MARSHALL

(1911–1967)

whose journey among men was too short by half

CONTENTS

LIST OF CONTRIBUTORS

Numbers in parentheses indicate the pages on which the authors' contributions begin.

DONALD S. FARNER (343), Department of Zoology, University of Washington, Seattle, Washington

HANS FISHER (431), Department of Nutrition, Rutgers—The State University, New Brunswick, New Jersey

ROBERT L. HAZELWOOD (471), Department of Biology, University of Houston, Houston, Texas

KJELL JOHANSEN* (157), Department of Zoology, University of Washington, Seattle, Washington

DAVID R. JONES (157), Department of Zoology, University of British Columbia, Vancouver, British Columbia, Canada

ROBERT C. LASIEWSKI† (287), Department of Zoology, University of California, Los Angeles, California

RALPH S. PALMER (65), New York State Museum, Albany, New York

ROBERT B. PAYNE (103), Museum of Zoology, University of Michigan, Ann Arbor, Michigan

VAUGHAN H. SHOEMAKER (527), Department of Life Sciences, University of California, Riverside, California

PETER STETTENHEIM (1), Plainsfield, New Hampshire

VINZENZ ZISWILER (343), Zoologisches Museum der Universität Zürich Künstlergasse, Zürich, Switzerland

*Present address: Department of Zoophysiology, University of Aarhus, Aarhus, Denmark.

†Deceased.

PREFACE

The birds are the best-known of the large and adaptively diversified classes of animals. About 8600 living species are currently recognized, and it is unlikely that more than a handful of additional species will be discovered. Although much remains to be learned, the available knowledge of the distribution of living species is much more nearly complete than that for any other class of animals. It is noteworthy that the relatively advanced status of our knowledge of birds is attributable to a very substantial degree to a large group of dedicated and skilled amateur ornithologists.

Because of the abundance of basic empirical information on distribution, habitat requirements, life cycles, breeding habits, etc., it has been relatively easier to use birds instead of other animals in the study of the general aspects of ethology, ecology, population biology, evolutionary biology, physiological ecology, and other fields of biology of contemporary interest. Model systems based on birds have played a prominent role in the development of these fields. The function of this multivolume treatise in relation to the place of birds in biological science is therefore envisioned as twofold. We intend to present a reasonable assessment of selected aspects of avian biology for those for whom this field is their primary interest. But we view as equally important the contribution of these volumes to the broader fields of biology in which investigations using birds are of substantial significance.

Only slightly more than a decade has passed since the publication of A. J. Marshall's "Biology and Comparative Physiology of Birds," but progress in most of the fields included in this treatise has made most of the older chapters obsolete. Avian biology has shared in the so-called information explosion. The number of serial publications devoted mainly to avian biology has increased by about 20% per decade since

1940, and the spiral has been amplified by the parallel increase in page production and by the spread of publication into ancillary journals. By 1964, there were about 215 exclusively ornithological journals and about 245 additional serials publishing appreciable amounts of information on avian biology (P. A. Baldwin and D. E. Oehlerts, *Studies in Biological Literature and Communications, No. 4. The Status of Ornithological Literature, 1964.* Biological Abstracts, Inc., Philadelphia, 1964).

These stark numbers reflect only the quantitative acceleration in the output of information in recent time. The qualitative changes have been much more impressive. Avifaunas that were scarcely known except as lists of species a decade ago have become accessible to scientific inquiry as a consequence of improved transportation and facilities in many parts of the world. Improved or new instrumentation has allowed the development of new fields of study and has extended the scope of old ones. Examples that come readily to mind include the use of radar in visualizing migration, of telemetry in studying the physiology of flying birds, and of spectrography in analyzing bird sounds. The development of mathematical modeling, for instance in evolutionary biology and population ecology, has supplied new perspectives for old problems and has created a new arena for the examination of empirical data. All of these developments—social, practical, and theoretical—have profoundly affected many aspects of avian biology in the last decade. It is now time for another inventory of information, hypotheses, and new questions.

Marshall's "Biology and Comparative Physiology of Birds" was the first treatise in the English language that regarded ornithology as consisting of more than anatomy, taxonomy, oology, and life history. This viewpoint was in part a product of the times, but it also reflected Marshall's own holistic philosophy and his understanding that "life history" had come to include the whole spectrum of physiological, demographic, and behavioral adaptation. This treatise is a direct descendent of Marshall's initiative. We have attempted to preserve the view that ornithology belongs to anyone who studies birds, whether it be on the level of molecules, individuals, or populations. To emphasize our intentions we have called the work "Avian Biology."

It has been proclaimed by various oracles that sciences based on taxonomic units (such as insects, birds, or mammals) are obsolete, and that the forefront of biology is process-oriented rather than taxon-oriented. This narrow vision of biology derives from the hyper-specialization that characterizes so much of science today. It fails to notice that lateral synthesis as well as vertical analysis are inseparable partners in the search for biological principles. Avian biologists of both stripes have together contributed a disproportionately large share of the information and thought that have produced contemporary principles in zoogeography, systematics, ethology, demography, comparative physiology, and other fields too numerous to mention.

In part, this progress results from the attributes of birds themselves. They are active and visible during the daytime; they have diversified into virtually all major habitats and modes of life; they are small enough to be studied in useful numbers but not so small that observation is difficult; and, not least, they are esthetically attractive. In short, they are relatively easy to study. For this reason we find gathered beneath the rubric of avian biology an alliance of specialists and generalists who regard birds as the best natural vehicle for the exploration of process and pattern in the biological realm. It is an alliance that seems still to be increasing in vigor and scope.

In the early planning stages we established certain working rules that we have been able to follow with rather uneven success.

1. "Avian Biology" is the conceptual descendent of Marshall's earlier treatise, but is more than simply a revision of it. We have deleted some topics and added or extended others. Conspicuous among the deletions is avian embryology, a field that has expanded and specialized to the extent that a significant review of recent advances would be a treatise in itself.

2. Since we expect the volumes to be useful for reference purposes as well as for instruction of advanced students, we have asked authors to summarize established facts and principles as well as to review recent advances.

3. We have attempted to arrange a balanced account of avian biology as its exists at the beginning of the 1970's. We have not only retained chapters outlining modern concepts of structure and function in birds, as is traditional, but have also encouraged contributions representing a multidisciplinary approach and synthesis of new points of view. Several such chapters appear in this volume.

4. We have attempted to avoid a parochial view of avian biology by seeking diversity among authors with respect to nationality, age, and ornithological heritage. In this search we have benefited by advice from many colleagues to whom we are grateful.

5. As a corollary of the preceding point, we have not intentionally emphasized any single school of thought, nor have we sought to dictate the treatment given to controversial subjects. Our single concession to conceptual uniformity is in taxonomic usage, as explained by Kenneth Parkes in the Note on Taxonomy.

We began our work with a careful plan for a logical topical sequence through all volumes. Only its dim vestiges remain. For a number of reasons we have been obliged to sacrifice logical sequence and have given first priority to the maintenance of general quality, trusting that each reader would supply logical cohesion by selecting chapters that are germane to his individual interests.

DONALD S. FARNER
JAMES R. KING

NOTE ON TAXONOMY

Early in the planning stages of "Avian Biology" it became apparent to the editors that it would be desirable to have the manuscript read by a taxonomist, whose responsibility it would be to monitor uniformity of usage in classification and nomenclature. Other multiauthored compendia have been criticized by reviewers for use of obsolete scientific names and for lack of concordance from chapter to chapter. As neither of the editors is a taxonomist, they invited me to perform this service.

A brief discussion of the ground rules that we have tried to follow is in order. Insofar as possible, the classification of birds down to the family level follows that presented by Dr. Storer in Chapter 1, Volume I.

Within each chapter, the first mention of a species of wild bird includes both the scientific name and an English name, or the scientific name alone. If the same species is mentioned by English name later in the same chapter, the scientific name is usually omitted. Scientific names are also usually omitted for domesticated or laboratory birds. The reader may make the assumption throughout the treatise that, unless otherwise indicated, the following statements apply:

1. "The duck" or "domestic duck" refers to domesticated forms of *Anas platyrhynchos*.
2. "The goose" or "domestic goose" refers to domesticated forms of *Anser anser*.
3. "The pigeon" or "domesticated pigeon" or "homing pigeon" refers to domesticated forms of *Columba livia*.
4. "The turkey" or "domestic turkey" refers to domesticated forms of *Meleagris gallopavo*.

5. "The chicken" or "domestic fowl" refers to domesticated forms of *Gallus gallus;* these are often collectively called *"Gallus domesticus"* in biological literature.

6. "Japanese Quail" refers to laboratory strains of the genus *Coturnix,* the exact taxonomic status of which is uncertain. See Moreau and Wayre, *Ardea* **56,** 209–227, 1968.

7. "Canary" or "domesticated canary" refers to domesticated forms of *Serinus canarius.*

8. "Guinea Fowl" or "Guinea Hen" refers to domesticated forms of *Numida meleagris.*

9. "Ring Dove" refers to domesticated and laboratory strains of the genus *Streptopelia,* often and incorrectly given specific status as *S. "risoria."* Now thought to have descended from the African Collared Dove (*S. roseogrisea*), the Ring Dove of today *may* possibly be derived in part from *S. decaocto* of Eurasia; at the time of publication of Volume 3 of Peters' "Check-list of Birds of the World" (p. 92, 1937), *S. decaocto* was thought to be the direct ancestor of *"risoria."* See Goodwin, *Pigeons and Doves of the World* **129,** 1967.

As mentioned above, an effort has been made to achieve uniformity of usage, both of scientific and English names. In general, the scientific names are those used by the Peters "Check-list"; exceptions include those orders and families covered in the earliest volumes for which more recent classifications have become widely accepted (principally Anatidae, Falconiformes, and Scolopacidae). For those families not yet covered by the Peters' list, I have relied on several standard references. For the New World I have used principally Meyer de Schauensee's "The Species of Birds of South America and Their Distribution" (1966), supplemented by Eisenmann's "The Species of Middle American Birds" (*Trans. Linnaean Soc. New York* **7,** 1955). For Eurasia I have used principally Vaurie's "The Birds of the Palaearctic Fauna" (1959, 1965) and Ripley's "A Synopsis of the Birds of India and Pakistan" (1961). There is so much disagreement as to classification and nomenclature in recent checklists and handbooks of African birds that I have sometimes had to use my best judgment and to make an arbitrary choice. For names of birds confined to Australia, New Zealand, and other areas not covered by references cited above, I have been guided by recent regional check-lists and by general usage in recent literature. English names have been standardized in the same way, using many of the same reference works. In both the United States and Great Britain, the limited size of the avifauna has given rise to some rather provincial English names; I have added appropriate (and often previously used) adjectives to these. Thus *Sturnus vulgaris* is "European Starling," not simply "Starling"; *Cardinalis cardinalis* is "North American Cardinal," not simply "Cardinal"; and *Ardea cinerea* is "Gray Heron," not simply "Heron."

Reliance on a standard reference, in this case Peters, has meant that certain species appear under scientific names quite different from those used in most of the ornithological literature. For example, the Zebra Finch, widely used as a laboratory species, was long known as *Taeniopygia castanotis*. In volume 14 of the Peters' "Check-list" (pp. 357–358, 1968), *Taeniopygia* is considered a subgenus of *Poephila*, and *castanotis* a subspecies of *P. guttata*. Thus the species name of the Zebra Finch becomes *Poephila guttata*. In such cases, the more familiar name will usually be given parenthetically.

For the sake of consistency, scientific names used in Volume I will be used throughout "Avian Biology," even though these may differ from names used in standard reference works that would normally be followed, but which were published after the editing of Volume I had been completed.

Strict adherence to standard references also means that some birds will appear under names that, for either taxonomic or nomenclatorial reasons, would *not* be those chosen by either the chapter author or the taxonomic editor. As a taxonomist, I naturally hold some opinions that differ from those of the authors of the Peters' list and the other reference works used. I feel strongly, however, that a general text such as "Avian Biology" should not be used as a vehicle for taxonomic or nomenclatorial innovation, or for the furtherance of my personal opinions. I therefore apologize to those authors in whose chapters names have been altered for the sake of uniformity, and offer as solace the fact that I have had my objectivity strained several times by having to use names that do not reflect my own taxonomic judgment.

KENNETH C. PARKES

CONTENTS OF OTHER VOLUMES

OBITUARY

Robert C. Lasiewski (1937–1971)

Robert C. Lasiewski, the author of Chapter 5, died on August 2, 1971, in Gainesville, Florida, where he was chairman of the Department of Zoology at the University of Florida. At the time of his death he was 34 years old.

As an undergraduate of Cornell University, Lasiewski's primary interests were in ecology. During his graduate years at the University of Michigan his interests became more and more physiologically oriented. After completing work for his doctorate on temperature regulation and energetics of hummingbirds under the supervision of Dr. William R. Dawson, he joined the faculty of the University of California at Los Angeles, where he remained until shortly before his death. His research interests were centered on the general problem of avian energetics. The topics to which Lasiewski devoted his most persistent attention were temperature regulation and dormancy, oxygen consumption in relation to temperature and body size, pulmocutaneous water loss with particular reference to high ambient temperatures, and, most recently, the establishment of allometric equations for the prediction of rates of avian physiological function over a wide range of body sizes. Most of Lasiewski's research was necessarily based on laboratory experimentation, but he had a keen interest in taking physiological investigations into the field. He participated in expeditions to South Africa, Madagascar and the Comorros, and New Guinea. Most of his research was devoted to birds, but his publications also include studies of crustaceans, reptiles, and mammals. In addition to his professional interests, Lasiewski found time to follow two of his principal avocations, deep sea fishing and skin diving. His death at a time in which his scientific powers were fully maturing is a substantial loss to avian environmental physiology.

Chapter 1

THE INTEGUMENT OF BIRDS

Peter Stettenheim

I.	Skin	2
	A. Gross Morphology and Functions	2
	B. Embryology	2
	C. General Histology	3
II.	Feathers	9
	A. Structure of Basic Types	9
	B. Embryonic Development of Feather and Follicles	17
	C. Structure of a Feather Follicle	20
	D. Regeneration of Feathers	24
	E. Chemical Composition of Feathers	28
	F. Color and Pattern	29
III.	Pterylosis	31
	A. Fundamentals	31
	B. Natal Pterylosis	32
	C. Adult Pterylosis	32
	D. Functional Significance of Pterylosis	36
	E. Amount of Feathers in Relation to Body Size	37
	F. Value of Pterylosis for Taxonomy	37
IV.	Muscles of the Skin	38
	A. Adjustment of Feather Position	38
	B. Subcutaneous Muscles	38
	C. Intrinsic Muscles	39
	D. Skin Muscles and Feather Movements	44
V.	Integumentary Derivatives Other Than Feathers	44
	A. Rhamphotheca	44
	B. Scales and Claws	47
	C. Glands of the Integument	50
References		54

I. Skin

A. Gross Morphology and Functions

The skin is the covering or integument of the body. It is notable in birds for its production of feathers and a variety of special external structures (e.g., horny covering of the beak, wattles, comb, scaly covering of legs and feet). The skin of birds is elastic, often translucent, and thinner than in mammals of equal size. It is attached to the body muscles at relatively few places. On the other hand, and in contrast with the conditions in other vertebrates, the skin of a bird has extensive attachments to the skeleton—on the skull and beak, the bones of the wrists and wing tips, the dorsal side of the pelvis, the tarsometatarsus and toes, and often on a distal portion of the tibiotarsus. Air chambers or inflatable air pouches, sometimes of great size, are present beneath the skin of the throat or neck in certain birds (Mayaud, 1950).

The general functions of the skin are much the same in birds as in other vertebrates. The skin is primarily a protective envelope for the body, keeping out injurious substances and organisms, and retaining vital fluids and gases. It also provides thermal insulation, but comparison of the thickness of the skin in birds and mammals shows that feathers are more effective for this purpose than are skin and subdermal fat. The skin also serves as a large sensory organ, having receptors for temperature, pressure, and vibration. Sebaceous or lipoid secretions are produced by glands at the base of the tail and in the external ear, and by the general skin itself (Section V,C,4).

B. Embryology

In birds, as in other vertebrates, the skin consists of two main layers, epidermis above and dermis (corium, cutis) below. Epidermis forms from ectoderm. By developmental stage 11 [40–45 hours (Hamburger and Hamilton, 1951)] in a chicken embryo, the original single layer of cells has begun to separate into an outer squamous layer, the periderm (epitrichium), and an inner layer of cuboidal cells (H. L. Hamilton, 1952, p. 546). The periderm later thickens and cornifies, but it is sloughed off at or before hatching. The inner cells gradually become columnar, and at about stage 39 (13 days) they start to multiply rapidly, forming a stratified squamous epithelium (McLoughlin, 1961). These cells become vacuolated, flattened, and keratinized as they are pushed to the surface, with the result that an outer corneous layer is formed. The process of epidermal cellular proliferation, differentiation, and sloughing off the surface of the skin continues throughout the life of a bird.

Dermis, of mesodermal origin, forms from mesenchyme that is in contact with skin ectoderm. The mesenchyme differentiates into connective tissues and muscles, starting at about stage 38 (12 days) in the chick. Blood vessels and nerves develop from other components of the germ layers and grow into the dermis.

The skin of a bird differentiates in various ways on different parts of the body. In each region, the particular derivative of the embryonic epidermis is determined at an early, sometimes brief, stage in ontogeny by reaction with the underlying embryonic dermis (Cairns and Saunders, 1954). Further differentiation depends on a series of reciprocal interactions between the dermis and the epidermis (Sengel, 1958, 1964). "The inductive capacity of the dermis and the response of the epidermis vary with developmental stage and with location on the body" (Rawles, 1963, p. 784). The epidermis remains able to respond to specific dermal stimuli for a relatively long time, after which, at least in certain regions, it becomes restricted in competence during a specific, rather brief period.

C. General Histology

The layers of the skin in birds have been variously subdivided and named by anatomists (e.g., Wodzicki, 1928; Lange, 1931). The treatment to be used here follows Lucas and Stettenheim (1972), who reviewed earlier ideas and added their own findings.

1. *Epidermis*

Avian epidermis is generally thin in areas that are covered by feathers, both in places where feathers arise (feather tracts) and in intervening bare areas (apteria). The epidermis is generally thick in exposed, featherless integumentary derivatives, such as the horny coverings of the beak (rhamphotheca) and the feet (podotheca). Especially in its corneous layer, the epidermis has many variations among birds. The fact that there is general similarity between the epidermis of such distantly related birds as a penguin and a chicken (Spearman, 1969) shows, however, that the same histological plan can be used throughout the class.

Avian epidermis is composed of two major layers—a superficial corneous layer (stratum corneum) and a deeper germinative layer (stratum germinativum). The germinative cells are all living and are mitotically producing new cells or becoming cornified. They are divisible into three subordinate layers. The basal layer (stratum basale) of the germinative layer usually consists of a single ply of cuboidal or columnar cells. Cells in this layer are dividing and do not show signs of keratinization.

Above the basal cells are larger polyhedral cells which constitute the intermediate or polyhedral layer (stratum intermedium). The thickness of this layer and the shape of its cells vary widely among birds and in parts of the body. Mitosis continues in this layer, but signs of keratinization and vacuolization appear. Where the layer is several plies thick, the more superficial cells are also somewhat flattened.

The intermediate cells grade outward into cells that are more flattened and vacuolated and that often have prominently staining outer membranes. They have been called transitional cells because they are transitional from the germinative to the corneous cells (Lucas and Stettenheim, 1972). These are not the same as the cells designated as transitional by Cane and Spearman (1967, p. 340), which are more superficial and are regarded by Lucas and Stettenheim as belonging to the deepest part of the corneous layer. There is no mitosis in the transitional layer, but keratinization is advanced.

A granular layer (stratum granulosum) and a clear layer (stratum lucidum), found in the epidermis of mammals, are probably absent from the epidermis of reptiles and birds (Lange, 1931; Spearman, 1966, 1969; Lucas and Stettenheim, 1972).

The corneous or horny layer is usually composed of highly flattened cells that are joined together mainly at their edges, forming thin lamellae. This layer varies widely, but it is generally thinnest on feathered skin and the hinges between scales, while it is thickest on the beak, spurs, claws, metatarsal pads, and digital pads. In most places, the lamellae are loosely arranged, probably an adaptation to reduce weight in flight rather than for flexibility (Spearman, 1966, p. 72). Keratinization is completed in the deepest part of the corneum (transitional layer of Cane and Spearman, 1967, p. 340), and the cells above are dead. Nuclei are often demonstrable with stain in the deeper cells but not in the superficial cells of the corneous layer (Lange, 1931, p. 401; Cane and Spearman, 1967, p. 340; Spearman, 1969, p. 362). Considering its microanatomy, its fluorescence with congo red, and the presence of bound phospholipid and cysteine, the corneous layer of avian body epidermis differs from that of mammals and resembles that on the neck of a tortoise (Spearman, 1966).

The superficial lamellae of the corneous layer are shed at times, either as thin pieces or as definite structures. This process probably takes place at about the same time as the molting of feathers (Spearman, 1966, p. 73). The superficial lamellae in the ordinary skin of the body and the linings of the feather follicles are shed in pigeons (*Columba livia*) as large thin pieces (Voitkevich, 1966, p. 56). This process may also occur in penguins (Spearman, 1969, p. 369), though it is not to be confused here with the molting of the feathers in sheets, owing to interlocking of the barbules.

Periodic replacement of definite epidermal structures other than feathers is known in a number of birds (reviewed by Mayaud, 1950; Harrison, 1964). Some of these structures are secondary sexual characters that are developed for each breeding season and shed afterward. Other epidermal structures are lost annually but have no apparent secondary sexual role. Finally, there are structures on the wings or feet that function only in nestling birds and that are lost about the time of leaving the nest.

In certain cases, corneous structures in birds appear to be lost because of the formation of a mechanically weak fission plane beneath them (Spearman, 1966). Such a layer occurs in the epidermis of lizards and snakes but not in that of tortoises, crocodiles, and mammals. A similar layer has not been found, however, in epidermis of chicks other than that of the tarsus, in the tarsal scales of adult fowls, or in the skin of a penguin (Cane and Spearman, 1967, p. 347; Spearman, 1969, p. 369). The absence of a morphologically distinct fission layer in these cases suggests that in birds, as in turtles, crocodiles, and mammals, the corneous layer is commonly lost in tiny flakes. Such dandruff-like particles can be seen on the surface of the skin of feather tracts and apteria. Abrasion causes the exfoliation of flakes on the surface of hard keratinous structures such as the rhamphotheca, scales, spurs, and claws; evidence of this is the irregular pattern of faint layering. Wear is in fact necessary to keep these continuously growing structures at their proper size.

Keratin is produced to a limited extent in the epidermis of certain bony fishes and amphibians, but it is the chief secretory product in the epidermis of reptiles, birds, and mammals. Indeed, "the epidermis has been likened to a holocrine gland producing dead keratin-filled cells" (Cane and Spearman, 1967, p. 338). Keratin is a fibrous protein characterized by its resistance to attack by proteolytic enzymes. Its durability is chiefly a result of close cross-bonding between keratin polypeptide chains, the most important of which is probably the disulfide bond of cystine (Spearman, 1966, p. 59). Three types of keratin have been found in vertebrates, of which the beta type ("feather keratin") occurs in reptiles and birds. The structure of the feather keratin molecule has received much study (reviewed by Dickerson, 1964), yet it is still in doubt.

Keratinization is a specific form of cell differentiation in which metabolically highly active cells undergo changes in their morphology and physiology while they reach the terminal stage and become filled with keratin (Matoltsy, 1962, p. 4). Two basic processes take place in keratinization—synthesis of keratin protein and breakdown of other cellular constituents by hydrolytic enzymes in the epidermal cells. The type of horny cell formed in a given case probably depends on which process occurs first (Spearman, 1966, p. 60).

Keratin synthesis has been described in the rhamphotheca and the feathers of domestic chick embryos (Bell and Thathachari, 1963; Fell, 1964). The process in birds resembles that in mammals (Matoltsy, 1962; Jarrett and Spearman, 1967) but differs in the orientation and size of the microfibrils (tonofibrils), and the absence of keratohyalin. A further difference is that in feathers, keratin protein synthesis, polymerization, and the initial stages of keratin bonding appear to take place simultaneously, whereas in hair these steps occur sequentially (Cane and Spearman, 1967, p. 349).

2. *Dermo-epidermal Interface*

a. Basement Membrane. A basement membrane is present between the epidermis and the dermis in birds as in other animals. As seen in domestic fowls of various ages, this layer is thin in the skin of the apteria, wattles, ear lobe, rhamphotheca, and oil gland, but often as thick as a small bundle of connective tissue in the skin of the feather tracts (Lucas and Stettenheim, 1972). The membrane follows closely the deep ends of the basal epidermal cells, dipping upward between them. The basement membrane is thought to serve for mechanical attachment of the epidermis to the dermis. It may also function in the transfer of materials from the dermis to the epidermis in the embryo (Kischer, 1963, p. 319).

b. Conformation. Where the histology of avian body skin has been investigated, the dermo-epidermal interface has been found as in mammals. Low wrinkles in the interface occur in feathered and unfeathered body skin of fowl (Cane and Spearman, 1967; Lucas and Stettenheim, 1972). In some places, such as the sternal apterium, these folds may conform to folds on the surface of the skin. Sharp ridges of upward-directed dermal cells are said to be characteristic of the feathered skin of ratites, especially the ostrich (*Struthio camelus,* Lange, 1931, pp. 429–430). A folded dermo-epidermal interface, representing papillae or ridges and grooves, has been reported in the rhamphotheca, rictus, ventral plate of the claws, and the metatarsal spurs of domestic fowl and in the cere of the domestic pigeon (Lucas and Stettenheim, 1972).

3. *Dermis*

The dermis is composed chiefly of connective tissues. Its layers are more difficult to identify than those of the epidermis because they vary so greatly among species, parts of the body, and individuals of different ages. The account that follows is based on conditions in the domestic fowl, the most thoroughly studied species (Böhm, 1962, 1964; Lucas and Stettenheim, 1972). The dermis in other birds has been investigated by Greschik (1916) and Lange (1931).

The superficial layer of the dermis (stratum superficiale dermi) consists of collagenic connective tissue that ranges from dense to loose. Blood vessels are present and sometimes abundant. Böhm (1962) found capillaries and larger vessels, and regarded them as constituting a distinct vascular layer (stratum vasculosum). Lucas and Stettenheim (1972), on the contrary, found only capillaries and regarded them as the bottom component of the superficial layer. In domestic fowl, the distinctness and thickness of this layer vary with age and the part of the body.

The deep layer of the dermis (stratum profundum dermi) has two principal components. Uppermost is the layer of compact connective tissue (stratum compactum dermi), a dense feltwork of collagenic fiber bundles and elastic fibers. This layer is usually denser than the superficial layer. In the skin of the apteria, as in the skin of reptiles, the collagenic bundles are arranged chiefly in horizontal layers penetrated by perpendicular bundles. The arrangement is less regular in the skin of the feather tracts, particularly in carinate birds (Lange, 1931, p. 422). Capillaries and larger vessels occur in the compact dermis.

The lower part of the deep dermis is a loose layer of connective tissues (stratum laxum dermi) (Lucas and Stettenheim, 1972). This is not the same as the layer termed "stratum laxum" by Lange (1931), which corresponds to the superficial dermis of Lucas and Stettenheim. In addition to the loose collagenic fibers, there may be present fat, capillaries and larger blood vessels, elastic fibers, smooth muscles, the bases of contour feather follicles, nerves, and sensory corpuscles.

Fat is usually present in the dermis of the feather tracts, but sparse or absent in that of the apteria. The fat cells are of the plurivacuolar and univacuolar types, as well as intermediate forms. Plurivacuolar cells develop into univacuolar cells (Clara, 1929; Eastlick and Wortham, 1947). These two types of avian dermal fat may be equivalent to the brown and the white fat of mammals, although the brown fat is not now considered to be a precursor of the white (Bloom and Fawcett, 1968, p. 165). The fat cells are larger in the feather tracts than in the apteria (Böhm, 1963).

The smooth muscles and their elastic tendons constitute a subordinate layer of the loose dermis known as the "musculo-elastic layer (stratum musculo-elasticum)." These muscles will be discussed in Section IV,C.

At the bottom of the dermis is sometimes a very thin layer of elastic fibers, the elastic lamina of the dermis (lamina elastica dermi). It is a horizontal lattice composed of one or more levels of fine, straight fibers, running generally parallel to each other in each level (Wodzicki, 1928).

4. *Subcutis*

The subcutis (subdermis, hypodermis, superficial fascia, tela subcutanea) is composed of loose collagenic fibers, often in horizontal sheets. Dense collagenic tissue is sometimes present. Blood vessels and nerves pass through the subcutis. Skeletal muscles are usually situated below this layer, but the subcutaneous muscles (Section IV,B) pass through it. In places where there is no subcutis, the dermis (or elastic lamina) rests on the skeletal muscles, (e.g., in the lateral caudal apterium), or on the periosteum of a bone (e.g., in tip of beak or claw).

Fat is an important part of the subcutis, occurring both as a layer and as discrete bodies that lie deeper, bound by fascia to the underlying muscles. The layer and the bodies together may be known as the panniculus adiposus, and probably correspond to the panniculus adiposus of mammals (Wodzicki, 1928). In domestic fowl, a subcutaneous layer of fat is well developed in the skin of the feather tracts, comb, wattles, earlobes, and claws; it is sparsely developed or absent in the apteria (Lucas and Stettenheim, 1972). The fat bodies occur in definite locations in a number of species of birds. There are sixteen such bodies in domestic fowl (Liebelt and Eastlick, 1954), and fifteen in the White-crowned Sparrow (*Zonotrichia leucophrys gambelii,* McGreal and Farner, 1956). Judging from their origin, growth, structure, and function, Liebelt and Eastlick stated that fat bodies should be regarded as organs rather than as simple tissues. The lipid content of these organs in domestic chick embryos has been reported by Feldman *et al.* (1962).

5. *Deep Fascia*

The deep fascia is properly a connective tissue but it is included here because it is commonly seen in microscopic sections of integument. It is usually a single dense layer of collagenous bundles and elastic fibers arranged in sheets. Deep fascia can be found overlying either skeletal muscle or bone.

6. *Summary of Histology*

The following outline, taken from that of Lucas and Stettenheim (1972), summarizes the layers that may be found in avian integument:

- Epidermis
  - Corneous layer
  - Germinative layer
    - Transitional layer
    - Intermediate layer (spinous-cell layer)
    - Basal layer
- Basement membrane
- Dermis
  - Superficial layer of dermis
  - Deep layer of dermis
    - Compact layer
    - Loose layer
      - Musculo-elastic layer
  - Elastic lamina
- Subcutis (superficial fascia)
  - Superficial layer of subcutis[1]
  - Deep layer of subcutis[1]
- Deep fascia

II. Feathers

A. Structure of Basic Types

Feathers, the characteristic skin covering of birds, are probably the most complex derivatives of the integument to be found in any vertebrates. Five main structural categories of feathers can be recognized, though there are intergrades between most of them. The classification of feathers and terminology of feather parts as used here mostly follow Chandler (1916).

1. *Contour Feathers*

a. Typical Body Feathers. The structure of feathers is most easily reviewed by examining the most familiar type of feathers, those that form the covering over most of the body of a bird. The base of the feather shaft, the calamus, is short, tubular, and implanted in a socket, the feather follicle. Beyond the calamus is a long, tapered section, the rachis, which bears branches, the barbs, on its sides. The division between the calamus and the rachis is marked by the lowermost barbs

[1]Panniculus adiposus may be present in one or both layers.

and by a pit on the underside of the shaft, the superior umbilicus. At the bottom of the calamus is an opening, the inferior umbilicus, through which the pulp entered a feather while it was growing. The rachis is composed of an outer thin, solid layer, the cortex, and an inner, thick, spongy core, the medulla or pith. Internal ridges of cortex and septa in the medulla stiffen the rachis against dorsoventral movement, yet allow sidewise flexibility (Rutschke, 1966a).

The barbs on each side of the rachis constitute a sheet known as a vane. The vanes of body feathers are symmetrical or nearly so; asymmetry is more pronounced in the feathers of the wings and all but the central tail feathers. The portion of the vanes at the proximal end of a feather has downy (plumulaceous) texture—soft, loose, and often fluffy. It is this part of contour feathers, as well as feathers that are wholly downy, that gives the feather covering of a bird its properties of thermal insulation and filling out the body contours. The distal portion of the vanes has a pennaceous texture—firm, compact, and closely knit. This, the exposed part of a feather, provides an airfoil, protects the downy parts of feathers, repels moisture, and has a visual function.

A barb consists of an axis, the ramus, and many closely spaced branches, the barbules, or radii (singular, radius). The ramus is a compressed filament that tapers in height from base to tip. Ramal cortex is usually uniform but is sometimes divided into a superficial layer of flattened cells (so-called "cuticle") and a deeper layer of thicker cells, the cortex proper (Strong, 1902; Auber and Appleyard, 1951; Olson, 1970). Three categories of medullary cells in the rami are known, depending on the size of their vacuoles; the smallest of these are responsible for most cases of blue color in feathers (Auber, 1957).

A barbule is essentially a stalk of single cells that are serially differentiated to some degree. It has a base of compressed, fused cells and a distal segment, the pennulum, of cylindrical, jointed cells. In a downy (plumulaceous) barbule, the base is short and straplike, and the pennulum is long and simple, resembling a stalk of bamboo. The distal ends of the pennulum cells are variously swollen or furnished with tiny prongs. These shapes are often characteristic of orders of birds, and hence can be used for identifying isolated feathers (Day, 1966).

Pennaceous barbules are often highly differentiated and bear diverse outgrowths, collectively termed barbicels. Proximal barbules, those facing the proximal end of a feather, have a pronounced flange along the dorsal edge of the base, and barbicels that are small, simple, or absent. Distal barbules, those facing the distal end of a feather, have little or no dorsal flange, but have various, well-developed barbicels. Distal barbules of each barb cross over proximal barbules of the next higher barb in the vane. Hooklet-shaped barbicels of the distal barbules grasp the dorsal flanges of the proximal barbules. Other parts of the barbules aid in making a flexible, self-adjusting interlocking mechanism (Sick, 1937). Since functional demands imposed on pennaceous barbules are variable, the barbules themselves vary within and among feathers.

Barbules of a third category are smaller and simpler than those already described. They show slight differentiation among the cells, faint markings or none at the ends of the cells, and have few or no outgrowths. Simplified pennaceous and simplified plumulaceous barbules have very reduced barbicels or nodes, respectively. Barbules with even less differentiation are known as "stylet barbules" (German, *Spiessradien*) in reference to their stiffness and slender pointed shape (Hempel, 1931). Stylet barbules are thought to resemble the primitive type of barbule from which both pennaceous and plumulaceous types evolved (Becker, 1959, p. 517).

Analysis of the structure of barbules and the proportions of feathers in juvenal and adult plumages of Ring-necked Pheasants (*Phasianus colchicus*) suggests that primitive feathers may have been contour feathers with a relatively large pennaceous portion and simple pennaceous barbules (Buri, 1967).

Feathers often bear outgrowths on the underside, attached to the rim of the superior umbilicus. There may be a fringe or a cluster of barbs, an auxiliary shaft with barbs on each side, or an intermediate structure. All of these should be regarded as forms of afterfeathers (Hempel, 1931; Ziswiler, 1962; Lucas and Stettenheim, 1972). The commonly used term "aftershaft" should be applied only to an auxiliary shaft, when such is present. The texture of an afterfeather, whatever its structure, is entirely downy except in cassowaries and emus, where it is coarse and lax. The barbules are somewhat smaller and simpler than the downy barbules on the main feather.

Afterfeathers vary in different parts of a bird's body, being relatively largest on feathers that are wholly or largely downy, and smallest or absent on the remiges and rectrices (Hempel, 1931; Ziswiler, 1962). Several forms of afterfeather have been found among birds. Some ornithologists (e.g., Van Tyne and Berger, 1959, p. 78) have regarded afterfeathers as useful for taxonomy, but others (Ziswiler, 1962, p. 305) disagree. The diversity in the presence and structure of afterfeathers suggests that they have evolved independently several times in birds.

The theory that feathers evolved from reptilian scales is supported by the recognition of homology between the two kinds of structures. As explained by Steiner (1917) and confirmed by Ziswiler (1962), the main part of a feather is homologous with the dorsal part of a scale and the afterfeather is homologous with the ventral half. The underside of the vanes does not correspond to the underside of a scale but to the innermost layer of its epidermis.

An important function of body contour feathers, especially in water birds, is that of shielding a bird against water. The water repellancy of feathers is determined mainly by their structure, particularly the diameter and spacing of the rami and barbules (Rutschke, 1960; Rijke, 1970). A film of uropygial gland secretion is also necessary (Section V,C,1,e).

b. Remiges and Rectrices. The flight feathers of the wings and tail—the remiges and the rectrices—are a type of contour feathers. They are large, stiff, and mostly pennaceous. These feathers are chiefly modified for their roles in flight. Such adaptations are both gross and microscopic, and are most pronounced in the remiges on the manus (primary remiges), the feathers that produce propulsive force. In many birds, the distal portions of certain primaries and rectrices are distinctly narrowed (notched, incised, emarginate). When the wings are spread, the tips of these feathers consequently do not overlap, but are separated by gaps or slots. This reduces the induced drag of the wing and allows the feather tips to twist and act as individual propeller blades (Graham, 1930; Cone, 1962).

The barbules of flight feathers are of the pennaceous type except at the lower end of the vanes, where short, simple downy barbules may occur. The pennaceous barbules are generally larger and have more highly developed barbicels than those of body contour feathers. They are closely adapted to the operation of the feathers, and hence vary considerably within a vane. A common adaptation are "friction barbules" (German, *Reibungsradien*), which are distal barbules modified by lobate dorsal barbicels (Sick, 1937, p. 213). Friction barbules occur in those portions of flight feathers that overlap when a wing or tail is spread. By rubbing against the rami of overlapping vanes, they supplement the action of muscles and ligaments in keeping the feathers from slipping too far apart (Graham, 1930; Oehme, 1963). Friction barbules and other flight adaptations of the outer vanes of remiges have been investigated by Rutschke (1965).

While the remiges and rectrices are adapted principally for locomotion, in some birds they have become modified for display (Brinckmann, 1958; Fry, 1969) or sound production (Heinroth and Heinroth, 1958; Koenig, 1962; W. J. Hamilton, 1965).

c. Ear Coverts. Surrounding the external opening of the ear in most birds are one to four rows of small, modified contour feathers known as ear coverts (auricular feathers, opercular feathers). The vanes of these feathers have open pennaceous texture owing to wide spacing between moderately stiff barbs and the small, simple character of the barbules. The coverts anterior to the ear project posteriorly across the opening, and in many birds their tips are propped up by the coverts posterior to the ear. Ear coverts appear to screen the external ear opening and to improve hearing ability (Ilychev and Izvekova, 1961; Il'ichev, 1962). These feathers are well developed in nocturnal owls, owlet-frogmouths, and certain parrots and hawks (Il'ichev, 1961).

2. *Semiplumes*

Semiplumes combine a large rachis with entirely plumulaceous vanes. They grade into contour feathers on the one hand, and into down feathers on the other. Semiplumes can be separated from down feathers by measuring the lengths of the rachis and the longest barb—the rachis of a semiplume is longer than the longest barb (Lucas and Stettenheim, 1972). The barbules are of the plumulaceous type, and in a given bird resemble those in the fluffy basal portions of body contour feathers. An afterfeather is present if it also occurs on the body feathers. Semiplumes can be found along the margins of contour feather tracts and in tracts by themselves. Mostly hidden beneath the contour feathers, they provide thermal insulation and fill out the contours of the body.

3. *Down Feathers*

Down feathers (downs, plumules) are wholly fluffy feathers in which the rachis is either shorter than the longest barb or absent.

a. Natal Downs. Natal downs (nestling downs, neossoptiles) are present on many birds at hatching or within a few days. Their barbules are of a plumulaceous type, but are smaller than those of adult feathers and lack characteristic processes at the nodes. The tips of the terminal barbs lack barbules, apparently a good mark for distinguishing between natal downs and unworn downs of later plumages (Schaub, 1912; Parkes, 1965).

Most natal downs precede contour feathers, semiplumes, or down feathers in apteria. They are generally alike on all parts of an individual, yet they differ widely among groups of birds. At one extreme, as in ducks, they have a calamus and a moderately long rachis bearing barbs. At the other extreme, as in shorebirds and passerines, the calamus is vestigial or absent and the natal down barbs continue into those of the next feather (Jones, 1907; Becker, 1959). Probably in all cases, natal downs can be regarded as a distinct feather generation. Feathers of each generation are joined to their successors (Watson, 1963), yet there is probably always a cessation of growth between the completion of one feather and the start of the next.

Birds with adult down feathers amidst the contour feathers, such as ducks and owls, also have a second type of natal downs preceding the adult downs. Ducklings even have a third type of natal downs preceding their filoplumes (Ewart, 1921).

Intermediate feathers are produced between the natal downs and the juvenal contour feathers in certain birds. They are a second set of downs in penguins and Barn Owls (*Tyto alba*), semiplumes in rheas, and abortive feathers in ducks. Lucas and Stettenheim (1972) suggested that all these feathers be classed as "mesoptiles," not just those with downy structure (A. L. Thomson, 1964, p. 455).

b. Definitive Downs. Down feathers that appear on the body after the natal downs or the mesoptiles are known as "definitive downs" (plumules). These feathers always have a true calamus. The rachis is sometimes extremely short and sharply tapered, scarcely more than the fused proximal ends of rami. Often it is somewhat longer and may approach the length of the longest barbs; in this way, down feathers grade into semiplumes. In a given bird, the barbs of the down feathers resemble those in the downy portion of contour feathers, but tend to be longer and have longer, more closely spaced barbules. An after-feather is commonly present, particularly in birds where this structure is present on the body contour feathers.

c. Uropygial Gland Downs. Most birds have a sebaceous gland, the uropygial gland, at the base of the tail (Section V,C,1). The papilla of the gland commonly bears a tuft of modified down feathers at or near its tip. The nature and arrangement of these feathers in many birds were reported by Paris (1913) and details from three domestic species were given by Lucas and Stettenheim (1972). The tuft aids in transferring the oily secretion from the gland to the bill. Oil probably enters the tuft as a result of pressure due to secretion by the glandular cells, pressure of the bird's bill, action of the muscles in the papilla, and capillary action in the feathers.

d. Powder Feathers. Modified feathers of many birds shed an extremely fine, white powder, and hence are known as "powder feathers." They will be discussed here because most often they have the structure of downs, but in some birds, such as pigeons, powder is also produced by special semiplumes and downy portions of contour feathers. The distribution, structure, and growth of powder feathers have been discussed by Schüz (1927).

In different birds, powder feathers may be found among the ordinary downs and contour feathers, in distinct patches by themselves, or both in patches and mingled with the ordinary feathers. The powder feathers sometimes grade into the ordinary feathers, making it difficult to determine exactly their distribution.

The powder is composed of granules of keratin approximately one micron in diameter. It is derived from modified cells on the surface and in the middle of each ridge of barb-forming tissue in a growing feather. At least in pigeons, it does not contain fragments of the barbs or the feather sheath. The powder is nonwettable and therefore is generally regarded as a waterproof dressing for the ordinary contour feathers. It also causes the inherent colors of these feathers to appear paler or dimmer than otherwise.

The structure of powder feathers in different birds tends to be modified in proportion to the amount of powder produced. The powder feathers of ordinary domestic pigeons generally shed small to moderate amounts of powder, and their structure is like that of ordinary downs, semiplumes, and contour feathers. They are molted more often than the ordinary feathers (Stettenheim and Fay, 1968). In herons and bitterns, on the other hand, copious amounts of powder are shed and the powder downs are highly modified, having vestigial rami, no rachis, and abortive barbules or none. These feathers maintain the powder supply by growing continuously.

Highly modified powder downs in certain varieties of domestic pigeons produce a fatty substance instead of powder (Eiselen, 1939). Similar feathers appear to be the source of a creamy secretion reported by Abdulali (1966) and Berthold (1967) in the Pied Imperial-Pigeon, *Ducula bicolor*. The resinous secretion on the back of certain woodpeckers (*Hemicircus* spp.) may also be derived from such feathers (Bock and Short, 1971).

4. *Bristles*

Bristles in birds are generally feathers with a stiff rachis and barbs only on the proximal portion, if at all. They usually grade structurally into nearby contour feathers and replace them in the pterylosis. Bristles occur most commonly around the base of the bill, around the eyes, and as eyelashes, but in various birds they can be found elsewhere on the head and, rarely (as in certain owls), on the toes.

In nearly all bristles the distal part of the rachis is very dark brown or black, owing to a dense concentration of a melanin pigment. Melanin not only colors feathers but also makes them firmer and more resistant to wear and photochemical changes.

The so-called "bristles" that make up the beard of a turkey are not feathers but solid horny fibers that grow from the surface of the epidermis (Boas, 1931; Lucas and Stettenheim, 1972).

Feathers that are intermediate between contour feathers and bristles occur on many birds. These have been termed "semibristles" by Lucas and Stettenheim (1972). The rachis of such feathers is often dark, its terminal barbs are far apart, and barbules are either very small or absent on most barbs. Semibristles can be found, for example, covering the nostrils of corvids, on the rictus and lores of owlet-frogmouths, on the lores and eyelids of some owls (such as *Bubo* spp.), and caudal to the eye in hawks.

Bristles and semibristles that cover the nostrils appear to screen out foreign particles. Large eyelashes may likewise protect the eyes and give a striking visual effect, especially if the skin around the eyes is bare and brightly colored. Bristles and sembristles on the rictus and lores appear to be tactile structures, analogous to the whiskers of a cat. The stiffening and strengthening provided by melanin probably improve the ability of bristles to transmit pressure changes.

5. *Filoplumes*

Filoplumes are hairlike feathers that, when full-grown, consist of a very fine shaft with a few short barbs or barbules at the tip. They are the most uniform of feathers and do not grade into other types of feathers. A calamus is present but often breaks off when a filoplume is pulled from its follicle because the rachis is narrowed at its base. While a filoplume is growing, it sometimes bears a cluster of long downy barbs on the rim of the superior umbilicus. This appears to represent an afterfeather, although it is lost before the feather is fully developed (Lucas and Stettenheim, 1972).

In some cormorants and bulbuls, filoplumes emerge over the contour feathers and contribute to the appearance of the feathering. Typically, however, they are not exposed. The filoplumes appear to transmit slight vibration or movement of the contour feathers to pressure and vibration receptors, the lamellar corpuscles (Vater-Pacini or Herbst corpuscles) (von Pfeffer, 1952). Sensory input of this kind is probably necessary for keeping the feathers in place and adjusting them for flight, insulation, or bathing. Filoplumes have abundant free nerve endings in their follicle wall and lamellar corpuscles around the follicles (Borodulina, 1966; Golliez, 1967). If this functional view is correct, filoplumes should not be regarded as degenerate but as highly specialized structures that aid the operation of other feathers.

B. EMBRYONIC DEVELOPMENT OF FEATHERS AND FOLLICLES

The embryogeny of feathers and follicles has been well described at the histological level, chiefly by Davies (1889), Hosker (1936), Watterson (1942), and Wessells (1965). Finer morphological details are being added with electron microscopy (reviewed by Kischer, 1968). Histochemical studies (reviewed by H. L. Hamilton, 1965) show that morphogenesis and final structure of embryonic feathers depend strongly on proper functioning of the enzyme alkaline phosphatase. Experiments by Cairns and Saunders (1954), Saunders and Gasseling (1957), Rawles (1963), and Sengel (1964) reveal the roles of the germ layers in determining the characteristics of integumentary derivatives. Studies (reviewed by Johnson, 1968) show that embryonic feathers respond clearly to several types of endocrine manipulation, suggesting that these structures will serve well for future investigations of endocrine regulation of development.

In the following account, the ages and equivalent stages when events occur refer to advanced feather germs on the dorsopelvic tract of White Leghorn chick embryos. The process differs in several respects in Pekin duck embryos (Hosker, 1936).

The first signs of feather primordia are clusters (placodes) of epidermal cells, formed on the fifth day of incubation (stage 28). Shortly thereafter (stages 29+ to 30), mesenchymal cells (presumptive dermis) congregate below each placode, and the combination constitutes the germ or rudiment of a feather and its follicle. Feather germs are formed in a regular arrangement and sequence, thereby laying out the feather tracts (Section III,A). The patterns of the tracts may have their genesis in a fibrous lattice of collagen in the mesenchyme (Stuart and Moscona, 1967; Sengel and Rusaouën, 1968).

The various inductions or tissue interactions in feather morphogenesis or feather cell specialization continue to receive attention (e.g., Sengel, 1958, 1964; Rawles, 1963; Wessells, 1965) yet their roles remain unclear. The dermis causes the differentiation of the general ectoderm into a typical epidermis, but the epidermis seems to initiate feather development and play the major role thereafter. Anchor filaments, processes that extend downward from the placodes into the mesenchyme (Fig. 1), appear to function in inductive tissue interactions at stages 28 and 29 (Wessells, 1965; Kallman *et al.*, 1967; Kischer, 1968). Neural substances seem to be necessary for the induction of feather development (Sengel, 1964; Sengel and Kieny, 1967a,b; Kischer, 1968).

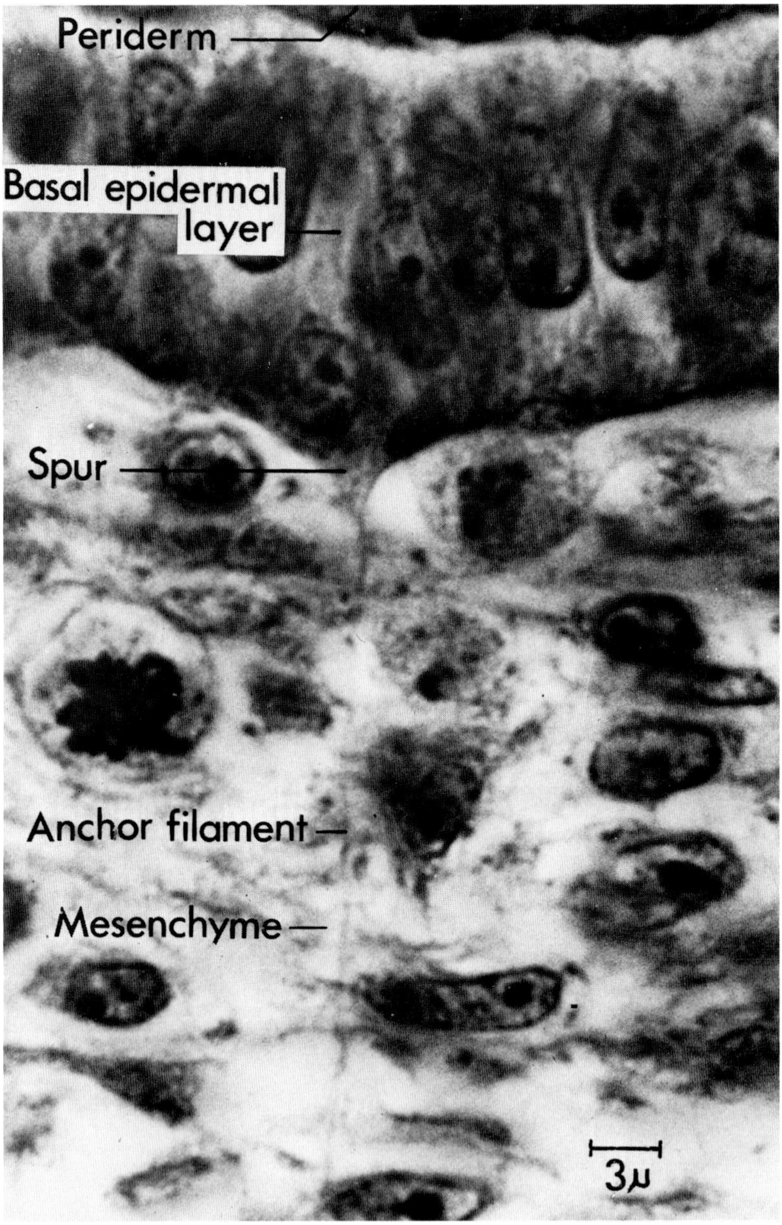

FIG. 1. Epidermal spur and anchor filament in chick embryo skin. Paraffin section stained by Masson's technique. × 2950. (Courtesy Kischer and Keeter, 1971.)

Feather germs start to push above the surface late in stage 33 (about 7.5 to 8 days) and soon lengthen and bend posteriorly, the first sign of their orientation. The germinative layer of the epidermis divides into intermediate and basilar layers (terms discussed by Lucas and Stettenheim, 1972).

Late in stage 36 (end of tenth day) the intermediate layer starts to differentiate into ridges parallel to the long axis of the feather germ. These formations are the rudiments of the barbs and hence are called barb ridges.

At stage 37 (eleventh day) a ring of epidermis at the base of a germ begins to invaginate, and the rudiment of the feather is separated from that of the follicle. Subsequent growth takes place chiefly by mitosis at the base of the follicle and feather. The follicle becomes a deep pit, and the feather germ a long cone protruding from it.

The barb ridges increase in number and length, and begin to differentiate, starting at the distal end, which is the oldest portion. Four components form lengthwise in each ridge—two barbule plates, which are precursors of the barbules, an axial plate between them, which mostly atrophies, and a ramogenic column, the precursor of the ramus. The barb ridges surround a core of pulp, derived from the dermis, but are separated from the pulp by a thin layer of basilar epidermis. A network of capillaries and larger vessels develops inside the pulp as described by Goff (1949). The barb ridges are enclosed by a thin sheath, which arises from the intermediate layer. Epitrichium covers the sheath outside the follicle.

When a feather grows, cell division takes place mostly at the base, and the parts differentiate as they move upward. The distal and peripheral parts arise first, followed by progressively more proximal and central parts. The barbs at the tip of the shaft are the first parts to be formed while the calamus is the last. Factors such as hormones that affect growing feathers exert their influence in this same sequence.

Embryonic keratin can first be detected on the thirteenth day (stage 39), long before the feather is fully formed (Bell and Thathachari, 1963). Pituitary hormones appear to be necessary for keratinization to take place (Yatvin, 1966).

By the sixteenth day (stage 42), a presumptive rachis and a presumptive hyporachis may have arisen on the dorsal and the ventral sides of the feather germ, respectively. Barb ridges have continued to form and differentiate. The pulp has reached its maximum length and started to regress (Goff, 1949). A new thick-walled tube of basilar epidermis forms around the pulp. After stage 42, development chiefly involves completion of a rachis, if any, and formation of a calamus.

Differentiation of form and keratinization are completed in the barbs by the nineteenth day (stage 45). The pulp regresses and is resorbed in the region of the calamus, where pulp caps form from the tube of basilar epidermis (Watterson, 1942; Goff, 1949).

Shortly after a chick hatches (stage 46), the feather sheath dries and flakes off. The barbs and their barbules spread out, transforming the bundle into a fluffy down feather. At the bottom of the follicle, there remains a small hump of dermal cells, the dermal papilla. It is covered by a thin layer of germinative epidermis, which is continuous with the epidermis of the follicle wall, and is overlaid by the last pulp cap in the calamus. The dermal papilla and its epidermal cap remain in a resting stage for two days or longer, before giving rise to a new feather.

C. Structure of a Feather Follicle

The follicle of a feather is a tubular invagination of the integument. It is slightly dilated at the base, where a dermal papilla projects up-

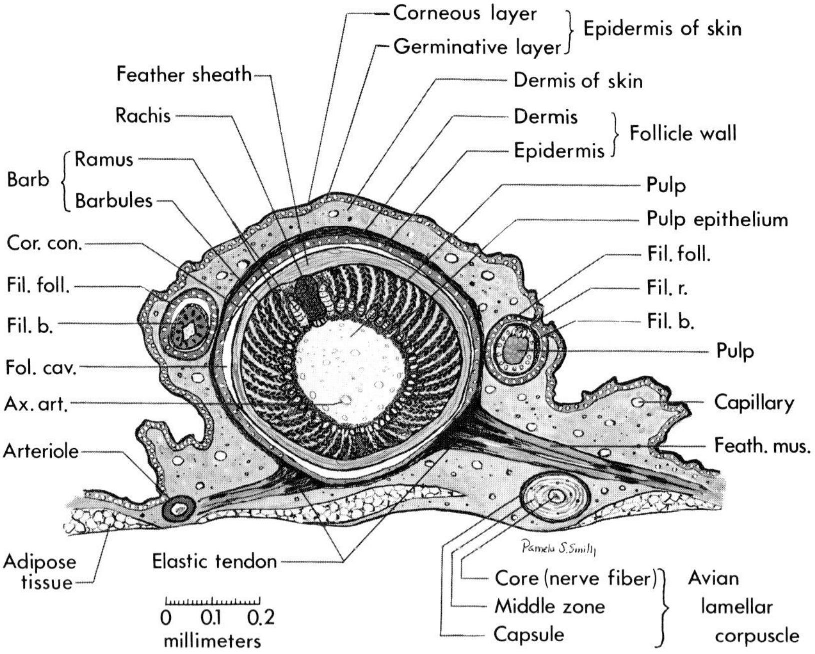

Fig. 2. Cross section of skin, a growing contour feather, and growing filoplumes from the crural tract of a 34-day-old domestic fowl. Bouin fixation, stained by Margolena and Dolnick's (1951) technique. Abbreviations: Ax. art., axial artery; Cor. con., corneous connection between feather sheath and follicle lining; Feath. mus., feather muscle; Fil. b., filoplume barbs; Fil. foll., filoplume follicle; Fil. r., filoplume rachis; Fol. cav., follicular cavity.

ward into the inferior umbilicus of a feather. Thin corneous sheets often bridge the follicular cavity, the space between the follicle lining and the feather sheath (Fig. 2) (Ostmann *et al.*, 1963). These sheets and the variable nature of the follicular cavity raise questions about the source of the follicle lining and the formation of the cavity (Lucas and Stettenheim, 1972).

A very thin keratinized layer of highly flattened cells lines a follicle. This is surrounded by a single ply of low cuboidal germinative epidermal cells. A basement membrane separates this layer from the surrounding dermis. Follicular dermis is composed chiefly of dense collagenic tissue, with elastic fibers arranged in roughly three concentric layers.

Bands of nonstriated muscles, the feather muscles, link adjacent follicles and cause feathers to move (Section IV,C). Numerous blood vessels and capillaries supply these muscles, the follicle itself, and the dermal papilla (Fig. 3) (Peterson *et al.*, 1965). The arteries and veins in the skin are branches of vessels to underlying skeletal muscles.

The dermal portion of a follicle, the feather muscles, and the surrounding dermis have a complex innervation that is clearly revealed (Fig. 4) by the technique of Tetzlaff *et al.* (1965). The wall of a follicle is richly innervated with general somatic afferent (sensory) fibers, which come from different sources at different levels of a follicle (Dreyfuss, 1937). The dermal papilla of a resting follicle contains a vasomotor system of autonomic ganglia and a network of efferent (motor) fibers around the axial artery. Short fibers that may have a sensory function are also present here. These nerves will grow and degenerate with the pulp itself each time a new feather is produced. Strictly speaking, a fully grown feather has no innervation.

Encapsulated sensory end-organs occur in the dermis of feather tracts and unfeathered parts of the body. These bodies, long known as Herbst's corpuscles, are now being called avian lamellar corpuscles (Malinovsky, 1967; Lucas and Stettenheim, 1972). They have been compared to the Vater-Pacini (Pacinian) corpuscles of mammals by Winkelmann and Myers (1961), and their embryology has been reported by Saxod (1967) and Golliez (1967). Avian lamellar corpuscles are approximately ellipsoidal and often bent in one or two places. Their abundance varies in different parts of the body (Stammer, 1961), but there are commonly one or two corpuscles beside each follicle. A corpuscle consists of a nerve tip and three concentric sets of fluid-filled lamellae around it. The axon is that of a general somatic afferent nerve which loses its sheath before entering the internal set of lamellae. Characteristically, the nuclei of the cells that form the internal

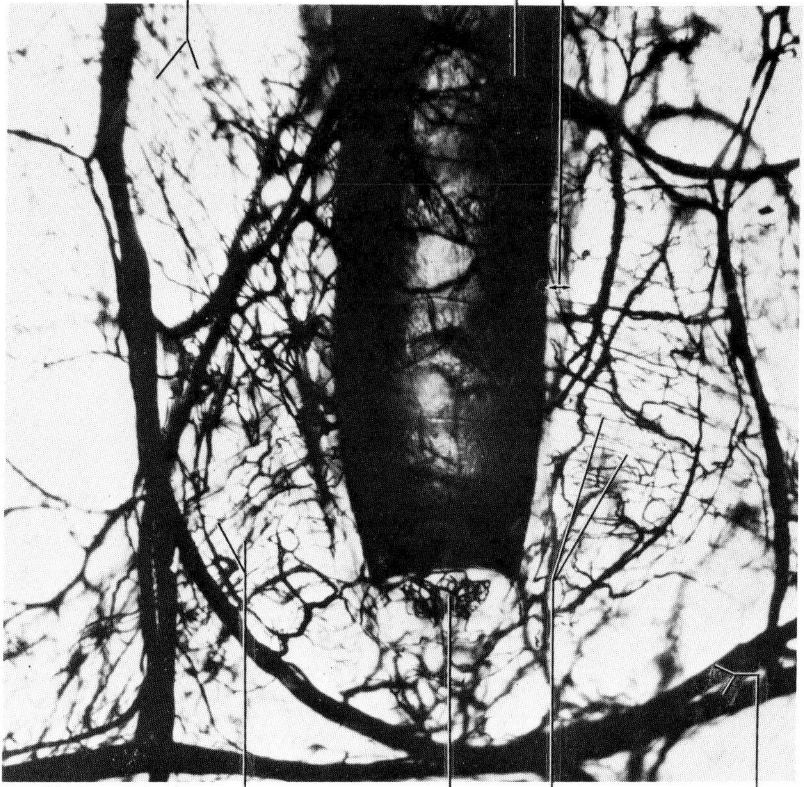

FIG. 3. Ink-perfused blood vessels in a feather follicle in a whole mount of skin from a domestic fowl. × 18. Courtesy Peterson *et al.*, *Stain Technol.*, ©1965, The Williams & Wilkins Co., Baltimore, Maryland.

lamellae (actually hemilamellae) are lined up in a row on each side of the nerve fiber. Variations in the morphology of the corpuscles appear to be related to different topographic conditions, such as thickness of the epidermis, and to sensitivity to different kinds of stimuli (Stammer, 1961; Malinovsky, 1967).

Various functions have been proposed for lamellar corpuscles in birds and mammals (Malinovsky, 1968), but findings in birds indicate that they are mechanoreceptors. These corpuscles are very sensitive to vibration (Schwartzkopff, 1955; Skoglund, 1960) and they may

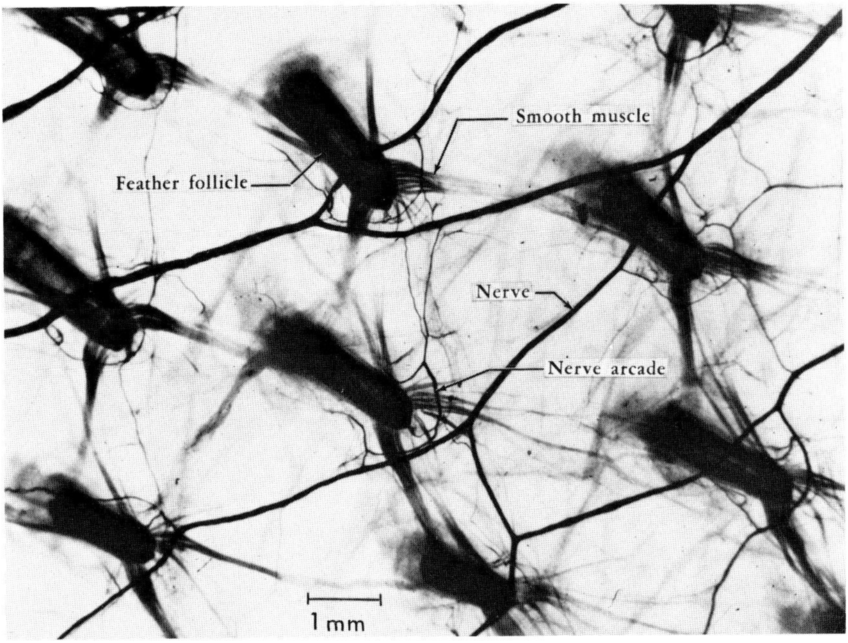

FIG. 4. Feather follicles and net of nerves in a whole mount of skin from the femoral tract of a domestic fowl. The nerves were stained by a phenylhydrazine-leucofuchsin method. Courtesy Tetzlaff *et al.*, *Stain Technol.*, ©1965, The Williams & Wilkins Co., Baltimore, Maryland.

respond to other pressure stimuli as well (Winkelmann, 1960). Both vibration and pressure changes are likely to occur in a follicle whenever a feather is moved by its muscles, air pressure, or some other force. The lamellar corpuscles beside the follicles appear to provide the sensory input necessary for proper positioning of the feathers for flight, thermoregulation, display, or other activity. Filoplumes seem to transmit delicate sensations of pressure changes from the contour feathers to the sensory corpuscles (Borodulina, 1966). Sensory corpuscles will be discussed further in Chapter 8, Vol. III.

Feathers are normally seated tightly in their follicles, whether they are growing or fully grown. Retention of feathers appears to depend on motor impulses from the brain (Ostmann *et al.*, 1964; Smith, 1964). This supports the idea that the feather muscles and their elastic tendons exert a constricting force on the follicle. Friction and the corneous sheets between the follicular lining and the feather may also aid in holding a feather in place.

D. Regeneration of Feathers

1. *The Occurrence of Regeneration*

A follicle ordinarily produces a series of feathers during the life of a bird. Regeneration of a new feather is the essential event in natural molting. The process of molting often begins with the initiation of growth of a new feather, which then pushes its predecessor out of the follicle (Watson, 1963). Feathers that are lost by plucking or by fright molt are also replaced, either promptly or at the next regular molt involving those follicles. The timing of such regeneration depends on whether the feathers that were lost were growing or fully grown, on the reproductive state of the bird, and probably other factors as well.

After a feather becomes fully grown, germinal activity ceases at the base of the follicle. The follicle returns to a resting stage, where it normally remains for a period ranging from 2 days to more than a year, depending on the molt schedule. The dermal papilla and its epidermal covering constitute the blastema or germ from which the next feather will develop.

2. *Activation of the Blastema*

Molting is triggered in some way by changes in photoperiod and hormone levels (reviewed by Voitkevich, 1966), but the actual stimuli at the feather follicles do not seem to be known. The control of molting is discussed in Chapter 3, this volume.

The epidermis of a blastema must be influenced by the dermal papilla before it can develop. The papilla not only activates the epidermis but also seems to induce the orientation and symmetry of a feather (Wang, 1943). A postembryonic dermal papilla exerts no tract-specific influence and the epidermis seems to have no tract specificity except that imposed on it by dermis surrounding the papilla (Cohen, 1965). Cohen proposed a two-stage mechanism, regulated by feedback, for induction of postembryonic follicles. According to Cohen (1965, p. 198), "the dermis maintains the regional character of its overlying epidermis *and* [italics his] of the ectoderm of the appendage follicles associated with it. This follicular ectoderm acts upon dermal papillae to regulate them and their function according to the kind of appendage being locally produced."

Regeneration of a feather normally begins with an increase in the height of a blastema. Cell division takes place at first uniformly throughout the blastema but soon becomes most active at the base. This causes the epidermis to thicken there, forming a collar around the dermal papilla. The collar eventually develops into a tube of epidermis, which in turn differentiates into a feather (Cohen and 'Espinasse, 1961).

3. *Development of a Blastema into a Feather*

The epidermal collar becomes thicker by continued cell division, and starts to differentiate into concentric layers. These layers become more distinct as they are pushed upward. In sequence from the innermost outward, these layers are as follows: basilar, intermediate, incipient transitional, and sheath. The incipient transitional cells show no morphological sign of keratinization close to the collar but become fully transitional at a higher level. In addition to proliferation in the collar, the intermediate cells increase in number at higher levels by continued division of themselves and perhaps also by proliferation from the basilar layer (Hosker, 1936; Eiselen, 1939).

As the tube of epidermis is pushed upward, the basilar and intermediate cells begin to rearrange themselves into barb ridges. This takes place first middorsally on the blastema and then progressively around the sides to the midventral region (Fig. 5) (Ziswiler, 1962). As new sites are added at the bottom of the barb ridges, they seem to organize their successors (F. R. Lillie and Juhn, 1938). The sites of organization shift dorsally around the collar, and the barb ridges are thus laid out on a spiral course. The lower end of each ridge eventually meets the rudiment of the rachis in the middorsal line, fuses with it, and moves distally. Meanwhile, new ridges arise in the ventral part of the collar, on both sides of a site known as the ventral locus. This process continues until the full complement of barb ridges has been formed. The original ridges, which were close to the middorsal line, join the rudiment of the rachis early and thus become barbs at the tip of the feather. Ridges of later origin take up progressively lower stations on the rachis, the last ridges to arise being those at the bottom of the vanes.

The rachis originates several days after feather growth has begun, by fusion of a few barb ridges in the middorsal line. As the bases of subsequent ridges arrive dorsally, they join an independent rudiment in the midline, the rachidial ridge.

The barb ridges start to differentiate as soon as they have formed. Each ridge begins as a wedge-shaped column of intermediate epidermis, bounded on its outer surface by incipient transitional cells, and elsewhere by a layer of basilar cells. The intermediate cells group themselves into a pair of barbule plates laterally and an axial plate between them. Development proceeds as the cells ascend; in addition the cells at the outer surface become more advanced than those further in. Later, the innermost axial cells will form the rudiment of a

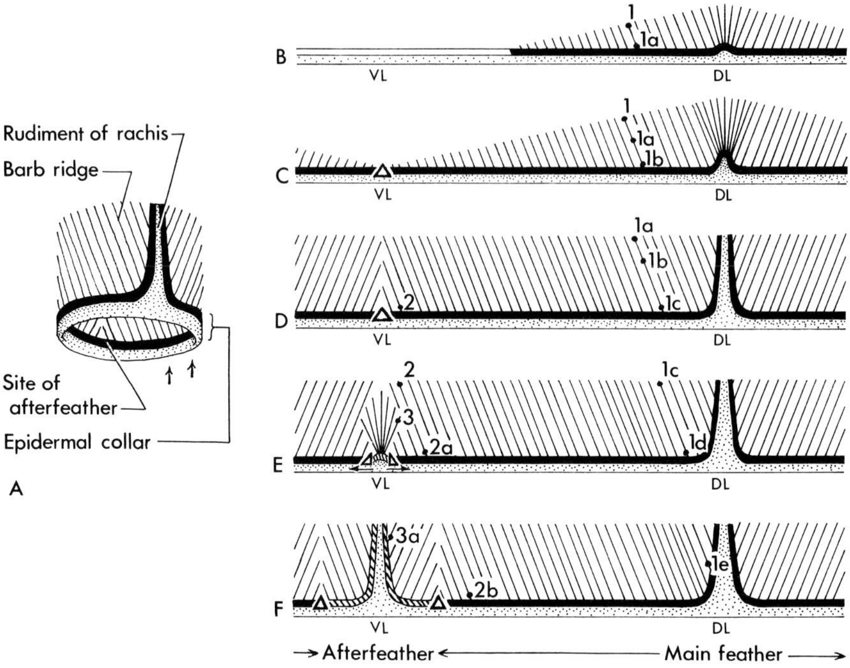

FIG. 5. Formation of main feather barbs and afterfeather barbs in regenerating feathers of carinate birds. (A) Proximal portion of feather tube at stage D. (B–F) Split and unrolled basal portions of feather tube at successive stages of growth. The vertical arrows in A mark the locations of the cuts (and the piece of tube removed between them) in B to F. 1, 1a, 1b, etc., mark successively formed points on a barb and show how the growth centers shift around the epidermal collar toward the rachis or the hyporachis. All such points at the same level (e.g., 2 and 1c) were formed at the same time. Abbreviations: DL, dorsal locus; VL, ventral locus. Regions of the epidermal collar: stippled, zone of most active cell proliferation; black, dorsal ramogenic zone; hatched, ventral ramogenic zone. The triangles in the ramogenic zone mark the ventral or lateral sites where growth centers of barbs arise. (Figures B to F based on Ziswiler, 1962, Abb. 10.)

ramus. The remaining axial cells and the basilar cells between ridges flatten and degenerate. The last cells to differentiate at any level are the basilar cells on the inner surfaces of the ridges. They fuse and form a tube around the pulp, the pulp epithelium (Schüz, 1927).

Development is most advanced at the tip and the periphery of a blastema, and follows gradients inward and toward the base. Hence, in the feather as a whole, in a barb, or even a barbule, the distal parts develop before the proximal parts. The cells of the barbule plates become aligned in columns which curve outward and upward. Individual cells lengthen, differentiate, and form their protuberances. The rami later differentiate from their rudiments and the barbules join them. The core of the rachis meanwhile develops from the rachidial ridge, and the sides are added by the progressive fusion of the bases of the rami. The growth of each barb is completed when it joins the rachis.

The afterfeather, if any, arises on the ventral side of the blastema after the main barbs and the rachis have begun. Rudiments of the afterfeather barbs form first, followed by that of the aftershaft, if any (F. R. Lillie and Juhn, 1938). The system of organization for the afterfeather is dominated by that of the main feather, and can manifest itself only after the influence of the main feather gradients has diminished (Ziswiler, 1962).

A feather grows inside a sheath of squamous epithelium that is thin and dense when completed. The tip of the sheath joins the bottom of the preceding feather. Morphogenesis and functions of the sheath were discussed by Lucas and Stettenheim (1972).

The tip of a feather completes its growth and extends well above the surface of the skin long before the lower parts have finished or even started forming. The upper barbs start emerging from the sheath soon after they are fully grown, keratinized, and dry. As the barbs are freed, they unfurl and assume their proper orientation.

After the full complement of barb ridges for the main vanes and the afterfeather has been formed, cell proliferation continues in the collar. The additional cells form a tube as they move upward, thus producing the calamus.

Keratinization—the process by which the cells become fused, dehydrated, and filled with keratin (Section I,C,1)—takes place while the cells are assuming their final shape (Cane and Spearman, 1967). Keratinization begins at the tip of the sheath and proceeds along the same gradients as those on which the feather parts were formed. The upper barbs may be fully keratinized before the lower barbs and the calamus have even arisen.

4. *Pulp and Pulp Caps*

After a feather blastema starts to grow, the dermal papilla ceases to act as an inducer and begins generating pulp, probably by mitosis. Pulp is a loose mesenchymal reticulum with a firm, elastic consistency. It forms a core inside the epidermal tube that will become a feather. The pulp grows from its base, while it dies and is resorbed at its apex. At first it lengthens at the same rate as the feather tube, but eventually it shortens and vanishes, as resorption proceeds at a faster rate than growth. The blood vessels and nerves in the pulp grow and degenerate with it every time a new feather is produced (Dreyfuss, 1937; F. R. Lillie, 1940). The pulp and its vascular system support a growing feather by their physical presence and by exerting turgor pressure. The vascular system also conveys nutrients to the epidermal cells and carries away their metabolic waste products. Lüdicke (1965) has reported on the movement of radioactive tracers in the pulp.

As the pulp is resorbed, a series of caps forms from the pulp epithelium at its apex, proceeding from the tip of the feather into the calamus. Resorption and cap formation in fowl take place during 12-hour periods, alternating with equal periods when the apical pulp, its blood vessels, and epidermal covering are reconstituted (F. R. Lillie, 1940).

After the last pulp cap has formed (in the inferior umbilicus), there remains a dermal papilla and a tiny hump of pulp with a covering of germinative epidermis. These constitute the start of a blastema for the next feather.

5. *Rate of Feather Growth*

Feathers grow at a rate that changes during the entire period of regeneration (F. R. Lillie, 1940). Growth rate also varies in a diurnal rhythm shown by regularly spaced transverse light and dark bands known as growth bars, seen on the vanes of feathers of many birds (Wood, 1950). These markings correspond with the pulp caps, indicating that both are reflections of the same metabolic cycle. The cycle has a short-phased primary rhythm that produces narrow, closely spaced bands and a long-phased secondary rhythm that produces broader, more widely spaced bands (Lüdicke and Geierhaas, 1963). The primary rhythm coincides with the rhythm of pulp cap formation (Lüdicke and Teichert, 1963). The portion of germinative epidermis (pulp membrane) on the apex of the pulp plays a major role in establishing the bands of the primary rhythm (Lüdicke, 1965). Both primary and secondary growth rhythms are closely related to the rhythm of melanin deposition (Lüdicke, 1967).

E. Chemical Composition of Feathers

Unhydrolyzed chicken feathers contain about 90.7% crude protein, 1.3% ether extract (lipids), and 7.9% moisture (McCasland and Richardson, 1966, p. 1233). The protein is mostly keratin. The amino-acid composition of the keratin differs among species of birds and even among parts of a feather (Schroeder and Kay, 1955; Harrap and Woods, 1967). Feathers and the corneous layer of the skin also contain small amounts of water-soluble proteins and other substances known collectively as nonkeratins (Gross, 1956). The free amino acids are thought to be precursors or by-products of keratinization, while the other nonkeratins are probably products of cellular breakdown (Spearman, 1966).

Feather lipids are unusual in their high content of cholestanol and large amount of free fatty acids (Bolliger and Varga, 1961). They are thought to develop in the feather as by-products of keratinization. If so, the secretion of fatty material by the feathers of certain pigeons (Eiselen, 1939; Berthold, 1967) may be simply an extreme manifestation of a normal process.

F. Color and Pattern

The display of colors and patterns is a major function of feathers, and hence has figured in their evolution (Portmann, 1963; Brush and Seifried, 1968). The causes of color and the formation of patterns in birds have been reviewed comprehensively by Rawles (1960) and Lucas and Stettenheim (1972). Accordingly, this section will be only an introduction to recent literature.

Melanin biology has been well reviewed by Fitzpatrick *et al.* (1967), and in a less advanced manner by McGovern and Lane Brown (1969). The chemistry of melanins has been discussed by Mason (1967) and R. D. Lillie (1969). A peculiar iron-containing phaeomelanin which is closely related to, or possibly identical with trichosiderin, a pigment of human red hair, has been found in feathers (Flesch *et al.*, 1966), but such pigments may be artifacts (Prota and Nicolaus, 1967). Melanic patterns in feathers and plumage have been classified, and their various mechanisms have been discussed by Cohen (1966). Langerhans cells have received much attention in mammalian pigment cell biology (e.g., Breathnach and Wyllie, 1967; Wolff and Winkelmann, 1967; Breathnach, 1969) but apparently have not been investigated in birds. These are nonkeratinizing dendritic cells with several morphological characteristics and the property of not synthesizing melanin. They are now thought to be independent of melanocytes, yet somehow involved in melanin pigmentation.

Recent studies of carotenoid pigments have considered their sources and metabolic pathways as well as their identification. Feeding experiments indicate that red pigments in the plumage of certain finches and flickers (*Colaptes*) are derived from yellow or orange carotenoids in the food (Völker, 1962; Test, 1969). Differences in the color of the flight feathers of "yellow-shafted" and "red-shafted" populations of flickers result chiefly from differences in the birds' carotenoid metabolism, not in their diet. Fox and Hopkins (1965) discovered several peculiar features of carotenoid metabolism in a flamingo, *Phoenicoparrus andinus*. Seasonal differences in the color of male Scarlet Tanagers *(Piranga olivacea)* appear to be caused by an oxidation and reduction of a xanthophyll carotenoid (Brush, 1967). The isozeaxanthin of (green) females and fall males is apparently converted into red-producing canthaxanthin at the time of the prealternate (spring) molt of males. The conversion is presumably mediated by a single enzyme and is undoubtedly under hormonal control. A study of the Gouldian Finch *(Chloebia gouldiae)* shows that "carotenoid-containing feathers, where they occur in specific patterns or in relatively small areas of intense coloration on monochromatic birds, are frequently modified in a manner that produces maximum exposure of the pigment" (Brush and Seifried, 1968, p. 423). In the facial feathers of this finch, the structure of feathers and the production of pigment are controlled by separate sets of genes. Brush and Seifried (1968) and Brush (1969) have considered the evolution of biochemical pathways for carotenoid metabolism. These vary in complexity among birds, but there does not seem to be any phylogenetic sequence in the occurrence in the class. Porphyrins in feathers have been treated by Thiel (1968).

Investigating the iridescence of hummingbird feathers, Greenewalt *et al.* (1960) found closely stacked layers of hollow melanin bodies in the outer keratin of the barbules. Iridescence in this case results from the interference of light waves reflected from the upper and lower surfaces of the gas-filled vacuoles. Studies of other iridescent birds have disclosed various arrangements of solid or hollow melanin bodies and sometimes gas vacuoles in the keratin between them (e.g., Durrer and Villiger, 1966, 1967; Rutschke, 1966b). The iridescent effects depend on extreme regularity (and precise variation thereof) in the refractive indices of the media, in the shape and dimensions of the pigment bodies, and in the spacing between the bodies. Since these factors also vary closely within and among feathers of a bird, and since the whole color system is repeated in subsequent, sometimes alternating plumages, the mechanism in control must be truly remarkable.

A brief review of hormonal effects on color systems in birds (Ralph, 1969) shows that effects on the synthesis and deposition of melanins are rather well established, but that effects on other pigments are largely unexplored. A study by Hall (1969) reveals new differences among birds in their response to melanizing hormones.

III. Pterylosis

A. Fundamentals

Most adult birds are covered with feathers except on the beak, the eyes, and the feet. The contour feathers are arranged in rows, which are in groups known as tracts or pterylae (singular, pteryla). Pterylae are separated by apteria (singular, apterium), areas of skin without contour feathers or bristles. Apteria are usually concealed by their surrounding feathers. Bare areas are visible on several parts of the body in various birds, most commonly on the head, neck, and legs. Ratites, penguins, screamers, and sometimes mousebirds, on the contrary, are often said to lack apteria, but these are present, even conspicuous, in embryos although much reduced in adults (Clench, 1970, p. 684).

Not only feathers but also their distribution, known as pterylosis, is a unique property of birds. Clench has pointed out that no other land vertebrates have an ordered grouping of epidermal derivatives, interrupted by "bare" spaces. The study of pterylosis, pterylography, was begun by C. L. Nitzsch in the early nineteenth century, and his monograph (1867) is still a fundamental reference in this field. Pterylography provides a system for designating the location of feathers, as is necessary for studies of feather morphogenesis and coloration, for describing the feathering (ptilosis), and for analyzing molts and plumages—in turn, a requirement for age determination. Pterylography is also a starting point for studying the functional and taxonomic significance of feather arrangement.

Pterylography must be based on the distribution of the follicles, for the ptilosis as seen externally does not reflect the actual pterylosis. The character and color of feathers sometimes aid in delimiting tracts, and Lucas and Stettenheim (1972) have shown that the patterns of the feather muscles can also reveal tract boundaries. Techniques for mapping feathers have been discussed by Humphrey and Butsch (1958), Naik (1965), Clench (1970), and Lucas and Stettenheim (1972).

The feather tracts are fully laid out in an embryo although in some birds, such as coraciiforms, piciforms, and passerines, many follicles do not produce feathers until some time after hatching. The rudiments of the contour feathers appear in a regular sequence within and among tracts (Holmes, 1935; Gerber, 1939; Hamburger and Hamilton, 1951; Burckhardt, 1954; Koecke and Kuhn, 1962; Golliez, 1967). In birds that have downs in the pterylae, the rudiments of these feathers appear later, while the rudiments of filoplumes are last of all, strictly located in relation to the contour feathers. In chick and duck embryos, the rudiments on the dorsal side of the body (spinal tracts) are arranged in metameric fashion corresponding to the somites (Verheyen, 1953; Koecke and Kuhn, 1962; Sengel and Mauger, 1967). This indication of an induction process is not evident, however, in other tracts.

B. Natal Pterylosis

When hatched, some birds are fully covered with down whereas others are scantily downy or even naked. The degree of feathering is related to the overall level of development and, in many cases, to the environment and exposure of the nest. Natal pterylosis is based on the presumptive pterylosis of a fully grown bird but there are often wide differences in the pattern of the tracts and the number of feathers. The pterylosis of newly hatched birds, as seen, tends to be more restricted than that in older birds, follicles being present but inactive. The study of natal pterylosis may thus reveal apteria that are later filled in. The younger pterylosis may be an evolutionary clue in some birds, being perhaps more primitive than the adult pattern (Clench, 1970). Baby ostriches, for example, show distinct tracts and apteria, whereas adults are fully feathered.

The process of filling in the tracts has been described for several nonpasserine birds by Gerber (1939) and Kuhn and Hesse (1957). In many passerines, numerous follicles, especially along the borders of the tracts, are very late in producing juvenal feathers (e.g., Naik and Andrews, 1966; Gwinner, 1969). This appears to be an adaptation that keeps the amount of feathering in proportion to the size of the body and spreads the energy cost of feather production over a longer period. A few birds, such as the Sacred Ibis (*Threskiornis aethiopica*), show an opposite change between natal and definitive pterylosis, the head being covered with down in nestlings but bare in adults.

Natal pterylosis of passerines varies to a small degree, chiefly in the number of downs and rarely in the presence or absence of tracts (Burckhardt, 1954; Wetherbee, 1957; Clark, 1967). Accordingly, it can sometimes be a useful taxonomic character in the classification of passerines. For example, differences in the natal pterylosis of tanagers support the view that this family is heterogeneous (Collins, 1963).

C. Adult Pterylosis

1. *Feather Tracts*

A look at Nitzsch's (1867) plates shows the diversity in adult pterylosis found in birds as a class. These differences are reflected in the various systems that have been proposed for classifying and naming the tracts and apteria. It is not only the landmarks and boundaries that differ, but also the views of pterylographers. Probably no one system of pterylography will be satisfactory for all kinds of birds. The outline of the tracts given below is based on the scheme of Lucas and Stettenheim (1972), which was worked out for fowl and other domestic birds. Pterylographers working with passerines may prefer to follow the system of Clench (1970), which was critically worked out on species of *Passer*. The chief difference between these two systems lies in the concept of a tract. Following Nitzsch and most successors, Clench stressed the idea of a few large tracts composed of smaller parts, which she called "elements" (equivalent to "regions" of many other pterylographers). Lucas and Stettenheim designated the parts themselves as "tracts."

a. Capital Tracts. A group of tracts (as many as thirteen) on the head, from the base of the bill posteriorly to the base of the skull. It is regarded by many workers as a single tract.

b. Spinal Tracts. A group of tracts along the vertebral column from the base of the head to the posterior end of the pelvis (usually just anterior to the oil gland). It is designated by Clench as the "dorsal tract."

c. Ventral Tracts. A group of tracts on the underside of the body from the bill to the base of the tail. Clench regarded this as a single tract and found that in passerines it begins on the upper breast; she considered the ventral head and neck feathering as part of the capital tract. The ventral tracts are paired for part or all of their length, always being divided by a midventral apterium. The tracts (or elements) are variously united or divided in different birds, but they are essentially continuous, except for the medium apterium, in many water birds, pigeons, and others.

d. Lateral Body Tract. A group of feathers on the side of the body below the axillary region. It is weakly developed in most passerines.

e. Caudal Tracts. All the feathers on the tail, including the large flight feathers (rectrices) at the posterior border, their coverts on the upper and under surfaces, feathers on the uropygial gland, down feathers associated with the rectrices, and others. The rectrices are typically in pairs, numbered from the midline outward. The median rectrices are bound to the sides of the pygostyle.

f. Wing Tracts. The pterylosis on the wings is subdivided into more tracts and named groups of feathers than that anywhere else on the body. Pterylography of the wing has been described in detail for the Mallard (*Anas platyrhynchos*) (Humphrey and Clark, 1961) and certain ploceids (Morlion, 1964).

i. Upper arm tracts. The humeral (scapulohumeral) tract overlies the scapula, the humerus, or some intermediate point. The subhumeral (axillar or subaxillar) tract occupies the ventral surface of the upper arm and the posthumeral tract occupies the posterior surface (Lucas and Stettenheim, 1972). The longest posthumeral feathers have commonly been called "tertiary remiges" but they differ from the true remiges in several respects.

ii. Remiges. Three series of mostly large feathers that extend from the trailing edges of the manus, the forearm, and the alula. The primary remiges are bound to the bones of the longest digit and are numbered from the wrist outward. There are ten primaries in nearly all families of functionally flying birds, though the tenth is very small in many passerines. Anterior to the tenth primary is sometimes found a small feather without a greater upper covert, attached to the second phalanx of digit II. Different interpretations of this feather, the "remicle," lead to different conclusions on the evolutionary history of the primaries (Stegmann, 1962; Stresemann, 1963).

The secondary remiges are anchored to the ulna and are numbered toward the elbow. Their number ranges from 6 to 40 among birds, though there are commonly 10 to 20. The number of secondaries may be a useful character for taxonomy of some groups of oscines (Stephan, 1965). A small feather, the carpal remex, is present in some birds on the posterior side of the wrist between the innermost primary and the outermost secondary or their upper major coverts. In many birds, there is no secondary remex beneath the fifth upper major secondary covert, leaving a gap, the "diastema," between the fourth secondary and the remex just proximal to it. Birds possessing such a gap are said to be "diastataxic," those without it, "eutaxic"; taxa of birds have been listed according to these conditions by Verheyen (1958). When present, the gap is numbered as the fifth position in the row of secondaries. The origin and significance of diastataxy have been studied most thoroughly by Steiner (1917, 1956) and Verheyen (1958); these findings have been summarized by Humphrey and Clark (1961).

The alular remiges are small- to medium-sized feathers that attach to the pollex. They range from two to six among birds and are commonly four in number. Some workers (e.g., Van Tyne and Berger, 1959; Humphrey and Clark, 1961) number them from proximal to distal, whereas others (e.g., Lucas and Stettenheim, 1972) follow the opposite order.

iii. Coverts. These are the rows of feathers that grade anteriorly from the remiges on the upper and undersides of the wing. The rearmost coverts correspond to the remiges, but anteriorly this relationship is lost. The rows also tend to be suppressed or shifted in the distal part of the wing. The major and median under secondary coverts are uniquely "reversed," that is, with their concave surfaces facing outward. This is supposedly because in ontogeny their rudiments shift from the dorsal side of the wing to the ventral side. There is much diversity among birds in the number, arrangement, and overlapping of the coverts, especially on the underside.

g. Hind Limb Tracts. i. Femoral tract. This is a group of feathers on the thigh, sometimes extending anteriorly on the side of the body. It is also known as the "lumbar" tract.

ii. Crural tract. This is a group of feathers encircling the leg from the knee downward, in many cases to the ankle (intertarsal joint) or even a short distance below.

iii. Metatarsal and digital tracts. The metatarsal tract is a group of feathers that arises anywhere on the metatarsus more than a short distance below the ankle. The digital tract is the group of feathers on the dorsal surface of the toes. These tracts are found only in birds with feathered feet—grouse, ptarmigans, some owls, and certain varieties of domestic pigeons and fowl.

2. *Apteria*

Apteria are areas of skin that are either bare or furnished with semiplumes or downs. In mapping the pterylosis of a bird, semiplumes can be included in either the apteria or the pterylae, according to the conditions found and the interpretation of the pterylographer. Apteria occur on all parts of the body, but they are generally small on the dorsal side of the wings and the tail. The most conspicuous apteria are those between tracts and between halves of a paired tract. Apteria also occur within tracts but it is sometimes hard to decide whether or not these are simply extra-large gaps between rows of feathers. Like the tracts, apteria should be named according to the body regions they occupy. Terminology and descriptions of apteria in domestic birds have been given by Lucas and Stettenheim (1972).

3. *Distribution of Down Feathers and Filoplumes*

Definitive down feathers are present in different birds in tracts amid contour feathers, in apteria, or in various intermediate patterns. There do not seem to be any consistent differences in the structure or musculature of downs as related to their location. Downs in the tracts are situated in a regular manner with respect to the contour feathers, but the governing factors are not known. Primordia for these downs form in an embryo after those for contour feathers and apterial downs (Gerber, 1939). Even birds that lack downs on the body tracts may have downs on the wings and tail beside the calami of the flight feathers.

Down feathers on the apteria may be sparse or numerous, and their distribution may be random or weakly organized. In the latter case, the feathers are widely spaced but in rows that are extensions of rows in the bordering tracts. The presence of weakly organized downs on the apteria suggests that the pterylosis has phylogenetically become restricted.

Filoplumes are situated only beside other feathers, never by themselves. They can be found with all other types of feathers, including bristles and downs on feather tracts. Usually one or two filoplumes accompany each "host" feather, but more are sometimes present, as in hawks and ducks. Remiges and rectrices tend to have more filoplumes beside them than do other feathers. Details on the number and arrangement of filoplumes have been reported by Fehringer (1912). In embryos, the primordia for filoplumes usually arise after those for contour feathers and tract downs (Gerber, 1939; Golliez, 1967). The filoplumes have a regular though varying placement in relation to their "hosts," showing the detail of the plan that governs the symmetry of the tracts.

D. Functional Significance of Pterylosis

Why do birds have feather tracts and apteria, and what is the significance of differences in the pattern among taxonomic groups? The low degree of intraspecific variation in pterylosis found in ploceids (Morlion, 1964; Clench, 1970) indicates that the pattern is functionally important, yet this topic seems to have been little investigated. Nitzsch (1867, p. 17) suggested that apteria reduce the total weight of the feathering, lessen the metabolic cost of producing feathers, and allow greater freedom of movement for the limbs. Apteria on the body may also sometimes accommodate underlying structures, although this idea has been refuted in the case of the apterium over the long hyoid apparatus in hummingbirds (Aldrich, 1956). Shortly before incubation in most groups of birds, special areas, the incubation patches, develop in the ventral apteria by loss of feathers, and by thickening and increased vascularity of the skin (Bailey, 1952). These patches aid the transfer of heat from the body of an incubating bird to the eggs.

Clench (1970) marshalled several lines of evidence to support the view that birds employ their pterylae and apteria to control their body temperature. Heat exchange is partly dependent on the ambient temperature. Clench showed that pterylosis seems to be affected by climate in some birds, but the type of reaction may differ among families. The number of feathers seems to vary seasonally in birds inhabiting regions of marked temperature change but not in those inhabiting regions of nearly constant temperature (Markus, 1963).

E. Amount of Feathers in Relation to Body Size

Counts of the number of contour feathers on many species of birds have been compiled by several authors (e.g., Van Tyne and Berger, 1959; Markus, 1963, 1965; Lucas and Stettenheim, 1972). These data indicate that large birds have more feathers than small birds, a conclusion that has been borne out by Turček (1966). Body size alone, at least in passerines as a whole, "does not necessarily influence the size and density of pterylae. Within a family, however, the larger members of a closely related group tended to have slightly longer or more heavily feathered tracts" (Clench, 1970, p. 681). In three varieties of domestic pigeon that differed in body size, however, Kuhn and Hesse (1957) found the number of feathers to be nearly constant, and that birds of a large variety have larger feathers than those of a small variety. In either case, the total weight of the feathers is heavier in large birds than in small ones. As a proportion of body weight, however, plumage weight decreases slightly with increasing body size (Turček, 1966).

F. Value of Pterylosis for Taxonomy

The organization of feathers into pterylae and apteria is a universal trait of birds, and the overall pattern seems to be very consistent and evolutionarily conservative within the class. There is no significant individual variation, at least in passerines. In the light of these considerations and a pterylographic survey of passerines, Clench (1970) concluded that pterylosis (and differences therein) can be highly useful for taxonomic studies.

Pterylosis varies on two levels, according to Clench. Minor differences, especially in feather counts, occur among closely related forms; such differences may be useful for taxonomic decisions within a family. Major differences, especially in patterns, occur at higher taxonomic levels, usually on the family level, or higher.

IV. Muscles of the Skin

A. Adjustment of Feather Position

Birds frequently adjust the orientation of their feathers and their degree of elevation above the body surface. Such positioning may take place all over the body or be confined to certain groups of feathers. The adjustment of feathers plays a role in thermoregulation, preening, clearing the vent before defecation, brooding, control of the wings and tail as airfoils, control of buoyancy in diving birds, and in displays for social signals (Morris, 1956). Many birds have in fact evolved individual feathers or groups of feathers with display functions.

Feathers are moved passively as a result of movement of parts of the body or inflation of air sacs or subdermal air pouches. Active movements, a result of muscle contraction, tend to be more specific and precise. The two kinds of movement are produced independently yet their effects may be combined. Three categories of muscles share in the active positioning of feathers: (1) striated muscles that insert on the bones of the wing and tail or on the follicles of flight feathers, (2) striated subcutaneous muscles, and (3) nonstriated intrinsic muscles of the skin.

B. Subcutaneous Muscles

The subcutaneous muscles are sheets or bands of striated fibers that attach to the underside of the skin (Fig. 6). They are often called "dermal" muscles, but this term is inappropriate because they really do not enter the dermis. As worked out by Osborne (1968), the subcutaneous muscles are Mm. constrictor colli and cucullaris, and components of Mm. latissimus dorsi, serratus superficialis, and pectoralis. M. ypsilotrachealis may be added to this list (George and Berger, 1966). A few of these muscles attach to the skin at both ends, but most of them originate on the skeleton or on body muscles. Of the latter, some are firmly bound to the skin for most of their length, but most of them are firmly bound only at their insertion. The attachments to the skin are sometimes beneath the follicles of feather tracts, but more often they are beneath the follicles at the borders of tracts. Subcutaneous muscles usually pass beneath apteria, although they are not confined to such areas. They occur only where the skin is loose (notably in the neck), never where it is tightly bound to the body.

Subcutaneous muscles are innervated by branches of cranial and spinal nerves; those muscles that are components of skeletal muscles receive branches of nerves to the main belly. Details of innervation are given by Osborne (1968).

Morphological and experimental evidences indicate that subcutaneous muscles serve principally to position the skin and to control its tension (Langley, 1904; Osborne, 1968). They may shift groups of feathers but do not affect their degree of elevation. Contraction of the muscles may wrinkle and shorten apteria, drawing together the borders of two tracts, thereby adjusting the body contours. Specific actions of these muscles have been suggested by Osborne.

C. Intrinsic Muscles

1. *Morphology and Action*

a. Feather Muscles. Adjacent feather follicles are linked by tiny bundles of nonstriated muscle known as feather muscles (pennamotor muscles, musculi pennati). These muscles are confined to the dermis and are associated with all types of feathers except filoplumes. Every follicle in a tract is generally connected by muscles to each of three or four surrounding follicles, and the repetition of these units forms a network across the tract (Fig. 6). The rows of muscles within this network may or may not correspond with the pattern of follicle rows. Certain muscle rows may be dominant over others, depending on the thickness of muscles within a row. The intersection of muscle rows between adjacent follicles forms quadrilateral or triangular figures, depending on the number of rows, their direction, and dominance. The gross pattern is similar in both sexes of a species, but in fowl and pheasants it is stronger in males, the sex that performs stronger feather displays (Osborne, 1968, p. 273).

In the sets of muscles between follicles, one muscle runs from the base of one follicle to a higher level on a more anteriorly or dorsally situated follicle (Fig. 7). Contraction of this muscle raises the base of the first follicle and possibly lowers the top of the second one. In either case, the feathers are depressed, so this is called a depressor muscle (Langley, 1904). The antagonistic muscle, an erector, runs from an upper level of one follicle to the base of a more anteriorly or dorsally situated follicle. The muscles are either entire or composed of several bundles, and these may interdigitate as they cross. In each set of muscles, the depressor is generally shorter and thicker than the erector, and has greater mechanical advantage for strength of pull. In some tracts, muscles pass at the same level between follicles. They are known as retractors because they appear to pull feathers together with little or no change in their elevation.

A feather muscle attaches to a follicle by way of a short, thick tendon that is composed largely of elastic fibers. These fibers ramify

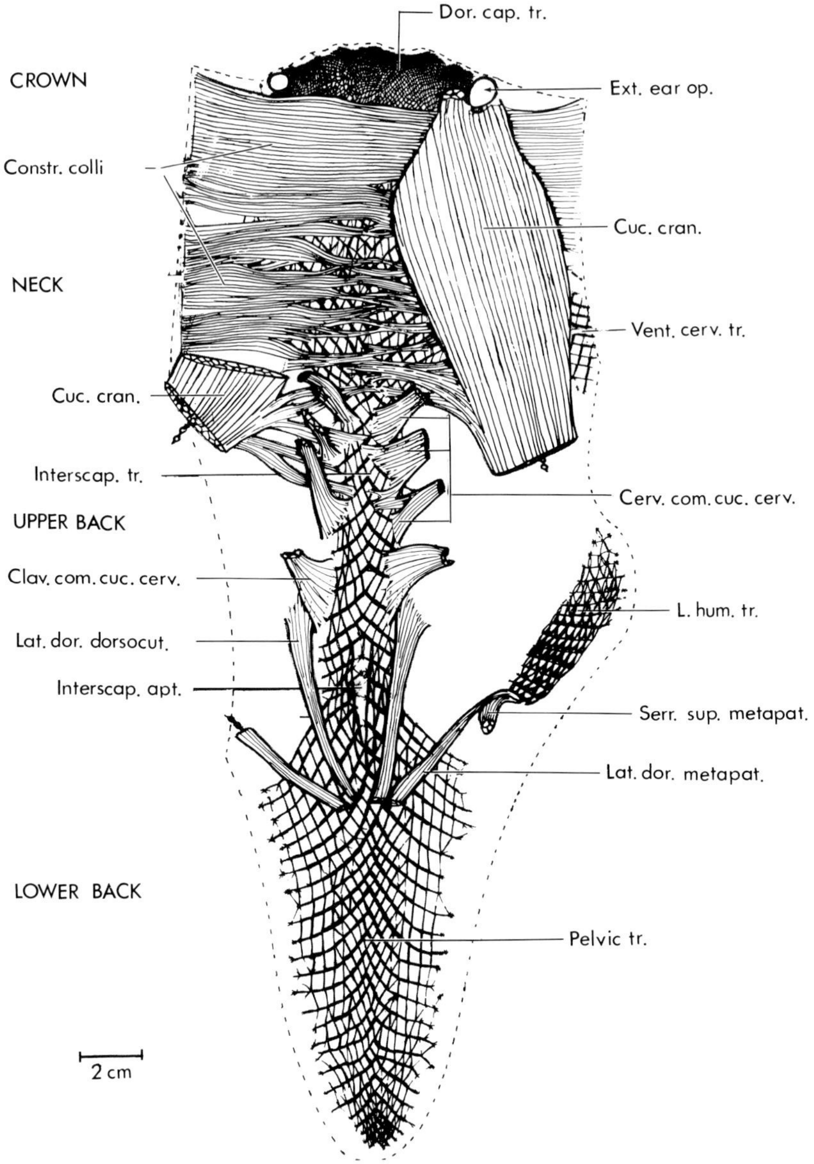

FIG. 6. Subcutaneous muscles and feather muscles as seen in an internal view of the skin from the dorsal side of a male Ring-necked Pheasant, *Phasianus colchicus*. The extreme anterior part of the dorsal capital tract is omitted. The criss-crossing lines within tracts represent the feather muscles, which are superficial to the subcutaneous muscles. The dashed line represents the boundary of the cut skin. Bouin fixation, unstained and partially cleared. Drawn from a photograph of a whole mount. Abbreviations: Cerv. com. cuc. cerv., cervical component of M. cucullaris pars cervicalis (cut); Clav. com. cuc. cerv., clavicular component of M. cucullaris pars cervicalis (cut); Constr.

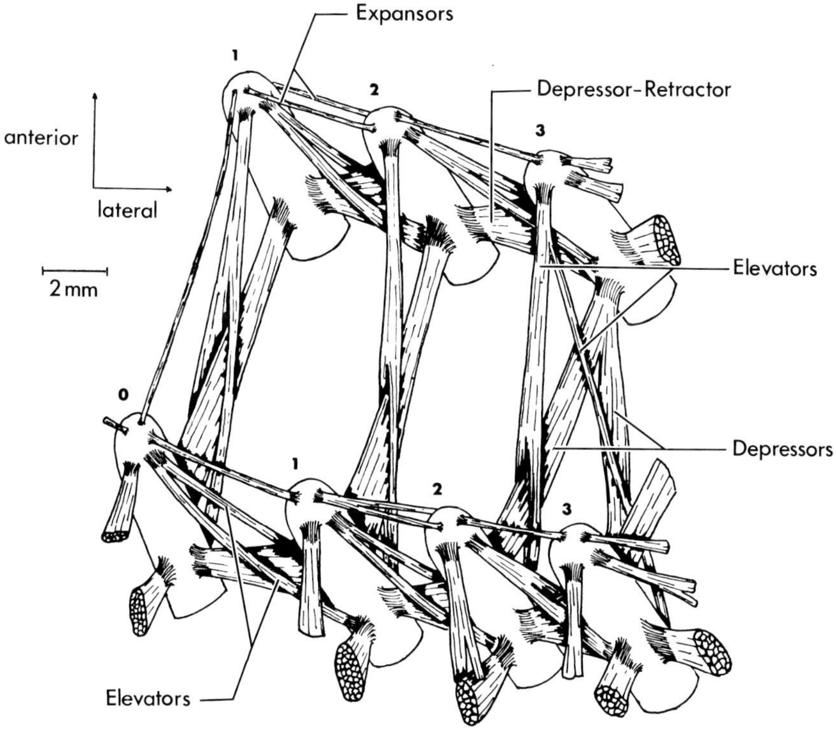

FIG. 7. Internal view of feather muscles associated with seven medial follicles of the interscapular tract of a male Ring-necked Pheasant, *Phasianus colchicus*. (Courtesy Osborne, 1968.)

into the outer layer of elastic fibers encircling a follicle (Ostmann *et al.*, 1963). Passing between follicles, the muscles are commonly interrupted by elastic tendons or even anastomoses of muscles and elastic tendons.

Feather muscles vary greatly in different parts of the body, as shown by Petry (1951), Stettenheim *et al.* (1963), Osborne (1968), and Lucas and Stettenheim (1972). In Ring-necked Pheasants, their volume varies within certain tracts in direct proportion to the length and weight of the feathers (Osborne, 1968, p. 165).

colli, M. constrictor colli; Cuc. cran., M. cucullaris pars cranialis (left side cut at cranial attachment and midcervical region, right side partially removed to show superficial M. constrictor colli); Dor. cap. tr., dorsal capital tract; Ext. ear op., external ear opening; Interscap. apt., interscapular apterium; Interscap. tr., interscapular tract; Lat. dor. dorsocut., M. latissimus dorsi pars dorsocutaneus (cut); Lat. dor. metapat., M. latissimus dorsi metapatagialis (cut); L. hum. tr., left humeral tract; Pelvic tr., pelvic tract; Serr. sup. metapat., M. serratus superficialis pars metapatagialis (cut); Vent. cerv. tr., ventral cervical tract. (Courtesy Osborne, 1968.)

In areas where feathers are widely spaced, the muscles tend to be randomly arranged, and their functions are no longer evident. They usually do not span apteria. At the borders of tracts, they radiate from the follicles and often ramify into the apterial muscles, to be described below. On the legs, feather muscles stop abruptly where the scales begin.

Down feathers in apteria have muscles that radiate from the follicles at all levels and anchor on the apterial muscles below them. Downs on the tracts, as seen in ducks, are linked to each other and to the contour feathers by a somewhat regular but delicate musculature (Lucas and Stettenheim, 1972).

Feather muscles can raise or lower feathers, draw them together, twist them, or combine these actions. The effect of their synergistic action depends on many variables of muscular morphology and mechanical arrangement. Because follicles are linked by a muscular net, feathers are moved or held in a graded or even manner within a tract.

b. Apterial Muscles. Apteria are sometimes devoid of smooth muscles throughout or in certain areas. Narrow median apteria formed by the separation of rows of follicles along the middorsal line of the body may be spanned by feather muscles. Commonly, however, the skin of apteria is partly or entirely furnished with a sheet of alternating smooth muscle bundles and elastic tendons arranged in an anastomosis. This tissue has been called a musculo-elastic net (Petry, 1951) or apterial muscle (Lucas and Stettenheim, 1972). Apterial muscles lie below any isolated down feathers or semiplumes in apteria, and may link with their muscles. At the borders of tracts, the apterial muscles typically grade into the feather muscles, but sometimes they pass beneath them for a short distance.

Apterial muscles appear to exert tension on the skin so that it does not sag, while allowing the parts of the body to move freely. The elasticity of apteria depends on their development of apterial muscles, which in turn may be related to their need for freedom of movement. The tension of the apterial muscles may also stabilize the boundaries of the tracts (Lucas and Stettenheim, 1972).

c. Tensor Muscles. Portions of the smooth musculature, often those that serve feathers involved in displays, sometimes become specially developed. These muscles have various forms and actions but are collectively known as tensors (Osborne, 1968). They attach to feather follicles at least at one end and have not been found with intermittent elastic segments. Their fibers are usually longer than those of typical feather muscles, and thus appear capable of contracting over a longer distance than other smooth muscles of the skin. Tensor muscles have been described in several galliform birds (Osborne, 1968; Lucas and Stettenheim, 1972) and will undoubtedly be found in many other birds that move their feathers in displays.

2. *Blood and Nerve Supply*

The skin and its smooth muscles are richly supplied by branches of vessels that supply the underlying skeletal muscles. As the vessels course beneath the skin, they do not follow the rows of feather muscles or follicles, but cut across the tracts. Anastomoses can be found among arteries and among veins, but not between arteries and veins. Blood vessels enter the muscle bundles at various points, but tend to be most numerous at the musculo-elastic junctions.

The arrangement of the cutaneous nerves and their relationship to their feather tracts in a chicken have been shown by Yasuda (1964). The nerves course without regard to the follicle rows, and commonly form anastomoses as shown by Tetzlaff *et al.* (1965) (Fig. 4). The main trunks consist of (1) somatic afferent fibers from the lamellar corpuscles and free nerve endings in the skin and subcutaneous muscles, (2) visceral afferent fibers from the walls of blood vessels, (3) somatic efferent fibers to the subcutaneous muscles, and (4) visceral efferent (i.e., autonomic) fibers to the intrinsic muscles and the blood vessels.

The smooth muscles of the skin are innervated principally by postganglionic fibers from the thoracolumbar (i.e., sympathetic) division of the autonomic system (Langley, 1904). This has been confirmed by the finding of evidence of alpha adrenergic transmission (Golenhofen and Petry, 1968). Evidence of the added presence of cholinergic transmission (Ostmann *et al.*, 1963; Peterson and Ringer, 1968) indicates that the muscles are also innervated by craniosacral (i.e., parasympathetic) fibers. In the muscles of the femoral feather tract of a fowl, the transmission sites are arranged "in such a manner as to cause feather shaft erection when the feather muscles are stimulated by cholinergic drugs and to cause the feather shafts to become depressed when adrenergic drugs are injected" (Peterson and Ringer, 1968, p. 497). Sensory receptors have been reported in the innervation of feather muscles of a grebe and a cormorant; they probably respond to the state of these muscles (Borodulina, 1966).

3. *Feather Muscles and Pterylosis*

Feather muscles appear in chick embryos at about stage 38 (twelfth day of incubation), several days after the formation of the feather rudiments (Lucas and Stettenheim, 1972). The muscles hence are probably laid out under the regulation of the follicles or the factors that determine the pattern of the follicles. Indeed, the pattern of the muscles generally agrees with the pterylosis, and gives a functional insight into it. Lucas and Stettenheim have shown that the two patterns differ in details of the rows, and have suggested that the musculature can also be helpful in determining the boundaries of tracts. Study of feather muscle patterns thus would be a useful approach to comparative pterylography. Muscle pattern differs from pterylosis on the dorsal surface of the prepatagium in fowl, and may have a bearing on the theory of diastataxy (Lucas and Stettenheim, 1972).

D. Skin Muscles and Feather Movements

There are two main muscular systems in the skin—the subcutaneous muscles, which move the skin and groups of feathers, and the intrinsic muscles, which adjust the tension of the skin and the posture of the feathers. In some birds, the subcutaneous muscles may be modified for better control of body contours, operation of the metapatagium, or for movement of display feathers, but this has not been shown. The feather muscles and tensor muscles, however, have clearly been shown to have developed adaptations for feather movements in flight and displays.

The two systems occur together in certain regions, particularly the head, neck, and anterior end of the trunk. In the skin of pigeons, Petry (1951, p. 529) even found junctions between striated and nonstriated muscles by way of elastic tendons. Since these muscles are innervated by somatic and autonomic fibers, respectively, the neural integration of their combined action—even where they are not linked—poses intriguing problems.

V. Integumentary Derivatives Other Than Feathers

The integument of birds is modified or formed into a variety of structures such as the horny coverings of the beak and feet, the wattles, comb, caruncles, ear lobes, claws, spurs, and uropygial gland. Limitations of space here permit discussion of only a few of these structures. Information and introductory references on those not discussed here can be found in Lange (1931), Mayaud (1950), A. L. Thomson (1964), and Lucas and Stettenheim (1972).

A. Rhamphotheca

1. Gross Morphology

The outer surface and part of the inner surface of the beak are covered with modified integument known as the rhamphotheca. It is hard and heavily cornified in most birds, but leathery except at the tip in flamingoes and anatids, and soft, especially at the tip of the upper jaw, in some scolopacids. The shape of the rhamphotheca is basically that of the underlying bone, but it is modified by local thickenings of the corneous layer. Various plates, knobs, ridges, and other projections occur on the beaks of birds. The rhamphotheca at the proximal end of the upper jaw is sometimes soft and thickened, this portion being known as a cere. In parrots, the cere is usually partly feathered, but in falconiformes, some cracids, pigeons, and owls, it is bare and often brightly colored. The histology of the cere in a pigeon and an owl has been described by Lucas and Stettenheim (1972).

2. *Growth and Loss*

After the breeding season, certain penguins (*Aptenodytes* spp.), male White Pelicans (*Pelecanus erythrorhynchos*), and many alcids shed plates or projections that form parts of the rhamphotheca. In ptarmigans (*Lagopus* spp.) and certain other grouse (*Tetrao* spp.), the rhamphotheca is shed intact after the breeding season. In most birds, however, the rhamphotheca continually grows toward the tip and the edges, where it is worn down. The corneous layer does not flow parallel to the germinative layer but flows obliquely outward, and keratinizes over the entire surface (Lüdicke, 1933, p. 531). Correspondingly, the surface of the rhamphotheca also sloughs or is worn off, as is obvious in Shoebills (*Balaeniceps rex*). If the jaws do not oppose each other properly, the rhamphotheca may grow unchecked and may hinder a bird (Pomeroy, 1962).

3. *Embryology and Natal Structures*

The first indication of the beak in a chick embryo can be seen at 5.5 days (stage 28). The ontogeny of the rhamphotheca was reported by Kingsbury *et al.* (1953). As far as is known, all birds develop a small protuberance, the egg tooth, on the distal end of the outer surface of either or both jaws. It forms from the germinative layer and the periderm (epitrichium) together. The cells become solidly keratinized but do not flatten as they do elsewhere (Lange, 1931, p. 403). The egg tooth functions only at hatching, when it cuts through the shell membranes and the shell. In most birds it disappears during the first week after hatching, but in some species it is kept longer (Clark, 1961).

In nestlings of many altricial species, the skin of the gape becomes temporarily enlarged, forming structures known as oral flanges. These are found mainly on the lower jaw in woodpeckers, wrynecks, and perhaps colies, whereas they pass around the corner of the gape in passerines and others (Clark, 1969). Sensory corpuscles are scarce or absent in the oral flanges of the European Starling (*Sturnus vulgaris*), an indication that the flanges lack tactile sensitivity (Wackernagel, 1954). Oral flanges appear to have a visual function in feeding, perhaps in releasing feeding behavior by the adults or guiding adults to the mouths of the nestlings. The flanges are pale or bright, and may supplement the bright colors (often in spots) on the palate and tongue of many altricial nestlings.

4. *Histology*

a. Layers of the Integument. The bones of the beak are covered by dermis that is not differentiated into layers as it is in the skin. Abundant collagenic and elastic fibers in the dermis run in various directions and join the periosteum, binding the rhamphotheca to the bones. The dermis is well supplied with blood vessels, nerves, and sensory corpuscles (discussed below). Blood vessels are more numerous in the palate than in the sides of the beak. Papillae and ridges of dermis invading the epidermis are commonly found in the rhamphotheca, palate, and tongue, especially in places that are not heavily cornified (Lange, 1931, p. 440; Lüdicke, 1940; Lucas and Stettenheim, 1972).

The epidermis is generally thickest on the tomial edges, thinnest on the inner surfaces, and of varying intermediate thickness on the outer surfaces. Except in the palate, the germinative layer is thinner than the corneous layer.

The corneous layer of the rhamphotheca is composed of many closely packed cornified lamellae. The superficial contour of these sheets is essentially that of the germinative layer, but it varies in parts of the beak owing to the differential flattening, accumulation, and wear of cells. Lüdicke (1933, 1940) has distinguished four main components of the corneous layer, originating on different surfaces of the beak.

b. Innervation. Nerve fibers with free endings in the epidermis have been found in the cere of a few birds (Stammer, 1961). Two kinds of sensory corpuscles, however, are common in the dermis—lamellar (Herbst) and Grandry's corpuscles. The lamellar corpuscles vary (Malinovský and Zemánek, 1969), yet they resemble those described (Section II,C) in the skin and joints. They tend to be most numerous at the tip of the beak and tongue, but they are distributed in various locations in different birds (Stammer, 1961, p. 120).

Grandry's corpuscles occur in ducks and owls. Typically they consist of a pair of subspherical sensory cells with large nuclei, sandwiching a disclike terminal network of neurofibrils. This group is surrounded by satellite cells and connective tissue. These corpuscles are oriented with their discs parallel to the surface of the beak. Malinovský (1967, p. 297) has discussed variation in their morphology. In the rhamphotheca of ducks, where lamellar and Grandry's corpuscles occur together, the Grandry type lie in the uppermost level of the dermis and the lamellar type are generally deeper (Malinovský, 1967, p. 292).

Both types of corpuscles are thought to be tactile receptors though their exact functions are not known. The lamellar corpuscles on the beak are probably sensitive to pressure and vibration as they are in feathered skin and in joints. Stammer (1961) has suggested that Grandry's corpuscles may play a role in the selection of food. Variations in both types of corpuscles appear to be adaptations for improving sensitivity to different kinds of stimuli. The specialization of these receptors makes the rhamphotheca not merely a covering for the beak but truly a tactile organ.

B. Scales and Claws

1. *Gross Morphology*

The feet of birds are at least partly covered with scales, the modified integument being known as the podotheca. In different kinds of birds, the podotheca begins on the tibiotarsus (shank), on the ankle, or at some more distal level; it extends to the tips of the toes. Feathers tend to diminish as scales arise and become prominent. Various relationships between feathers and scales occur on different parts of the foot and among birds. Feathers may emerge between scales, through a notch in the margin, or even through the interior of a scale.

The podotheca is generally horny but it is softer and more flexible in ducks and other water birds. Scales are flat, rounded, or conical raised areas of highly keratinized epidermis, separated by inward folds of thinner, less keratinized epidermis. They vary in size, shape, amount of overlap, and degree of fusion, even on parts of the same foot. They tend to be larger, more rectangular, and regularly arranged on the anterior surface of the tarsus and dorsal surface of the toes, and smaller, rounder, and more irregular on the other surfaces. Scales on the tarsus are often fused, especially those on the anterior surface. The pattern of tarsal scales has long been used as a taxonomic character at the ordinal or familial level. As applied to the song birds (oscines), however, Rand (1959) concluded that, except in the larks

(Alaudidae), the pattern is of limited importance for indicating relationships within the suborder, or for separating oscines from nonoscines in all cases. He also judged that the scutellar pattern in these birds has no adaptive significance. Variations among scales on the same foot appear to be related to local needs for freedom of movement, and hence to the use of the feet for locomotion or grasping.

On the underside of the toes, the podotheca is formed into thick pads separated by furrows or interpad spaces—areas of less elevated skin. The pads underlie the joints and extend beneath the phalanges, whereas the spaces are only beneath the phalanges (Lucas and Stettenheim, 1972). The arrangement of the digital pads varies, presumably in relation to the length of the toes and their need for support and flexibility. On both the pads and the interpad spaces, the scales are tiny papillae, usually flattened from wear.

Specializations of the podotheca include webs between the toes, heel pads, and tarsal spurs. Webs vary in size and in the number of toes they connect; large webs are present in most aquatic birds and small ones are present in some terrestrial birds. Heel pads are thickenings of the loose integument on the posterior side of the ankle. They are found in nestlings of certain species of parrots (Canella, 1959), owls, trogons, coraciiforms, toucans, barbets, woodpeckers, and wrynecks, and are reduced or lost after the birds leave the nest. Spurs consist of a bony core and a hard pointed, horny sheath. They are situated on the posterior and medial surfaces of the tarsi of male pheasants, peacocks, and turkeys. Lucas and Stettenheim (1972) have described the growth and histology of spurs in fowl.

Claws or nails (flattened claws) are present on the terminal phalanges of all toes, although they are vestigial on the outer toes of ostriches. The integument of a claw is formed of a dorsal plate (unguis) that curves downward on the tip and sides, and a ventral plate (subunguis) that fills the space between the sides on the bottom.

2. *Embryology of Scales*

In chick embryos, the first scale primordia appear at stages 36½ to 37 (10 to 11 days), arising as thickenings of the basal epidermis and the overlying periderm (J. L. Thomson, 1964, p. 208). These grow outward, and furrows form between them. In stage 41 (15 days) the definitive shape has been attained, and by stage 43 (17 days) keratinization is completed (Wessells, 1961).

Scales and feathers are morphologically and histochemically similar in many aspects of their differentiation, particularly during the first 2 days of their growth (Rawles, 1963, p. 771; J. L. Thomson, 1964, p.

211). In order to investigate the factors that control differentiation of scales and feathers, Rawles analyzed the results of many reciprocal exchanges of epidermis and dermis from prospective scaled and feathered regions at stages before and during formation of primordia. She found that the inductive capacity of the dermis and the competence of the epidermis vary with their stage of development and location on the body. The dermis of a scaled region (tarsus) is not only later than that of a feathered region (middorsum) in acquiring demonstrable inductive properties but also "weaker" in the intensity of its stimulus. Depending on the ages of the epidermis and the dermis in a given combination, middorsal skin can produce scales, and tarsal skin can produce feathers. Tarsal dermis is unique in that it can induce both scales and feathers, in normal chicks as well as in grafts. "The remarkable ability of epidermis to alter its course of differentiation at relatively late developmental stages, suggests that the morphological similarities between the scale and the feather may be more than superficial. It is possible that relatively few metabolic differences exist between them, thus permitting one to be converted into the other with relative ease" (Rawles, 1963, p. 785).

3. *Histology*

a. Dorsal Scales. The following account is based on the description of scales and claws in fowl by Lucas and Stettenheim (1972). The epidermis of the dorsal scales has a moderately thick, dense corneum composed of closely packed laminae of hard keratin. In the folds (sulci) between scales, this layer grades into a thicker, spongier corneum of soft keratin. The dermis consists of a superficial layer of delicate collagenic fibers, elastic fibers, and blood vessels. A subcutis of dense collagen and elastic tissue lies between the elastic lamina and the bones and cartilages of the foot.

b. Ventral Scales. The corneum of scales on the ventral side of the toes appears, from different staining reactions, to have a thin outer layer of hard keratin and a thick deeper layer of presumably softer keratin. The germinative layer is thicker than that in the dorsal scales and it interdigitates with dermal papillae. The sulci are either narrow open grooves or solid masses of folded corneum, never spongy. The dermis consists of coarse collagenic bundles, parallel to the surface, and delicate elastic fibers. Unlike the dermis of dorsal scales, it is raised into primary papillae beneath entire scales, and secondary papillae, which are fingerlike projections from the primary papillae (Lange, 1931, p. 428).

c. Claw. Claws on the toes have a thick corneum that is arranged in layers in the ventral plate but is without visible structural organization in the dorsal plate. The dermis is a thin layer of dense collagen with few elastic fibers. In the ventral plate, but not in the dorsal plate, the dermis forms papillae that invade the epidermis. Lamellar sensory corpuscles are numerous in the dermis on the top and sides of a claw, but there are only a few on the bottom. A subcutis is absent.

4. Growth and Loss

Certain grouse and passerines are said to molt their scales in summer (Mayaud, 1950). A possible explanation for this process has been suggested by Spearman (1966, p. 73), who found in 5- to 9-day old chickens a thin unkeratinized layer between two layers of horny scales. The unkeratinized layer apparently functions as a fission plane, indicating that, as in snakes and lizards, keratinization in birds takes place intermittently, and that the outer layer is sloughed off.

The claws and horny fringes of the toes in certain grouse are molted in the spring. In other birds, however, the claws are believed to grow constantly and to be worn down by friction.

C. Glands of the Integument

1. Uropygial Gland

a. Anatomy. The uropygial gland (oil gland, preen gland) is a bilobed sebaceous gland between the dorsal skin and the muscles at the base of the tail. Fundamental references on this organ are the monograph by Paris (1913) and the review by Elder (1954). Each lobe of the gland is composed of numerous acinar tubules arranged radially around a central cavity, which collects the secretion from the tubules. Ducts convey the secretion caudally to the surface where they open, in some birds at the tip of a papilla. The total number of orifices varies among birds from 1 in the Hoopoe (*Upupa epops*) to 18 in a pelican (*Pelecanus* sp.). The skin over the gland may be bare or feathered, but the papilla is usually bare except at the tip, where there is commonly a tuft of down feathers, the uropygial wick. In any case, the location of the gland is concealed by the body feathers except when a bird exposes it in preening. The gland is enclosed in a capsule of connective tissue and sometimes embedded in fat. It receives its blood supply from branches of the caudal artery, and is drained by branches to the caudal vein. The gland is innervated by the first pair of caudo-spinal nerves and by sympathetic fibers. Stimulating the sympathetic fibers is thought to cause dilation of the sphincter muscles around the external openings of the ducts.

The gland is present in most birds and is relatively large in many aquatic species. The most thorough survey of its condition in various birds is that of Paris (1913).

b. Embryology. Paired invaginations of the embryonic epidermis of the lower rump region appear on the eighth day of incubation in common pigeons, ninth to tenth days (stages 35 to 36) in fowl, and the tenth day in Pekin ducks. Properties of the gland are already determined several days earlier, however, in the undifferentiated epidermis of duck embryos (Gomot, 1959). The epidermis sinks inward and gives rise to solid buds of cells, which become covered by dermal connective tissue. The buds subdivide and acquire a lumen, eventually forming tubules around a central cavity. A papilla develops and secretion droplets appear in the tubules a few days before hatching.

c. Histology. The tubules are arranged in simple branched fashion and their walls form anastomoses, projecting as a network of trabeculae into the central cavity. Each tubule is composed of stratified epidermis which is basically like that of ordinary skin of the body (Lucas and Stettenheim, 1972). This resemblance is especially clear in young birds. There is a single ply of flattened basal cells, an intermediate layer, and a secretory layer that is equivalent to the transitional layer of ordinary integument. In the glands of fowl and ducks, histochemical study shows that each gland has a peripheral sebaceous zone and an inner glycogen zone (Cater and Lawrie, 1950). In the sebaceous zone, new cells develop lipoid droplets when they reach the secretory layer. As the cells move inward they enlarge and then disintegrate, leaving corneous laminae. These fragments, along with the liberated secretion, are pushed into the narrow lumen of the tubule. High esterase activity in this zone also indicates rapid production of lipids. Golgi apparatus in the secretory cells has been reported by many workers (e.g., Das and Ghosh, 1959), but Kanwar (1961) found no evidence for its presence. The glycogen zone of a gland also contains lipids, but is chiefly marked by the presence of glycogen and acid phosphatase, substances found in the secretion but not in the cells of the sebaceous zone (Cater and Lawrie, 1950). These differences in properties strongly suggest that the zones have different functions in the production of the secretion. Lucas and Stettenheim (1972) histologically distinguished a third zone at the inner ends of the tubules and the confluent secondary cavities in the uropygial glands of domestic fowl.

The gland is enclosed in a capsule of collagenic, reticular, and elastic fibers. Smooth muscles have been observed, mostly near the caudal end (Lucas and Stettenheim, 1972), but they do not ensheath the gland. Lucas and Stettenheim reported finding in the capsule a peculiar system of ducts that were unrelated to the sebaceous secretory system of the gland proper. From the capsule, septa of connective tissue extend between the tubules, enclosing and supporting them. These septa, sometimes very thin, carry nerves and small blood vessels. Elder (1954, p. 8) stated that they have smooth muscle fibers, but these were not observed by Paris (1913) or Lucas and Stettenheim. The lobes of the gland are often separated by an interlobular septum, and they are always joined at their caudal ends by a band of tissue known as the isthmus. The papilla arises from the isthmus. All these parts are composed of dense collagen, and the papilla also contains interwoven bundles of smooth muscles. Avian lamellar corpuscles are clustered in the papilla.

d. Composition of the Secretion. Apandi and Edwards (1964) reviewed the literature on the chemistry of the secretion. Although the composition varies among species, approximately half of the secretion is ether-soluble (Cater and Lawrie, 1950, p. 232). In an analysis of the secretion in 14 species of birds, E. Haahti *et al.* (1964) found the lipids in most cases to be mainly aliphatic waxes, but in 3 galliform species they were esters of dihydric alcohols with fatty acids (further analyzed by E. O. A. Haahti and Fales, 1967). Other lipids found are cholesterol esters, triglycerides, and phospholipids or substances with similar characteristics (Apandi and Edwards, 1964). The gland seems to be able to synthesize certain fatty acids, but at least in fowl the composition of these compounds can be influenced by the composition of the diet.

Of the nonlipids, several proteins and inorganic compounds were named by Elder (1954); polysaccharides, ascorbic acid, alkaline phosphatase, and acid phosphatase were reported by Das and Ghosh (1959). Iron is present in the secretion of many species of birds but is thought not to be responsible for the rust color occasionally seen on feathers (Fretzdorff *et al.*, 1966).

e. Function. When preening, many birds rub their bill and head on the uropygial gland papilla or its wick, annointing them with the secretion, which is then released, perhaps by nervous reflex. The secretion is transferred to the feathers on the body and wings or to the scales on the feet. Apparently, the secretion keeps the feathers, the rhamphotheca, and the podotheca from drying, thereby helping to keep them supple and in good condition. The lipids in the secretion are hydrophobic and are a necessary supplement to the meshwork of the barbs in making feathers water repellant (Rijke, 1970). Studies ascribing water repellancy chiefly to feather structure (Fabricius,

1959) may have overlooked the very thin film of secretion that remains on the feathers long after the uropygial gland has been removed. The feather lipids themselves have also been neglected as a possible factor in water repellancy.

Experiments by Hou (1929) indicate that the uropygial gland secretion contains a precursor of vitamin D, and that the vitamin forms when the secretion on the feathers is subjected to sunlight. The vitamin may be ingested when a bird preens itself. Removal of the gland caused young domestic chickens and ducks to show symptoms of rickets in spite of normal diet and exposure to sunlight. These findings have not been confirmed by some later workers, possibly due to species differences in the threshold of reaction to vitamin D deficiency.

Removal of the gland from young birds was followed by growth at a decreased rate in certain ducks (Elder, 1954), but at an increased rate in fowl (E. Haahti *et al.*, 1964).

The secretion is odorous in petrels, Musk Ducks (*Biziura lobata*), and incubating female and nestling Hoopoes. It may serve as a sexual attractant for the ducks or as a repellant for defense of the Hoopoes. The gland probably does not have a scent function in most birds, however, because the secretion generally does not smell strongly (to humans) and most birds do not have a keen sense of smell (A. L. Thomson, 1964, p. 552).

2. *Outer Ear Glands*

The wall of the outer ear contains a number of small, holocrine sebaceous glands in at least some birds. As seen in domestic fowl, these are sac-shaped organs that each extend lengthwise into the wall and open by a short duct directly to the outer ear canal (Lucas and Stettenheim, 1972). The embryology and structure of these glands were described by Glimstedt (1942). The ear glands secrete wax but differ greatly from the uropygial gland, as in the fact that their cells do not disintegrate completely and the lumen of each gland is filled with discharged cells.

3. *Anal Glands*

Birds of most orders have glands in the integument around the anal openings and along the anal canal. These are not to be confused with the cloacal glands because they lie outside the cloacal chambers and the true cloacal glands of the cloacal walls. Quay (1967) reported that anal glands differ widely among birds in their location, structure, and staining reaction.

These glands appear always to originate from, and to secrete onto, the relatively thick stratified squamous epithelium of the anal canal or the anal region. The secretory portions of the glands are either simple or branched sacs or tubules of columnar secretory epithelium. Histology and staining reactions indicate that the glands secrete mucus.

The function of anal glands is unknown, but in Japanese Quail (*Coturnix*) the secretion appears to be associated with the mechanics of internal fertilization (Coil and Wetherbee, 1959).

4. Secretory Activity of the Skin Epidermis

The skin of birds has long been thought to differ from that of mammals in lacking a general secretory function. Lipoid secretion in the epidermis of the toe pads of fowl was discovered, however, by Varićak (1938). Lucas (1968) and Lucas and Stettenheim (1972) reported evidence of secretion in the integument on many parts of the body in fowl.

These findings indicate that the entire skin of a bird is a sebaceous secretory organ, of which the uropygial gland and the ear glands are specialized parts. Lucas suggested that the secretion of the uropygial gland is applied primarily to the feathers and that the secretion of the body epidermis is the chief source of fatty material needed by the skin itself. The secretion probably helps maintain the pliability of the corneum, serves as a moisture barrier, and performs some of the other functions of human sebum. Varićak observed that it also serves as a carrier solvent for carotenoid pigments in the epidermis.

REFERENCES

Abdulali, H. (1966). On the "creaminess" in the plumage of *Ducula bicolor* (Scopoli). *Bull. Brit. Ornithol. Club* **86**, 162–163.

Aldrich, E. C. (1956). Pterylography and molt of the Allen Hummingbird. *Condor* **58**, 121–133.

Apandi, M., and Edwards, H. M., Jr. (1964). Studies on the composition of the secretions of the uropygial gland of some avian species. *Poultry Sci.* **43**, 1445–1462.

Auber, L. (1957). The structures producing "non-iridescent" blue color in bird-feathers. *Proc. Zool. Soc. London* **129**, 455–486.

Auber, L., and Appleyard, H. M. (1951). Surface cells of feather barbs. *Nature (London)* **168**, 736–737.

Bailey, R. E. (1952). The incubation patch of passerine birds. *Condor* **54**, 121–136.

Becker, R. (1959). Die Strukturanalyse der Gefiederfolgen von *Megapodius freyc. reinw.* und ihre Beziehung zu der Nestlingsdune der Hühnervögel. *Rev. Suisse Zool.* **66**, 411–527.

Bell, E., and Thathachari, Y. T. (1963). Development of feather keratin during embryogenesis of the chick. *J. Cell Biol.* **16**, 215–223.

Berthold, P. (1967). Zur Creme-Färbung von *Ducula bicolor* (Scopoli). *J. Ornithol.* **108**, 491–493.

Bloom, W., and Fawcett, D. W. (1968). "A Textbook of Histology," 9th ed. Saunders, Philadelphia, Pennsylvania.

Boas, J. E. V. (1931). Federn. *In* "Handbuch der vergleichenden Anatomie der Wirbeltiere" (L. Bolk *et al.*, eds.), Vol. 1, pp. 565–584. Urban & Schwarzenberg, Berlin.

Bock, W. J., and Short, L. L., Jr. (1971). "Resin secretion" in *Hemicircus* (Picidae). *Ibis* **113**, 234–236.

Böhm, R. (1962). Beitrag zu den postnatalen Veränderungen der Lederhaut des Haushuhns. *Acta Univ. Agr., Brno, Fac. Vet.* **10**, 95–100.

Böhm, R. (1963). Das Fettgewebe in der Haut des Haushuhns. *Acta Univ. Agr., Brno, Fac. Vet.* **11**, 473–485.

Böhm, R. (1964). Zum Bau der Lederhaut beim Haushuhn. (In Czech with German abstract.) *Acta Univ. Agr., Brno, Fac. Vet.* **12**, 351–362.

Bolliger, A., and Varga, D. (1961). Feather lipids. *Nature (London)* **190**, 1125.

Borodulina, T. L. (1966). On the morphology of filoplumes. *In* "Flight Mechanisms and Orientation of Birds" (S. E. Kleinenberg, ed.), pp. 113–145. Akad. Nauk SSSR, Inst. Morfologii Zhivotn., Moscow (in Russian).

Breathnach, A. S. (1969). Normal and abnormal melanin pigmentation of the skin. *In* "Pigments in Pathology" (M. Wolman, ed.), pp. 354–394. Academic Press, New York.

Breathnach, A. S., and Wyllie, L. M.-A. (1967). The problem of the Langerhans cells. *Advan. Biol. Skin* **8**, 97–113.

Brinckmann, A. (1958). Die Morphologie der Schmuckfeder von *Aix galericulata* L. *Rev. Suisse Zool.* **65**, 485–608.

Brush, A. H. (1967). Pigmentation in the Scarlet Tanager, *Piranga olivacea. Condor* **69**, 549–559.

Brush, A. H. (1969). On the nature of "cotingin." *Condor* **71**, 431–433.

Brush, A. H., and Seifried, H. (1968). Pigmentation and feather structure in genetic variants of the Gouldian Finch, *Poephila gouldiae. Auk* **85**, 416–430.

Burckhardt, D. (1954). Beitrag zur embryonalen Pterylose einiger Nesthocker. *Rev. Suisse Zool.* **61**, 551–633.

Buri, A. (1967). Das Juvenilgefieder von *Phasianus Colchicus* [sic] L., ein Beitrag zur Kenntnis dieser Altersetappe des Gefieders. *Rev. Suisse Zool.* **74**, 301–387.

Cairns, J. M., and Saunders, J. W., Jr. (1954). The influence of embryonic mesoderm on the regional specification of epidermal derivatives in the chick. *J. Exp. Zool.* **127**, 221–248.

Cane, A. K., and Spearman, R. I. C. (1967). A histochemical study of keratinization in the domestic fowl *(Gallus gallus). J. Zool.* **153**, 337–352.

Canella, M. F. (1959). Il "torulo del tallone" dei nidiacei di *Melopsittacus undulatus* (Shaw) e di altri uccelli. *Ann. Univ. Ferrara, Sez. 3* **2**, 1–45.

Cater, D. B., and Lawrie, N. R. (1950). Some histochemical and biochemical observations on the preen gland. *J. Physiol. (London)* **111**, 231–243.

Chandler, A. C. (1916). A study of the structure of feathers, with reference to their taxonomic significance. *Univ. Calif., Berkeley, Publ. Zool.* **13**, 243–446.

Clara, M. (1929). Bau und Entwicklung des sogenannten Fettgewebes beim Vogel. *Z. Mikrosk.-Anat. Forsch.* **19**, 32–113.

Clark, G. A., Jr. (1961). Occurrence and timing of egg teeth in birds. *Wilson Bull.* **73**, 268–278.

Clark, G. A., Jr. (1967). Individual variation in natal pterylosis of Red-winged Blackbirds. *Condor* **69**, 423–424.

Clark, G. A., Jr. (1969). Oral flanges of juvenile birds. *Wilson Bull.* **81,** 270–279.
Clench, M. H. (1970). Variability in body pterylosis, with special reference to the genus *Passer. Auk* **87,** 650–691.
Cohen, J. (1965). The dermal papilla. *In* "Biology of the Skin and Hair Growth" (A. G. Lyne and B. F. Short, eds.), pp. 183–199. American Elsevier, New York.
Cohen, J. (1966). Feathers and patterns. *Advan. Morphog.* **5,** 1–38.
Cohen, J., and 'Espinasse, P. G. (1961). On the normal and abnormal development of the feather. *J. Embryol. Exp. Morphol.* **9,** 223–251.
Coil, W. H., and Wetherbee, D. K. (1959). Observations on the cloacal gland of the Eurasian Quail, *Coturnix coturnix. Ohio J. Sci.* **59,** 268–270.
Collins, C. T. (1963). The natal pterylosis of tanagers. *Bird-Banding* **34,** 36–38.
Cone, C. D., Jr. (1962). Thermal soaring of birds. *Amer. Sci.* **50,** 180–209.
Das, M., and Ghosh, A. (1959). Some histological and histochemical observations on the uropygial gland of pigeon. *Anat. Anz.* **107,** 75–84.
Davies, H. R. (1889). Die Entwicklung der Feder und ihre Beziehungen zu anderen Integumentgebilden. *Morphol. Jahrb.* **15,** 560–645.
Day, M. G. (1966). Identification of hair and feather remains in the gut and faeces of stoats and weasels. *J. Zool.* **148,** 201–217.
Dickerson, R. E. (1964). X-ray analysis and protein structure. *In* "The Proteins" (H. Neurath, ed.), 2nd ed., Vol. 2, pp. 603–778. Academic Press, New York.
Dreyfuss, A. (1937). L'innervation de la plume. *Arch. Zool. Exp. Gen.* **79,** 30–42.
Durrer, H., and Villiger, W. (1966). Schillerfarben der Trogoniden. Eine elektronenmikroskopische Untersuchung. *J. Ornithol.* **107,** 1–26.
Durrer, H., and Villiger, W. (1967). Bildung der Schillerstruktur beim Glanzstar. *Z. Zellforsch. Mikrosk. Anat.* **81,** 445–456.
Eastlick, H. L., and Wortham, R. A. (1947). Studies on transplanted embryonic limbs of the chick. III. The replacement of muscle by "adipose tissue." *J. Morphol.* **80,** 369–389.
Eiselen, G. (1939). Untersuchungen über den Bau und die Entstehung von Schmalzkielen bei Tauben. *Z. Wiss. Zool.* **152,** 409–438.
Elder, W. H. (1954). The oil gland of birds. *Wilson Bull.* **66,** 6–31.
Ewart, J. C. (1921). The nestling feathers of the mallard, with observations on the composition, origin, and history of feathers. *Proc. Zool. Soc. London* 609–642.
Fabricius, E. (1959). What makes plumage waterproof? *Annu. Rep. Wildfowl Trust (Slimbridge, Gloucester, England)* **10,** 105–113.
Fehringer, O. (1912). Untersuchungen über die Anordnungsverhältnisse der Vogelfedern, insbesondere der Fadenfedern. *Zool. Jahrb., Abt. Syst., Geogr. Biol.* **33,** 213–248.
Feldman, G. L., Churchwell, L. M., Culp, T. W., Doyle, F. A., and Jonsson, H. T. (1962). The lipid content of the subcutaneous fat organs of the chick embryo. *Poultry Sci.* **41,** 1232–1240.
Fell, H. B. (1964). The experimental study of keratinization in organ culture. *In* "The Epidermis" (W. Montagna and W. C. Lobitz, Jr., eds.), pp. 61–81. Academic Press, New York.
Fitzpatrick, T. B., Miyamoto, M., and Ishikawa, K. (1967). The evolution of concepts of melanin biology. *Advan. Biol. Skin* **8,** 1–30.
Flesch, P., Esoda, E. J., and Katz, S. (1966). The iron pigment of red hair and feathers. *J. Invest. Dermatol.* **47,** 595–597.
Fox, D. L., and Hopkins, T. S. (1965). Exceptional carotenoid metabolism in the Andean Flamingo. *Nature (London)* **206,** 301–302.
Fretzdorff, A. M., Schüz, E., and Weitzel, G. (1966). Über den Eisengehalt von Vogel-Bürzeldrüzen. *J. Ornithol.* **107,** 225–228.

Fry, C. H. (1969). Structural and functional adaptation to display in the Standard-winged Nightjar *Macrodipteryx longipennis*. *J. Zool.* **157**, 19–24.

George, J. C., and Berger, A. J. (1966). "Avian Myology." Academic Press, New York.

Gerber, A. (1939). Die embryonale und postembryonale Pterylose der Alectoromorphae. *Rev. Suisse Zool.* **46**, 161–324.

Glimstedt, G. (1942). Über Morphogenese, Histogenese und Bau der Gehörgangdrüsen bei einigen Vögeln." Gleerupska Univ. Bokhandeln. Lund. (Not seen.)

Goff, R. A. (1949). Development of the mesodermal constituents of feather germs of chick embryos. *J. Morphol.* **85**, 443–481.

Golenhofen, K., and Petry, G. (1968). Physiologische Untersuchungen an der innervierten glatten Muskulatur der Vogelfedern (*Mm. pennarum*). *Experientia* **24**, 1137–1138.

Golliez, R. (1967). Beitrag zur Pterylose von *Melopsittacus undulatus* Shaw mit besonderer Berücksichtigung der Filoplumae. *Verh. Naturforsch. Ges. Basel* **78**, 315–364.

Gomot, L. (1959). Sur la détermination de la glande uropygienne des oiseaux. *Arch. Anat., Histol. Embryol.* **42**, 245–281.

Graham, R. R. (1930). Safety devices in wings of birds. *Brit. Birds* **24**, 2–21, 34–47, and 58–65.

Greenewalt, C. H., Brandt, W., and Friel, D. D. (1960). Iridescent colors of humming-bird feathers. *J. Opt. Soc. Amer.* **50**, 1005–1016.

Greschik, E. (1916). Zur Histologie der Vogelhaut. Die Haut des Kernbiessers und Haussperlings. *Aquila* **22**, 89–110.

Gross, R. (1956). Water soluble compounds (non-keratins) associated with the plumage of the pigeon (*Columba livia*). *Aust. J. Exp. Biol.* **34**, 65–70.

Gwinner, E. (1969). Untersuchungen zur Jahresperiodik von Laubsängern. *J. Ornithol.* **110**, 1–21.

Haahti, E., Lagerspetz, K., Nikkari, T., and Fales, H. M. (1964). Lipids of the uropygial gland of birds. *Comp. Biochem. Physiol.* **12**, 435–437.

Haahti, E. O. A., and Fales, H. M. (1967). The uropygiols: Identification of the unsaponifiable constituent of a diester wax from chicken preen glands. *J. Lipid Res.* **8**, 131–137.

Hall, P. F. (1969). Hormonal control of melanin synthesis in birds. *Gen. Comp. Endocrinol., Suppl.* **2**, 451–458.

Hamburger, V., and Hamilton, H. L. (1951). A series of normal stages in the development of the chick embryo. *J. Morphol.* **88**, 49–62.

Hamilton, H. L. (1952). "Lillie's Development of the Chick. An Introduction to Embryology," 3rd ed. Holt, New York.

Hamilton, H. L. (1965). Chemical regulation of development in the feather. *In* "Biology of the Skin and Hair Growth" (A. G. Lyne and B. F. Short, eds.), pp. 313–328. American Elsevier, New York.

Hamilton, W. J. III (1965). Sun-oriented display of the Anna's Hummingbird. *Wilson Bull.* **77**, 38–44.

Harrap, B. S., and Woods, E. F. (1967). Species differences in the proteins of feathers. *Comp. Biochem. Physiol.* **20**, 449–460.

Harrison, J. M. (1964). Moult. *In* "A New Dictionary of Birds" (A. L. Thomson, ed.), pp. 484–489. Nelson, London.

Heinroth, O., and Heinroth, K. (1958). "The Birds." Univ. of Michigan Press. Ann Arbor.

Hempel, M. (1931). Die Abhängigkeit der Federstruktur von der Körperregion, untersucht an *Xantholaema rubricapilla*. *Jena. Z. Naturwiss.* **65**, 659–737.

Holmes, A. (1935). The pattern and symmetry of adult plumage units in relation to the order and locus of origin of the embryonic feather papillae. *Amer. J. Anat.* **56,** 513–537.

Hosker, A. (1936). Studies on the epidermal structures of birds. *Phil. Trans. Roy. Soc. London, Ser. B* **226,** 143–188.

Hou, H.-c. (1929). Relation of the preen gland (glandula uropygialis) of birds to rickets. *Chin. J. Physiol.* **3,** 171–182.

Humphrey, P. S., and Butsch, R. S. (1958). The anatomy of the Labrador Duck, *Camptorhynchus labradorius* (Gmelin). *Smithson. Misc. Collect.* **135,** 1–23.

Humphrey, P. S., and Clark, G. A., Jr. (1961). Pterylosis of the Mallard Duck. *Condor* **63,** 365–385.

Il'ichev, V. D. (1961). Morphological and functional details of the external ear in crepuscular and nocturnal birds. *Dokl. Biol. Sci.* **137,** 253–256.

Il'ichev, V. D. (1962). Additional vanes in the pterylosis of the avian ear, their structure and function. *Dokl. Biol. Sci.* **144,** 532–535.

Ilychev, V. D., and Izvekova, L. M. (1961). Some peculiarities of the function of the external portion of auditory analizer in birds (in Russian). *Zool. Zh.* **40,** 1704–1714.

Jarrett, A., and Spearman, R. I. C. (1967). Keratinization. *Dermatol. Dig.* **6,** 43–53.

Johnson, L. G. (1968). Endocrine influences on growth and pigmentation of embryonic down feather cells. *J. Embryol. Exp. Morphol.* **20,** 319–327.

Jones, L. (1907). The development of nestling feathers. *Oberlin Coll. Lab. Bull.* No. 13.

Kallman, F., Evans, J., and Wessells, N, K. (1967). Anchor filament bundles in embryonic feather germs and skin. *J. Cell Biol.* **32,** 236–240.

Kanwar, K. C. (1961). Morphological and histochemical studies on the uropygial glands of pigeon and domestic fowl. *Cytologia* **26,** 124–136.

Kingsbury, J. W., Allen, V. G., and Rotheram, B. A. (1953). The histological structure of the beak in the chick. *Anat. Rec.* **116,** 95–115.

Kischer, C. W. (1963). Fine structure of the developing down feather. *J. Ultrastruct. Res.* **8,** 305–321.

Kischer, C. W. (1968). Fine structure of the down feather during its early development. *J. Morphol.* **125,** 185–204.

Kischer, C. W., and Keeter, J. S. (1971). Anchor filament bundles in embryonic skin: Origin and termination. *Amer. J. Anat.* **130,** 179–194.

Koecke, H. U., and Kuhn, O. (1962). Die embryonale Pterylose und ihre entwicklungsphysiologischen Vorbedingungen bei der Hausente (*Anas boschas domestica*). I. Die Entstehung der primordialen Federanlagen in den Körperregionen und die entwicklungsphysiologischen Probleme. *Z. Morphol. Oekol. Tiere* **50,** 651–686.

Koenig, O. (1962). Der Schrillapparat der Paradieswitwe *Steganura paradisaea. J. Ornithol.* **103,** 86–91.

Kuhn, O., and Hesse, R. (1957). Die postembryonale pterylose bei Taubenrassen verschiedener Grösse. *Z. Morphol. Oekol. Tiere* **45,** 616–655.

Lange, B. (1931). Integument der Sauropsiden. *In* "Handbuch der vergleichenden Anatomie der Wirbeltiere" (L. Bolk *et al.*, eds.), Vol. 1, pp. 375–448. Urban & Schwarzenberg, Berlin.

Langley, J. N. (1904). On the sympathetic system of birds, and on the muscles which move the feathers. *J. Physiol. (London)* **30,** 221–252.

Liebelt, R. A., and Eastlick, H. L. (1954). The organ-like nature of the subcutaneous fat bodies in the chicken. *Poultry Sci.* **33,** 169–179.

Lillie, F. R. (1940). Physiology of development of the feather. III. Growth of the mesodermal constituents and blood circulation in the pulp. *Physiol. Zool.* **13,** 143–175.

Lillie, F. R., and Juhn, M. (1938). Physiology of development of the feather. II. General principles of development with special reference to the after-feather. *Physiol. Zool.* **11**, 434–448.

Lillie, R. D. (1969). Histochemistry of melanins. *In* "Pigments in Pathology" (M. Wolman, ed.), pp. 327–351. Academic Press, New York.

Lucas, A. M. (1968). Lipoid secretion in the avian epidermis (Abstract). *Anat. Rec.* **160**, 386–387.

Lucas, A. M., and Stettenheim, P. R. (1972). "Avian Anatomy. Integument." Agriculture Handbook 362, U.S. Dept. Agr., Washington, D.C.

Lüdicke, M. (1933). Wachstum und Abnutzung des Vogelschnabels. *Zool. Jahrb., Abt. Anat. Ontog. Tiere* **57**, 465–534.

Lüdicke, M. (1940). Aufbau und Abnutzung der Hornzähne und Hornwülste des Vogelschnabels, *Z. Morphol. Oekol. Tiere* **37**, 155–201.

Lüdicke, M. (1965). Die Beziehungen des ventralen Coriumraumes der wachsenden Flugfeder zu den radioaktiven Querbändern nach Applikation von ^{35}S- Sulfat- und ^{35}S-DL-Cystinlösungen. *Biol. Zentralbl.* **84**, 273–297.

Lüdicke, M. (1967). Der Einfluss des Pigments auf das radioaktive Muster der Flugfedern vom Mäusebussard (*Buteo buteo* L.) nach intramuskulärer Injektion von ^{35}S-Sulfatlösungen. *Z. Morphol. Oekol. Tiere* **58**, 429–442.

Lüdicke, M., and Geierhaas, B. (1963). Über das Ablagerungsmuster des radioaktiven Schwefels in der wachsenden Konturfeder nach Applikation von ^{35}S-DL-Cystinlösungen. *J. Ornithol.* **104**, 142–167.

Lüdicke, M., and Teichert, H. (1963). Über den Ort der Aufnahme des radioaktiven Schwefels in der Federanlage nach Injektion von ^{35}S-DL-Cystin- und ^{35}S-Natriumsulfatlösungen. *Naturwissenschaften* **50**, 737.

McCasland, W. E., and Richardson, L. R. (1966). Methods for determining the nutritive value of feather meals. *Poultry Sci.* **45**, 1231–1236.

McGovern, V. J., and Lane Brown, M. M. (1969). "The Nature of Melanoma." Thomas, Springfield, Illinois.

McGreal, R., and Farner, D. S. (1956). Premigratory fat deposition in the Gambel White-crowned Sparrow: Some morphologic and chemical observations. *Northwest Sci.* **30**, 12–23.

McLoughlin, C. B. (1961). The importance of mesenchymal factors in the differentiation of chick epidermis. I. The differentiation in culture of the isolated epidermis of the embryonic chick and its response to excess vitamin A. *J. Embryol. Exp. Morphol.* **9**, 370–384.

Malinovský, L. (1967). Die Nervenendkörperchen in der Haut von Vögeln und ihre Variabilität. *Z. Mikrosk.-Anat. Forsch.* **77**, 279–303.

Malinovský, L. (1968). Types of sensory corpuscles common to mammals and birds. *Folia Morphol. (Prague)* **16**, 67–73.

Malinovský, L., and Zemánek, R. (1969). Sensory corpuscles in the beak skin of the Domestic Pigeon. *Folia Morphol. (Prague)* **17**, 241–250.

Margolena, L. A., and Dolnick, E. H. (1951). A differential staining method for elastic fibers, collagenic fibers, and keratin. *Stain Technol.* **26**, 119–121.

Markus, M. B. (1963). The number of feathers in the laughing dove *Streptopelia senagalensis* [*sic*] (Linnaeus). *Ostrich* **34**, 92–94.

Markus, M. B. (1965). The number of feathers on birds. *Ibis* **107**, 394.

Mason, H. S. (1967). The structure of melanin. *Advan. Biol. Skin* **8**, 293–312.

Matoltsy, A. G. (1962). Mechanism of keratinization. *In* "Fundamentals of Keratinization," Publ. No. 70, pp. 1–25. *Amer. Ass. Advance. Sci.*, Washington, D.C.

Mayaud, N. (1950). Téguments et phanères. *In* "Traité de Zoologie" (P.-P. Grassé, ed.), Vol. 15, pp. 4–77. Masson, Paris.

Morlion, M. (1964). Pterylography of the wing of the Ploceidae. *Gerfaut* **54,** 111–158.

Morris, D. (1956). The feather postures of birds and the problem of the origin of social signals. *Behaviour* **9,** 75–113.

Naik, R. M. (1965). The pterylography of the House Swift *Apus affinis*, as revealed by a new staining technique. *Pavo* **3,** 89–96.

Naik, R. M., and Andrews, M. I. (1966). Pterylosis, age determination and moult in the Jungle Babbler. *Pavo* **4,** 22–47.

Nitzsch, C. L. (1867). "Pterylography" (W. S. Dallas, transl.; P. L. Sclater, ed.). Ray Soc., London.

Oehme, H. (1963). Flug und Flügel von Star und Amsel. *Biol. Zentralbl.* **82,** 413–454 and 569–587.

Olson, S. L. (1970). Specializations of some carotenoid-bearing feathers. *Condor* **72,** 424–430.

Osborne, D. R. (1968). "The Functional Anatomy of the Skin Muscles in Phasianinae." Ph.D. Dissertation, Michigan State University, East Lansing.

Ostmann, O. W., Ringer, R. K., and Tetzlaff, M. (1963). The anatomy of the feather follicle and its immediate surroundings. *Poultry Sci.* **42,** 958–969.

Ostmann, O. W., Peterson, R. A., and Ringer, R. K. (1964). Effect of spinal cord transection and stimulation on feather release. *Poultry Sci.* **43,** 648–654.

Paris, P. (1913). Recherches sur la glande uropygienne des oiseaux. *Arch. Zool. Exp. Gen.* **53,** 139–276.

Parkes, K. C. (1965). Personal communication.

Peterson, R. A., and Ringer, R. K. (1968). The effect of feather muscle receptor stimulation on intrafollicular pressure, feather shaft movement and feather release in the chicken. *Poultry Sci.* **47,** 488–498.

Peterson, R. A., Ringer, R. K., Tetzlaff, M. J., and Lucas, A. M. (1965). Ink perfusion for displaying capillaries in the chicken. *Stain Technol.* **40,** 351–356.

Petry, G. (1951). Über die Formen und die Verteilungen elastisch-muskulöser Verbindungen in der Haut der Haustaube. *Morphol. Jahrb.* **91,** 511–535.

Pomeroy, D. E. (1962). Birds with abnormal bills. *Brit. Birds* **55,** 49–72.

Portmann, A. (1963). Die Vogelfeder als morphologisches Problem. *Verh. Naturforsch. Ges. Basel* **74,** 106–132.

Prota, G., and Nicolaus, R. A. (1967). On the biogenesis of phaeomelanins. *Advan. Biol. Skin* **8,** 323–328.

Quay, W. B. (1967). Comparative survey of the anal glands of birds. *Auk* **84,** 379–389.

Ralph, C. L. (1969). The control of color in birds. *Amer. Zool.* **9,** 521–530.

Rand, A. L. (1959). Tarsal scutellation of song birds as a taxonomic character. *Wilson Bull.* **71,** 274–277.

Rawles, M. E. (1960). The integumentary system. *In* "Biology and Comparative Physiology of Birds" (A. J. Marshall, ed.), Vol. 1, pp. 189–240. Academic Press, New York.

Rawles, M. E. (1963). Tissue interactions in scale and feather development as studied in dermal-epidermal recombinations. *J. Embryol. Exp. Morphol.* **11,** 765–789.

Rijke, A. M. (1970). Wettability and phylogenetic development of feather structure in water birds. *J. Exp. Biol.* **52,** 469–479.

Rutschke, E. (1960). Untersuchungen über Wasserfestigkeit und Struktur des Gefieders von Schwimmvögeln. *Zool. Jahrb. Abt. Syst.* **87,** 441–506.

Rutschke, E. (1965). Funktionell-morphologische Studien an der Aussenfahne von Schwungfedern. *Mitt. Zool. Mus. Berlin* **41,** 59–80.

Rutschke, E. (1966a). Untersuchungen über die Feinstruktur des Schaftes der Vogelfeder. *Zool. Jahrb., Abt. Syst. Geogr. Biol.* **93**, 223–288.

Rutschke, E. (1966b). Die submikroskopische Struktur schillernder Federn von Entenvögeln. *Z. Zellforsch. Mikrosk. Anat.* **73**, 432–443.

Saunders, J. W., Jr., and Gasseling, M. T. (1957). The origin of pattern and feather germ tract specificity. *J. Exp. Zool.* **135**, 503–528.

Saxod, R. (1967). Histogenèse des corpuscules sensoriels cutanés chez le Poulet et le Canard. *Arch. Anat. Microsc. Morphol. Exp.* **56**, 153–166.

Schaub, S. (1912). Die Nestdunen der Vögel und ihre Bedeutung für die Phylogenie der Feder. *Verh. Naturforsch. Ges. Basel* **21**, 131–182.

Schroeder, W. A., and Kay, L. M. (1955). The amino acid composition of certain morphologically distinct parts of white turkey feathers, and of goose feather barbs and goose down. *J. Amer. Chem. Soc.* **77**, 3901–3908.

Schüz, E. (1927). Beitrag zur Kenntnis der Puderbildung bei den Vögeln. *J. Ornithol.* **75**, 86–223.

Schwartzkopff, J. (1955). Schallsinnesorgane, ihre Funktion und biologische Bedeutung bei Vögeln. *Acta 11th Int. Ornithol. Congr., 1954* pp. 189–208.

Sengel, P. (1958). La différenciation de la peau et des germes plumaires de l'embryon de poulet en culture *in vitro*. *Année Biol.* **62**, 29–52.

Sengel, P. (1964). The determinism of the differentiation of the skin and the cutaneous appendages of the chick embryo. *In* "The Epidermis" (W. Montagna and W. C. Lobitz, Jr., eds.), pp. 15–34. Academic Press, New York.

Sengel, P., and Kieny, M. (1967a). Production d'une ptéryle supplementaire chez l'embryon de Poulet. I. Etude morphologique. *Arch. Anat. Microsc. Morphol. Exp.* **56**, 11–30.

Sengel, P., and Kieny, M. (1967b). Production d'une ptéryle supplementaire chez l'embryon de Poulet. II. Analyse expérimentale. *Develop. Biol.* **16**, 532–563.

Sengel, P., and Mauger, A. (1967). La métamerie de la ptéryle spinale, étudiée chez l'embryon de Poulet à l'aide d'irradiations localisées aux rayons X. *C. R. Acad. Sci., Ser. D* **265**, 919–922.

Sengel, P., and Rusaoüen, M. (1968). Aspects histologiques de la différenciation précoce des ébauches plumaires chez le Poulet. *C. R. Acad. Sci., Ser. D* **266**, 795–797.

Sick, H. (1937). Morphologisch-funktionelle Untersuchungen über die Feinstruktur der Vogelfeder. *J. Ornithol.* **85**, 206–372.

Skoglund, C. R. (1960). Properties of Pacinian corpuscles of ulnar and tibial location in cat and fowl. *Acta Physiol. Scand.* **50**, 385–386.

Smith, C. J. (1964). "The Role of the Central Nervous System in the Feather Release Mechanism of Young Chickens." Ph.D. Dissertation, University of Maryland, College Park.

Spearman, R. I. C. (1966). The keratinization of epidermal scales, feathers, and hair. *Biol. Rev. Cambridge Phil. Soc.* **41**, 59–96.

Spearman, R. I. C. (1969). The epidermis and feather follicles of the King Penguin (*Aptenodytes patagonica*). *Z. Morphol. Tiere* **64**, 361–372.

Stammer, A. (1961). Die Nervenendorgane der Vogelhaut. *Acta Biol. (Szeged)* **7**, 115–131.

Stegmann, B. K. (1962). Die verkümmerte distale Handschwinge des Vogelflügels. *J. Ornithol.* **103**, 50–85.

Steiner, H. (1917). Das Problem der Diastataxie des Vogelflügels. *Jena. Z. Naturwiss.* **55**, 1–276.

Steiner, H. (1956). Die taxonomische und phylogenetische Bedeutung der Diastataxie des Vogelflügels. *J. Ornithol.* **97,** 1–20.

Stephan, B. (1965). Die Zahl der Armschwingen bei den Passeriformes. *J. Ornithol.* **106,** 446–458.

Stettenheim, P., and Fay, J. A. (1968). The powder feathers of common pigeons, *Columba livia. 86th Meet. Amer. Ornithol. Union, College Alaska* Paper 21.

Stettenheim, P., Lucas, A. M., Denington, E. M., and Jamroz, C. (1963). The arrangement and action of the feather muscles in chickens. *Proc. 13th Int. Ornithol. Congr., 1962* pp. 918–924.

Stresemann, E. (1963). Variations in the number of primaries. *Condor* **65,** 449–459.

Strong, R. M. (1902). The development of color in the definitive feather. *Bull. Mus. Comp. Zool., Harvard Coll.* **40,** 147–185.

Stuart, E. S., and Moscona, A. A. (1967). Embryonic morphogenesis: Role of fibrous lattice in the development of feathers and feather patterns. *Science* **157,** 947–948.

Test, F. H. (1969). Relation of wing and tail color of the woodpeckers *Colaptes auratus* and *C. cafer* to their food. *Condor* **71,** 206–211.

Tetzlaff, M. J., Peterson, R. A., and Ringer, R. K. (1965). A phenylhydrazine-leucofuchsin sequence for staining nerves and nerve endings in the integument of poultry. *Stain Technol.* **40,** 313–316.

Thiel, H. (1968). Die Porphyrine der Vogelfeder: Untersuchungen über ihre Herkunft und Einlagerung. *Zool. Jahrb., Abt. Syst., Geogr. Biol.* **95,** 147–188.

Thomson, A. L. (1964). "A New Dictionary of Birds." Nelson, London.

Thomson, J. L. (1964). Morphogenesis and histochemistry of scales in the chick. *J. Morphol.* **115,** 207–224.

Turček, F. J. (1966). On plumage quantity in birds. *Ekol. Pol., Ser. A* **14,** 617–633.

Van Tyne, J., and Berger, A. J. (1959). "Fundamentals of Ornithology." Wiley, New York.

Varićak, T. D. (1938). Neues über Auftreten und Bedeutung von Fettsubstanzen in der Geflügelhaut (speziell in der Epidermis). *Z. Mikrosk.-Anat. Forsch.* **44,** 119–130.

Verheyen, R. (1953). Oiseaux. *Explor. Parc Nat. Upemba Miss. G. F. de Witte,* **19.** Inst. Parcs Nat. Congo Belge, Bruxelles.

Verheyen, R. (1958). Note sur l'absence de la cinquième rémige secondaire (diastataxie) dans certains groupes d'oiseaux récents et fossiles. *Gerfaut* **48,** 157–166.

Voitkevich, A. A. (1966). "The Feathers and Plumage of Birds." Sidgwick & Jackson, London.

Völker, O. (1962). Experimentelle Untersuchungen zur Frage der Entstehung roter Lipochrome in Vogelfedern. *J. Ornithol.* **103,** 276–286.

von Pfeffer, K. (1952). Untersuchungen zur Morphologie und Entwicklung der Fadenfedern. *Zool. Jahrb., Abt. Anat. Ontog. Tiere* **72,** 67–100.

Wackernagel, H. (1954). Der Schnabelwulst des Stars (*Sturnus vulgaris* L.). *Rev. Suisse Zool.* **61,** 9–82.

Wang, H. (1943). The morphogenetic functions of the epidermal and dermal components of the papilla in feather regeneration. *Physiol. Zool.* **16,** 325–350.

Watson, G. E. (1963). The mechanism of feather replacement during natural molt. *Auk* **80,** 486–495.

Watterson, R. L. (1942). The morphogenesis of down feathers with special reference to the developmental history of melanophores. *Physiol. Zool.* **15,** 234–259.

Wessells, N. K. (1961). An analysis of chick epidermal differentiation *in Situ* and *in Vitro* in chemically defined media. *Develop. Biol.* **3,** 355–389.

Wessells, N. K. (1965). Morphology and proliferation during early feather development. *Develop. Biol.* **12,** 131–153.

Wetherbee, D. K. (1957). Natal plumages and downy pteryloses of passerine birds of North America. *Bull. Amer. Mus. Natur. Hist.* **113,** 339–436.

Winkelmann, R. K. (1960). "Nerve Endings in Normal and Pathologic Skin. Contributions to the Anatomy of Sensation." Thomas, Springfield, Illinois.

Winkelmann, R. K., and Myers, T. T., III. (1961). The histochemistry and morphology of the cutaneous sensory end-organs of the chicken. *J. Comp. Neurol.* **117,** 27–35.

Wodzicki, K. (1928). Beitrag zur Kenntnis der Haut und des Fettansatzes bei Vögeln. *Bull. Int. Acad. Cracovie, Cl. Sci. Math. Nat., Ser. B* pp. 667–685.

Wolff, K., and Winkelmann, R. K. (1967). Nonpigmentary enzymes of the melanocyte-Langerhans cell system. *Advan. Biol. Skin* **8,** 135–167.

Wood, H. B. (1950). Growth bars in feathers. *Auk* **67,** 486–491.

Yasuda, M. (1964). Comparative and topographical anatomy of the fowl. XXXIV. Distribution of cutaneus [sic] nerves of the fowl (in Japanese). *Jap. J. Vet. Sci.* **26,** 241-248.

Yatvin, M. (1966). Hypophyseal control of genetic expression during chick feather and skin differentiation. *Science* **153,** 184–185.

Ziswiler, V. (1962). Die Afterfeder der Vögel. Untersuchungen zur Morphogenese und Phylogenese des sogenannten Afterschaftes. *Zool. Jahrb., Abt. Anat. Ontog. Tiere* **80,** 245–308.

Chapter 2

PATTERNS OF MOLTING

Ralph S. Palmer

I. Introduction

This chapter presents an approach to the subject of molting and resultant feather generations that is applicable to all birds. That is, the concepts and terminology relate to feather generations—"plumages" in the restricted sense of Humphrey and Parkes (1959)—and the antecedent molting. This seems preferable to summarizing known patterns of molting in avian families, since so complex a subject does not lend itself readily to condensation into a single chapter. The examples used, however, are drawn from various families, although waterfowl are singled out for treatment in some detail because (1) their molting has been investigated quite thoroughly, and (2) known variations and complexity of patterns may be used to illustrate many facets of the general subject.

The two prime characteristics of the vestiture of birds are (1) feathers, and (2) "the constant property of the skin with respect to formation of new feathers, which comes into effect periodically as molting" (Voitkevich, 1966, p. 88). That is, the integumentary derivative undergoes at most only a very few periods of succession or "renewal" in the course of a cycle of feather generations in the normal life of a bird of any size or age. Generally speaking, after one to several developmental generations of pennaceous feathers (sometimes called "immature plumages"), the pattern of molts and feather generations becomes repetitive and stabilized, usually with one or two generations per cycle, and the cycle is repeated thereafter throughout the life of the bird. To state it another way, the germinal tissue in a follicle that gives rise to a feather is activated only once or twice per cycle, in a few known cases three times, in one case (in an arctic grouse, *Lagopus lagopus*) reportedly four times (see p. 73). Furthermore, a follicle may be activated during one period of molting and not at another, so that one feather generation has a different number of feathers in a particular area than does another.

Even though feathers and their functions are so varied, the phenomenon of renewal common to all birds consists of such a limited number of occurrences that it seems to have some common basis throughout the class Aves (Fig. 1). Almost all of the present chapter

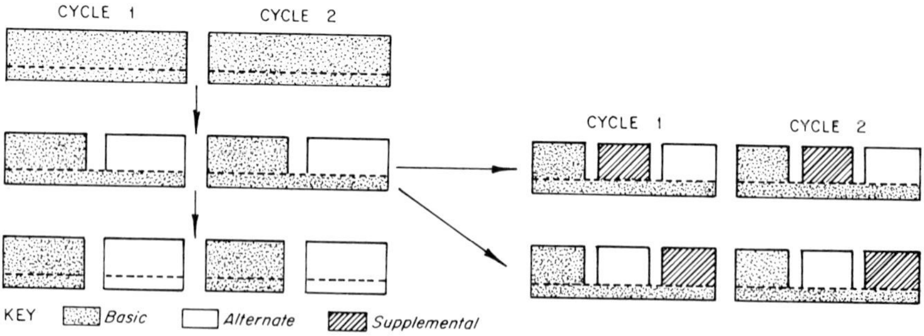

FIG. 1. Ways in which increasing complexity in cycles of feather generations may have evolved. Wing portion of feathering is below dashed line. (After Humphrey and Parkes, 1959.)

is restricted to matters relating to renewal of pennaceous feathers, without reference to periodic renewal of other feathers or structures (such as tarsal scutellae), which have been discussed in Chapter 1. Furthermore, various topics relating to feathers—such as their anatomy, differentiation into various kinds, distribution, and even the sequence and rates of growth of feathers in certain series—have their own specialized terminology. A minimum of terminology, mostly relating to the gross features of molting and the resulting feather generations, is defined and used in the present chapter.

The topical development of this chapter begins with some essential definitions and then proceeds with a presentation of the concepts of homology and functional equivalence in plumage cycles. This is followed by a brief consideration of the ways in which social factors may influence molt cycles, and the chapter is concluded with an extensive series of examples contrasting arbitrary and functional systems for describing patterns of molting.

II. Definitions

A. Cycle of Molting or of Feather Generations

The events occurring within a span of time until a particular event (such as onset of a certain molt, presence of a particular feather generation, etc.) is repeated constitute a cycle. Cycles of definitive feathering have a 12-month duration in most temperate-zone birds. Well-demonstrated cases of molting (hence also of feather generation) cycles that are shorter than 12 months are mainly among tropical and oceanic species. The cycle may be longer than a year in some species in various latitudes on the globe, but demonstrable cases evidently are few (Fig. 2). A bird has no more molts during a single cycle than the maximum number of times any feather follicle normally is activated (Humphrey and Parkes, 1959, p. 7).

There appear to be some cases of more or less continuous (acyclic) "breeding" and/or more or less continuous molting in some warm-climate species. There is a dearth of precise information. This subject is touched upon in Chapter 3.

B. Feather Generation

The new feathering acquired by a molt is a feather generation. An entire generation may consist, say, of only a few feathers (i.e., occur on only part of the bird) or may be so extensive as to include the entire feathering. Furthermore, it may be acquired (and lost) during a brief, or a protracted, or an interrupted molt.

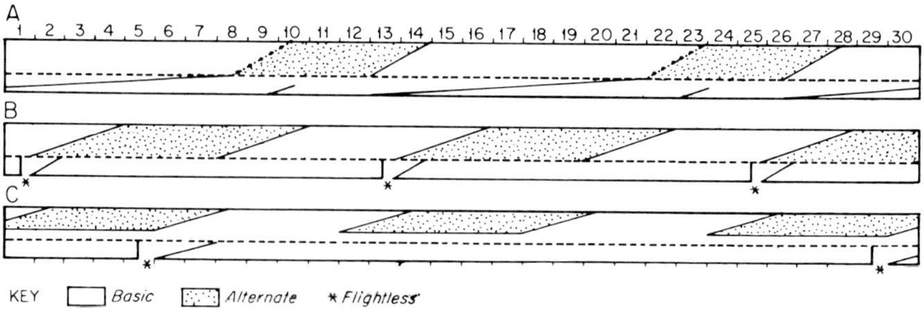

FIG. 2. Some reported variations in length of definitive cycle. Scale at top represents months; the species have two feather generations per cycle; wing is below dashed line in each diagram. (A) Sooty Tern (*Sterna fuscata*), shorter than 12 months; wing starts molting, then is interrupted, then all the flight feathers of wing are renewed gradually. (B) Common Loon *(Gavia immer)*, 12-month cycle with wing molted in winter. (C) European Crane (*Grus grus*), reported to molt the wing only every other year, as diagrammed (but in captive *Grus americana* the wing is molted annually).

C. Feathering

All or part of the vestiture of a bird may be referred to as feathering. The totality of feathers (pennaceous feathers, the down, filoplumes, etc.) is included, but authors commonly mean by it only the more "obvious" feathers, i.e., pennae. Feathering or "total feathering" refers to the covering of the bird regardless of feather generation(s) present and irrespective of whether the bird is molting. Humphrey and Parkes (1959, pp. 4–5) used the word "aspect" for the appearance of the total feathering at a specified time.

D. Definitive Feathering

After the developmental or "ripening" feather generation(s) have occurred, the shapes, proportions, colors, and other attributes of the feathers of each generation and the timing of occurrence of each (via molting) become relatively stabilized and, thereafter, each corresponding molt and resultant generation is essentially similar, i.e., definitive. Also, the antecedent molts may be referred to as definitive molts.

As long as any portion of the Juvenal generation, which is noncyclic, is retained, that portion of the feathering obviously cannot be definitive. If, for example, the head-body-tail is renewed twice before the Juvenal wing is lost, and these added generations approximate their later counterparts in appearance, the question arises as to whether or not they may be termed definitive. They seldom are on the same time schedule as corresponding later feather generations, however. The opposite situation occurs when a bird has so many developmental generations that there is, say, a full cycle beyond the retention of any portion of the Juvenal feathering, in which shapes, pattern, or color have not attained the stabilized condition. Thus the number of developmental generations varies. As long as inclusive feathering and its timing are not on a duplicating schedule, it is best to regard the generations as predefinitive.

In ornithological literature there is also the term "definitive feather," meaning the "final" feather, the "true" or pennaceous feather, i.e., a feather of the Juvenal or subsequent generations, as opposed to the down(s). Voitkevich (1966, in translation) used "definitive feather" and, for the same collectively, "definitive plumage."

A nonrecurring feather generation is not necessarily absolutely identical on all individuals in a species, and a recurring one is not necessarily absolutely identical with its counterpart in other cycles on the same individual. Many factors cause minor variation, including nutrition, weather, and disturbance of a bird while it is forming new feathers.

The concept of definitive thus is based on whether certain major kinds of change have been terminated. It is independent of (and often does not coincide with) attainment of reproductive maturity, and it also applies to individuals that live for years without ever breeding. Though possibly somewhat arbitrary, it is a very useful concept when describing the sequence of molts and feather generations of a species. Only one apparently satisfactory case of notable change in definitive feathering has come to my attention. In the Southern Giant Fulmar or Petrel (*Macronectes giganteus*), dark-phase birds at first are very dark; by degrees, through a series of molts, eventually they become nearly white on head and down onto the breast. They begin breeding within the span of this change. If a distinction must be made, because of some variation in definitive condition, it might be termed a "variant" or possibly "advanced."

E. Molting

The process of natural feather loss ("shedding") and replacement is called molting. It is not merely the loss, but also—and more importantly—the replacement of part of the integument, as in various reptiles. In terms of feather generations, molting may be continuous (within a brief span of time or protracted) or discontinuous (interrupted). A molt can be extended for a longer time than a calendar year, but not over more than one cycle. The replacement of a feather generation that consists of only a portion of the feathering is the entire molting of that generation (but is "partial" in terms of the entire feathering). Also a generation may, for example, consist of the entire feathering (as the Juvenal) and, in the aggregate, its loss may be accomplished during more than one period of molting. Thus part of the Juvenal may be succeeded by Basic I and, at the next period of molting, all the Basic I plus the remaining portion of the Juvenal may be lost. This illustrates the additional point that more than one feather generation (in this example, all of one and part of another) may be lost in a single span of molting.

A concept of "complete molt" has been so widely used as to include situations that are not exactly comparable. This may be explained by expanding the Juvenal-Basic I schematic example just mentioned. (1) the first generation of true feathers (Juvenal) consists of total feathering. (2) Soon the head-body-tail molt into another generation (Basic I) and, still later, head-body-tail again molt (into Alternate I), while the Juvenal wing remains. (3) At the next period of molting the entire feathering is renewed. Thus the loss of the Juvenal wing is greatly offset in time, in this instance not occurring until the third period of molting after acquisition of the entire Juvenal feathering. In this third period the entire feathering, in Dwight's (1900) words, is "entirely swept away." (Dwight wrote of molting in terms of loss of the "old," rather than acquiring the "new," feathering.) The cycle recurs, now minus the noncyclic Juvenal, and consists of (1) entire new feathering (Basic), a subsequent period of molting in which head-body-tail is renewed (Alternate), and (3) still later, the cycle is completed with a period of molting during which all feathering (Alternate head-body-tail plus Basic wing) is succeeded by the next-incoming Basic (entire feathering).

It should be noted that the pattern of having the "loss" of the Juvenal wing greatly offset in time is repeated by offset loss of the Basic wing in subsequent cycles. This "offset" feature, which has many variations (and will be discussed again later), occurs in a vast number of birds of all sizes and various taxonomic groupings that have at least two feather generations per cycle. In all of these, when all feathering is renewed in a single period of molting, it is somewhat misleading to refer to this as a "complete molt" when it actually includes loss of one entire feather generation plus part of another. But from the point of view of emphasizing *incoming* feathers, it is logical because all feathers are renewed regardless of their relative generation age.

In birds having more than one feather generation per cycle, if one tries to use a period of molting during which all feathering is "swept away" (Dwight) as a "landmark," without regard to how many feather generations are involved, sometimes the result is confusing. The so-called "landmark molt" actually is divided into two separate periods of molting in certain waterfowl, discussed later on.

The naming of a molt for the incoming feather generation was proposed formally by Humphrey and Parkes (1959, p. 10). That is, the Basic feather generation is acquired by a Prebasic molt, the Alternate by a Prealternate molt, and so forth. This has a developmental or physiological basis, well stated independently by Voitkevich (1966, p. 54): "[the feathers] are pushed out by the anlage of the new growing feather[s]. The development of the anlage begins before the old feather falls from its follicle." The falling out is a by-product of the event, just as discarding the scaly outer covering of various reptiles occurs after the new vestiture has formed underneath. The general scheme of this, in birds, seems to have been understood and described, more or less independently, at least several times. For example, Lowe (1933) included a diagram of "new feathers thrusting out old" and Voitkevich (1966, pp. 54, 55) referred to it by citing a 1934 paper by himself. Recent authors include Humphrey and Parkes (1959) and Watson (1963). The concept, however, has been implicit in numerous papers on integumentary derivatives of domestic fowls and ducks that have appeared in technical physiological journals over the past 40 years. The names of molts and resultant feather generations used by Humphrey and Parkes, and in this chapter, are independent of seasonal, reproductive, chronological age, or other phenomena (or combinations, as seasonal plus reproductive).

A bird's energy is "budgeted"—that for molting being fitted into rather short periods or, in other cases, distributed over a considerable period of time. (The anatomical and physiological data are treated in the preceding and following chapters.) Molting is done at least cost to the bird in maintaining itself. Prolonged and gradual molting must require some energy drain, but many birds evidently can continue to function satisfactorily without all feathers being present or fully developed. Heavy molting during briefer periods must be more of a handicap; that is, efficiency to perform other functions is reduced, or other functions may be omitted, but only temporarily. There are species that have part of their molting prolonged and part of it much more rapid. Prolonged molting, which may be derived secondarily, quite often is limited to the wing and then is a means of renewing the flight feathers without losing ability to fly.

In certain Northern Hemisphere ducks, the ventral body feathering, acquired in spring, is so worn and ragged on nesting females that it then hardly can be a satisfactory covering. On the other hand, in birds in which the feathers still are intact when normally molted, they really are not as good as new: for example, they may be brittle from drying out or exposure to sunlight. In still other cases, in which a feather generation (usually not the total feathering) is worn only a short time, it may be "exhausted" at the time it is lost in the sense that it has fulfilled some special function while still being structurally as good as new.

F. Names of Feather Generations

The names of feather generations are capitalized, all but the first one listed below being ordinary words that are being used in a special sense.

1. *Juvenal*

Juvenal is the first covering of "true" ("vaned" or "pennaceous") feathers. It has no counterpart later on; i.e., it precedes the cyclic feather generations. It is acquired long before sexual maturity. No species is known to breed when entirely in Juvenal feathering. [Unlike "juvenile," Dwight's (1902) term Juvenal is precise and unambiguous.]

2. *Basic*

Basic is the name of one feather generation in a cycle. If the cycle consists of a single generation, then the bird molts from Basic to Basic; if there is more than one, then one of them is Basic.

3. *Alternate*

Alternate is the name of a feather generation when there are two or more per cycle. That is, two generations per cycle consist of Basic and Alternate.

The naming relates to how many times germinal tissue is activated per cycle, being independent of season, sex, age, and other variables. Neither Basic nor Alternate necessarily is vividly colored or otherwise elaborate; both sexes may be somber in Basic or Alternate or in both. The idea that the "male looks most conspicuous and ornate during mating time" (Stresemann, 1963, p. 2), which is true for relatively few species, is irrelevant to the number of feather generations per cycle.

4. *Supplemental*

Supplemental is a generation that occurs in addition to Basic and Alternate in a cycle. It may be acquired during a period of molting into either Basic or Alternate, or possibly be offset from Prebasic or Prealternate molt. Not much is known about it. In the Long-tailed Duck or Oldsquaw (*Clangula hyemalis*) the molt into Supplemental begins before and continues into the span of molting from Basic into Alternate; then the Supplemental is succeeded by Alternate still within the span when other parts of the bird are acquiring Alternate. There seem to be very few cases of three feather generations per cycle. Four generations (Basic, Alternate, Supplemental A and B) are said to occur in some populations of Willow Ptarmigan (*Lagopus lagopus*) (Johnsen, 1929). A feather follicle must be activated twice per cycle to have two feather generations, three times to have three, and so on. Thus, when a molt is interrupted in, say, a two-generation cycle, that portion that grows after the pause is not a Supplemental generation.

III. Homology

A physical structure (including a generation of feathers), a display, a physiological process (including molting), a call note, or other attribute that occurs in two or more organisms and that resemble each other to such a degree—as determined by careful study—that they may be called by the same name are homologous. Determination of homology requires subjective assessment of resemblance and origin; then one can predict occurrence in additional organisms.

In a phylogenetic sense (common ancestry), avian wings are homologous. So are feathers. "It is, or course, impossible to be certain that plumage [feather generation] sequences which appear to be exactly equivalent in various groups of birds are truly homologous in the phylogenetic sense; however, we believe it is not only useful but even necessary to treat such equivalence provisionally as homology . . ." (Humphrey and Parkes, 1959, p. 2). Homology is implicit in any comparative study of molting and of feather generations. With this as a starting point, the problem is to describe seemingly endless variations and to try to unravel intrinsic and extrinsic causal factors (also possible convergence) and their interrelations. This is a core problem in avian biology and one of its major fascinations. "Many complexities in homological relations come to mind when we review the facts of anatomy, variation, genetics, and experimental embryology, and the uncertainties may be accentuated in studying closely related kinds of animals. Such irregularities encumber almost all biological concepts, but destroy neither their validity nor their usefulness" (Hubbs, 1944, p. 306).

From an evolutionary viewpoint, the "offset" principle is one of the most fundamental features of the avian integumentary system. This may be illustrated schematically as follows. If the ancestral avian pattern consisted of a simple cycle of one feather generation (entire feathering), then a cycle of two or more generations (hence two or more periods of molting) can be arrived at by evolving added generations that replace all or part of the single one—as diagrammed in Fig. 1. Thus, if one or more feather generations replace only part (say head-body-tail) in accordance with need to fulfill particular periodic functions, and the wing portion is retained throughout the entire cycle (so that ability to fly is least affected), the various evolutionary events required to achieve this have resulted in a more elaborate cycle. Then, in terms of feather generations present at, or just prior to, the time of "postponed" molting of the wing, it can be said that the wing molting is "offset." This approach to the evolution of cycles, which was discussed at length by Humphrey and Parkes (1959), fits in with (1) the limited number of molts and feather generations per cycle throughout the class Aves, and (2) a broad concept of homology. The great variation in modes of living that have evolved in birds includes such great variation in the nature, timing, etc., of molts and resultant feather generations, in whole or in part, as to appear to mask any underlying pattern. For the "offset" feature, see especially Fig. 3, but also Figs. 4, 5, and 8.

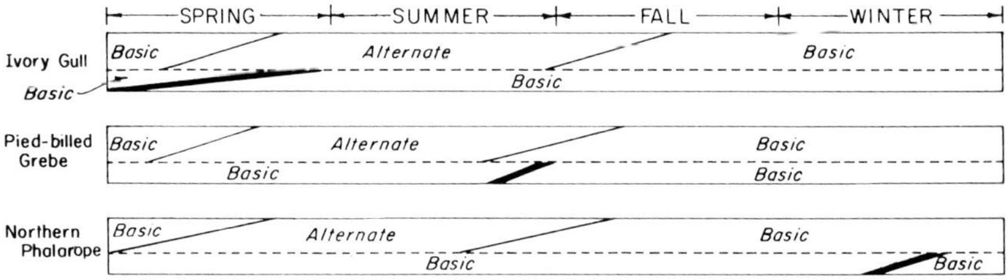

FIG. 3. Temporally "offset" compared with "normal" timing of wing molt. These species—Ivory Gull *(Pagophila eburnea)*, Pied-billed Grebe *(Podilymbus podiceps)*, and Northern Phalarope *(Phalaropus lobatus)*—have two feather generations per cycle and they molt head-body at about the same time. Wing is below the dashed line; the heavy diagonal line indicates molting. Reading down, the wing is molted before nesting, soon after ("normal"), and in winter.

IV. Functional Equivalency

(This section, and the one following, are included to indicate the breadth of the base of functional relationships of molting and its equivalency to other facets of avian biology.)

In many ways birds "extend" their capacity to function even while not having the "necessary" means. The jerking of the head from one stationary viewing point to another, by fowls and pigeons, achieves—with one eye—the equivalent of binocular vision. Some birds "extend their vocabulary" by producing nonvocal sounds; some even have feathers that are modified for this purpose. Some, as a result of abrasion of their feathering, achieve the equivalent of molting.

The European Starling (*Sturnus vulgaris*) has one feather generation per cycle. After the gray-brown Juvenal feathering, the sexes molt (in late summer) into light-tipped dark head-body feathering (Basic I) and this "spangled" appearance is renewed at each subsequent Prebasic molt. The light tips of the fully grown Basic I wear off during the winter and the birds are in highly iridescent dark feathering before they breed in spring. In the latitude of central New York state, after the multiple-brood season, the sexes (at time of minimal gonad activity) molt again into spangled feathering (Basic II). Then, in late summer, birds of both sexes (but especially males) visit nest holes, perch and sing very close by, and occasionally a female will defend a nesting territory; see Kessel (1957) for details. At this season, on approach of another Starling, the "owner" has a special display which includes fluffing out the spangles. The site may be tenanted all winter or the owners may join a social flock and be absent for a period. In spring, after the spangles have worn off at their line of demarcation, the male displays in quite another manner, generally from a higher perch or from near the nest hole—the song differs, the wings are fluttered vigorously, the throat-feathers are vibrated, and the fluffing out that was done in fall is minimal or omitted. A single feather generation accommodates both fall and spring situations.

The Starling's Basic feather generation is homologous with the Basic of the Scarlet Tanager (*Piranga olivacea*); in the latter species there is an added generation—the head-body portion of the Basic being succeeded by red Alternate in spring (Humphrey and Parkes, 1959, p. 9). Stresemann (1963, p. 2), however, stated that "if one wants to compare biologically, one will have to equate the brilliant spring plumage of the Starling . . . with the bright red plumage of the Scarlet Tanager, obtained by molt . . . ." Without suggesting here why an activity apparently relating to next year's nesting is "set back" into the preceding late summer, or even homologizing the two conditions of the feathering of the Starling with the seasonally corresponding two feather generations of the Scarlet Tanager, it is very evident that (1) the Starling has the equivalent of two generations per cycle with both equivalents relating to the reproductive cycle, and

(2) in the tanager, all facets of the reproductive cycle occur after an intervening molt (into Alternate head-body).

An icterid, the Bobolink (*Dolichonyx oryzivorus*), has two feather generations per cycle. In the definitive cycle of the male, the Basic generation (in fall) is quite buffy, i.e., femalelike; after molting into Alternate (in spring), the male again appears buffy. In due course, however, the buffy ends of the Alternate feathering wear off and the male becomes black and white with buff nape; see color plate in Chapman 1932, opp. p. 482). There are several interesting points about this case. At each period of molting, from loss of first Alternate onward, the entire feathering is renewed (as in the hypothetical ancestral bird having one feather generation per cycle, discussed earlier, also at top left in Fig. 1); molting of the wing is not offset from that of head-body. The recurrence of buffy coloration, after molting from Basic to Alternate, appears to be the functional equivalent of retaining the Basic for a longer period than it is actually worn. From the point of view of intrinsic factors, such as hormonal levels, it seems as though a molt of all feathering has been offset in time.

Another icterid, the Orchard Oriole (*Icterus spurius*), has two feather generations per cycle and in the male the new feathers of Definitive Basic (in fall) have wide buffy distal margins; these margins wear off by spring, in which season there is also a molt into Alternate (restricted to black feathering of the head) before onset of the breeding season, and the male is mostly reddish brown and black; see color plate in Chapman (1932, opposite p. 488). In still another icterid, the the Rusty Blackbird (*Euphagus carolinus*), the "rusty" margins of the Basic feathering wear off and then the bird is glossy black. (There are other icterids in which the feathering is all dark from the beginning.) This "margin" phenomenon occurs also in some birds that are neither icterids nor temperate-zone nor migratory; the neotropical Blue-black Grassquit (*Volatinia jacarina*) is a conspicuous example.

The Snow Bunting (*Plectrophenax nivalis*) has two generations per cycle; the Alternate may be disregarded for present purposes, being a molting in spring into white or very light feathering on part of the head. In the new Definitive Basic (in fall), both sexes have various feathers (especially on the breast and head) white with brown margins and others (on the mantle) black with white margins. The margins wear off by spring, so that the male becomes essentially white and black. The structural differences between the black and white portions of a white-margined feather were discussed and depicted by Stone (1896, p. 119). The Snow Bunting is rather unusual in that it has white feathers that lose nonwhite margins. The condition of two feather generations per cycle, with the Alternate more or less limited to the head and the Basic being the equivalent of two generations as a result of abrasion, is prominent also in the Snow Finch (*Montifringilla nivalis*), Lapland Longspur (*Calcarius lapponicus*), and various *Emberiza* species (e.g., Rustic Bunting, *E. rustica*).

The three icterids, the Snow Bunting, Lapland Longspur, and Rustic Bunting—highly migratory species—acquire the equivalent of an added feather generation without expending part of their energy budget on molting. The essentially nonmigratory Snow Finch, in its harsh habitat, does likewise. But the situation is different in arctic grouse (*Lagopus*), which seldom are more than locally migratory, and which have seasonal change from "white" to "brown" via molting.

The phenomenon of differently colored margins or ends of feathers, generally simplified in structure, and worn for less time than the remainder of the feather, is very widespread in birds. There are many obvious examples among anatids, turdids, icterids, emberizids, fringillids, and others. Among shorebirds (waders), the light margins often extend into the sides of the feathers as a series of indentations, as in curlews (*Numenius*); after abrasion, the feathers are essentially unicolor and are notched or scalloped in outline. In Golden Plovers (*Pluvialis apricaria, P. dominica*), the "golden" in the margins wears off well in advance of the time that the feathers drop out. In various ducks, some with and others without iridescent head coloring, color and even pattern are more or less obscured very briefly when the new Alternate feathering comes in; the feathers have differently colored tips (in some species buffy, in others darkish) which break off readily. This is so fleeting a condition in the drake Mallard (*Anas platyrhynchos*) that it is seldom noticed; it is more lasting in the drake Baikal Teal (*Anas formosa*), the head-neck having a hoary appearance for perhaps 3 months, so that the highly patterned, partly iridescent condition of the head is not evident until well into winter. Except for this teal (for which no explanation is suggested here), occurrence of different appearing feather ends or margins in the birds just mentioned does not seem to be quite the full functional equivalent of an added feather generation. The matter probably is complicated, with a different explanation required for different groups or species. In the shorebirds mentioned, however, it would appear that the margins, spots, and other markings may be functionally useful (as in pair formation) rather briefly and so can be dispensed with at a time when the remaining portions of these same feathers still are needed for other functions (such as thermoregulation).

The concept of equivalent includes the use of "characters" of feathering that are "detached" from the bird and are part of its physical environment. Certain male bowerbirds in the genus *Amblyornis* that lack the colorful adornment characteristic of other bowerbirds "compensate" for this morphological inequity by adding colorful ornaments to their bowers (though they are not the only ones to do this). This transfer to the environment having been made, natural selection operates so as to favor protective coloration—the males become more femalelike. See Gilliard (1956) for details.

V. Social Factors

The condition of one bird, as reflected by its behavior, may influence the time of molting of another. That is, it may promote a more nearly synchronous molting of mates. In various geese and swans of the Northern Hemisphere, the time when the female begins to lay varies with latitude, also locally depending on season, and other factors. There are a number of fairly constant time intervals, for example, as follows in the Lesser Snow/Blue Goose (*Anser c. caerulescens*): 4–5 days for laying, 22–23 days for incubation and, at approximately 21 days posthatching, the goose drops her wing quills. After an added 3–4 days, the gander drops his quills. If the goose begins molting after a fairly constant time span measured from a variable (within limits) date of clutch initiation, and the gander (who guards but does not otherwise assist the goose) goes into molt soon afterward, then timing of onset of molting of the gander quite likely is triggered some time beforehand by visual and/or auditory cues from the goose. That is, some signal initiates the preparatory period, and the time when the gander begins molting may be adjustable in some measure, depending on when the signal is received. "If, under the influence of a nerve-impulse, the old feather is to some extent loosened within the follicle, favourable conditions are created for the development of the previously dormant feather germ" (Voitkevich, 1966, p. 253). Perhaps the timing or nature of any cue is variable since, if nesting is disrupted, molting becomes a dominant energy expenditure at an earlier time in both sexes. There is no information on interval (if there is one) between molting of mates under these circumstances.

The situation is not the same in various Northern Hemisphere ducks, in part because (1) more than one feather generation is involved in the corresponding period of molting, and (2) drakes generally take leave of their mates for the season at some rather variable time during the incubation period. (The question of why the time of the drake's departure varies has not been investigated; the duck may discourage his presence!) The presence or activities of the duck may have some influence on the drake's schedule. This has some support from the observation that "bachelor" drakes (unmated males of breeding age) begin molting at another time (earlier) in the season.

According to the Fraser Darling hypothesis, in colonial birds a social factor tends to bring the reproductive cycles of the various individuals into synchrony. It may be noted here that there is an underlying common influence on molting as well as on reproduction; a tendency to synchrony applies to both. However, in the colonial Sooty Tern (*Sterna fuscata*), reproduction and molting can be too far out of phase for this to apply. This species (having two feather generations per cycle) nests on marine islands scattered within and beyond the tropics. The cycle of molting, as studied at certain terneries, is less than a year long and appears to be a more constant periodic phenomenon than is reproduction—the latter can "interrupt" the former; an individual can nest again after as brief an interval as 7 months. See Ashmole (1963) for details. There appear to be quite comparable situations among tropical land birds.

Also in regard to the Fraser Darling hypothesis, it must be kept in mind that both positive and negative factors exist in social situations. In certain seabirds, also swans and geese that nest in view of one another, the amount of optimal nesting habitat is limited. The various individuals or pairs that "own" a site where the egg(s) are incubated (1) interact among themselves, as mentioned above, and (2) also tend to interact with their young so as to inhibit breeding of the latter. That is, they keep younger birds away from optimal habitat and so prevent prebreeders from becoming breeders. The result is "in" birds and "out" birds and interactions within each category and between them. In various swans and geese the pattern is as follows. Established breeders have the optimum habitat divided among themselves and use it year after year. Their offspring—prebreeders of more than one cohort (age class)—are driven away by the established owners. Evidently this occurs at a sensitive time in the annual cycle when reproduction can be inhibited. Some of the former eventually find vacancies and nest there; others may nest in less than optimal habitat thereabouts; others may remain in peripheral areas, engage in nest building, and even copulate, but do not produce eggs (i.e., they have an incomplete breeding cycle); and still others—the majority being of the youngest cohort—engage only in mere fragments of reproductive behavior, soon form in flocks, and depart on molt migration. In large races of Canada Geese (*Branta canadensis*) and the Trumpeter Swan (*Cygnus buccinator*), the inhibiting influence is known to result in some young birds continuing as prebreeders for several years beyond the age when they are potential breeders physiologically. Trumpeter Swans, transported to an area remote from the presence of any others of their kind, formed a stable pair bond when 20 months old and began their first successful nesting just before they were 3 years old (Monnie, 1966). On the other hand, there are individual wild Canada Geese in which an inhibitory influence seems to have become permanent; they never become breeders. In the different categories and cohorts the molting schedule is as follows: prebreeders (and permanent non-breeders?) begin first; the timing is intermediate in birds having an incomplete breeding cycle and in breeders whose reproduction is disrupted (failed breeders); it is latest, being "postponed" into the rearing period, in successful breeders. Also, a very tardy season, i.e., unfavorable environment, can inhibit breeding even in long-established breeders; they then correspond to failed breeders.

In the cases cited in which periodicity of breeding and molting are not vastly out of phase, when the social influence on breeding is positive it also promotes a tendency toward synchronous molting in a breeding unit; when it is negative, the molting that ordinarily occurs afterward gains ascendancy and occurs earlier.

VI. Describing Molts and Feather Generations

A. Methods

For studying molts and feather generations on the exterior bird, about the only instrumentation needed is a thin-bladed tool, such as a scalpel, for parting short feathers, and a hand lens, for examining the margins and surfaces of feathers. A useful technique consists of dampening feathers with alcohol, especially for looking under coverts at the bases of flight feathers, when the coverts might fall out if disturbed. Signs of molting include empty feather follicles, partly ensheathed feathers, small "pinfeathers" and others fully ensheathed, and (as seen on the flesh side of the skin) dark follicles of growing feathers.

There are various ways, which have both advantages and disadvantages, of studying live birds; or one may examine museum series of skins. If birds can be captured alive in quantity, examined for molting and identification of whatever generation(s) of feathers may be present, and suspected key specimens are sacrificed (for checking sex if not obvious beforehand, any internal age characters, flesh side of skin), the resulting data should indicate the span of molting in the entire sample and when most molting occurred. The data are communicated best if they can be analyzed to the extent of describing an individual which represents the mode for a sex or age class for that locality and time interval—that is, a modal individual—and then adding to the description such variables as were found in the sample. (See Fig. 8 for modal Gadwalls.)

The "population" method naturally works best for studying definitive molts and feather generations of single-brooded species; there are comparatively few variables in the cycles of the various individuals. The picture is much more complex in predefinitive molts and feather generations of multiple-brood social species (such as certain icterids) in the interval after the long breeding season and before the end of the migration that follows. Flocks contain early- to late-hatched individuals and, since their variation in age is reflected by differences in molting and other functions, the general picture for the sample may mask the desired information about an individual.

In working with museum bird skins, trying to assemble specimens of both sexes in each and every molt is a constant problem. If it can be done, a continuum can be worked out from these static points of reference. Sometimes a handicap of inadequate material can be minimized by investigating several closely related species at the same time. The nature of a "gap" in data on one species may be inferred (via homology) from material that fits the span corresponding to this gap in another species. This at least will serve tentatively until further material can be studied in another collection. One must always be prepared, however, for a species to have characteristics peculiar to itself, which require revision of any tentative conclusions. The study of museum material was discussed by Parkes (1963, pp. 123–125).

B. Examples

1. *Agelaius*

Two birds were collected, a male and female (probably mates), in mid-August in Peru, and described by Short (1969) as a new species of *Agelaius* (Icteridae). Male—feathering soft in texture, black with some blue-green gloss. Female—as male except less glossy, a single breast feather had a narrow brown tip; there was a patch of newly emerging feathers in the dorsal tract (believed to be a replacement of accidental loss, since there were no other signs of molting).

The feathering might be designated in terms of season (in Southern Hemisphere), or reproduction (both birds had somewhat enlarged gonads; the male was in song), or feather generation(s)—tentatively, via homology with better-known members of the genus. However, with so little material available, nothing is gained by invoking conceptual terminology; it should be omitted, as in the present instance.

2. European Blackbird (Turdus merula)

The molts and feather generations of this thrush (Turdidae) are well known; the following is based in the main on Witherby *et al.*, (1938, pp. 139–140):

European Blackbird (*Turdus m. merula*)—sexes much alike in appearance; one feather generation per cycle or possibly two (some body feathers may be molted twice); part of the Juvenal generation is retained and worn with Basic I.

♂ Definitive Basic (entire feathering), acquired within the period August–October and retained until within the following August–October. All feathering glossy black, except proximal margin of flight feathers of wing grayish. Notes: (1) some individuals acquire some new body feathering in spring; (2) the wings are brownish when worn and abraded; (3) a considerable number of individuals have more or less white feathering. An increase in the amount of white in the feathering of an individual in succeeding years "appears to be due to accumulating deficiencies"—probably dietary. The initial abnormal white may begin at any age and increase from there. "Senility may be a cause of increasing white in old birds and the two causes may work together as the bird ages" (Rollin, 1964).

♀ Definitive Basic (entire feathering), timing as ♂. Coloring mostly a dark umber (the tail nearest black). Head slightly patterned—lores freckled paler and with grayish line above, ear coverts dark with off-white shaft lines, chin and throat pale grayish (latter streaked dark); remainder of underparts variously brownish, often with rufous-tawny tinge, and terminal parts of the feathers (especially on breast and sides) conspicuously dark; undertail area umber, streaked pale; wing—upper coverts match the mantle; flight feathers have outer webs deep umber, inner webs margined grayish; wing lining brownish, the feathers with pale shafts and often rufous tips and edging. Note: The feathering often is much abraded by summer, the dorsum then browner; the venter loses any rufous tinge and many dark spots, hence becomes grayer.

♂ and ♀ Nestling. The down is scanty, fairly long, and pale gray-buff.

♂ Juvenal (entire feathering), evidently worn for some weeks after it is fully acquired and then, within the span August–October (or sometimes to November or December), replaced by Basic I, except the Juvenal tail, flight feathers of wing, primary coverts, and some distal greater coverts, which are retained into the following late summer or later. General appearance: brownish, with darker wing and tail, feathers of dorsum have a light central streak (along the shaft), ventral surface with rather blended pattern of darkish spots on medium brownish background. Head patterned as ♀ Definitive Basic, but browner and with paler chin and throat. Often a rufous tinge on upper tail coverts. Wing: upper surface—flight feathers and primary coverts almost as brown as ♀ Definitive Basic, median and lesser coverts have pale reddish-brown central wedge-shaped marks (base of wedge at outer end of feather), broader than in ♀ Definitive Basic; wing lining and axillars medium reddish brown.

♀ Juvenal (entire feathering), timing as ♂ and the same feathers are retained and worn with Basic I. Differs from ♂ Juvenal as follows: dorsum not as dark; tail browner; wing browner, with covert markings usually less rufescent (but individuals vary greatly), and underparts also variable.

♂ Basic I (all feathering except tail, flight feathers of wing, primary coverts, and some distal greater coverts—which are retained Juvenal), acquired in late summer or fall (August–October usually, sometimes later) of year of hatching. Differs from ♂ Definitive Basic as follows: generally not as dark, being blackish brown; feathers of crown and dorsum often have inconspicuous brownish edging (it wears off); feathers of chin and throat generally are conspicuously edged with gray, occasionally are gray with dark streaks (quite like ♀ Definitive Basic); most feathers of underparts have clearly evident brownish or grayish margins and often pale shaft streaks.

♀ Basic I (inclusive feathering as ♂ Basic I), timing as ♂. Often difficult to distinguish from ♀ Definitive Basic; upperparts commonly not quite as dark. Retained Juvenal feathering (tail, part of wing) becomes somewhat bleached and abraded; in the wing, therefore, some contrast may be evident between the old (Juvenal) and new (Basic I) in the covert area.

The following are some points of interest about the above description. (a) The sequence is arranged arbitrarily—first the definitive cycle of the ♂, then ♀, then stages from posthatching up to definitive. (b) The inclusive feathering of each feather generation is stated. (c) The periods of molting are given. (d) The spans of molting are based on examination of many individuals and are inclusive for all; the actual duration of molting of an individual is within each span. (e) The "offset" feature consists of part of the Juvenal generation being retained and worn with Basic I. (f) In describing total feathering, the wing is described last. (In order to make descriptions consistent with one another in presentation, this practice quite commonly is extended to all descriptions. That is, cases with and without the "offset" feature arbitrarily are described in the same order.) (g) Comparisons all are within the species; the various feather generations of each sex are diagnosed and compared with one another and with those of the other sex. (h) Since all necessary data on the feathering are given, one knows what an individual bird looks like at any time of year, whether breeding or not, and so forth. (For present purposes, the description is limited to diagnostic features of feathering; nothing is included on seasonal coloring of unfeathered parts, duration of the down, development of the Juvenal feathering, or whether there are similar-appearing feather generations in related species.) (i) The description is of one subspecies only and includes individual variation; other subspecies then might be discussed comparatively as to timing of events, coloring of feathers, and other relevant matters. (j) From an evolutionary viewpoint, one might speculate that this thrush is in the process either of adding or eliminating a feather generation since, in the definitive cycle, there is limited molting on the body of some individuals in spring.

3. *Various Ducks*

The presentation here is different from that of previous examples. First there is a background of concepts—derived from study of live birds, birdskins, and published information—and then follow terse descriptions which are a product of applying all relevant ideas and concepts given below and elsewhere in this chapter.

a. Concepts. i. Evolutionary change. The "Mallard group" in the genus *Anas* is especially interesting; the following is set forth as a working hypothesis regarding the nature of certain present-day species. The ancestral Mallard stock was not sexually dimorphic and it dispersed widely, from an Old World place of origin to various continents and islands. Those reaching North America were, subsequently, at a time when much of the continent was not habitable for them, isolated (as though on islands) in (a) the North Atlantic coastal area, (b) Gulf coastal area, and (c) northern Mexico or thereabouts. Each of these populations developed certain habitat preferences and evolved certain morphological characters: (a) The Black Duck (*A. rubripes*) became darkest, (b) Mottled Duck (*A. fulvigula*) smaller and brownish, and (c) Mexican Duck (*A. diazi*) large and pale. None of these developed marked sexual dimorphism. In the Old World in the meantime, the Mallard stock evolved into a highly sexually dimorphic bird (*A. platyrhynchos*)—the drake becoming green-headed and maroon-breasted. Subsequently, when interior and northern North America became habitable, the green-headed birds invaded and occupied the prairies and other interior areas. The latest major event not influenced by man was northward occupation, by green-headed birds, of part of Greenland; the present morphological uniqueness of Greenland birds (*A. p. conboschas*) consists mainly of larger size.

The wide-ranging Gadwall (*Anas strepera*) (not in the Mallard group), at a fairly recent time and after its dimorphic pattern was fully developed, colonized Washington Island (Fanning group) in the southwest Pacific. Judging from the two specimens of this now-extinct population, the birds merely became a diminutive form of the ordinary Gadwall.

Thus, well after sexual dimorphism was firmly established, green-headed Mallards went north and became larger; Gadwalls went south and became smaller.

All members of the "Mallard group" that have been investigated are found to have the same molts and feather generations regardless of whether or not they are sexually dimorphic. That is, homologous molts and feather generations are older throughout the group than either (a) present distribution of members of the group or (b) attainment of sexual dimorphism by certain of its members.

In the "Shoveler group" there appears to have been considerable sexual dimorphism in spatulate-billed parent stock prior to wide dispersal to various continents and islands. This dimorphism probably is relatively old. Later evolution of a dimorphic pattern seems to have been greatest in the presently holarctic birds, while the various other geographical populations appear (to the human eye, at least) not to have proceeded as far along the path of sexual dimorphism. In all Shovelers, evidently, Basic I in the drake (all feathering except most of wing) is distinct in pattern and, for an *Anas* species, is long retained. Again, the homologous molts and feather generations antedate, and have been more stable than, development of sexual dimorphism.

The above hypothesis in regard to some *Anas* species (others, such as Pintails, could have been included) is contrary to the usual thinking that, for example, Mallards were green-headed before they dispersed widely and, via natural selection, secondarily lost most of the sexual dimorphism in geographical areas where it was not needed.

ii. Some definitive cycles compared. In various Northern Hemisphere ducks the female has a spring molt into Basic head-body-tail, the wing portion of this molt being "postponed" or "offset" into the time when the duck has a preflight brood and then it may continue and "overlap" the next (Prealternate) molt. In the female North American Wood Duck (*Aix sponsa*) the Prebasic molt (of at least head-body) begins somewhat later, about the time of the first attempt to nest, and may continue between (and to some extent during) successive nesting attempts. In a northern stifftail, the Ruddy Duck (*Oxyura jamaicensis rubida*), the entire cycle is simpler and similar in both sexes (see Fig. 4); they acquire Basic (entire feathering) in late summer–early fall and wear this until spring, when they acquire Alternate head-body-tail and retain the Basic wing. [In the Southern Hemisphere, captive *O. maccoa* of both sexes, and wild male *O. australis*, are known to molt the wing twice per cycle (Siegfried, 1970), but no evidence of this has been found in the Northern Hemisphere *O. j. rubida*.] In the northern population of the Ruddy, display is not a prominent activity until spring, so the Alternate is both pair formation and "breeding" feathering; in various other waterfowl, pair formation occurs in fall–winter (Alternate head-body-tail) and the females (but not the drakes) have an intervening molt (into much of Basic) in spring before they "breed" (nest); see Mallard in Fig. 4.

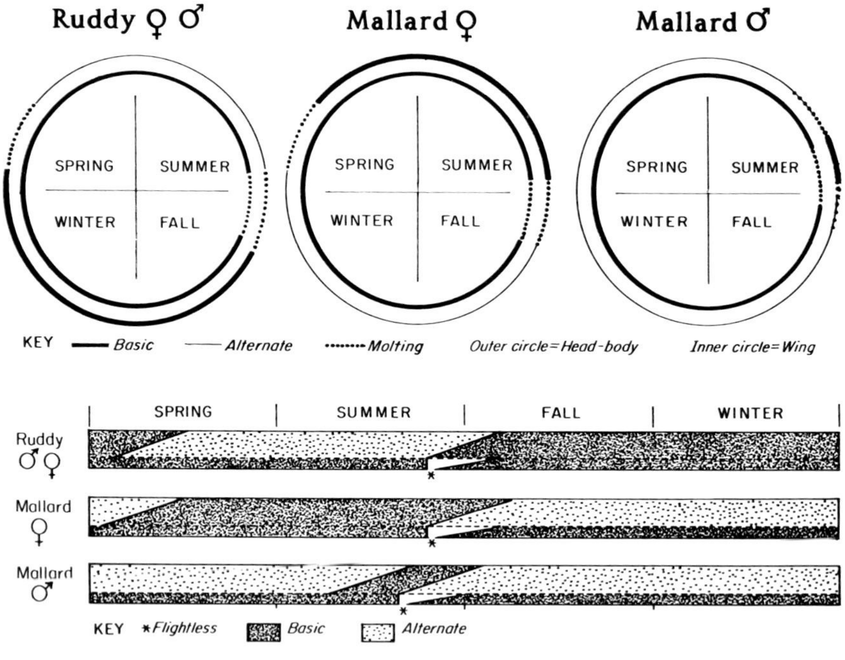

FIG. 4. Definitive cycles (of two feather generations) of northern population of Ruddy Duck *(Oxyura jamaicensis rubida)* and of Mallard *(Anas p. platyrhynchos)*. Ruddies acquire Alternate ("display") head-body-tail in spring; the ♀ Mallard acquires Basic head-body-tail in spring, but the drake wears Alternate head-body-tail most of the year and his two molts are strung together so that intervening Basic head-body-tail is worn briefly in late summer. The same information is shown in circular and in linear diagrams. Also see text for additional details.

Thus both Alternate and Basic generations are directly involved in the reproductive cycle and, furthermore, it is known that some of these waterfowl actively engage in pair-formation activity even while molting. [*Note added in proof:* Northern Ruddies (*O. j. rubida*) now have been found to have an added wing molt in captivity in winter.]

iii. Periodic phenomena. Molting, fat deposition, gonad development, and migratory drives are not necessarily on the same schedule in relation to each other in members of a species that nest, say, in the latitude of Missouri and also in the subarctic Mackenzie Delta. Within certain limits, there are adjustments depending on length of warm season. (The Canada Goose is a more striking example, since it includes southerly breeders that are essentially nonmigratory.)

iv. Components of certain feather generations. In writing descriptions, "body," or "head-body," and "wing" are used somewhat arbitrarily. In various waterfowl having more than one feather generation per cycle, several innermost secondaries and some of their overlying coverts are renewed each time the head-body is renewed. That is, the actual dividing line between "head-body" and "wing" occurs a short distance out on the latter. The innermost secondaries ("tertials") often are elongated and strikingly patterned in Alternate feathering; They grow fairly early during the span when the body is acquiring Alternate and usually attain full length earlier than the scapulars. (The tail is renewed each time head-body is renewed in some genera, as *Anas,* but only each time the wing is renewed in others, as *Clangula*).

This linkage of innermost wing feathers with the body occurs widely in birds. Examples: certain grouse (Galliformes), various gulls, terns, plovers, and sandpipers (Charadriiformes), and many Passerines.

v. The "offset" phenomenon. This is typical of wing molting of waterfowl having two or more feather generations per cycle, as in Mallard in Fig. 4. The Juvenal wing is retained and worn with Basic I and Alternate I head-body; in the female, which has a spring molt into Basic head-body, it is also retained beyond the molting into that feathering. In the definitive cycle the molting into and out of Basic ("eclipse") has not been understood and may be clarified as follows. First, it is important to recall that, in swans and geese (one feather generation per cycle), molting begins with loss of flight feathers of the wing; this seems to be a very old pattern. In at least many Northern Hemisphere duck species having two generations per cycle, the drake in spring wears Alternate head-body plus retained Basic wing; (a) in early summer the Alternate is replaced over a period of time by Basic head-body (the "molt into eclipse"); (b) then the wing quills ("offset" —that is, retained from the previous Basic) are dropped, and thereafter (c) the newly acquired Basic head-body is succeeded, in the same prolonged molting, by Alternate head-body concurrently with, or some time after, growth of Basic wing quills. The duck differs from the drake mainly by exchanging Alternate head-body for Basic during an earlier, separate, period of molting in spring. Thus much of the duck's Prebasic molt occurs well prior to nesting (but not in *Aix*). The combination of unusual cycles, differing in the sexes, plus an extended period of molting, plus the "offset" phenomenon, have resulted in endless confusion in the ornithological literature. Compare Figs. 3, 4, and 8.

vi. "Order of molt." The feather tracts are activated in a certain order. Example: A duckling needs protection and waterproofing ventrally before it needs equipment for flying; a Passerine needs flight feathers early. In waterfowl, the order of activation of tracts and relative rate of growth of areas of the Juvenal feather generation tend to be reflected in other periods of molting throughout the life of the individual. It does not follow that a tract that starts early also ends early; some require more time than others. Also, more than one tract may begin at approximately the same time. Generally speaking, molting on head-neck precedes the body, and the tail is molted gradually or late; the wing is a separate matter. (The order of molt within tracts is not discussed in this chapter.)

vii. Pressure of time. Many waterfowl seem to be "in a hurry" to get through the early, i.e., developmental, feather generations. Various swans and geese go from Juvenal into Basic I head-body surprisingly early and rapidly (Fig. 5). Various ducks go through Juvenal, then Basic I, then Alternate I head-body rapidly. There is continuous, or nearly continuous, molting and the feather generations come in like overlapping waves. An entire "wave" may not occur on the individual at any time, since early acquired feathers may be lost

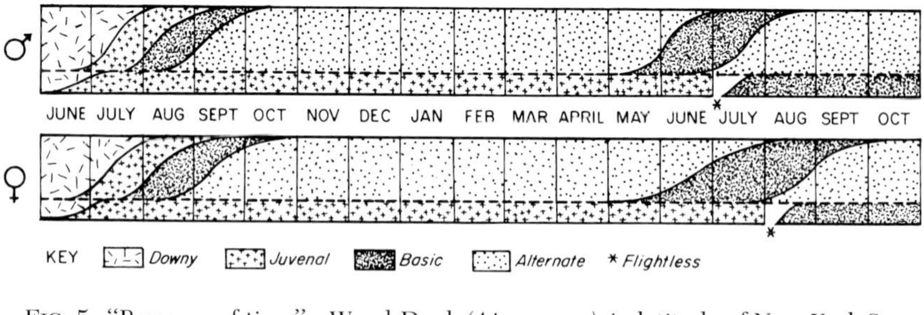

FIG. 5. "Pressure of time"—Wood Duck *(Aix sponsa)*, in latitude of New York State. The early feather generations succeed each other so rapidly that each is not worn in entirety at any time. Wing is below the dashed line.

before the late ones grow; a particular feather generation is only part of a continuum. From a physiological viewpoint, it may be noted that so much energy is devoted to this continuum that these waterfowl have very little fat from about the time they attain flight until at least well into winter. In telescoping the sequence of molts so that feather generations overlap, (a) the Juvenal generation may begin to fall out before all its feathers are fully grown and (b) Basic I is the most variable from species to species in extent of inclusive feathering and in time span retained. One might say that any "slack" is "taken up" or "let out" via this feather generation.

There is a limited "corrective effect" in that late-hatched waterfowl broods may grow their Juvenal feathering more rapidly than early-hatched ones. This has been demonstrated for flight feathers of the Redhead (*Aythya americana*) by Smart (1965) and Dane (1965), but the latter observed no difference in late-hatched Blue-winged Teal (*Anas discors*).

Various Northern Hemisphere waterfowl that first breed at an age of approximately 1 year—even with such "telescoping" as their growth processes will allow—do not "catch up" (get into synchrony) with older cohorts until some time after their first breeding season. That is, they are not equipped with the appropriate feather generation for pair formation until well along in the appropriate season (usually winter), nor ready to nest as early in the following summer. In some species the peak period of clutch initiation tends to be late. Therefore, since reproduction and molting are directed by a common influence, their period of molting after breeding comes later. Thus the schedule of the youngest breeding cohort may differ somewhat from that of older ones.

The utilitarian imperative, or pressure of time, results in the telescoping of feather generations of various birds other than waterfowl. It occurs widely in Galliformes, for example.

viii. Momentum of molting. After the initial threshold is past, molting gains momentum or intensity and, after a period of much feather growth ("heavy molt," "peak of molt"), tapers off again. The process might be diagrammed as a flatter curve in certain predefinitive as compared to certain definitive molts. In relation to a particular feather generation (not necessarily total feathering), the latter molts seem to be more "compact" events, but the pattern is upset easily—for example, by repeated disturbance of molting birds, poor nutrition, and other factors.

ix. Corresponding molts in the sexes. In waterfowl that have been studied in detail, a feather generation that is present in one sex in a species is present in the other also. However, they may not correspond in timing or extent of inclusive feathering. In birds in general, the exceptions to this rule may be rather few.

x. Feather shapes. In waterfowl, feathers tend to be narrow in Juvenal, wide with rather truncated (squarish) ends in Basic, and longest, often tapering, and with rounded (even pointed) ends in Alternate generations (see Fig. 6). The downy basal portion of pennaceous feathers is increasingly smaller, in proportion to the vane, in successive feather generations up to the stabilized condition of definitive cycles. In definitive cycles there is a slight difference in amount of downy structure between Basic and Alternate (more in the former) but, when comparing feathers from a series of individuals, this tends to be obscured by several variables.

xi. Feather markings. In many waterfowl, feathers having longitudinal markings are more typical of Alternate and transverse of Basic; sometimes, however, molting into a feather generation evidently straddles a period of change in hormonal or some other influence and the resulting pattern differs on early- and late-acquired feathers. It even varies on the same feather.

xii. Similarity of feather generations. In various waterfowl, and especially in females, the Juvenal and Basic I feathering often are not markedly different in general appearance; this has resulted in a mixture of the two, or even a preponderance of the latter, being described as Juvenal. (There seldom is difficulty in recognizing Basic I after much of it has appeared.) In females of various *Anas* species in which the venter has a streaked pattern, there is considerable difference (such as width of markings) between Alternate and Basic but, due to overlap in variation, it is easy to mistake the two for individual variation in a single feather generation. The process of molting from one generation into another that closely resembles it might be compared to cinematographic lap dissolving of one image into another quite similar one.

xiii. Morphs. If there is a white morph, as in Blue/Snow Goose, it is easier to work with the colored one and then compare with the other. The times of molting, inclusive feathering, shapes of feathers, and other characteristics are the same in both. (Certain swans are not difficult in off-white early stages; adventitious staining plus feather shapes and wear help after they become all white.)

xiv. Nest down. The down that is used by waterfowl to form the nest and cover the eggs in the sitter's absence has special characteristics, such as a cohesive quality. Swans and geese produce a smallish to large quantity of nest down, depending on the species, with deposition continuing from within the laying span into incubation. Whist-

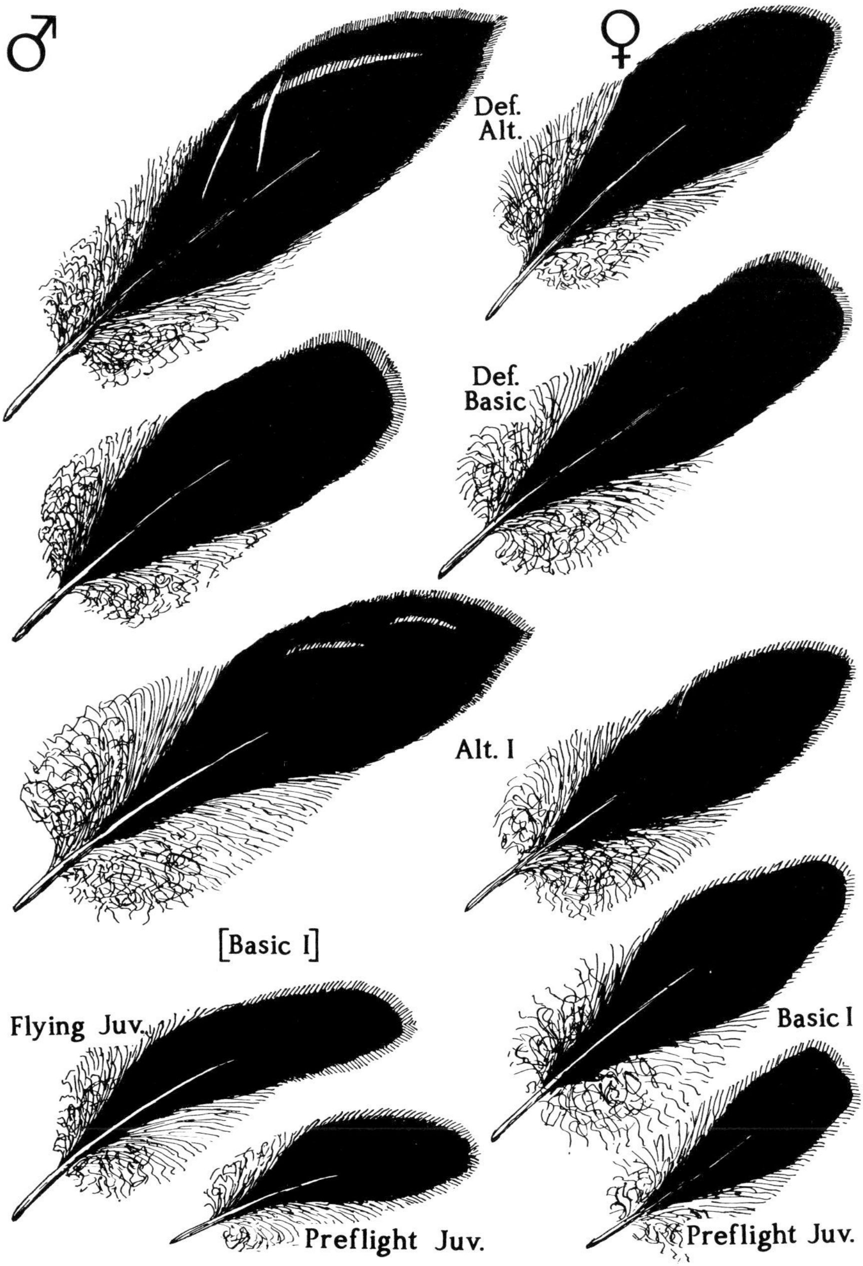

FIG. 6. Black Duck *(Anas rubripes)*—shapes and other details of long scapulars. The Basic I generation in the drake is limited in extent and evidently does not include the scapulars.

ling (also called Tree) ducks (*Dendrocygna* spp.), which are warm-climate species, apparently produce none; at least none is found with clutches. In Northern Hemisphere waterfowl having two or three feather generations per cycle, the nest down, which is produced abundantly, grows toward the end of or possibly after the spring molting of contour feathering. Rather interestingly, certain drakes (e.g., Mallard, Gadwall), when rapidly producing a series of predefinitive feather generations, have Basic I as a mere fragment—the feathers grow rapidly, are attached loosely, and very soon fall out. This same phenomenon of rapid growth and rather loose attachment is characteristic of the production of nest down on the venter of females. Generally speaking, the initial nest-down production antedates rapid development of eggs within the bird, thus spreading out the combined energy drain. Then the down is added to the nest beginning within the laying period and ending within the incubation period. In cases of re-nesting, the duck again produces eggs, but evidently little (or no) nest down. The North American Wood Duck, a cavity nester, does not produce abundant down even with her initial clutch of the season; thereafter the duck may make repeated attempts to re-nest; i.e., clutch production may continue for a long time after nest-down production is terminated. The northern population of the Ruddy Duck is interesting in that production of nest down occurs while the female is in Alternate contour feathering (not Basic as in Anatinae); this situation probably applies to all stifftails (*Oxyura*).

xv. Castration. After a period of adjustment, de-sexing results in only male-type Alternate head-body-tail being acquired at succeeding molts by both sexes of the Mallard; both continue, however, to have two feather generations per cycle. In various larids, the result of de-sexing is Basic feathering in both sexes, also Basic in both male and female domestic fowl, and reportedly no change in either sex in the House Sparrow *(Passer domesticus).*

xvi. Captivity. This tends to upset the timing, including duration, of molting in anatids. Some species adjust fairly well to captivity—e.g., various *Anas*, *Aix*—but it is common to see captives (especially drakes) of these species still largely in Basic feathering much later than is normal in the wild. Captivity apparently can have the same effect as castration in *Aythya;* a drake Canvasback *(A. valisineria)* kept captive for a number of years eventually molted from Alternate to Alternate head-body (Hochbaum, 1944, p. 112). It may be concluded, then, that the natural situation is at least blurred in captivity, or may be profoundly altered, so that observations of captives must be treated with caution.

b. Examples. The various concepts and principles of molt analysis now will be applied to three kinds of ducks. Lengthy descriptions of the feather generations are omitted entirely here and emphasis is restricted to inclusive feathering and patterns of molting.

i. Mallard (Anas p. platyrhynchos). The Mallard has two feather generations per cycle in both sexes; Basic I in the male is very limited and fleeting; predefinitive feather generations of the female commonly are not distinguished because of their relative similarity. The downs are omitted and the order of listing is from Juvenal through the definitive cycle (the latter is diagrammed in Fig. 4).

♂ Juvenal (entire feathering), reaches fullest development usually within the span 55–70 days posthatching, i.e., soon after flight is attained, then is lost quite rapidly, except most of the wing is retained through winter into the following summer.

♀ Juvenal (entire feathering), timing essentially as ♂, including long retention of most of the wing.

♂ Basic I (so far as known, limited to portions of head, breast, scattered feathers on mantle and upper sides, and usually the tail; apparently individuals may differ in the amount of this feather generation that they acquire). The feathers come in during the continuum of molting in fall, generally in October, when the Juvenal contour feathers are largely gone and Alternate I is appearing. The Basic I feathers very soon fall out.

♂ Alternate I (inclusive feathering as Definitive Alternate, described beyond), acquired beginning at about 65–80 days posthatching and continuing into fall, sometimes slowly or perhaps with a pause (into early winter); then it is worn, along with retained Juvenal wing, through winter into the next summer (when approximately 1 year old).

♀ Basic I (apparently all of head-body-tail and innermost feathers of wing) succeeds corresponding Juvenal rather rapidly, beginning before or soon after the latter is fully developed; most of the Juvenal wing is retained and worn with it.

♀ Alternate I (all feathering except, in wing, only a few innermost feathers), succeeds Basic I usually well along in fall or from then into early winter, then retained until some time in spring. The Juvenal wing not only is retained and worn with it, but also with the Basic feathering acquired in spring.

♂ Definitive Basic (entire feathering). In the prolonged summer molting period the drake gradually acquires Basic head-body, then comes simultaneous loss of flight feathers of wing (of the preceding Basic in older drakes, of the Juvenal in the case of Basic II) and loss of tail. While the new Basic wing ("offset" from the rest of that generation) is growing, with flightless period of about 25 days, new Alternate head-body-tail-inner wing feathers begin to replace the corresponding portions of the comparatively briefly worn Basic. The Basic wing then is retained and worn with the Alternate generation described next. (Note: Captive and feral birds tend to acquire entire Basic, including all of the wing, and wear it for a considerable time—even many weeks—before head-body-tail begin molting back into Alternate.)

♂ Definitive Alternate (head-body-tail and some innermost feathers of wing), acquired usually during late summer (occasionally well into fall) and retained into the following early summer; the Basic wing is retained and worn with it. Usually the last feathers to grow fully are long scapulars and part of the tail.

♀ Definitive Basic (entire feathering), head-body-tail and innermost feathers of wing are acquired in spring, as is also the nest-down; the wing worn after that is either Juvenal or Basic (depending on age of the duck) and the new Basic wing that goes with the spring-acquired feathering does not grow until well along in summer (in successful breeders, when the duck is rearing her brood; in failed breeders, earlier). At the time it is growing, the Basic contour feathering is being succeeded by Alternate. From this it follows that the duck at no time is entirely in Basic feathering of the same generation—the wing being "offset."

♀ Definitive Alternate (inclusive feathering as ♂ Definitive Alternate), worn from approximately early fall through winter into spring, along with retained Basic wing.

ii. Redhead (Aythya americana). The Redhead has two feather generations per cycle in both sexes; molting and feather generations, sex for sex, as in Mallard—except that the drake's Basic I is extensive and not fleeting. The early feather generations of both sexes blend from one into the next in a continuum of molting, hence—especially in the ♀ whose early generations do not differ strikingly—a transition is not very obvious.

iii. Wood Duck (Aix sponsa). The Wood Duck has two feather generations per cycle in both sexes, although Heinroth (1910) stated, as translated, that "the female molts only once annually and continually wears the same feathers"—and authorities have perpetuated this error down to the present. (The sequence of feather generations is diagrammed in Fig. 5 and the head feathering of all generations is illustrated by Fig. 7.) Basic I, the earliest white-bellied stage, evidently has been misconstrued as incoming Alternate I; i.e., this intervening generation has been overlooked. In the female the various

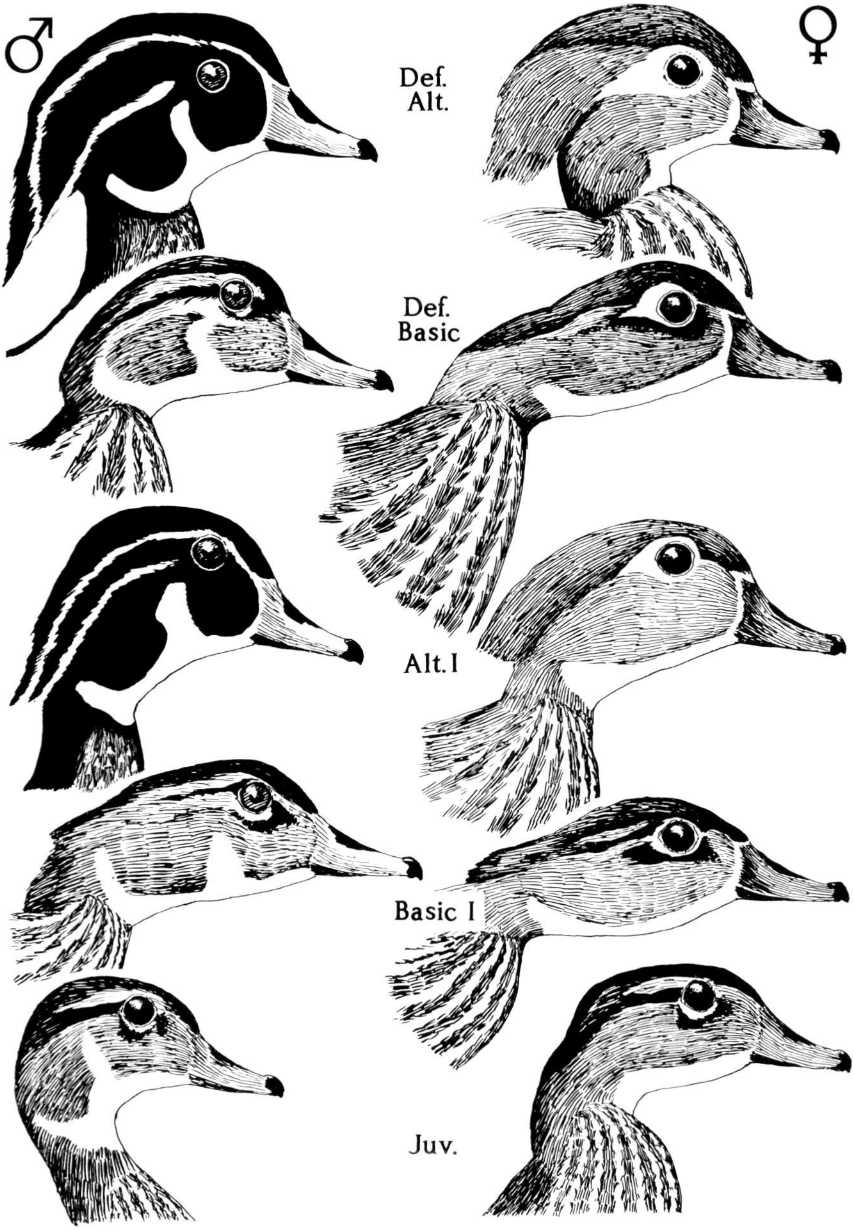

FIG. 7. Wood Duck *(Aix sponsa)*—all feather generations. Basic I in both sexes and Definitive Basic in the female have been overlooked or unrecognized. (See Fig. 5 for timing of these feather generations.)

feather generations after Juvenal are not, unlike the drake, strikingly different; furthermore, some of the differences are masked slightly by overlap in variation in individuals, and gradual molting has the effect of blurring the distinctions shown in Fig. 7. The picture also is obscured by geographical variation in hatching dates (which include those from re-nesting) which extend, say, from December in Cuba until the following August in New England. Thus a stage (as Basic I) occurring at about 85 days posthatching is found somewhere within the total range of the species for at least three quarters of the year. The timing of events of the definitive cycle also varies with latitude.

It is evident that, although there are differences in timing and in amount of feathering in certain feather generations, the periods of molting and resultant generations of the three waterfowl just described are homologous.

C. Arbitrariness

Assuming that a goal is to describe the molting and resultant feather generations of a species, a logical approach would be to begin with the earliest relevant event in the life of the individual and then, for each sex, continue in chronological order up to and through the definitive cycle (as done, from Juvenal onward, for the Mallard in this chapter). Even relatively technical ornithological treatises do not follow this sequence, however. Generally speaking, the definitive cycle is best known, commonly regarded as most "diagnostic" of the species, most abundant in collections, and so—arbitrarily—it is described first (as done for the European Blackbird in this chapter). Again, within this cycle, it would make sense from a comparative viewpoint to start with the molting ("landmark molt") by which the Basic generation is acquired, then describe that generation, then (if two generations per cycle) Alternate, then (if three per cycle) Supplemental. In each case, if any feathering was "offset" (retained) from previously acquired generation(s), this would have to be added to account for total feathering at any particular time. (The Supplemental also may occur in another chronological sequence; see Fig. 1.) Treatment by authors has varied, however, even when all feathering is renewed in a single period of molting and there is only a single generation per cycle. The feathering of some Northern Hemisphere birds of this sort has been described as, say, "spring" or "nuptial" or "breeding plumage," plus "fall" or "resting plumage" which, of course, is re-describing the same feather generation but placing the molting between the descriptions. The picture becomes more scrambled in cases of two feather generations per cycle, with part of one retained and worn with the other, say in spring. Many descriptions begin with the latter "added" feather generation and include the retained Basic wing without noting that it is of an earlier generation. In the waterfowl discussed at length earlier in this chapter (also see diagram of Gadwall, Fig. 8) there are special problems. The "landmark molt" is not a unified event, portions of it being widely separated in the female, and the wing is "offset" from the remainder of even its own Basic generation. In seasonal terms, the "winter plumage" is really what would be "spring plumage" in many Northern Hemisphere birds. The "spring" or "breeding plumage" of the drake is the same as his "winter plumage," but in the duck the "spring" or "breeding plumage" actually is homologous with the "resting" or "eclipse plumage" of the drake. Particularly in cases as complicated as this, treatment based on season or reproduction obscures the picture; it can be clarified, however, regardless of any particular labeling, by starting at the beginning and describing events in chronological order. Any arbitrary departure from this sequence should be indicated clearly.

A person who examines various ornithological treatises finds that, even assuming that approximately similar data are available to the authors, there is no consistent pattern of labeling or even in arbitrary departure from chronological sequence among them. There is, however, more or less of a tradition of treating certain kinds of information in roughly the same way.

VII. Concluding Remarks

The key to this chapter is Fig. 1 which shows, schematically, some possible ways in which increasingly elaborate cycles of molting and resultant feather generations may have evolved. Even in the most complex known cases, there are only a few periods of molting and a small number of feather generations, most of which are recurring, on an individual bird. Even so, the problems of working out the details in particular groups or species, discovering their causal influences and complicated integration in the life cycle, and defining the seemingly infinite number of variations within the general framework make the analysis of molts and Plumages one of the more difficult areas of descriptive biology.

Part of the problem stems from the fact that end products of adaptation may give the appearance of masking, or at least obscuring, the fundamental pattern. There is an indication of this in the present chapter, for example in variations in "offset" phenomena, objects in

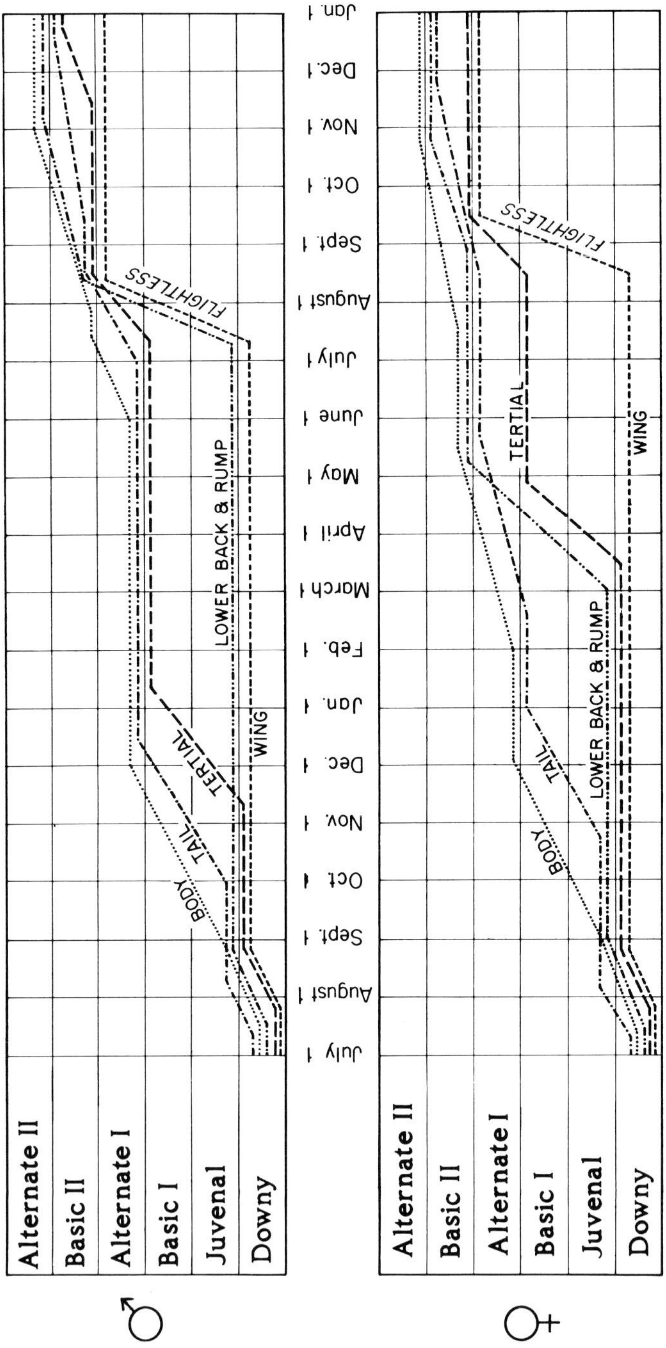

FIG. 8. Modal Gadwalls (*Anas strepera*). Corresponding portions of feathering are not acquired or lost at the same time by both sexes. (After Oring, 1968.)

the environment being functionally equivalent to characteristics of feathering, social influences on time of molting, and so on. To indicate complexity within a group, certain Anatidae are prime examples.

In writing descriptions, the trend, which comes with increasing experience on the part of the describer, is to shift emphasis from feather generations toward patterns of molting. The "description" of the Mallard, earlier in this chapter, is restricted arbitrarily to molting; it would be at least five times as long if details of appearance of the exterior bird and individual variations were added. On the other hand, having mastered the molting, one can re-convert to more accurate definitions of feather generations and then add appropriate details. The latter approach has broader appeal, but is very different from authoring descriptive works without having the aforementioned background on molting.

If the periods of molting, the feather generations, and temporal relation of these, were known for both sexes in all birds, there still would be the problem of communicating this information. Various "systems" of description and terminology were compared by Humphrey and Parkes (1959) and need not be dwelt on at length here. It should be pointed out, however, that describing feather generations and their antecedent molts has no constant equivalent in any "system" other than the one these authors presented and advocated. Dwight's (1900, 1902) terminology was based on temperate-zone passerines primarily; if, in addition, he had studied tropical birds, he would have had to modify his concepts or abandon them. On the other hand, Miller (1961) studied the Rufous-collared Sparrow *(Zonotrichia capensis)* near the equator in the Andes of Colombia. Its definitive cycle consists of only a single feather generation, renewed by a molt (of all feathering) at approximately 6-month intervals. That is, it molts from Basic to Basic (it also has two complete reproductive cycles per year; i.e., the reproductive cycle also is compressed into a half-year). Miller could not dissociate molting and reproduction; he reasoned that, since Northern Hemisphere *Zonotrichia* breed in Alternate, therefore *capensis* must be in Alternate when breeding, hence there is no Basic Plumage! A reasonable hypothesis to explain the difference between two species of *Zonotrichia* is diagrammed in Fig. 9. The Gadwall is another interesting case (see Fig. 8); the sexes may be said to be wearing "spring plumage" on May 1, but the duck is molting on that date, the drake is not, and they are not wearing corresponding (homologous) feather generations.

Describing each period of molting and the resultant feathering is not new; it was done over 40 years ago by Miller (1928) for the Log-

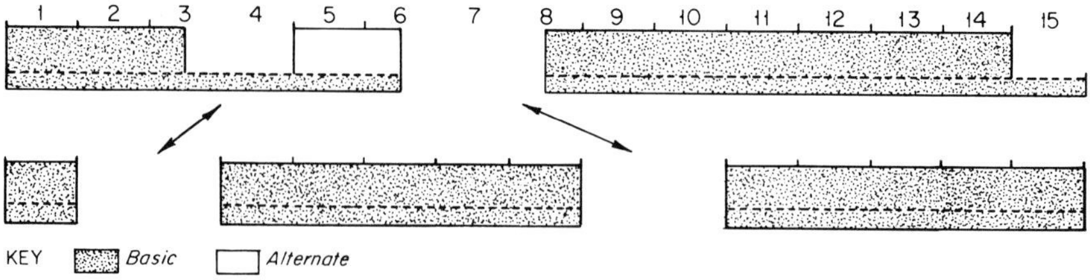

FIG. 9. Two *Zonotrichia* species compared. Time scale in months (1 = Jan. 1); empty areas represent molting; wing is below the dashed line. (Above) Temperate-zone *Z. leucophrys gambelii* has two head-body feather generations in a single cycle lasting 12 months and one of its two periods of molting includes all feathering. (Below) Tropical *Z. capensis costaricensis* has the equivalent time span divided into two cycles, each consisting of a single feather generation plus molt of all feathering. One of these patterns may have evolved from the other.

gerhead Shrike *(Lanius ludovicianus).* This approach can be applied to all birds and the host of particular adaptations encountered is a secondary matter. For example, the vast range of adaptations of molting of the wing—to maintain ability to fly, to accommodate flightless periods in some birds and to fit molting into the energy budget of migrants—is only a facet of the subject. A lengthy study of wing molt did not provide evidence for judgment of phylogenetic relationships, as Stresemann and Stresemann (1966) demonstrated. At the level of investigating the order in which wing feathers are replaced, one is dealing with possible homology of sequences that are of a limited nature within a broader concept of homology of molts and feather generations. More fundamental questions concern such topics as whether homologies can be traced throughout or beyond avian families and orders, and the nature of the evolutionary process whereby a feather generation is added to or perhaps subtracted from a cycle.

Here are some fairly obvious kinds of situations at the species level. In the Red-breasted Nuthatch (*Sitta canadensis*) scattered individuals apparently have a limited Prealternate molt, i.e., two feather generations per cycle (Banks, 1970); also see *Turdus merula,* earlier in this chapter. In the Red-winged Blackbird *(Agelaius phoeniceus)* there is geographical variation not only in time of molting but also in extent of Prealternate molt on male individuals (Parkes, 1972). In the Willow Ptarmigan the geographically isolated population (*Lagopus lagopus scoticus)* of the British Isles (the "Red Grouse") apparently has one less feather generation per cycle than is reported from Scandinavia. A host of additional situations—not only at the species level—await discovery and explanation.

REFERENCES

Ashmole, N. P. (1963). Molt and breeding populations of the Sooty Tern *Sterna fuscata*. *Postilla* No. 76.

Banks, R. C. (1970). Molt and taxonomy of Red-breasted Nuthatches. *Wilson Bull.* **82,** 201–205.

Chapman, F. M. (1932). "Handbook of Birds of Eastern North America," 2nd rev. ed. Appleton, New York.

Dane, C. D. (1965). "The Influence of Age on the Development and Reproductive Capability of the Blue-winged Teal (*Anas discors* Linnaeus)." Unpublished Ph.D. dissertation, Purdue University, West Lafayette, Indiana.

Dwight, J., Jr. (1900). The sequence of plumages and moults of the passerine birds of New York. *Ann. N. Y. Acad. Sci.* **13,** 73–360.

Dwight, J., Jr. (1902). Plumage-cycles and the relation between plumages and moults. *Auk* **19,** 248–255.

Gilliard, E. T. (1956). Bower ornamentation versus plumage characters in bower-birds. *Auk* **73,** 450–451.

Heinroth, O. (1910). Beobachtungen bei einem Einbergerungs-versuch mit der Brautente (*Lampronessa sponsa* (L.)). *J. Ornithol.* **58,** 101–156.

Hochbaum, H. A. (1944). "The Canvasback on a Prairie Marsh," Amer. Wildl. Inst., Washington, D.C.

Hubbs, C. L. (1944). Concepts of homology and analogy. *Amer. Natur.* **78,** 289–307.

Humphrey, P. S., and Parkes, K. C. (1959). An approach to the study of molts and plumages. *Auk* **76,** 1–31.

Humphrey, P. S., and Parkes, K. C. (1963). Comments on the study of plumage succession. *Auk* **80,** 496–503.

Johnsen, S. (1929). Draktskiftet hos lirypen (*Lagopus lagopus* Lin.) i Norge. *Bergen Mus. Aarb.* No. 1.

Kessel, B. (1957). A study of the breeding biology of the European Starling (*Sturnus vulgaris* L.) in North America. *Amer. Midl. Natur.* **58,** 257–331.

Lowe, P. R. (1933). On the primitive characters of the penguins, and their bearing on the phylogeny of birds. *Proc. Zool. Soc. London* **103,** 483–538.

Miller, A. H. (1928). The molts of the Loggerhead Shrike, *Lanius ludovicianus* Linnaeus. *Univ. Calif., Berkeley, Publ. Zool.* **30,** 393–417.

Miller, A. H. (1961). Molt cycles in equatorial Andean Sparrows. *Condor* **63,** 143–161.

Monnie, J. B. (1966). Reintroduction of the Trumpeter Swan to its former prairie breeding range. *J. Wildl. Manage.* **30,** 691–696.

Oring, L. W. (1968). Growth, molts, and plumages of the Gadwall. *Auk* **85,** 355–380.

Parkes, K. C. (1963). The contribution of museum collections to knowledge of the living bird. *Living Bird* **2,** 121–130.

Parkes, K. C. (1972). Personal communication.

Rollin, N. (1964). Abnormal plumage. *Bird Res.* **2,** No. 1, 1–44. (World Bird Research Sta., Glanton, Northumberland, England.)

Short, L. L. (1969). A new species of blackbird *(Agelaius)* from Peru. *La. State Univ., Occas. Pap. Mus. Zool.* No. 36.

Siegfried, W. R. (1970). Double wing-moult in the Maccoa Duck. *Wildfowl* **21,** 122.

Smart, G. (1965). Development and maturation of primary feathers of Redhead ducklings. *J. Wildl. Manage.* **29,** 533–536.

Stone, W. (1896). The molting of birds with special reference to the plumages of the smaller land birds of eastern North America. *Proc. Acad. Nat. Sci. Philadelphia* **48,** 108–167.

Stresemann, E. (1963). The nomenclature of plumages and molts. *Auk* **80,** 1–8.

Stresemann, E., and Stresemann, V. (1966). Die Mauser der Vögel. *J. Ornithol.* **107,** Suppl.

Voitkevich, A. A. (1966). "The Feathers and Plumage of Birds." Sidgwick & Jackson, London (Orig. publ.: "Pero Ptitsy" ["The Bird's Feather"]. USSR Acad. Sci. Publ. House, Moscow, 1962.)

Watson, G. E. (1963). Feather replacement in birds. *Science* **139,** 50–51.

Witherby, H. F., Jourdain, F. C. R., Ticehurst, N. F., and Tucker, B. W. (1938). "The Handbook of British Birds," Vol. 2, pp. 136–141. Witherby, London.

Chapter 3

MECHANISMS AND CONTROL OF MOLT

Robert B. Payne

I. Introduction

The control of the time and rate of molting is important in the lives of birds. Molting, breeding, and migration commonly are events that are essentially mutually exclusive in time; many fascinating adaptive compromises have developed between molt and other phases of the annual cycle (Stresemann and Stresemann, 1966; Stresemann, 1967; Ashmole, 1968). The adaptive significance of periodic molt is related to the wear and tear of the plumage, as birds usually molt before their feathers are excessively worn. The importance of molt in keeping a functional plumage for insulation and flight is evident in species that are subject to unusually severe feather abrasions. Some New World sparrows and wrens and their Old World warbler counterparts that live in coarse grass molt their flight feathers as well as the body plumage twice each year whereas others in less abrasive vegetation molt the flight feathers only once (Stresemann and Stresemann, 1966). Within a species the molting schedules are often closely timed to occur when breeding or migration are not taking place, when food is abundant, or when further abrasion of the plumage would be harmful. In addition to this practical requirement of replacing plumage, the alternation of breeding and nonbreeding plumages is a means by which many birds change their appearance between a conspicuous aspect for display and attracting a mate in the breeding season and a cryptic aspect for passing unobserved in the off-season.

The physiological bases of the control of the time and rate of molting and the integration of molt into the annual cycle are of considerable interest, and their discovery remains as a challenge to the comparative physiologist. The events of the replacement of a feather and the mechanisms that result in the orderly progression of molt along the feather tracts are poorly known. A central problem is to uncover the physiological changes that may be the initiating factors in seasonal molt. The relation of molting to other processes involved in breeding and in migration provides some insight into the ways in which molt is integrated into the complete annual cycle, and it suggests ways in which the molt may be controlled through different hormones. A final field of interest is in the amount of energy involved in the growth of a new plumage.

Molt is the periodic replacement of feathers. The shedding of the old feathers and the growth of the new ones are closely related events in avian molt, as the old feather is pushed out by the new. The old feather for a time remains in the resting follicle but is eventually dropped as the follicle grows and produces a new feather. Often the new feather emerges from the follicle while the old feather is still attached to its tip. The replacement of feathers thus differs from that of hair in mammals, since in birds the new structure develops directly beneath and pushes away the old (Dwight, 1900; Spearman, 1966).

While molt usually is discussed in terms of feather loss and replacement, other keratinized epithelial parts are also molted. These include the stratum corneum of the skin, the claws, the tarsal scales, and in some birds horny plates of the bill. Usually these structures are shed at the same time as the feathers, but in certain birds they may be shed weeks earlier (Gullion, 1953; Spearman, 1966). The stratum corneum is sloughed along with the feathers in response to thyroxine (Kobayashi *et al.*, 1955), suggesting that the loss and replacement of all of these integumentary derivatives may be under common hormonal control.

Many of the experimental investigations on the hormonal control of molt have been made on domestic birds which, through domestication, have been selected for physiological characters related to molt. The White Leghorn chicken, selected for its ability to lay eggs, now lays almost continuously, and its laying and molting schedules overlap, in contrast to its undomesticated relatives. Comparative studies of the hormonal control of molt in wild birds are clearly desirable. The diversity of molt–breeding and molt–migration relationships in different birds allows us to think in terms of multiple solutions in the hormonal control of molt in different species.

II. Mechanisms of the Pattern of Molt

The loss and replacement of the feathers of molting birds proceed in a regular sequence on each feather tract. (See Chapter 2.) A pattern of molt is most conspicuous in the flight feathers, in which molt of the primaries commonly progresses from the wrist outward, and only one or two feathers grow simultaneously (Stresemann and Stresemann, 1966). Detailed studies of the body pterylae also show a regular pattern of molt with definite waves of feather loss and growth passing from the midline outwards and sweeping caudally. The sequential loss of feathers in a definite pattern involves two aspects. First, molt begins at certain definite points (foci, loci) in a feather series. Second, molt progresses in a wave of feather replacement from each of these points through a series of neighboring feathers, sometimes in two directions and sometimes in only one as in the outer flight feathers of the wing. Few experimental investigations of the pattern of molt have been reported. The sequence and relative timing of the molt of neighboring feathers is of considerable adaptive significance. An example is the retention of the outer primaries while the inner are growing, since this pattern allows the bird to fly during molt and provides protection to the growing feathers.

A. The Beginning of Molt

The onset of molt at each locus (rather than at any other follicle in the same tract) probably results from a lower threshold of response at the locus than elsewhere in the feather series. In most birds that replace all feathers in a series during a molt the first feather to be molted is the one that was first replaced in the previous molt. A time-dependent state of each feather follicle may make each follicle prepared to molt only after a long period of preparation on the level of the cells of the follicle. A long period of obligate inactivity or preparation within a follicle could then act as a refractory period, preventing the growth of each follicle until a certain time had passed after the previous feather had been shed (Ashmole, 1968). Each follicle may have its own rhythm of molting; adjacent follicles may have the same periodicities but may be out of phase.

The first feather to drop in a series may be the one with an increased blood supply (A. H. Miller, 1941). A follicle that is highly vascularized would perhaps have a lowered threshold to molt, receive more of the hormone that acts peripherally on the follicles and initiates molt, or receive more oxygen and nutrients than its neighbors. Vascularization of one follicle might also cause it to deprive its neighbors of these substances and thereby inhibit their molt. In the wing of some birds, molt begins in the part of the wing least crowded by feathers, and the last part of the middle wing to be molted is in the more densely feathered region of secondaries 4 or 5; perhaps a region of restricted blood supply (A. H. Miller, 1941). The local regional differences in vascularization of the dermis surrounding each feather follicle have not been studied, and microanatomical examination is desirable to determine whether structural features of vascular or other tissue systems are correlated with the pattern of molt in a feather tract.

B. The Sequential Pattern of Molt

Once the molt has begun in the initial locus the other feathers in a series are replaced in a regular sequence. A feather is replaced only after its neighbor has partly regrown. The local physiological mechanisms responsible for this pattern are not well understood; several have been suggested.

A state of differing phase relationships among the population of follicles composing a feather series may account for the regular pattern of molt in some birds (Ashmole, 1968), each follicle growing only after it has completed a time-dependent change. This mechanism may occur in birds with prolonged molts, because here the phase relationships of the follicles would be pronounced. However, in birds such as *Zonotrichia leucophrys gambelii* in which molt is compressed into a short period and as many as seven sequentially shed primaries may be growing at the same time (Morton *et al.*, 1969) the time difference in their phase relationships is negligible, as the time between molts is long. Similarly among the 11 families of birds known to have a simultaneous molt of the flight feathers (Stresemann and Stresemann, 1966) the wing feathers are all dropped on nearly the same day. Here the vagaries of time change in the feathers suggest a common response of the feathers to a timing mechanism (perhaps a hormonal change) which is external to the feather series. In some birds, feathers accidently lost between molts are regrown at once, and these feathers are again replaced in the normal sequence at the following molt, indicating no prolonged refractory period of the individual feather follicle. In birds with one complete and one incomplete molt in a year some of the follicles grow new feathers twice each year and they would have to have preparatory periods of half the duration of the feathers replaced once a year. The importance of an automatic timer at the level of the individual follicle has yet to be shown experimentally.

Another possible mechanism involved is a differential hormone sensitivity of the follicles in a series. The molt loci may be sensitive to low doses of molt-inducing hormones, and the last-molting feathers are those with the highest thresholds that are reached through an increasing output of molt-promoting hormones by the relevant endocrine glands (Ashmole, 1968). No endocrine glands are known with a closely graded, continually increasing output through the time of molt, however.

A third mechanism, suggested by A. H. Miller (1941), assumes a more or less constant hormonal state throughout molt. Growth of one feather may stimulate the molt of the adjacent one, and molt then may continue until the feather series has been molted. If the shedding of the inner primaries improves the vascularization of the adjacent follicles, one might expect to find a process of local control that produces an orderly molt sequence. The interaction of a molting follicle and its neighbors would provide a more precise mechanism to account for the regular sequence of molt typical of most of the tracts of most birds than would the "internal cycle" model. If a long period of follicle preparation occurs in birds with protracted molts, the follicle interaction may function as a mechanism that coordinates the molting of individual follicles into a temporally patterned molt.

The development of the avian brood patch suggests differential local hormone sensitivities in the integument, and although the mechanism of feather loss in brood patch formation does not involve the growth of the underlying feather follicles as in molt, the first stages of brood patch development do involve local changes in vascularization. An ordered arrangement of feather loss in female birds with certain hormones results in defeathering in the ventral tract but not in other feather tracts (Jones, 1969). A comparably topographic response of feather follicles may occur on a smaller scale within a feather tract in the process of molt. In the domestic fowl the various feather tracts exhibit "orderly and characteristic differences in threshold of reaction to known doses of hormones (thyroxine and estrogen)" (Rawles, 1960, p. 230). If the pattern and rate of molt are determined by local differences in sensitivity to certain hormones, then minor changes in the threshold may be involved in the proximate control of different patterns of molt in different kinds of birds. Simple changes in the local responsiveness of neighboring follicles could prolong the molt of some feather tracts or could lead to an increased rate of molt through several tracts at once. Species differences in molting patterns may have similar physiological bases. The responses of the ventral tract in brood patch formation suggest a comparable difference among species as some icterids respond to certain hormones whereas other species are insensitive at the same hormonal levels (Selander and Kuich, 1963).

III. Molting and Breeding Schedules

The periodic molt of birds usually occurs at the end of the breeding season after reproductive activities have been completed. In birds with two molts per year the second one also often takes place outside of the breeding season. The separate scheduling of molt and breeding (and of migration, in birds that fly long distances between their breeding grounds and wintering areas) has been recognized as an ecological adaptation permitting these events to coexist in the annual cycle with a minimum of energetic stress (Kendeigh, 1949; Farner, 1964). The majority of birds in the tropics and in higher latitudes have molting schedules that do not overlap with breeding activity (A. H. Miller,

1962, 1963; Ashmole, 1963, 1968; Snow and Snow, 1964; Stresemann and Stresemann, 1966; Stresemann, 1967). Nevertheless the occurrence of active molt in breeding birds has been reported for more than 100 species (Dorward, 1962; Immelmann, 1963a; Ingolfsson, 1970; Lofts and Murton, 1968; Keast, 1968; A. H. Miller, 1955b, 1963; Moreau *et al.*, 1947; Payne, 1969a; Serventy and Marshall, 1957; Snow and Snow, 1964; Stresemann and Stresemann, 1966; Wagner and Müller, 1963). Overlap in some of these is known only from circumstantial evidence of molting activity during the time of breeding or from a few molting specimens collected with enlarged gonads, and some of these may not involve molt of nesting individuals. However, nearly half of these species appear to molt regularly while their gonads are active and while they are actually breeding.

The regular overlap in time of breeding and molting in these birds occurs mainly in two ecological circumstances. In habitats where food is apparently available all year and not so seasonally variable in abundance as in most temperate land regions the molt is often prolonged and may overlap breeding. When the feathers are replaced over a long period of time, relatively little energy per day is required to grow a new plumage. As a result, in some tropical land and marine environments there appears to be no real restriction of energy resources to a single physiological process. On the other hand birds such as some arctic gulls and some sandpipers that exploit a more highly seasonal food than do the typical temperate region birds may have a surplus of food in summer but a severe shortage in the rest of the year, and sufficient food for the completion of both molt and nesting is apparently available in one brief period. Irregular overlap of breeding and molt may occur in other forms in a facultative manner in highly unpredictable, aperiodic environments such as deserts where rain may fall at any time of the year, and though birds molt seasonally they may be physiologically prepared to breed whenever the rains come. Molt is regular every year, breeding is not. Some birds interrupt their molt when they breed in response to rains whereas others continue to molt while they breed (Snow, 1966; Keast, 1968). The time of molt here may be quite independent of the time of breeding, sometimes coinciding, sometimes overlapping, and sometimes alternating, rather than a regular seasonal overlap. The various ecological adaptations in these instances of overlap in certain species have been discussed elsewhere (Stresemann, 1967; Immelmann, 1971).

From the physiological aspect the timing of molt and breeding is of interest, particularly as different molt–breeding patterns may be associated with different hormonal control systems of molting. Separation of molting and breeding schedules in some birds suggests an inhibitory role of molt by the reproductive hormones, whereas this would be unlikely in birds with simultaneous molt and nesting. Investigation of species with overlapping schedules may prove rewarding in experimental, comparative analyses of the physiology of molt and reproduction.

A. Molt in Sexual Inactivity

The most common molt schedule in birds is alternation with breeding. Most birds in temperate regions begin the main annual molt shortly after breeding has been completed and the young are independent. Molt begins after the gonads have regressed to a small size and the interstitial cells of the testes have lost their lipoidal activity, indicating the loss of androgens (Marshall and Coombs, 1957; Payne, 1969b; Fig. 1). This timing of molt schedules with changes in reproductive condition suggests a possible causal relationship between the decrease in reproductive hormones and the onset of molt. Control of molting by a decrease in reproductive hormones is also suggested in unpaired or unsuccessful birds that begin molt earlier than successful later breeders, as molt begins only after the gonads have regressed. Individuals that nest last in the breeding season may have young during the time when others have begun the molt, but these birds may delay molt until after the young are independent (Putzig, 1938; Michener and Michener, 1940; Kramer, 1950; A. H. Miller, 1961; Keast, 1968; Payne, 1969b). In the European Bullfinch, *Pyrrhula pyrrhula*, molt may be delayed as long as 12 weeks (Newton, 1966). Along the Pacific coast of North America, populations of White-crowned Sparrows, *Zonotrichia leucophrys*, differ in the time of development of the gonads in spring; the southern birds breed 2 months earlier than the northern birds and they lack a prenuptial molt, probably because of inhibition of a spring molt by early spring surges of reproductive hormones (Mewaldt *et al.*, 1968). In these birds probably the mechanism that separates the processes of breeding and molting directly involves the reproductive hormones.

B. Interrupted Molt

In birds with seasonal breeding or prolonged periods of molt, the molt may be temporarily arrested during a period of nesting. Feathers that have started to grow complete their growth, but the neighboring feathers are not dropped until breeding has been completed. Interrupted molt occurs in several species of dry-country birds that cease

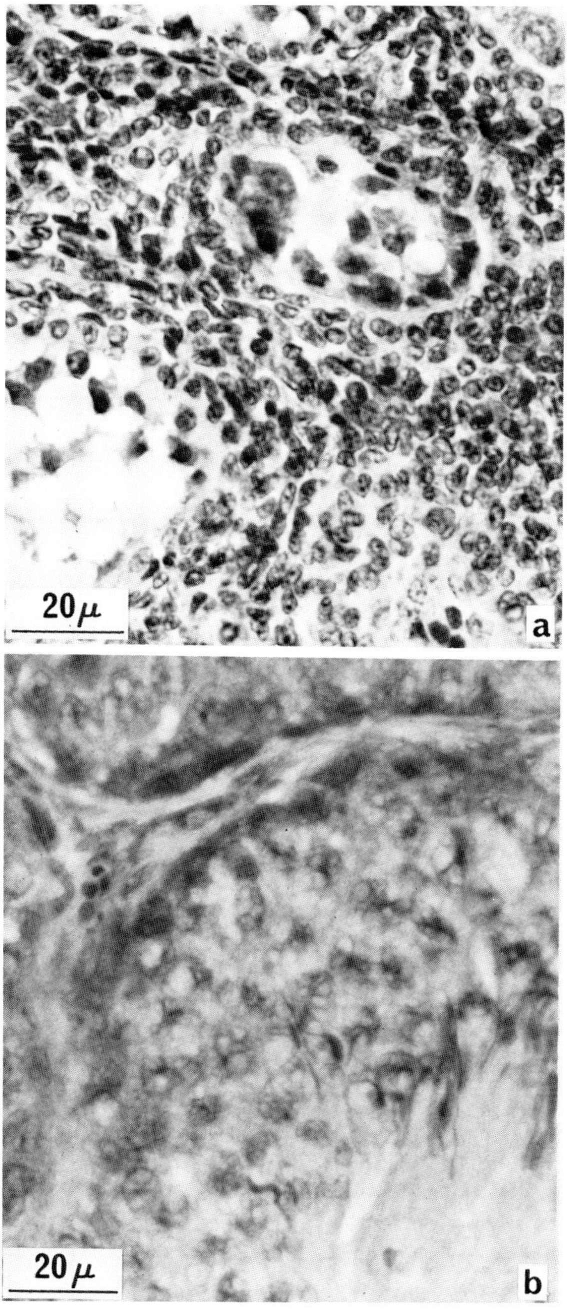

FIG. 1. Species differences in the gonadal activity of molting birds. (a) Testis photomicrograph of a bird which molts in sexual inactivity, the Tricolored Blackbird (*Agelaius tricolor*). (b) Testis photomicrograph of a bird which molts during gonadal activity, the Greater Honeyguide (*Indicator indicator*).

molting when rains fall and breeding ensues, notably in the Galapagos finches (Snow, 1966) and in some tropical land birds with prolonged periods of molt (A. H. Miller, 1961; Snow and Snow, 1964). Tropical seabirds with schedules of breeding and molting that do not follow the calendar of land environments may regularly interrupt the molt when they nest (Dorward, 1962; Ashmole, 1963, 1965, 1968). The alcid *Ptychoramphus aleuticus*, on the other hand, regularly begins molt off California while it breeds, but molt is slowed or sometimes completely stopped while the adults are feeding the more advanced young (Payne, 1965). A duck (*Anas strepera*) in Canada has a prolonged molt which is interrupted in the female during the nesting period (Oring, 1968). In all of these birds the molt is at times interrupted by breeding, and the increased levels of reproductive hormones at this time may be the cause of the decreased molt.

Molting also may be interrupted during migration, especially during long flights over seas or deserts (Stresemann, 1967). By molting some flight feathers before migration the wings may be more efficient in flight than worn wings would be, but because of early cold weather and food shortage the birds may have insufficient time to complete the molt before migration. The physiological changes involved in the interruption of molt in migrating birds are not known; some birds in captivity interrupt their molt at the time when wild birds of the same species are migrating (Stresemann, 1967).

C. Molt During Gonadal Development

Birds that molt immediately before nesting may molt while the gonads are seasonally enlarging and hence while gonadal steroids are being produced. Domestic pigeons may molt while the gonads are growing (Kobayashi, 1953a). The hummingbird *Calypte anna* begins to enlarge the gonads in autumn during the molt, and the periods of gonadal development and molt are largely coextensive (Williamson, 1956). Rooks, *Corvus frugilegus*, sometimes undergo a period of spermatogenesis in autumn, and these birds continue and complete their annual molt at the same time as nonbreeding fall birds (Marshall and Coombs, 1957). *Quiscalus mexicanus*, an icterid that normally molts after breeding in spring, does not appear to control its molt directly through reproductive hormones, as molt may be in progress during autumn gonadal recrudescence and as the prenuptial molt of first-year males occurs at the same time as the maximum rate of enlargement of the gonads in spring (Selander, 1958). Several Australian birds molt while the gonads are enlarging (Immelmann, 1963a), and chickens may do the same (Assenmacher, 1958). Some African viduine and estrildid finches undergo seasonal gonadal enlargement and molt at the same time. Molt and gonadal growth coincide from onset to completion in these birds. All of these birds are largely free from the energy strains imposed by breeding activities during the molt, but from the physiological point of view the molts overlap with periods of probable high levels of reproductive hormones, in contrast to birds such as *Zonotrichia leucophrys* and *Junco hyemalis* (Farner, 1964; King *et al.*, 1966; Wolfson, 1966).

D. Molt During the Periodic, Seasonal Breeding Cycle

This group includes a number of species many of which are birds of higher latitudes as well as some tropical forms. They may be considered with respect to the stage of nesting wherein molt commences.

1. Molt in Laying Birds

Few birds regularly begin molt at the time of laying. The arctic sandpiper *Calidris alpina* has a foreshortened molt on the breeding grounds; molt begins at the time of egg-laying and is completed before the southward migration (Holmes, 1966). Other species with such a schedule are *Corvus corax* (Stresemann and Stresemann, 1966) and *Nucifraga columbiana* (Mewaldt, 1958), though in the latter molt may be more directly related to season than to nesting as some birds molt before the eggs are laid. In addition to these high-latitude birds there are several species that may begin the molt at the time of laying, but which do not necessarily begin molt at this time. The levels of sex hormones of laying birds in this last group may be facultative in permitting a molt but probably do not cause molt.

The onset of molt in birds at the time of maximum gonadal enlargement and sexual behavior as a regular annual event clearly departs from the pattern of alternating molt and breeding schedules. Not only is breeding an energy-demanding period, but also the circulating levels of sex hormones and gonadotropins very likely are high at these times.

Some New World quail (Watson, 1962; Raitt and Ohmart, 1966) occasionally molt during the laying period. Breeding tropical African cuckoos with an egg in the oviduct have sometimes been found to be molting (Stresemann and Stresemann, 1966), though these cuckoos usually do not molt the flight feathers while breeding (Payne, 1969a). Most individuals of the tropical tern *Gygis alba* cease molting when they nest, though birds may start breeding before completion of the growth of the flight feathers which were shed shortly before breeding. Further loss of feathers is postponed until after breeding has been completed (Ashmole, 1968).

2. *Molt in Incubating Birds*

Females of several large birds of prey begin the molt during incubation. In *Falco peregrinus* and in *Accipiter gentilis* and *A. nisus* the molting female remains at the nest incubating and caring for the young, whereas the hunting males do not molt until later. Other birds of prey that may molt soon after egg-laying are the hawks *Buteo platypterus* and *B. lagopus* and the relatively large northern owls *Nyctea scandiaca* and *Surnia ulula*. Females of most tropical hornbills that incubate and brood the young in the confines of a partially sealed tree cavity also molt the wing and tail feathers immediately after egg-laying. Males bring food to the females and later to the young as well and do not molt until after nesting (Amadon, 1966, Stresemann and Stresemann, 1966). The female Giant Fulmar, *Macronectes giganteus*, also begins wing molt soon after egg-laying (Warham, 1962).

3. *Molt in Birds Feeding Young*

The molt begins during the period of parental care in the Northern Crag Martin, *Ptyonoprogne rupestris* (Prenn, 1937), in northern populations of the Chaffinch, *Fringilla coelebs* (Dolnik and Blyumental, 1967), and perhaps in the jay *Cyanocitta stelleri* (Pitelka, 1958). Brooke (1969) notes that it is usual for nonmigratory swifts to begin the wing molt immediately after egg-laying and to continue molt at a leisurely rate through the remainder of nesting and for a few months afterwards. In addition, the tern *Gygis alba* (Ashmole, 1968), owls (Stresemann and Stresemann, 1966), and several songbirds (Newton, 1966) may occasionally begin molt while they are feeding young, though most of these birds usually begin molt after the young are independent.

E. Molt Independent of Breeding

In several groups the molting schedule appears to begin each year on a definite seasonal schedule independent of breeding activity. Birds in the arid interior of Australia often have seasonal molts during which time the gonads remain somewhat active, and the birds may breed whenever the irregular rains break the prolonged droughts (Serventy and Marshall, 1957; Immelmann, 1963a,b; Farner and Serventy, 1960; Keast, 1968). A few of these species show a delayed molt when they breed, but in many there is no evidence of molt being slowed or interrupted when the rains come or when breeding occurs. Keast (1968) emphasizes that breeding is somewhat seasonal and tends to occur in many birds prior to molting, especially when rains do fall in summer, the time of their greatest probability. Independence of molting from the reproductive hormones may occur also in the arid-country African ploceid *Sporopipes squamifrons*. Body and wing molt had begun in a laying female taken at its nest shortly after rains had broken a long summer drought (Payne, 1969a).

In several seabirds the major periods of molt are highly seasonal and may overlap with the breeding schedules of an individual bird. Procellariids may molt the head and body during the nesting period, though molt of the flight feathers is delayed until later in the migratory species (Marshall and Serventy, 1956; Lockley, 1961; Stresemann and Stresemann, 1966). Northern Gannets (*Morus bassanus*) complete the molt at the breeding grounds (Nelson, 1964) and in this case the molt may be more closely related to seasonal changes than to breeding activities inasmuch as nonbreeding birds molt at the same time. A few tropical boobies and terns have overlapping schedules, though most commonly the schedulings of molt and breeding in the tropical seabirds are mutually exclusive (Ashmole, 1962; Dorward, 1962; Dorward and Ashmole, 1963). Independent control of breeding and molt may occur in the Masked Booby, *Sula dactylatra,* where breeding is periodic. In this species molt and breeding are usually carried out at different times but birds breeding out of phase with the majority of the populations sometimes molt while breeding. The Greater Golden Plover, *Pluvialis apricaria,* begins wing molt in Iceland in June, about the same time as laying (Stresemann, 1963). A few arctic and subarctic gulls, e.g., *Pagophila eburnea, Larus hyperboreus,* and *L. marinus,* overlap considerably in the times of breeding and molt within a species, although the timing of molt and breeding in individual birds has not been studied (Johnston, 1961; Stresemann, 1963; Ingolfsson, 1970). *Larus pipixcan,* a gull of temperate latitudes, also is reported to breed while molting (Stresemann, 1963).

Alcids of temperate and higher latitudes have closely defined molting schedules and individuals may breed and molt at the same time. The seasonal breeding and seasonal molting overlap in *Cyclorrhynchus psittacula* (Stejneger, 1885). Cassin's Auklets (*Ptychoramphus aleuticus*) commonly begin molt in the breeding season, and they may also begin a new nesting cycle while molting (Payne, 1965). The independence of breeding and molting schedules in this bird permits the molt to be completed even when the birds nest late into the summer, and high levels of reproductive hormones in the nesting period apparently do not cause molt to be interrupted as a rule. Occasionally molt is interrupted, however, and if a bird breeding in summer interrupts its molt while it is feeding its young, the old plumage may become so severely worn that the adult becomes flightless and may starve. A flightless adult was observed in late summer for several days in a bay off South Farallon Island by A. S. Thoresen. This bird had just begun to replace its body and flight feathers several weeks later than most auklets in this population had started the molt. The primaries were worn, probably from abrasion in the nesting burrow in the rocks or sand, to such a degree that the bird was apparently incapable of flight (Fig. 2). If Cassin's Auklets generally deferred their molt until the breeding season was past, their feathers might be as worn as they were in this bird. The adaptive significance of a seasonal molt that is not curtailed by prolonged breeding activity is clear in this species.

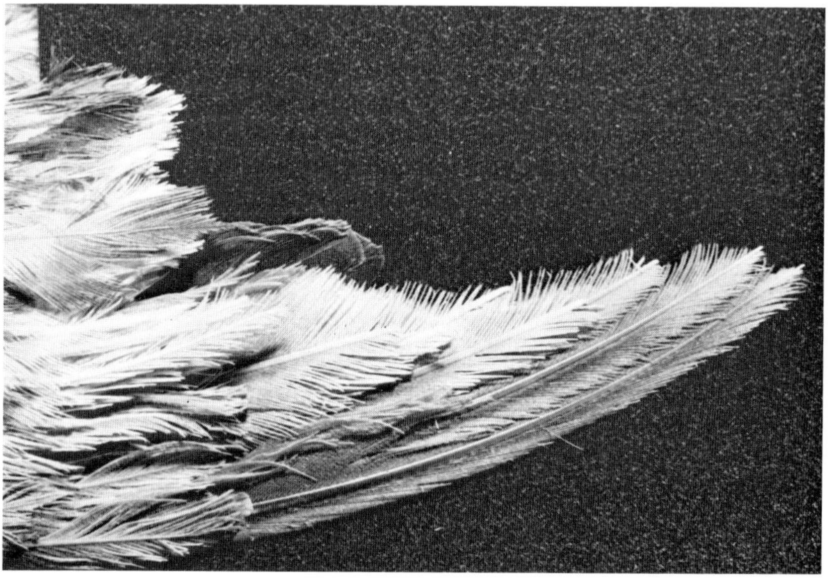

FIG. 2. Worn plumage in a late-molting, flightless adult Cassin's Auklet (*Ptychoramphus aleuticus*). The black wing feathers are the innermost primaries which have just grown. The other primaries are faded, the tips of the rachises are abraded, and the terminal barbs are missing or separated.

Independence of molt from the hormonal changes associated with breeding is likely in the Rock Ptarmigan, *Lagopus mutus,* in which the spring molt may precede, coincide with, or follow testicular recrudescence (Salomonsen, 1939). Similarly, breeding and molt sometimes overlap in quail, and so they may be in or out of phase in a single species (Raitt and Ohmart, 1966).

An independent control of breeding and molt is suggested in several birds (primarily tropical forms) with prolonged molts. Many parrots, pigeons, and doves, some with molts lasting as long as 8 or 9 months, continue to molt during the breeding season (Colquhoun, 1951; Wagner and Müller, 1963; Stresemann and Stresemann, 1966; Lofts and Murton, 1968) though in other doves the molt may be interrupted during nesting (Snow and Snow, 1964). Tropical species of mouse-birds (*Colius*) appear to be in perpetual molt and do not interrupt it when they breed (Moreau *et al.*, 1947). In temperate Africa, molting and breeding tend to occur at different seasons in *Colius* but individuals still may breed and molt synchronously (Rowan, 1967). Molt is prolonged in the Jungle Babbler, *Turdoides striatus*, of tropical India and overlaps with breeding (Naik and Andrews, 1966).

The time relations between molt and breeding are not well known in many tropical birds, and the tendency for some tropical forms to retain enlarged testes throughout the year makes it difficult to evaluate the breeding condition of molting birds (Moreau *et al.*, 1947; A. H. Miller, 1955b).

F. Physiological Integration of Molting and Breeding

The various patterns of timing of molting and breeding in birds suggest two possible main types of control of molt involving the reproductive hormones. First, in those birds that typically have separated schedules that do not overlap, the field evidence suggests an inhibition of molt by the reproductive hormones or an enhancement of molt by a decrease in those substances at the end of the breeding season, either of these giving a means of integrative control of molt into the annual cycle. Similarly in species with regularly overlapping schedules, especially in birds with a closely synchronized timing of laying or of incubation with the onset of molt, the onset of molt may involve a change in levels of reproductive hormones. These coincidences of the onset of molt with events of the breeding cycle all suggest a pattern of control of the time of molt by the levels of gonadal hormones, but there is no strong evidence to challenge another control hypothesis, that gonadal hormones are not *directly* involved in molt. On the other hand, independence of the initiation or arrest of molt by the reproductive hormones is indicated in those birds with only occasional overlap or with molt beginning at various times in the nesting cycle. In birds such as the dry-country passerines of Australia and the Pacific auklets the onset of molt is probably independent of changing levels of gonadal steroids or hypophysial gonadotropins.

No single pattern of control or of independence of molt by the reproductive condition of birds pertains to all kinds of birds—this is the main physiological implication of the variety of patterns of timing of molt and breeding in wild birds.

IV. Hormonal Control of Feather Loss and Replacement

Molt has traditionally been viewed within the domain of the endocrinologist. Many discussions of the relative importance of thyroxine, androgens, estrogens, and progesterone in the control of molt assume that birds molt when they have reached a certain hormonal condition (Assenmacher, 1958; Voitkevich, 1966). There appears to be an interplay between the local, developmental preparation of the feather follicle on the one hand and the endocrine status of the bird on the other hand. While thyroid gland secretions have often been regarded as responsible for molt, these may be more permissive in allowing a normal development of the feather follicles than coercive in bringing about an immediate reaction to increased levels of hormones. A difficulty in the study of the effects of the internal secretions on molt is the fact that many different kinds of treatment will all be followed by molt in laboratory birds, as shown most dramatically by Tanabe and Katsuragi (1962). Domestic fowl are generally unresponsive to hormonal changes and do not alter their molting schedule while they are actually in molting condition. Within 6 months after molt has been completed, fowl molt again in response to thyroxine, progesterone, and gonadotropins (PMS), whereas after this time they molt in response not only to these substances but also to deoxycorticosterone acetate, Enheptin, and prolactin, as well as to starvation. With this wide responsiveness of birds to various hormones it is evident that molt response to a chemical in the laboratory is not necessarily the same as the seasonal response to this substance in a wild bird.

A. Sex Hormones

1. *Castration*

The main source of sex hormones, the gonads, has been removed experimentally to determine the effect of these substances on molt. Castrated birds sometimes have little or no alteration in their molting schedules. Birds showing little effect include the House Sparrow, *Passer domesticus* (Keck, 1934), Chaffinch, *Fringilla coelebs* (Assenmacher, 1958), Golden-crowned Sparrow *Zonotrichia atricapilla* (Morton and Mewaldt, 1962), and canaries (Takewaki and Mori, 1944), while in pigeons castrates may molt several weeks earlier than sexually active individuals (Voitkevich, 1966). Domestic chickens and ducks may enter a more or less continuous state of molt after castration (Assenmacher, 1958; Voitkevich, 1966), suggesting that molt is inhibited by the presence of high levels of sex hormones, a conclusion supported by the response to exogenous hormones. The continuous molt of the capon may include peaks of feather loss and replacement in spring and in autumn which parallel the natural peaks of molt in some breeds of domestic fowl (Assenmacher, 1958). The absence of altered molt schedules following castration in some wild birds suggests no direct role in the initiation of molt in the annual cycle by changing levels of sex hormones, but rather an inhibition of molt in the breeding season. This inhibition apparently is relaxed in birds that molt while their gonads are active.

2. *Sex Hormones (Androgens, Estrogens)*

The administration of exogenous androgens and estrogens to intact birds slows or arrests the normal course of molt in many birds including pigeons (Kobayashi, 1954), canaries (Takewaki and Mori, 1944), House Sparrows (Vaugien, 1955), and others (Wagner and Müller, 1963). Circumstantial evidence of an inhibitory effect of estrogen is noted in domestic fowl, which usually molt after laying has ceased and the ovary has diminished in activity (Tanabe *et al.*, 1957). Molt of the fowl is inhibited by injection of sex hormones and by synthetic compounds with estrogenic activity (Assenmacher, 1958; Tanabe and Katsuragi, 1962; Wagner, 1962). A suppression of the replacement of feathers lost accidentally during the breeding season in House Sparrows (Clench, 1970) also indicates an inhibitory effect of the reproductive hormones on molting. The separation in time of breeding and molting in most wild birds similarly suggests an inhibitory action of the sex hormones on molt. In general the results of the sex hormone experiments suggest inhibition of molting by androgens and estrogens, and as the species tested are birds that usually show separation of breeding and molting it is likely that breeding activity may inhibit molt in these birds through the action of the sex steroids.

3. *Progesterone*

Progesterone treatment in chickens is followed by molt (Shaffner, 1955; Adams, 1956; Juhn and Harris, 1956, 1958; Tanabe and Katsuragi, 1962). Laying hens cease laying eggs directly upon progesterone treatment, but the molting response to single doses does not occur until several days after treatment. The amount of progesterone required to initiate molt varies and is less shortly before the normal molting period (Harris and Shaffner, 1957). This temporal change in sensitivity parallels the response to exogenous thyroxine. As laying hens have more progesterone than nonlaying fowl but do not molt while laying, the molt in response to progesterone appears not to be physiologically the same as a normal molt. A direct effect of progesterone on the feather papillae may occur in plucked fowls. Local treatment on one side of a plucked fowl with progesterone speeds the development of the regrowing feathers on that side (Shaffner, 1954). Progesterone effects are enhanced with thyroxine treatment and appear to be antagonistic to androgens and estrogens. Progesterone may act indirectly through increased thyroid activity inasmuch as thiouracil blocks molt in progesterone-treated birds (Harris and Shaffner, 1957), and progesterone may increase the rate of uptake of ^{131}I by the thyroid (Himeno and Tanabe, 1957). However, thiouracil-treated laying hens may molt as do progesterone-treated fowl upon thiouracil administration (Shaffner, 1955; Himeno and Tanabe, 1957). Another possible indirect effect of progesterone may involve inhibition of gonadotropin secretion and a subsequent decrease in sex hormones (Smith *et al.*, 1957; Assenmacher, 1958; van Tienhoven, 1968).

For several reasons it appears that progesterone probably plays little part in molt in wild birds. Most species tested do not molt upon treatment with progesterone (Kobayashi, 1958; Wagner and Müller, 1963). There are no known changes in progesterone in the blood of wild birds or of domestic fowl that would suggest a maximum before or during the natural molts. No such changes are to be expected, particularly in the prenuptial molt of birds in sexual inactivity or during the seasonal development of the gonads. The effect of progesterone may be permissive through inhibition of the secretion of sex hormones. Also, castrated birds may still molt on schedule, although these birds have lost their major source of progesterone. These data suggest no primary role of progesterone in the control of natural molt in most wild birds.

B. Gonadotropins

The role of the hypophysial gonadotropins in the control of molt is largely indirect. The gonadotropins are probably involved in molt mainly in their control of the secretion of gonadal hormones. Gonadotropins are effective also in regulating the development of feathers in some molting birds. The effect of hypophysial hormones on the molting process has been investigated mainly through hypophysectomy and administration of gonadotropins of mammalian origin.

1. *The Effect of Gonadotropins on Molt*

The results of hypophysectomy are similar to those of castration. Hypophysectomized domestic fowl undergo an intense molt about two weeks following removal of the pituitary, and the birds remain in a state of perpetual molt (Assenmacher, 1958). The time of the beginning of molt and the continual loss of feathers over many months resemble the results of some castration studies of domestic fowl, suggesting that decrease in secretion of the gonadal sex hormones is effective in precipitating molt. The form of the new generation of feathers is similar to that of capons in pattern, but in shape and degenerated barbules the feathers indicate also a hypothyroid condition resulting from the loss of thyrotropic hormone. Chemical treatment of intact fowl with 2-amino-5-nitrothiazol (Enheptin), a drug that inhibits pituitary secretion, sometimes has an effect like that of hypophysectomy. Treated birds stop laying, undergo regression of the gonads and other structures dependent on sex hormones, and enter a continuous state of molt. The molt induced in chickens by this drug begins when the gonad and comb have regressed, about three weeks after treatment. Administration of mammalian PMS or of fowl pituitary extract restores gonadal activity and laying. The blockage of hypophysial secretion by this drug appears to be restricted to gonadotropin secretion and seems not to involve a decrease in TSH secretion (Assenmacher, 1958). Here also the molt may be traced to a decrease in the secretion of sex hormones by the gonads.

Histochemical and assay correlations of pituitary gonadotropin and the phases of the annual cycle may clarify the role of hormones in molt (Gourdji, 1964; King *et al.*, 1966; Tixier-Vidal *et al.*, 1968; Matsuo *et al.*, 1969). Even in the birds that have been best studied, however, no pattern of hypophysial activity which is exclusive or even common to the two annual periods of molt is evident.

The application of crude pituitary extract to castrates is followed by an early molt in canaries with thyroids but not in thyroidectomized canaries (Takewaki and Mori, 1944), suggesting that TSH has more effect (acting through the thyroid) than do the gonadotropins on molt. Treatment of fowl with exogenous FSH or LH has an irregular effect on molting. Gonadotropin treatment is followed by an interruption of laying and by molt in intact fowl, but molt also occurs in castrates (Juhn and Harris, 1956, 1958).

These results lead to the conclusion that molt usually accompanies a decrease in the levels of sex hormones in most birds. No direct effect of gonadotropic activity on molt is evident although gonadotropic control of sex hormone secretion may indirectly inhibit molt in some birds.

2. *The Effect of Gonadotropins on Feather Character*

The only clearly direct effect of gonadotropins on molt is on the character state of the growing feathers in certain birds. Males of some ploceid finches have a seasonal alteration of brown, sparrowy plumage in the nonbreeding season and brightly colored plumage, including black feathers, in the breeding season. Finches of the African genera *Euplectes*, *Quelea*, and *Vidua* (including "*Steganura*") all show response to injections of mammalian luteinizing hormone (LH) by growing black feathers in plucked regions which are black in breeding birds (Witschi, 1961; Ralph, 1969). In the absence of LH activity these feathers are brown. The black feather response to exogenous LH or to gonadotropic complexes with LH activity occurs in castrates of either sex as well as in intact male birds. Castrated Baya Weavers, *Ploceus philippinus*, and Red Avadavats, *Amandava amandava*, also undergo normal seasonal plumage cycles (Thapliyal and Saxena, 1961; Thapliyal and Tewary, 1961) suggesting a direct effect of seasonally varying gonadotropin levels on the male breeding plumage. The differential staining technique of Gourdji (1964) shows a seasonal variation in the presumptive LH- (or ICSH-) secreting cells in the adenohypophysis of *Euplectes orix* and these gamma cells are most active prior to the breeding season during the period of the prenuptial molt, indicating that in the natural annual cycle the plumage changes are in fact naturally mediated through cyclic changes in the levels of this gonadotropin.

C. Prolactin

Prolactin has integumentary effects in all vertebrate classes and growth-stimulating effects in the tetrapods (Bern and Nicoll, 1968). Two effects of prolactin are of special relevance to molting in birds: exogenous prolactin has an effect similar to that of thyroxine in increasing the rate of epidermal sloughing (molt) in lizards, and prolactin is involved in the formation of the brood patch of birds. In addition, prolactin may be involved in the molt of birds.

An inhibition of molting by prolactin was first noted for birds in the domestic fowl (Podhradsky *et al.*, 1937). Experiments by Kobayashi (1953b) were made on domestic pigeons which normally had a prolonged molt overlapping with the breeding season. Molt was often arrested while adults were feeding young up to two weeks of age, but resumed while they were feeding older nestlings. Using pituitary extract with prolactin activity the effects on molting pigeons late in the molting season were observed in two birds. Feather loss on the body was interrupted, and one bird ceased molting the wing feathers also. Other experiments were completed after the normal molting season, and although the results were inconclusive, Kobayashi (1953b) concluded that prolactin inhibits molt. Pigeon crop-gland activity (a measure of prolactin activity) is high just before hatching and for a few days thereafter. The timing of crop-gland activity thus indicates a peak of prolactin secretion in the pigeon at the time when molting is interrupted. Perhaps the simultaneous inhibition of molt and stimulation of crop-gland activity in nesting pigeons by the same substance, prolactin, is the mechanism controlling the separation of these two processes, both of which may impose considerable nutritional demands upon the parents.

A diversity of results have been reported from other birds. Prolactin had no effect on the molt of several wild passerines tested by Wagner and Müller (1963). On the other hand, exogenous prolactin induces molting in domestic fowl (Juhn and Harris, 1956, 1958; Tanabe and Katsuragi, 1962). In hypophysectomized pigeons prolactin increases hemopoiesis and serum protein levels (Höcker, 1969a,b), and these increases in the activity of the vascular system may be involved in molt.

Prolactin may be an important regulator of the annual cycle, including molt. An antigonadal response, perhaps mediated through the gonadotropins, occurs in some migratory songbirds (Meier, 1969b), and so by depressing other hormones prolactin may affect molting. Prolactin induces fat deposition and nocturnal restlessness associated with migration in *Zonotrichia leucophrys gambelii*. As these events are separated in time from molt the high levels of endogenous prolactin in birds undergoing fat deposition, nocturnal restlessness, brood patch formation, and possibly parental care may be effective in inhibiting molt. The effect of prolactin is complex, and the responsiveness of birds to prolactin varies with the time of day. In *Z. albicollis* prolactin promotes fat deposition at the beginning of a light period, but fat depletion results when prolactin is given in the middle of a period of light, and conversely it evokes nocturnal restlessness only when given at midday (Meier, 1969a). Midday injection of prolactin can check the postnuptial molt of *Z. l. gambelii*. The time-dependence of these responses to prolactin may involve a diurnal cycle of pituitary secretion and release. In addition, the diurnal rhythm of pituitary prolactin is different during the periods of molting and nonmolting (Meier *et al.*, 1969). The time of release of prolactin and also a temporal pattern of sensitivity of the integument to hormones provide possible mechanisms for the integration of molt into the annual cycle.

D. Thyroid Hormone

The "generative ferment in the blood" often regarded as most closely linked to molt in birds is thyroid hormone (Voitkevich, 1966). More than 50 years ago Zawadowsky (1927) found that chickens molted when they were fed extracts of thyroid glands, and several other workers have induced molt under similar experimental conditions. Whether the natural seasonal molt of wild birds is brought on by increased thyroid activity is less certain.

1. Histophysiological Correlation of Thyroid and Molt

About 24 species of wild birds have been examined to determine whether the thyroid activity increases at the time of molt. In some birds the thyroid appears to be most active in the weeks before molt and in certain others activity is high during molt. Most reports of increased activity of the thyroid related to molt have involved small numbers of individual birds, and although some birds may undergo increased thyroid activity, none of the studies demonstrates a clear statistically significant increase in thyroid activity during molt or in the immediately preceding weeks. When thyroid increases do occur, in the annual cycle, they do not regularly occur at the time of molt. Birds in which some seasonal increase in activity of the thyroid around the time of molt has been noted are discussed by Andrews and Naik (1966), Bigalke (1956), Davis and Davis (1954), Gál (1940), Höhn (1949), Küchler (1935), Oakeson and Lilley (1960), Schildmacher (1956), Takewaki and Mori (1944), Vaugien (1948), and Voitkevich (1966). On the other hand, some species exhibit no increase in thyroid activity with molt (Elterich, 1936; D. S. Miller, 1939; Wilson and Farner, 1960; Kendeigh and Wallin, 1966; John and George, 1967; Ljunggren, 1968; Erpino, 1968; Raitt, 1968; Payne and Landolt, 1970). Until detailed studies of large samples have been completed with more species, it will be unclear whether any species of wild bird regularly shows in nature a specific increase of thyroid activity at the onset of molt.

The interpretation of thyroid activity in these studies depends upon the validity of histological data. In laboratory animals a close correlation occurs between radioiodine and histological measures of thyroid activity (Biellier and Turner, 1957; Wilson and Farner, 1960; Ringer, 1965; von Faber, 1967), but it remains possible that thyroid activity in birds may vary without evident histological change.

Little physiological evidence is available on thyroid activity in relation to molt. An increase in metabolic rate often occurs in molt and may be related to increased thyroxine output, as metabolic rates in most endothermic vertebrates are elevated by increased levels of thyroxine. Bioassay tests indicate seasonal variation in thyroxine activity in domestic fowl, and high rates occur near the end of molt (Reineke and Turner, 1945). Tanabe *et al.* (1955), however, using more reliable techniques, compared the thyroid uptake of radioiodine in molting and nonmolting fowls and found no difference. Himeno and Tanabe (1957) found an increase in thyroid activity measured by radioiodine uptake before molt, but when molting chickens are kept warm enough, there is no change in the thyroxine output in molt (Tanabe and Katsuragi, 1962).

Studies of pituitary cells thought to produce thyrotropic hormone (TSH) have involved histochemical techniques of cell identification. In some birds the presumptive TSH-secreting cells are most active at the time of molt. During the molting period of fowl the pituitary basophils are mainly cells identified as thyrotrophs (the delta cells), whereas in the breeding period the majority of basophils are gonadotrophs (Perek *et al.*, 1957). However, bioassay analysis shows no increased TSH content in the pituitaries of laying hens (Tanabe *et al.*, 1958). In the Pekin duck the rate of thyroxine secretion is high in the early stages of molt in May and June and then decreases in the second half of molt in July. This pattern of thyroid activity parallels but lags slightly behind the cycle of the activity of the delta cells of the duck (Tixier-Vidal *et al.*, 1962), suggesting that the delta cells undergo a change in TSH content and that this change alters the rate of thyroxine secretion. The delta cells in *Euplectes orix* hypertrophy and are cytologically more active at the time of the prenuptial molt (Gourdji, 1964), and here also molt may coincide with increased thyroid activity. Nevertheless, studies using these techniques for the differentiation of pituitary cells have not established unambiguously the cellular source of TSH in the avian hypophysis, nor has it been possible to establish a clear rise in the delta cell activity before or during molt in all species (Matsuo *et al.*, 1969).

2. *Thyroidectomy*

Removal of the thyroid glands generally inhibits the process of molt. The effect depends upon the time of thyroidectomy. In a series of experiments with European Starlings (*Sturnus vulgaris*) the birds were thyroidectomized 1, 2, and 3 months before their normal season of molt, and they were examined subsequently for more than a year (Voitkevich, 1966). Birds thyroidectomized three months prior to the molt season did not molt at all. Starlings thyroidectomized two months before the molting season began molt several weeks late. Birds thyroidectomized 1 month before the molting period began to molt at about the same time as controls and completed the molt more rapidly. Thyroidectomy of molting birds generally does not bring about an abrupt arrest of molt; rather the rate of feather renewal is retarded. Other species likewise usually are inhibited or arrested in their molt by experimental thyroidectomy. Experiments in which thyroidectomy had no apparent effect on molt have also been reported (Höhn, 1949), and Paredes (1956) found molt in domestic fowl to recur at the normal season for 3 years after thyroidectomy, although possibly in these studies some thyroid tissue had regenerated after incomplete removal of the organs.

Diminished thyroid hormone secretion resulting from reported antithyroid chemicals such as sulfonamides and thioureates also may affect molt. Thiouracil delays the onset of molt and slows the rate of molt in domestic fowl (Glazener and Jull, 1946; Domm and Blivaiss, 1948). It inhibits molt in some passerines as well (Wagner and Müller, 1963). On the other hand Sulman and Perek (1947) and Himeno and Tanabe (1957) found no inhibition in domestic fowl, and Kobayashi (1952) found no inhibition in the pigeon. The thyroid histology of thiouracil-treated birds shows large cells and small follicles. The interpretation of inhibited secretion in the thiouracil-treated birds is questionable (Assenmacher, 1958), and measured thyroxine secretion rates are in fact higher in thiouracil-treated domestic fowl (Tanabe and Katsuragi, 1962).

Other changes result from thyroidectomy. Feathers of hypothyroid adult birds are often lanceolate in shape and lack many barbules, and the melanin content is altered. Thiourea treatment of embryos has similar effects on the growing down feathers (Johnson, 1968). Young birds thyroidectomized in juvenile plumage often fail to molt; they also are stunted in their general growth. Some juvenile behavior patterns are retained and adult behavior fails to develop, and a high mortality rate occurs in thyroidectomized birds (Voitkevich, 1966). These various effects suggest that the failure of some thyroidectomized birds to molt normally may not be a specific response to levels of thyroid hormones.

3. *Effects of Thyroid Treatment on Molt and Plumage Development*

Treatment of birds with thyroxine is usually followed by molt. Response to thyroxine is rapid. The old feathers are shed and new ones emerge from the skin in 4–7 days after treatment (Takewaki and Mori, 1944; Wagner, 1961; Voitkevich, 1966). Treatment with thyroxine is followed by molt in birds that had not molted for a long time after thyroidectomy. Thyroid hormone treatment brings on a cessation of laying and an early molt in domestic fowl and other galliform birds, in pigeons, and in some passerines, whereas ducks and geese respond only to high doses of thyroid hormone. Among the passerine birds and birds of prey are species that apparently do not molt at all in response to thyroid hormone (Assenmacher, 1958; Wagner and Müller, 1963; Voitkevich, 1966).

The pattern of feather loss and regrowth in thyroxine-induced molt may depart from the orderly sequence of feather replacement in natural molt. Molting canaries become almost naked after a week of treatment with large doses (Takewaki and Mori, 1944). The simultaneous molt in nearly all feather tracts is markedly unlike the sequential replacement of feathers in natural molt. In domestic fowl low doses of thyroid extract have no effect on molt while larger doses result in complete molt, and the largest doses lead to nearly complete denudation. In experiments with physiologically appropriate dosage levels, however, pigeons molt their feathers in the same sequence as in natural seasonal molt (Assenmacher, 1958). Feathers grown by hyperthyroid birds are usually normal in appearance, but Voitkevich (1966) notes that often feathers grown in such conditions are different in structure, pigmentation, and pattern from feathers grown in natural seasonal molt. The downy base of feathers of pigeons and chickens is larger than in normal feathers and the barbules of the vane are poorly developed. Some birds show an unusually high degree of melanization while others, even breeds of the same species, undergo a lesser than normal amount of melanization (Lillie and Juhn, 1932; Assenmacher, 1958; Voitkevich, 1966). In plucked fowl the dose of thyroxine required to cause molt was lower than that which brought about an abnormal, hyperthyroid form of feather in the experiments of Juhn (1963), suggesting to her that surges of thyroid activity do not cause normal molt.

The effectiveness of thyroxine varies with time in relation to molt. The changing sensitivity of birds to thyroxine makes it unclear whether the reported species differences in responsiveness are real, or whether they result from testing birds in unlike physiological states. Higher doses of thyroxine are necessary to precipitate a new molt at the termination of an earlier molt. The hormone may be ineffective when it is administered a few weeks after the bird has molted, and it is most effective several months after molt has been completed (van der Meulen, 1939; Tanabe and Katsuragi, 1962; Wagner and Müller, 1963; Voitkevich, 1966). Repeated treatment with thyroxine may, nevertheless, bring on a series of molts in quick succession (Assenmacher, 1958). The refractoriness of most birds at the end of their molting periods may indicate a state of preparation of the feather follicle.

Histochemical tests show that thyroxine treatment increases the acid phosphatase activity in the feather tracts (Kobayashi *et al.*,

1955); this enzyme activity occurs in the skins of birds that are growing feathers (Kobayashi and Koscak, 1957) and is located in the follicle sheath (Spearman, 1966). The skin of birds in molt induced by exogenous thyroxine shows a wave of mitoses in the feather papilla within a few days. The period of a few days in which no external sign of molt is evident is a time when the new feather is growing rapidly below the surface of the skin and pushes out the old feather. The rapid cell division and growth of the feather papilla in thyroxinized birds suggest a direct response of the feather follicle to the thyroid hormone (Voitkevich, 1966).

A local effect of triiodothyronine on molt in hypophysectomized pigeons is the very local growth of new feathers from the plucked follicles (Höcker, 1967). The local response to minute quantities of thyroid hormone indicates a direct action of the hormone on feather follicle growth.

4. *Thyroid Activity and Molt*

In general, thyroid activity seems to be related to molt, but the apparent role of the thyroid in molt differs from species to species. In some species thyroid activity may increase before molt or in the early stages of seasonal molt. Thyroidectomy inhibits molt if the thyroid gland is removed several weeks prior to the onset of molt, but otherwise molt is begun and may be completed. Exogenous thyroxine or thyroid extract in high doses usually results in molt in intact birds as well as in thyroidectomized birds, but high doses may result in a molt abnormal in the pattern of replacement or in the structure and pigmentation of the feathers. Thyroxine may act directly upon the feather follicles. The absence of histological hyperactivity of the thyroid in molt in some species, the lack of response to thyroidectomy in molting birds or in birds within a few weeks of molt, and the irregularities of molt associated with experimental hyperthyroidism, however, all cast doubt on the role of thyroid hormone as a simple means of the control of molt in birds.

E. Gonad–Thyroid Interactions

In most birds the period of molt occurs only after the gonads have regressed, and in some of these the thyroid activity apparently has increased. The complementary timing of the activity of the gonads and the thyroid suggests an interaction between these two endocrine organs in the control of molt. Gonadal activity is enhanced by a normal thyroid activity, as shown by the regression of the gonads in thyroidectomized birds and the gonadal and secondary sex character development in sexually inactive birds given thyroxine treatment (Assenmacher, 1958). On the other hand, the sex hormones may inhibit thyroid activity. Castration in the pigeon causes an increase in the thyrotropic content of the pituitary (Schooley, 1937). The cytological thyrotropic activity of the adenohypophysis of photostimulated Japanese Quail is greater in castrates than in intact birds according to Tixier-Vidal *et al.* (1968), who commented that "the activation of thyrotropic function by castration appears to be a unique feature of avian endocrinology." This effect is blocked by androgens in castrated birds. One must be cautious here because of the problems of identification of TSH cells by cytological and cytochemical techniques. If the cell identifications are valid, the results suggest a mechanism by which a possible increase may be effected in thyroid activity at the time of post-breeding regression of the gonads immediately before the summer molt, since birds with regressing gonads are like castrates in having less androgens and estrogens. As a result, their TSH and thyroid activities may increase. The reverse, however, is likely in the prenuptial molt. Assuming that natural molts are associated with high thyroxine levels, the fact that the sex-specific plumage patterns and pigmentation in some birds are controlled by sex hormones or gonadotropins (Witschi, 1961) suggests that both thyroxine levels and sex hormone levels may be high in some natural molts. Neither a comparative review of molt schedules nor the results of earlier experimental work clarifies the importance of inhibition between the reproductive hormones and the thyroid hormone.

F. Other Hormonal Effects

In a search for correlations between seasonal hormonal changes and the events of the annual cycle in birds several workers have investigated the adrenal gland. Some birds have increased interrenal activity immediately before molt (Fromme-Bouman, 1962; Hall, 1968; Raitt, 1968; Thybusch, 1965), suggesting a possible effect of interrenal hormones on molt. The one adrenal hormone known to induce molt is deoxycorticosterone; its administration in the domestic fowl is followed by an arrest of laying and dropping of the feathers (Juhn and Harris, 1956; Herrick and Adams, 1957), although the pattern of molt is irregular. The interrenal cycle of *Zonotrichia leucophrys gambelii* shows increased cortical cellular volume and sudanophilia before and during the prenuptial molt but a decrease in sudanophilia associated with the postnuptial molt (Lorenzen and Farner, 1964), indicating no general correlation of secretory activity and the process of the seasonal molts. The occurrence of molt when birds are disturbed ("fright molt") is possibly related to unusually high levels of adrenal hormones as a response to a stressful situation, but the rapidity with which a fright molt can occur suggests a neural releaser.

Experimental work with exogenous hormones or other substances of the parathyroids, ultimobranchial body, glycogen body, and neurohypophysis in birds has not shown them to have an effect on molt, nor are any seasonal changes in the activity of these organs known to be associated with the timing of molt.

G. Hormonal Control of Molt

Endocrines clearly affect the launching and the course of molt in experimental birds, and hormonal changes most likely are important in the control of the seasonal molts of wild birds as well. Our present understanding of the role of the hormonal changes associated with molt is poor indeed, and in this field there is an opportunity for carefully designed experimental work. The old idea of a simple control of molt by the thyroid is probably not appropriate, because most experimental studies with thyroid hormones result in a molt unlike the naturally occurring seasonal molts of wild birds, and because most wild birds that have been relatively well studied do not show an increase of thyroid activity in molt. Both the experimental findings and the separation of breeding and molting schedules in most wild birds suggest that the sex hormones have an inhibitory effect on molt. A physiological control mechanism that prevents the expenditure of energy on molting during the period of nesting directly by the inhibitory influence of the sex hormones which control breeding itself is a process that is of readily apparent adaptive significance in those birds which have separate molting and breeding schedules. But the continuation of molting in progress in wild songbirds after castration reveals this mechanism to be one more easily deduced from field observations than it is demonstrable in experimental conditions. An interaction between the reproductive hormones and thyroid activity suggests that the collapse of reproductive condition at the end of the breeding season removes the effect of the inhibition of the sex hormones on pituitary TSH, and the postnuptial molt of seasonally breeding birds may result from the consequent surge of thyroid activity. Neither a direct inhibition of the sex hormones on molt nor this proposed sex hormone–thyroid interaction are to be expected in all birds, however, especially in species that breed and molt at the same time, as these birds presumably have high levels of sex hormones while they are molting. Other hormones that may be involved in molt in birds are progestins and prolactin, and in some experiments these substances have an effect upon molt. The hormonal balance of birds is probably different during the prenuptial and postnuptial molts in forms that molt twice a year.

Even within a single species, the domestic fowl, a wide variety of hormones can be used experimentally to initiate a molt. The nonspecific responsiveness of a bird to a diversity of hormones makes difficult the uncovering of the specific substance or substances involved in the molt of birds under natural conditions. Assay of endocrine activity in the annual cycle and molt of wild birds is desirable if we are to corroborate the effects of experimental changes as being relevant to the changes in natural molt.

It is probable that different hormonal control schemes are active in different groups of birds. Multiple physiological controls are suggested by the differences in the molt–breeding and molt–migration schedules of different kinds of birds and also by the species differences in responsiveness of birds in experimental conditions.

V. Direct Effects of the Environment on Molt

The proximate responses of molt schedules to changes in the environment have been noted by keeping birds on constant conditions and by altering the light schedule, temperature, and food. The steps through which these factors affect molt are unclear, though the response of the hypothalamo-hypophysial system to the daily photoperiod is known to affect both the thyroid and the gonads. All of these factors probably act indirectly on molt in wild birds. Abnormal molts can be induced by the same factors in captives, and in some cases these molts shed some light on the normal physiological control of molt.

A. Photoperiodic Control

The annual cycles of temperate-region birds are controlled in large measure by seasonally changing photoperiods. Effects of daylength have been studied in greater detail in the development of the gonads than in molt, though photoperiodic effects upon both gonadal development and molt are presumably mediated through the release of hypophysial hormones. The effect of daylength on molt is complex, and in some birds the phase relationships between gonadal enlargement and molt in experimentally prolonged daylengths are different from those of the natural annual cycle.

Two main types of response in molt occur in birds maintained on long photoperiods after the short days of winter. Birds that in nature molt shortly before their breeding season may undergo a molt simultaneous with their gonadal enlargement on long daylengths. This overlap occurs in captive white-eyes, *Zosterops* (Miyasaki, 1934; Keast, 1953), ploceid finches (Brown and Rollo, 1940), and the crowned sparrows *Zonotrichia albicollis* (Lesher and Kendeigh, 1941), *Z. atricapilla* (A. H. Miller, 1954), and *Z. leucophrys* (A. H. Miller, 1954; Farner and Mewaldt, 1955). In photoresponsive experimental birds the processes of molt, fat deposition, and gonadal enlargement may all coincide even though they are scheduled at different times in wild birds (Farner and Mewaldt, 1955; Farner, 1964; King, 1968). The resulting overlap makes more obscure the hormonal mechanisms involved in experimental molt studies. The molt of the sparrows occurs only in some individuals, and it occurs both in birds that are reproductively photorefractory in autumn and in photoresponsive birds in mid-winter. Another irregularity of photo-induced prenuptial molt in the crowned sparrows is the occasional molting of the flight feathers (A. H. Miller, 1954), which are replaced only after the breeding season in wild birds. Long days also advance the time of molt in Clark's Nutcracker, *Nucifraga columbiana,* which normally has overlapping breeding and molting schedules, although in this case it is the "postnuptial" (i.e., prebasic or complete) molt that precedes gonadal development in late winter (Mewaldt, 1958).

In contrast to the crowned sparrows, birds that normally have only a single annual molt generally do not molt in response to experimentally prolonged daylengths. An absence of an immediate molt on long experimental daylength has been noted in Japanese Quail (Follet and Farner, 1966; Follett and Riley, 1967), *Agelaius phoeniceus* and *A. tricolor* (Payne, 1969b), and *Junco hyemalis* (Wolfson, 1966); it therefore seems unlikely that molt is caused by increased gonadotropins or sex steroids in these birds. The hypothalamic response to long daylengths in the Japanese Quail includes an increase in gonadotropins and also in thyrotropin indicated by increased activity of the thyroid (Follett and Riley, 1967). Although the thyroid activity is quickly elevated by long daylengths it is again depressed as the growing gonads enlarge. The thyrotropic–thyroid inhibition resulting from the increasing levels of sex hormones from the gonads in long-day birds may be responsible for the failure of photoperiodically induced molt to be completed in crowned sparrows and other birds. The prenuptial molt of other species (Ploceidae) takes place during the time of an increasing gonadal activity. The effect of daylength on molt evidently differs among groups of birds.

Experimental shifts of periods of long and short photoperiods can alter the number of times a bird molts, and sometimes molting and gonadal cycles can be made to go out of phase. Miyazaki (1934) induced three molts in a year by alternating short and long photoperiods in *Zosterops japonica*, a species that normally has only one annual molt. Wolfson (1966) has brought about through alterations of light as many as five periods of gonadal enlargement and regression in a single year in *J. hyemalis*. Molt occurred only twice in the year and was not closely associated with gonadal condition (or, presumably, gonadotropins). The results may be explained by appreciating the period of unresponsiveness of the feather follicles after each molt.

Other experimental studies with photoperiod indicate an effect of reproductive hormones. When birds with large testes are maintained on short days, the gonads quickly regress and a molt similar to the natural postnuptial molt follows (Assenmacher, 1958). Similarly birds kept on long daylengths usually show a regression of the gonads after several months and molt follows (Wolfson, 1966), perhaps due to diminution of the reproductive hormones. Finally, molting birds may undergo a prolonged molt on long days but may complete the molt more quickly if the daylengths are shortened (Assenmacher, 1958). In nature the control of molt by short daylengths may be an adaptation for late-nesting birds of high latitudes to complete the molt rapidly before the time for migration or winter fattening.

It would be most useful to learn whether shortening the photoperiod would induce molt in seabirds. Each year thousands of loons, grebes, cormorants, ducks, and alcids are found barely alive with their plumage badly oiled from oil pollution at sea. The only cases in which birds have survived a heavy oiling and have been returned successfully to the wild again have been the ones in which birds have been cleaned by hand and held in captivity through the annual molt. The oil and the cleaning process destroy the waterproofing of the oiled feathers, and only a new plumage can waterproof a bird once oiled. Preliminary photoperiodic experiments on some alcids suggest that molt may be only slightly advanced by short hours of light (Aldrich, 1970). Whether or not a drop in sex hormones and an increase in thyroid hormones causes molt naturally in wild birds, a therapy based on hormonal injections together with photoperiodic manipulation may become an effective technique in restoring oiled seabirds back to nature.

B. Persistent Cycles of Molt in Constant Conditions

The hypothesis that autochthonous annual cycles independent of environmental changes may drive the molting rhythms can be tested by keeping birds under constant conditions. Any resemblance between the molt of such birds deprived of exogenous cues of the time of the year and the seasonal molts of wild birds may be ascribed to an internal periodicity. Such a cycle, however, may be dependent on other physiological conditions. As in wild birds of temperate regions, the processes of molting, migration, and breeding follow one another in regular manner, the hormonal or metabolic changes wrought by one event may precipitate the next, and molt may be causally linked to these events. Molt cycles cannot validly be divorced from the interrelated changes of reproductive cycles, at least in intact birds. King (1968) has clearly stated the problem in terms of the lipid cycle: ". . . the annual cycle of fat deposition and depletion may be the expression of an autonomous sequence of neuroendocrine events that is induced or set in phase, or both, by the increasing vernal photoperiod." The independence of the periodicity of molting seasons from environmental changes in intact birds is, nevertheless, of interest, as molt is the most readily observed aspect of the entire annual cycle in nonbreeding, captive birds.

Birds held on long-term short daylengths (9 hours or less of light per day) often show an interruption of the timing of molt. Several temperate-region birds (*Erithacus rubecula, Junco hyemalis, Zonotrichia leucophrys*) may go for more than a year without molting (Merkel, 1963; Wolfson, 1959; King, 1968). Depending upon the month in which the birds are initiated to short days some may molt irregularly. In a comparative study Merkel (1963) found captive *Sylvia communis* to maintain its natural molt schedule on short days for more than a year, whereas *Erithacus rubecula* did not molt at all over a period of more than 2 years. *S. communis* is highly migratory, wintering south of the equator and breeding in northern Europe, and *E. rubecula* is largely a year-round European resident. Merkel suggests that in migratory species which are subjected to long daylengths all year (as they "winter" in the local summer of southern Africa) the annual cycles (including molt) are in large part independent of the cues of daylength.

A regular pattern of persistent molt cycles in birds on constant conditions of tropical or spring photoperiods occurs in three Old World warblers (Wagner and Müller, 1963; Gwinner, 1967, 1968). Birds held on LD 12:12 (12 hours of light alternating with 12 hours of darkness) for more than 2 years molted at approximately the same time each year as do the wild warblers, which are highly migratory, breeding in northern Europe and wintering in Africa south of the equator. The same regularity on LD 12:12 was noted for a year-round European resident thrush, *Turdus iliacus,* which is as timely in its successive molts as birds on natural daylength.

Other birds on constant spring or summer photocycles (12 hours or more of light) may maintain their periodicity of molting for a year and then later molt irregularly (Wolfson, 1966; Lofts, 1964; Zimmerman, 1966; King, 1968). This pattern, the one most frequently observed in captive birds, indicates that autochthonous annual cycles alone—without external timers—do not account for the periodic changes in the physiological condition of birds. Environmental effects are important in most birds. However, cycles as in northern temperate-region Red Crossbills (*Loxia curvirostra*) are highly independent of daylength. Birds kept for more than a year all molted normally at the same time regardless of whether their photocycles included 8, 12, or 16 hours of light each day (Tordoff and Dawson, 1965).

In Fig. 3 are shown the molting cycles of tropical African finches (Ploceidae) on LD 12:12 and on natural daylength at 34° N latitude in Oklahoma for more than a year. Male *Vidua chalybeata* and *V. purpurascens* [these species are discussed in Payne, 1972] undergo two annual molts in Africa and have distinct breeding (summer) and nonbreeding (winter) plumage. When six birds from Rhodesia were imported and then maintained on northern daylength they regularly underwent a shift in their breeding phase, and within 7 months all were in the plumage appropriate for the northern season. A male maintained on natural daylengths for 2 years in a greenhouse re-

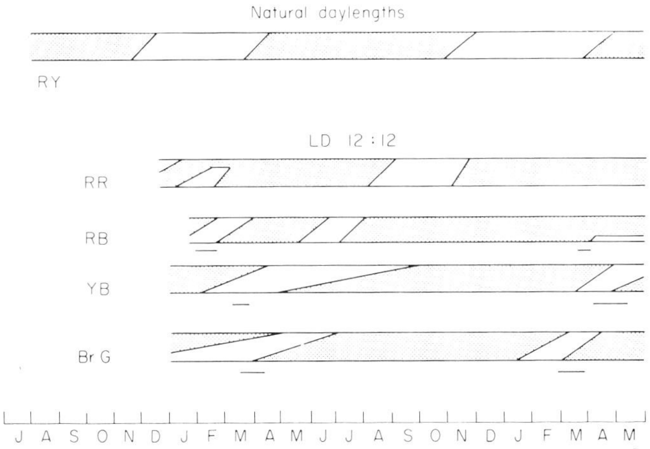

FIG. 3. Molt cycles of tropical ploceid finches on natural temperate daylength and on long continued LD 12:12. RY, RR, and RB = *Vidua purpurascens*, YB and Br G = *Vidua chalybeata*; all birds are males. Shaded areas represent breeding plumage and unshaded areas represent nonbreeding plumage. Molt of the primaries is indicated by a horizontal bar. Periods of molt correspond to the slanted lines between the plumage generations.

mained in breeding plumage through late summer, during the months when it would have been in winter plumage in southern Africa. Males kept on a constant LD 12:12 at 27°C showed less regularity in their molts. The time of molt varied a few weeks from year to year in a single bird and some molts were not completed in the usual time of a few weeks. Two birds in the second year began the prenuptial molt before the postnuptial molt had been completed, and these birds were almost continually in breeding plumage. In some molts a bird began to grow nonbreeding plumage, but within a week the feather development shifted to production of breeding plumage, producing some individual feathers that were brown at the tip and black at the base. Neither the time of molt nor the character state of the new feathers remained in phase with the seasonal wild birds. The time of molt of the primaries was more regular. The tropical finches on LD 12:12 do have alterations of breeding and nonbreeding plumage, but the timing is not very precise, and the finches do not have autochthonous annual rhythms that are completely independent of environmental timers and are also responsible for the regular seasonal changes of plumages in the wild birds.

The molting schedules of birds on constant conditions are diverse, and some of the differences may reflect real physiological differences among species. It is difficult to separate the causes of periodic molt from other seasonally recurring events of the annual cycle. The birds studied on prolonged short daylengths include species that require long daylengths for gonadal development, including *Zonotrichia leucophrys* and *Z. albicollis* (Farner and Wilson, 1957; Weise, 1962). If postnuptial molt in these sparrows is merely a consequence of a sequence of events set off by the long photocycles of spring, one might expect an absence of molt in birds held on prolonged short days as the result of the failure of gonadal development, yet under these conditions one species molts and the other does not (Weise, 1962; King, 1968). These results suggest that external events in addition to the daylengths of early spring are directly involved in timing the annual cycles of wild birds.

C. Temperature

Changes in temperature are generally regarded as of little importance as primary regulating factors in the annual cycle of birds, though indirectly the cycle is adapted to the ecological changes of the seasons. Some birds such as House Sparrows have fewer feathers in summer than in the colder seasons. This seasonal change in number of feathers involves not the development of an additional winter coat through seasonally unequal molts but rather the loss of many of the body feathers during spring. These feathers are lost by preening and abrasion rather than by the force of growing feather papillae. The gradual loss of feathers in the breeding season may be an adaptation for thinning the plumage in warm weather (Clench, 1970). Temperature may occupy a minor role in molting in captive birds. In hot weather caged birds sometimes shed their feathers in great masses at once, not in the sequence of normal seasonal molt. An African finch (*Vidua macroura*) which rode in my car across the desert near Needles, California, in August, dropped about half of its feathers during the next few days in Sacramento, apparently from the shock of heat.

Less often observed is molt in birds exposed to cold, and in captivity molt may be interrupted if the birds become chilled (Keast, 1953). Molt is nonadaptive in cold weather as the loss of feathers would permit greater loss of body heat at an inopportune time. The nonresponsiveness of the feather follicles to cold temperatures is interesting as both molt and cold temperatures may be associated with increased thyroid activity (Wilson and Farner, 1960). If both molting and thermal acclimatization to cold were mediated through increased thyroid activity, then cold birds with increased thyroid activity might molt at the coldest time of the year. There is generally less evidence for increased thyroid activity in birds in cold temperatures than for activity associated with molt, and there may be no simple, common thyroid function in these processes (Wilson and Farner, 1960).

D. Food

In captive and domestic birds an interruption of molt or untimely molt may result from changes or deficiencies in the food supply. Domestic fowl can be induced to molt by a sharp restriction of their food intake. Fowl in this period of "forced molt" generally stop laying eggs until the food intake is again increased. The cessation of laying at the beginning of forced molt suggests that withdrawal of ovarian hormones is involved in molting (Tanabe and Katsuragi, 1962). Deficiencies in the diet of caged birds are sometimes accompanied by chronic inhibition of molt. Protein shortages in the diets of House Sparrows inhibit molting (Kendeigh, in Tollefson, 1969). Similar effects may cause molt to be slowed or to be interrupted in wild birds, and Ashmole (1962) has reported delay in the onset of molt in wild seabirds related to an unseasonal shortage of food. An abnormal defeathering ("French molt") involving the breakage of feathers in young Budgerigars (*Melopsittacus undulatus*) is apparently related to diet. The incidence of French molt is increased in Budgerigars on a diet high in vitamin A and deficient in vitamins E and K (Taylor, 1969). Presumably the feathers are structurally unsound when they are first formed in the follicles. A sudden molt in captive songbirds sometimes follows a shift to a generous feeding of mealworms or fresh ant eggs (Wagner and Müller, 1963).

E. Fright Molt and Feather Plucking

Birds may shed their feathers when they are disturbed at the nest or are handled in captivity. Poultry farmers find that laying hens may molt when they are disturbed by changes in feeding, housing, or lighting (van der Meulen, 1939). The rapid dropping of feathers in this fright molt (Schreckmauser) does not follow the sequential pattern of feather loss in the seasonal molts, but rather large numbers of feathers are dropped at once. The rectrices are most often shed; body feathers are sometimes dropped. No correlations are evident between fright molt and breeding condition, molting seasons, age, or sex (Dathe, 1955). Perhaps a neurally mediated relaxation of the muscles between feather follicles could allow the follicle to dilate slightly, resulting in the loosening and falling out of the feathers (Stettenheim, 1972). This process is different from the loosening and pushing out of the feathers by the actively growing papillae in seasonal molt. The central nervous system, acting through the feather muscles, is not known to alter the state of integumentary tonus in the normal molting periods. Perhaps related to fright is the molt suffered by birds caught and shipped in small boxes (Wagner, 1962). The disturbances of the normal patterns of feather retention and molt by handling make more difficult the experimental study of molt.

Behavioral abberations may bring about the loss of feathers in parrots not through shedding but through feather plucking. Captive parrots sometimes pluck themselves naked. Excessive feather plucking often results from the sensory deprivation or boredom of parrots in small, unadorned cages, and it may occur when a parrot overpreens its mate (Dilger, 1969; Buckley, 1969).

F. Environmental Changes and Molt Schedules

Birds in seasonal temperate regions are generally unresponsive to temporary changes in the environment, inasmuch as unseasonal periods of hot weather or rains or temporary changes in food supply do not appear to cause molt. Molt schedules, like breeding schedules, are indirectly adapted to the changing seasons in periodic environments, and the role of seasonal photoperiodic responses in timing the entire annual cycle is more important than the responses to other environmental changes. The adaptive significance of responding to photoperiod and at the same time of remaining unresponsive to most other environmental features is that daylength is a reliable predictor of average future conditions, whereas other features vary irregularly and are poor predictors (Farner, 1964). If birds molted directly in response to hot temperatures, they would probably not have time to complete the molt before the weather again became cold, and they would also probably expend a great deal more energy by molting more than once or twice each year, because hot periods may occur frequently. The molting process is similar to the seasonal gonadal development which precedes breeding as both require several weeks for completion. The specificity of the molt response to seasonal daylengths serves as a safety mechanism in a manner similar to that through which the gonadal response is regulated. Nevertheless, the hormones involved in the molts, even when they are controlled through a photoresponsive hypothalamo-hypophysial system, are not necessarily the same ones involved in gonadal cycles.

VI. Energetics and Molt

The temporal separation of molt, breeding, and migration in the annual cycle is often interpreted as an adaptation that minimizes the day-to-day strain on the energy resources of a bird (Kendeigh, 1949; Farner, 1964). This concept has been developed from the widespread observation that the annual cycle of many species is, in fact, split into non-overlapping processes; this and the use of a terminology (Dwight, 1900) of "prenuptial" and "postnuptial" molts that are separated in name as well as time from the breeding season has apparently convinced many avian biologists of the general occurrence of separate seasons. Indeed, most species that have been studied do have separate timing schedules and in these the molts are appropriately named for purposes of describing the annual cycle by the terminology of Dwight. However, many species have overlapping schedules, and clearly in these birds the events of breeding and molting are energetically compatible. It is of interest to consider the more direct evidence bearing on the problem of whether molt does significantly increase the energy requirements and the nutritional demands of birds.

Both body temperatures and metabolic rates of molting birds suggest that molt involves increased energy demands. An elevated level of energy metabolism is evident in the body temperature of molting birds, as molting European Bullfinches average 0.8–1.0°C higher than birds not in molt (Newton, 1968a). The metabolic rates also are higher in molting birds than in nonmolting birds. Estimates of the increased metabolic rates of molting birds have been made for the molt following the breeding season—the molt that is generally complete and hence involves the greatest cost in energy. In most species of songbirds studied the metabolic rate increases from 5 to 30% above the rate of nonmolting birds, though the figure differs among species and investigators (Blackmore, 1969; Dolnik, 1965; Helms, 1968; Wallgren, 1954; West, 1960). In the Chaffinch, *Fringilla coelebs*, Koch and De Bont (1944) found metabolic rate to increase by 25% before the peak of molt, whereas in other species examined the peak of energy increase coincided with the molt. Domestic fowl increase their standard metabolic rate (SMR) by 30% during molt (Perek and Sulman, 1945).

Dolnik (1965) has compared the total energetic cost of molt in several Palearctic species by measuring metabolic rates through the period of molt. In Table I are the energetic costs of the complete postnuptial molt for the Fringillidae and for *Passer* determined by Dolnik, and a comparison of this energy with the SMR that has been calculated from mean body size from other sources, as Dolnik did not indicate directly the normal SMR for these species. An increase in energy expenditure of approximately 34%, on the average, in Palearctic songbirds is suggested from these data, a figure twice that of most other investigators. The duration of molt in Dolnik's birds is longer than in most wild birds or in captives of the same species (110 days vs. 73–82 days for *Pyrrhula pyrrhula*, 120 days vs. 43–56 days for *Acanthis flammea*) (Newton, 1966, 1968a; Evans, 1966, 1969; Evans *et al.*, 1967), and Dolnik's birds may have been studied under some abnormal conditions.

The duration of molt varies considerably among species, and several northern temperate-region birds that migrate replace their plumage in a shorter time than do resident species (Newton, 1968b; Snow, 1969). In the resident species in which molt is spread out over a longer time, the daily increase in the energy required for molt is proportionately less than in the migratory forms. A resident bunting, *Emberiza citrinella*, increases its daily energy requirements 14% during molt, whereas a closely related migratory form, *E. hortulana*, has a 26% daily energy increase in molt (Wallgren, 1954). The species in Table I suggest a similar trend, with the migratory *Carduelis chloris, Acanthis cannabina, A. flammea, Fringilla coelebs*, and *F. montifringilla* having larger daily increases in energy metabolism during molt than the less migratory or resident *Pyrrhula pyrrhula, Passer domesticus*, or

TABLE I
ENERGETIC COST OF MOLTING IN CERTAIN PALEARCTIC FINCHES AND SPARROWS

Species	Body weight (gm)[a]	Duration of molt (days)[b]	Energy expenditure for molt (kcal)[b]	Cumulative SMR during molt period (kcal)[c]	Increase in energy in molt (%)
Coccothraustes coccothraustes	55.0	100	127.8	1590	8.0
Carduelis chloris	29.0	100	374.8	1008	37.2
Carduelis carduelis	15.5	120	434.4	758	57.2
Carduelis spinus	12.0	130	145.0	684	21.2
Acanthis cannabina	18.5	130	366.6	933	39.3
Acanthis flammea	11.5	120	288.5	622	46.4
Pyrrhula pyrrhula	23.5	110	202.9	1056	19.2
Fringilla coelebs	23.5	110	335.4	1056	31.8
Fringilla montifringilla	25.0	80	267.2	714	37.4
Passer domesticus	30.5	80	205.5	824	24.9
Passer montanus	23.3	80	181.9	673	27.0

[a] Body weights are from Newton (1967), Myrcha and Pinowski (1970), and specimens in the University of Michigan Museum of Zoology.
[b] After Dolnik (1965).
[c] Calculated from equation (e) in Lasiewski and Dawson (1967).

P. montanus, though the migratory *Carduelis spinus* was rather low as well.

The increase in energy metabolism in molting birds comes at a time when wild birds are remarkably inactive and retiring in behavior, indicating that molt itself rather than activity may be responsible for the metabolic increase. The basis of the increase in metabolic rate during molt is complex and appears to involve both the cost of growing the feather and the thermoregulatory response to decreased insulation of the molting bird. King and Farner (1961) have calculated the energy involved in the keratinization of the total body plumage to be about 7.6% of the daily requirements of the bird, and this value is similar to the increase measured in terms of respiration for molting sparrows at summer temperatures (Blackmore, 1969). Most attempts to partition the energetic cost of growing feathers, of changes in thermoregulation, and any other changes due to altered daylengths have been mainly mathematical rather than experimental (Helms, 1968; Blackmore, 1969). Small differences in plumage insulation itself during molt are probably not important (King and Farner, 1961; Irving, 1964), but the growing feathers are highly vascularized and increase the surface area exposed to loss of heat at cool temperatures, and a molting bird may increase its heat production to maintain its body temperature. By measuring the oxygen consumption rates of molting and nonmolting Brown-headed Cowbirds (*Molothrus ater*), Lustick (1970) found higher metabolic rates and increased thermal conductance in molting birds at all temperatures over a range of 5–30°C. Molting birds in the zone of thermoneutrality had a 13% higher metabolic rate than did nonmolting birds, and at colder ambient temperatures molting birds increased their metabolic rates to 24% more than that of nonmolting birds at the same cool temperatures. In this species the 13% increase in metabolic rate is apparently due directly to the formation of the feathers in molt, and increased thermogenesis occurs only at temperatures below 20°C. Most songbird species usually molt during the warm season, and at this time of year the energetic cost of molt is at its lowest.

Body weights may increase in molting birds, and the increase may occur just before or during the time when the largest number of feathers are growing (King *et al.*, 1965; Holmes, 1966; Nisbet, 1967; McNeil, 1970; Newton, 1968a; Evans, 1969; Blackmore, 1969; Haukioja, 1969; Bell, 1970; Myrcha and Pinowski, 1970; Payne, 1969b; Fig. 4). The increase is associated with the growth of feather papillae, and it is due to increases in lean dry weight, fat, water content of the growing feathers and of other compartments, and blood volume (King *et al.*, 1965; Newton, 1968a; Evans, 1969). Molting birds may have enlarged thymus glands, apparently in response to a need for increased proliferation of lymphocytes for the increased blood volume (Ward and D'Cruz, 1968).

In relation to the energetic cost of molt and its impact on the resources of molting birds, it is notable that the stored body fat of some species shows no significant decrease during molt and in fact in some birds it increases at this time. No change is evident in the Alaskan populations of *Zonotrichia leucophrys gambelii* which have a brief period (34 days) of intensive molt in late summer (King *et al.*, 1965; Morton *et al.*, 1969). A net deposition of fat is evident in birds with a more protracted molt and has been described in the incomplete prenuptial molt of *Z. l. gambelii* on the wintering grounds and in the complete postnuptial molts of migratory *Acanthis flammea* and resident *Pyrrhula pyrrhula*, blackbirds (*Agelaius* spp.), and *Passer montanus* (King and Farner, 1966; Evans, 1966, 1969; Newton, 1966, 1968a; Payne, 1969b; Fig. 4; Myrcha and Pinowski, 1970). Fat is not usually deposited during the weeks when the greatest numbers of feathers are growing, but rather it is deposited early or late in the molt. The increase in body fat content during molt sug-

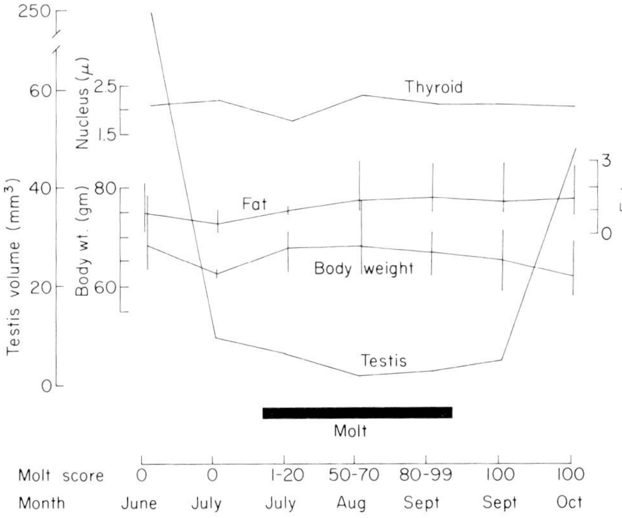

FIG. 4. Summary of the time relationships of molt score (percentage of molt which had been completed) with mean nuclear height of the thyroid, mean testis size, mean body weight, and mean body fat in adult male Tricolored Blackbirds (*Agelaius tricolor*). Vertical lines give the maximum and minimum values. Fat categories: 0, no fat; 1, little fat; 2, moderate fat; 3, fat. The data are original in part and in part are after Payne (1969b) and Payne and Landolt (1970).

gests that molt does not create an exhaustive strain on the energy resources in these birds.

In addition to its relation to the caloric energy resources, molting may impose a metabolic stress upon birds through a requirement for specific nutrients. In the European Bullfinch, Newton (1968a) found a significant increase in the overnight loss of lean dry weight in birds during peak molt. The rate of nitrogen metabolism in sparrows is greatest at the time when the largest number of feathers are growing, and the curve of nitrogen metabolism in time parallels that of metabolic rate (Blackmore, 1966). The amount of protein may be limiting. In molting Canada Geese (*Branta canadensis*) the muscles undergo a decrease in protein content which is not recovered until the molt has been almost completed, suggesting a long-term protein deficit with a shift of amino acids from the muscles into the growing feathers (Hanson, 1962). Bird feathers are composed almost entirely of the protein keratin. The amino acids of keratin as the building materials for the feather protein may be available in limited quantity. The proportions of the amino acids in feather keratin are similar to those in other tissues, with a major exception being cystine, which is important in the formation of disulfide bonds in keratin (Spearman, 1966). Radiolabeled S-containing amino acids are readily incorporated into growing feathers in experimental studies (Lüdicke, 1964). The calculated amounts of S-containing amino acids needed for feather growth do not approach the limits of these substances in the diet of birds whose molt is spread out over several weeks, however (Newton, 1968a).

From the physiological point of view, molting does not seem to cause a great strain on the energy or nutritional resources of birds. The energy requirements of molt are rather small compared to the increase in metabolic rates in thermoregulation, migration, and breeding, events that may more than double the energy requirements of small birds (West, 1960; Helms, 1968; Brisbin, 1969). From the ecological point of view, however, an increase in energy requirements in molting birds may be significant, and the 10–20% of the energy "saved" during the breeding period by scheduling molt at a different time may make the difference between nesting success and failure. Separation of breeding and molting of most temperate-region land birds may be a major adaptive tactic of their annual cycle. On the other hand, those birds with overlapping schedules are often species with a protracted molt, which energetically is not very expensive from day to day. Molting may be an ecologically demanding period in terms of survival as well as energetics, as the period of molt coincides with a high rate of mortality among young Reed Buntings, *Emberiza schoeniclus* (Haukioja, 1969), and late molting European Bullfinches, *Pyrrhula pyrrhula*, suffer higher losses than birds molting earlier (Newton, 1966). To appreciate the impact of molting on the energetic requirements of birds and on the various adaptive annual cycles among wild birds, we need accurate evaluations of the availability of food to individual birds and of the energy required for breeding. Hopefully, future studies of birds both in the field and in controlled experimental conditions will help make clear the adaptive significance of molt and breeding schedules in birds.

VII. Summary

Molt, the periodic replacement of the plumage in birds, involves the pushing out of the old feathers by the developing new ones. The mechanism of molt as well as its possible hormonal control differs from other means of integumentary change in birds such as seasonal abrasion or the development of a brood patch. The mechanism of feather replacement is similar in various groups of birds. The gradual replacement of the feathers in a feather series may involve mechanical or chemical interactions between neighboring follicles. In birds with gradual, prolonged molt the feathers in a series may be out of phase with each other in their development, and this phase difference may result in the spacing out of the molt so that no more than one feather grows at a time. In birds with a simultaneous molt or a graded but rapid molt the time differences in phase relationships of the follicles are slight, and an integrated pattern of molt probably results from the common responses of the follicles to changing hormonal levels at a certain stage in the annual cycle.

Molt offers more opportunities today for new studies of adaptations of the annual cycle and of comparative physiology than it does answered hypotheses of physiological control of the integration of the various phases of the annual cycle of birds. A few trends are evident. Molt in nature is not a simple mass sloughing of the plumage in response to a hormonal change (although this can be made to occur in the laboratory). Rather, local events of developing feather follicles and their interactions lead to a characteristic sequence of feather replacement. Follicles vary through time in their responsiveness to hormones, and while at the time just before a molt a wide variety of hormonal and other treatments can result in the replacement of feathers, during natural molt and shortly thereafter the integument remains insensitive to the same substances. The hormonal condition of birds is probably quite different during prenuptial and postnuptial molts, and also birds that begin the molt within various stages of the nesting cycle may have different means of hormonal control of molt. Physiological changes involved in molt and perhaps ultimately associated with its control include the energetic cost of molting (between 5 and 30% of total daily energy in most molting birds) and the nutritional demands for the protein of which feathers are composed. Probably the hormonal control of molt involves the preparation of the feather follicles and perhaps also their blood supply well in advance of the molting season. Thyroxine may act in this manner. In birds with separate breeding and molting schedules probably androgens, estrogens, and prolactin are involved in the physiological integration of molt and the breeding schedules by inhibiting the development of the new plumage. Molting overlaps with breeding in more than 100 species of birds. Both a comparison of molting and breeding schedules in birds and also a review of experimental work suggest that the hormonal control of molt varies among avian species, perhaps hand in hand with their particular versions of the annual cycle.

If there is a common scheme among birds for the physiological control of molt, no one has come close to finding it. The scarcity of appropriate generalities on the control mechanisms of molt makes all the more suggestive the variability among the natural annual cycles of different birds as resulting from different control schemes. There is a strong possibility that the control of molt is a case of multiple solutions with evolutionary divergences among related avian groups.

REFERENCES

Adams, J. L. (1956). A comparison of different methods of progesterone administration to the fowl in affecting egg production and molt. *Poultry Sci.* **35,** 323–326.

Aldrich, J. W. (1970). Review of the problem of birds contaminated by oil and their rehabilitation. *U. S., Bur. Sport Fish. Wildl. Resource Publ.* **87,** 1–23.

Amadon, D. (1966). Avian plumages and molts. *Condor* **68,** 263–278.

Andrews, M. I., and Naik, R. M. (1966). The body weight and the thyroid and gonadal cycles of the Jungle Babbler. *Pavo* **4,** 48–57.

Ashmole, N. P. (1962). The Black Noddy *Anous tenuirostris* on Ascension Island. Part I. General biology. *Ibis* **103b,** 235–273.

Ashmole, N. P. (1963). Molt and breeding in populations of the Sooty Tern *Sterna fuscata*. *Postilla* **76,** 1–18.

Ashmole, N. P. (1965). Adaptive variation in the breeding regime of a tropical sea bird. *Proc. Nat. Acad. Sci. U.S.* **53,** 311–318.

Ashmole, N. P. (1968). Breeding and molt in the White Tern (*Gygis alba*) on Christmas Island, Pacific Ocean. *Condor* **70,** 35–55.

Assenmacher, I. (1958). La mue des oiseaux et son déterminisme endocrinien. *Alauda* **26,** 242–289.

Bell, B. D. (1970). Moult in the Reed Bunting—a preliminary analysis. *Bird Study* **17,** 269–281.

Bern, H. A., and Nicoll, C. W. (1968). The comparative endocrinology of prolactin. *Recent Progr. Horm. Res.* **14,** 681–720.

Biellier, H. V., and Turner, C. W. (1957). The thyroid hormone secretion rate of domestic fowls as determined by radio-iodine techniques. *Mo., Agr. Exp. Sta., Bull.* **622,** 1–96.

Bigalke, R. C. (1956). "Über die zyklischen Veränderungen der Schilddrüse und des Körpergewichtes bei einigen Singvögeln im Jahresablauf." Thesis, Johann Wolfgang Goethe University, Frankfurt am Main.

Blackmore, F. H. (1966). "The Energy Requirements and Nitrogen Balance of the House Sparrow, *Passer domesticus,* During Molt." Ph.D. Thesis, University of Illinois, Urbana.

Blackmore, F. H. (1969). The effect of temperature, photoperiod and molt on the energy requirements of the House Sparrow, *Passer domesticus. Comp. Biochem. Physiol.* **30,** 433–444.

Brisbin, I. L., Jr. (1969). Bioenergetics of the breeding cycle of the Ring Dove. *Auk* **86,** 54–74.

Brooke, R. K. (1969). Taxonomic and distributional notes on *Apus acuticauda. Bull. Brit. Ornithol. Club* **89,** 97–100.

Brown, F. A., and Rollo, M. (1940). Light and molt in weaver finches. *Auk* **57,** 485–498.

Buckley, P. A. (1969). Disruption of species-typical behavior patterns in F_1 hybrid *Agapornis* parrots. *Z. Tierpsychol.* **26,** 737–743.

Clench, M. H. (1970). Variability in body pterylosis, with special reference to the genus *Passer. Auk* **87,** 650–692.

Colquhoun, M. K. (1951). "The Wood-Pigeon in Britain." HM Stationery Office, London.

Dathe, H. (1955). Über die Schreckmauser. *J. Ornithol.* **96,** 5–14.

Davis, J., and Davis, B. S. (1954). The annual gonad and thyroid cycles of the English Sparrow in southern California. *Condor* **56,** 328–345.

Dilger, W. C. (1969). Behavioral aspects. *In* "Diseases of Cage and Aviary Birds" (M. L. Petrak, ed.), pp. 19–21. Lea and Febiger, Philadelphia, Pennsylvania.

Dolnik, V. R. (1965). Bioenergetika linki byurovykh ptits kak adaptatsiya k migratsii. (Bioenergetics of the molts of finches as adaptations to migration.) *Novosti Ornithol.* **4,** 124–126.

Dolnik, V. R., and Blyumental, T. I. (1967). Autumnal premigratory and migratory periods of the Chaffinch (*Fringilla coelebs coelebs*) and some other temperate-zone passerine birds. *Condor* **69,** 435–468.

Domm, L. V., and Blivaiss, B. B. (1948). Plumage and other sex characteristics in the thiouracil-treated brown leghorn fowl. *Amer. J. Anat.* **82,** 167–202.

Dorward, D. F. (1962). Comparative biology of the White Booby and the Brown Booby *Sula* spp. at Ascension. *Ibis* **103b,** 174–220.

Dorward, D. F., and Ashmole, N. P. (1963). Notes on the biology of the Brown Noddy *Anous stolidus* on Ascension Island. *Ibis* **103b,** 447–457.

Dwight, J., Jr. (1900). The sequence of plumages and moults of the passerine birds of New York. *Ann. N. Y. Acad. Sci.* **13,** 73–360.

Elterich, C. F. (1936). Über zyklische Veränderungen der Schilddrüse in den einzelnen Geschlechtsphasen der Taube. *Endokrinologie* **18,** 31–37.

Erpino, M. J. (1968). Aspects of thyroid histology in Black-billed Magpies. *Auk* **85,** 397–403.

Evans, P. R. (1966). Autumn movements, moult and measurements of the Lesser Redpoll *Carduelis flammea cabaret. Ibis* **108,** 183–216.

Evans, P. R. (1969). Ecological aspects of migration, and pre-migratory fat deposition in the Lesser Redpoll, *Carduelis flammea cabaret*. *Condor* **71**, 316–330.

Evans, P. R., Elton, R. A., and Sinclair, G. R. (1967). Moult and weight changes of Redpolls, *Carduelis flammea*, in north Norway. *Ornis Fenn.* **44**, 33–41.

Farner, D. S. (1964). The photoperiodic control of reproductive cycles in birds. *Amer. Sci.* **52**, 137–156.

Farner, D. S., and Mewaldt, L. R. (1955). The natural termination of the refractory period in the White-crowned Sparrow. *Condor* **57**, 112–116.

Farner, D. S., and Serventy, D. L. (1960). The timing of reproduction in birds in the arid regions of Australia. *Anat. Rec.* **137**, 354.

Farner, D. S., and Wilson, A. C. (1957). A quantitative examination of testicular growth in the White-crowned Sparrow. *Biol. Bull.* **113**, 254–267.

Follett, B. K., and Farner, D. S. (1966). Pituitary gonadotropins in the Japanese quail (*Coturnix coturnix japonica*) during photoperiodically induced gonadal growth. *Gen. Comp. Endocrinol.* **7**, 125–131.

Follett, B. K., and Riley, J. (1967). Effect of the length of the daily photoperiod on thyroid activity in the female Japanese quail (*Coturnix coturnix japonica*). *J. Endocrinol.* **39**, 615–616.

Fromme-Bouman, H. (1962). Jahresperiodische Untersuchungen an der Nebennierenrinde der Amsel (*Turdus merula* L.). *Vogelwarte* **21**, 188–198.

Gál, G. (1940). A madarak pajzsmirigyszerkezetének ciklikus változásai. *Mat. És Természettudományi Értesitö* **59**, 360–378.

Glazener, E. W., and Jull, M. A. (1946). Effects of thiouracil on naturally occurring molt in the hen. *Poultry Sci.* **25**, 533–535.

Gourdji, D. (1964). "La préhypophyse de l'ignicolore male *Pyromelana franciscana* au cours du cycle annuel." Thesis, University of Paris, Paris.

Gullion, G. W. (1953). Observations on molting of the American Coot. *Condor* **55**, 102–103.

Gwinner, E. (1967). Circannuale Periodik der Mauser und der Zugunruhe bei einem Vogel. *Naturwissenschaften* **54**, 477.

Gwinner, E. (1968). Circannuale Periodik als Grundlage des jahreszeitlichen Funktionswandels bei Zugvögeln. *J. Ornithol.* **109**, 70–95.

Hall, B. K. (1968). The annual interrenal tissue cycle within the adrenal gland of the Eastern Rosella, *Platycercus eximus* (Aves: Psittaciformes). *Aust. J. Zool.* **16**, 609–617.

Hanson, H. C. (1962). The dynamics of condition factors in Canada Geese and their relation to seasonal stresses. *Arctic Inst. N. Amer. Tech. Pap.* **12.**

Harris, P. C., and Shaffner, C. S. (1957). Effect of season and thyroidal activity on the molt response to progresterone in chickens. *Poultry Sci.* **36**, 1186–1193.

Haukioja, E. (1969). Weights of Reed Buntings (*Emberiza schoeniclus*) during summer. *Ornis Fenn.* **46**, 13–21.

Helms, C. W. (1968). Food, fat, and feathers. *Amer. Zool.* **8**, 151–167.

Herrick, R. B., and Adams, J. L. (1957). The molting action of dehydrocortiscosterone acetate in single comb white leghorn pullets. *Poultry Sci.* **36**, 1125.

Himeno, K., and Tanabe, Y. (1957). Mechanism of molting in the hen. *Poultry Sci.* **36**, 835–842.

Höcker, W. (1967). Über lokale bzw. lokalisierte Trijodthyronin-wirkungen auf das Wachstum aktivierter Federkeime. *Naturwissenschaften* **54**, 207.

Höcker, W. (1969a). Über die Wirkung von Prolaktin auf das rote Blutbild hypophysektomierter Tauben. *Endokrinologie* **54**, 153–161.

Höcker, W. (1969b). Über die Wirkung von Prolaktin auf die Serumproteine hypophysektomierter Tauben. *Endokrinologie* **54**, 17–25.

Höhn, E. O. (1949). Seasonal changes in the thyroid gland and effects of thyroidectomy in the Mallard in relation to molt. *Amer. J. Physiol.* **158**, 337–344.

Holmes, R. T. (1966). Molt cycle of the Red-backed Sandpiper (*Calidris alpina*) in western North America. *Auk* **83**, 517–533.

Immelmann, K. (1963a). Tierische Jahresperiodik in ökologischer Sicht. *Zool. Jahrb., Syst., Geogr. Biol.* **91**, 91–200.

Immelmann, K. (1963b). Drought adaptations in Australian desert birds. *Proc. 13th Int. Ornithol. Congr., 1962* pp. 649–657.

Immelmann, K. (1971). Ecological aspects of periodic reproduction. *In* "Avian Biology" (D. S. Farner and J. R. King, eds.), Vol. 1, pp. 341–389. Academic Press, New York.

Ingolfsson, A. (1970). The moult of remiges and rectrices in Great Black-backed Gulls *Larus marinus* and Glaucous Gulls *L. hyperboreus* in Iceland. *Ibis* **112**, 83–92.

Irving, L. (1964). Terrestrial animals in cold: Birds and mammals. *In* "Handbook of Physiology" (D. B. Dill, E. F. Adolph, and C. G. Wilber, eds.), Sect. 4, pp. 361–377. Amer. Physiol. Soc., Washington, D.C.

John, T. M., and George, J. C. (1967). Certain cyclic changes in the thyroid and parathyroid glands of migratory wagtails. *Pavo* **5**, 19–28.

Johnson, L. G. (1968). Endocrine influences on growth and pigmentation of embryonic down feather cells. *J. Embryol. Exp. Morphol.* **20**, 319–327.

Johnston, D. W. (1961). Timing of annual molt in the Glaucous Gulls of northern Alaska. *Condor* **63**, 474–478.

Jones, R. E. (1969). Hormonal control of incubation patch development in the California Quail, *Lophortyx californicus*. *Gen. Comp. Endocrinol.* **13**, 1–13.

Juhn, M. (1963). An examination of some interpretations of molt with added data from progesterone and thyroxine. *Wilson Bull.* **75**, 191–197.

Juhn, M., and Harris, P. C. (1956). Responses in molt and lay of fowl to progestin and gonadotropins. *Proc. Soc. Exp. Biol. Med.* **92**, 709–711.

Juhn, M., and Harris, P. C. (1958). Molt of capon feathering with prolactin. *Proc. Soc. Exp. Biol. Med.* **98**, 669–672.

Keast, A. (1953). The physiology of the avian moult as evidenced by a study of the bird *Zosterops lateralis*. M.Sc. Thesis, University of Sydney, Sydney.

Keast, A. (1968). Moult in birds of the Australian dry country relative to rainfall and breeding. *J. Zool.* **155**, 185–200.

Keck, W. N. (1934). The control of the secondary sex characters in the English Sparrow *Passer domesticus*. *J. Exp. Zool.* **67**, 315–347.

Kendeigh, S. C. (1949). Effect of temperature and season on the energy resources of the English Sparrow. *Auk* **66**, 113–127.

Kendeigh, S. C., and Wallin, H. E. (1966). Seasonal and taxonomic differences in the size and activity of the thyroid gland in birds. *Ohio J. Sci.* **66**, 369–379.

King, J. R. (1968). Cycles of fat deposition and molt in White-crowned Sparrows in constant environmental conditions. *Comp. Biochem. Physiol.* **24**, 827–837.

King, J. R., and Farner, D. S. (1961). Energy metabolism, thermoregulation, and body temperature. *In* "Biology and Comparative Physiology of Birds" (A. J. Marshall, ed.), Vol. 2, pp. 215–288. Academic Press, New York.

King, J. R., and Farner, D. S. (1966). The adaptive role of winter fattening in the White-crowned Sparrow with comments on its regulation. *Amer. Natur.* **100**, 403–419.

King, J. R., Farner, D. S., and Morton, M. L. (1965). The lipid reserves of White-crowned Sparrows in the breeding ground in central Alaska. *Auk* **82**, 236–252.

King, J. R., Follett, B. K., Farner, D. S., and Morton, M. L. (1966). Annual gonadal cycles and pituitary gonadotropins in *Zonotrichia leucophrys gambelii*. *Condor* **68,** 476–487.

Kobayashi, H. (1952). Studies on molting in the pigeon. V. Oxygen consumption of the brooding pigeon and of the thiourea-treated pigeon. *Annot. Zool. Jap.* **25,** 371–376.

Kobayashi, H. (1953a). Studies on molting in the pigeon. III. Observations on normal process of molting. *Jap. J. Zool.* **11,** 1–9.

Kobayashi, H. (1953b). Studies on molting in the pigeon, VII. Inhibitory effect of lactogen on molting. *Jap. J. Zool.* **11,** 21–26.

Kobayashi, H. (1954). Studies on molting in the pigeon. VIII. Effects of sex steroids on molting and thyroid gland. *Ann. Zool. Jap.* **27,** 22–26.

Kobayashi, H. (1958). On the induction of molt in birds by 17α-oxyprogesterone-17-capronate. *Endocrinology* **63,** 420–430.

Kobayashi, H., and Koscak, M. (1957). Activity changes of phosphatases of bird skin during feather regeneration. *Physiol. Zool.* **30,** 350–359.

Kobayashi, H., Maruyama, K., and Kambara, S. (1955). Effect of thyroxine on the phosphatase activity of pigeon skin. *Endocrinology* **57,** 129–133.

Koch, H. J., and De Bont, A. F. (1944). Influence de la mue sur l'intensité de métabolisme chez le pinson, *Fringilla coelebs coelebs* L. *Ann. Soc. Zool. Belg.* **75,** 81–86.

Kramer, G. (1950). Über die Mauser, insbesondere die sommerliche Kleingefiedermauser beim Neuntöter (*Lanius collurio* L.). *Ornithol. Ber.* **3,** 15–22.

Küchler, W. (1935). Jahreszyklische Veränderungen im histologischen Bau der Vogelschilddrüse. *J. Ornithol.* **83,** 414–461.

Lasiewski, R. C., and Dawson, W. R. (1967). A re-examination of the relation between standard metabolic rate and body weight in birds. *Condor* **69,** 13–23.

Lesher, S. W., and Kendeigh, S. C. (1941). Effect of photoperiod on molting of feathers. *Wilson Bull.* **53,** 169–180.

Lillie, F. R., and Juhn, M. (1932). The physiology of development of feathers. I. Growth-rate and pattern in the individual feather. *Physiol. Zool.* **5,** 124–184.

Ljunggren, L. (1968). Seasonal studies of Wood Pigeon populations. I. Body weight, feeding habits, liver and thyroid activity. *Viltrevy, Swed. Wildl., Uppsala* **5,** 435–504.

Lockley, R. M. (1961). "Shearwaters." Amer. Mus. Natur. Hist., New York.

Lofts, B. (1964). Evidence of an autonomous reproductive rhythm in an equatorial bird (*Quelea quelea*). *Nature (London)* **201,** 523–524.

Lofts, B., and Murton, R. K. (1968). Photoperiodic and physiological adaptations regulating avian breeding cycles and their ecological significance. *J. Zool.* **155,** 327–394.

Lorenzen, L. C., and Farner, D. S. (1964). An annual cycle in the interrenal tissue of the adrenal gland of the White-crowned Sparrow, *Zonotrichia leucophrys gambelii*. *Gen. Comp. Endocrinol.* **4,** 253–263.

Lüdicke, M. (1964). Das Integument der Vögel. *Stud. Gen.* **17,** 390–406.

Lustick, S. (1970). Energy requirements of molt in cowbirds. *Auk* **87,** 742–746.

McNeil, R. (1970). Hivernage et estivage d'oiseaux aquatiques nord-américains dans le Nord-Est du Vénézuéla (mue, accumulation de graisse, capacité de vol et routes de migration). *Oiseau Rev. Fr. Ornithol.* **40,** 185–302.

Marshall, A. J., and Coombs, C. J. F. (1957). The interaction of environmental, internal and behavioural factors in the Rook. *Corvus f. frugilegus* Linnaeus. *Proc. Zool. Soc. London* **128,** 545–589.

Marshall, A. J., and Serventy, D. L. (1956). The breeding cycle of the Short-tailed Shearwater, *Puffinus tenuirostris* (Temminck), in relation to transequatorial migration and its environment. *Proc. Zool. Soc. London* **127,** 489–509.

Matsuo, S., Vitums, A., King, J. R., and Farner, D. S. (1969). Light-microscope studies of the cytology of the adenohypophysis of the White-crowned Sparrow *Zonotrichia leucophrys gambelii. Z. Zellforsch. Mikrosk. Anat.* **95,** 143–176.

Meier, A. H. (1969a). Diurnal variations of metabolic response to prolactin in lower vertebrates. *Gen. Comp. Endocrinol., Suppl.* **2,** 55–62.

Meier, A. H. (1969b). Antigonadal effects of prolactin in the White-throated Sparrow, *Zonotrichia albicollis. Gen. Comp. Endocrinol.* **13,** 222–225.

Meier, A. H., Burns, J. T., and Dusseau, J. W. (1969). Seasonal variations in the diurnal rhythm of pituitary prolactin content in the White-throated Sparrow *Zonotrichia albicollis. Gen. Comp. Endocrinol.* **12,** 282–289.

Merkel, F. W. (1963). Long-term effects of constant photoperiods on European Robins and Whitethroats. *Proc. 13th Int. Ornithol. Congr., 1962* pp. 950–959.

Mewaldt, L. R. (1958). Pterylography and natural and experimentally induced molt in Clark's Nutcracker. *Condor* **60,** 165–187.

Mewaldt, L. R., Kibby, S. S., and Morton, M. L. (1968). Comparative biology of Pacific coastal White-crowned Sparrows. *Condor* **70,** 14–30.

Michener, H., and Michener, J. R. (1940). The molt of House Finches of the Pasadena region, California. *Condor* **42,** 140–153.

Miller A. H. (1941). The significance of molt centers among the secondary remiges in the Falconiformes. *Condor* **43,** 113–115.

Miller, A. H. (1954). The occurrence and maintenance of the refractory period in crowned sparrows. *Condor* **56,** 13–20.

Miller, A. H. (1955a). The expression of innate reproductive rhythm under conditions of winter lighting. *Auk* **72,** 260–264.

Miller, A. H. (1955b). Breeding cycles in a constant equatorial environment in Colombia, South America. *Acta 11th Int. Ornithol. Congr., 1954* pp. 495–503.

Miller, A. H. (1961). Molt cycles in equatorial Andean Sparrows. *Condor* **63,** 143–161.

Miller, A. H. (1962). Bimodal occurrence of breeding in an equatorial sparrow. *Proc. Nat. Acad. Sci. U. S.* **48,** 396–400.

Miller, A. H. (1963). Seasonal activity and ecology of the avifauna of an American equatorial cloud forest. *Univ. Calif., Berkeley, Publ. Zool.* **66,** 1–78.

Miller, D. S. (1939). A study of the physiology of the sparrow thyroid. *J. Exp. Zool.* **80,** 259–281.

Miyazaki, H. (1934). On the relation of the daily period to the sexual maturity and to the moulting of *Zosterops palpebrosa japonica. Tokohu Univ. Sci. Rep., Sendai, Japan, Biol. Ser.* **4,** 183–203.

Moreau, R. E., Wilk, A. L., and Rowan, W. (1947). The moult and gonad cycles of three species of birds at five degrees south of the equator. *Proc. Zool. Soc. London* **117,** 345–364.

Morton, M. L., and Mewaldt, L. R. (1962). Some effects of castration on a migratory sparrow (*Zonotrichia atricapilla*). *Physiol. Zool.* **35,** 237–247.

Morton, M. L., King, J. R., and Farner, D. S. (1969). Postnuptial and postjuvenal molt in White-crowned Sparrows in central Alaska. *Condor* **71,** 376–385.

Myrcha, A., and Pinowski, J. (1970). Weights, body composition, and caloric value of postjuvenal molting European Tree Sparrows (*Passer montanus*). *Condor* **72,** 175–181.

Naik, R. M., and Andrews, M. I. (1966). Pterylosis, age determination and moult in the Jungle Babbler. *Pavo* **4,** 22–49.

Nelson, J. B. (1964). Factors influencing clutch-size and chick growth in the North Atlantic Gannet *Sula bassana*. *Ibis* **106**, 63–77.

Newton, I. (1966). The moult of the Bullfinch *Pyrrhula pyrrhula*. *Ibis* **108**, 41–87.

Newton, I. (1967). The adaptive radiation and feeding ecology of some British finches. *Ibis* **109**, 33–98.

Newton, I. (1968a). The temperatures, weights, and body composition of molting Bullfinches. *Condor* **70**, 323–332.

Newton, I. (1968b). The moulting seasons of some finches and buntings. *Bird Study* **15**, 84–92.

Nisbet, I. C. T. (1967). Migration and moult in Pallas's Grasshopper Warbler. *Bird Study* **14**, 96–103.

Oakeson, B. B., and Lilley, B. R. (1960). Annual cycle of thyroid histology in two races of White-crowned Sparrow. *Anat. Rec., Suppl.* **136**, 41–57.

Oring, L. W. (1968). Growth, molts, and plumages of the Gadwall. *Auk* **85**, 355–380.

Paredes, J. R. (1956). Causas y mecanismo de la muda en las gallinas. *Rev. Patron. Biol. Anim.* **2**, 95–175.

Payne, R. B. (1965). The molt of breeding Cassin Auklets. *Condor* **67**, 220–228.

Payne, R. B. (1969a). Overlap of breeding and molting schedules in a collection of African birds. *Condor* **71**, 140–145.

Payne, R. B. (1969b). Breeding seasons and reproductive physiology of Tricolored Blackbirds and Redwinged Blackbirds. *Univ. Calif., Berkeley, Publ. Zool.* **90**, 1–115.

Payne, R. B. (1972). Behavior, mimetic songs and song dialects, and relationships in the parasitic indigobirds (*Vidua*) of Africa. *Orn. Monogr.* 11.

Payne, R. B., and Landolt, M. (1970). Thyroid histology in the annual cycle, breeding, and molt of Tricolored Blackbirds (*Agelaius tricolor*). *Condor* **72**, 445–451.

Perek, M., and Sulman, F. (1945). The basal metabolic rate in molting and laying hens. *Endocrinology* **36**, 240–243.

Perek, M., Eckstein, B., and Sobel, H. (1957). Histological observations in the anterior lobe of the pituitary gland in molting and laying hens. *Poultry Sci.* **36**, 954–958.

Pitelka, F. A. (1958). Timing of molt in Steller Jays of the Queen Charlotte Islands, British Columbia. *Condor* **60**, 38–49.

Podhradsky, J., Wodzicki, K., and Golabeck, Z. (1937). Die Bedeutung der Schilddrüse für die Mauser bei Hühnern. *Ann. Tschechosl. Akad. Landwirtsch.* **12**, 604–613.

Prenn, F. (1937). Beobachtungen zur Lebensweise der Felsenschwalbe (*Riparia rupestris*). *J. Ornithol.* **85**, 577–586.

Putzig, P. (1938). Der Fruhwegzug des Kiebitzes (*Vanellus vanellus*). *J. Ornithol.* **86**, 123–165.

Raitt, R. J. (1968). Annual cycle of adrenal and thyroid glands in Gambel Quail of southern New Mexico. *Condor* **70**, 366–372.

Raitt, R. J., and Ohmart, R. D. (1966). Annual cycle of reproduction and molt in Gambel Quail of the Rio Grande Valley, southern New Mexico. *Condor* **68**, 541–561.

Ralph, C. L. (1969). The control of color in birds. *Amer. Zool.* **9**, 521–530.

Rawles, M. E. (1960). The integumentary system. *In* "Biology and Comparative Physiology of Birds" (A. J. Marshall, ed.), Vol. 1, pp. 189–240. Academic Press, New York.

Reineke, E. P., and Turner, C. W. (1945). Seasonal rhythm in the thyroid hormone secretion of the chick. *Poultry Sci.* **24**, 499–504.

Ringer, R. K. (1965). Thyroids. *In* "Avian Physiology" (P. D. Sturkie, ed.), pp. 592–648. Cornell Univ. Press, Ithaca, New York.

Rowan, M. K. (1967). A study of the colies of southern Africa. *Ostrich* **38**, 63–115.

Salomonsen, F. (1939). Moults and sequences of plumages in the Rock Ptarmigan (*Lagopus mutus* Montin). *Vidensk. Medd. Dansk Naturhist. Foren.* **103**, 1–491.

Schildmacher, H. (1956). Physiologische Untersuchungen am Grünfinken *Chloris chloris* (L.) im künstlichen Kurztag und nach "hormonaler Sterilisierung." *Biol. Zentralbl.* **75**, 327–355.

Schooley, J. P. (1937). Pituitary cytology in pigeons. *Cold Spring Harbor Symp. Quant. Biol.* **5**, 115–119.

Selander, R. K. (1958). Age determination and molt in the Boat-tailed Grackle. *Condor* **60**, 355–376.

Selander, R. K., and Kuich, L. L. (1963). Hormonal control and development of the incubation patch in icterids, with notes on behavior of cowbirds. *Condor* **65**, 73–90.

Serventy, D. L., and Marshall, A. J. (1957). Breeding periodicity in western Australian birds; with an account of unseasonal nestings in 1953 and 1955. *Emu* **57**, 99–126.

Shaffner, C. S. (1954). Feather papilla stimulation by progesterone. *Science* **120**, 345.

Shaffner, C. S. (1955). Progesterone induced molt. *Poultry Sci.* **34**, 840–842.

Smith, A. H., Bond, G. H., Ramsey, K. W., Reck, D. G., and Spoon, J. E. (1957). Size and rate of involution of the hen's reproductive organs. *Poultry Sci.* **36**, 346–353.

Snow, D. W. (1966). Moult and the breeding cycle in Darwin's finches. *J. Ornithol.* **107**, 283–291.

Snow, D. W. (1969). The moult of British thrushes and chats. *Bird Study* **16**, 115–129.

Snow, D. W., and Snow, B. K. (1964). Breeding seasons and annual cycles of Trinidad land-birds. *Zoologica* **49**, 1–39.

Spearman, R. I. C. (1966). Keratinization of epidermal scales, feathers, and hairs. *Biol. Rev.* **41**, 59–96.

Stejneger, L. (1885). Results of ornithological explorations in the Commander Islands and in Kamschatka. *U. S., Nat. Mus., Bull.* **29.**

Stettenheim, P. (1972). The integument of birds. *In* "Avian Biology" (D. S. Farner and J. R. King, eds.), Vol. 2, pp. 1–63. Academic Press, New York.

Stresemann, E. (1963). Zeitraum und Verlauf der Handschwingen-Mauser palaearktischer Möwen, Seeschwalben und Limicolen. *J. Ornithol.* **104**, 424–435.

Stresemann, E. (1967). Inheritance and adaptation in molt. *Proc. 14th Int. Ornithol. Congr., 1966* pp. 75–80.

Stresemann, E., and Stresemann, V. (1966). Die Mauser der Vögel. *J. Ornithol (Sonderheft)* **107**, 1–447.

Sulman, F., and Perek, M. (1947). Influence of thiouracil on the basal metabolic rate and on moulting in hens. *Endocrinology* **41**, 514–517.

Takewaki, K., and Mori, H. (1944). Mechanism of molting in the canary. *J. Fac. Sci., Imp. Univ. Tokyo, Sect. 4 Zool.* **6**, 547–575.

Tanabe, Y., and Katsuragi, T. (1962). Thyroxine secretion rates of molting and laying hens, and general discussion on the hormonal induction of molting in hens. *Nogyo Gijutsu Kenyusho Hokoku G* **21**, 49–59.

Tanabe, Y., Mozaki, H., and Makino, K. (1955). Studies on the thyroid function and the uptake of I^{131} in molting and laying hens. *Igaku To Seibutsugaku* **37**, 179–184.

Tanabe, Y., Himeno, K., and Nozaki, H. (1957). Thyroid and ovarian function in relation to molting in the hen. *Endocrinology* **61**, 661–666.

Tanabe, Y., Abe, T., Himeno, K., Kaneko, T., Mogi, K., Saeki, Y., and Hosoda, T. (1958). Thyrotrophic hormone content of the pituitary of molting and laying hens. *Endocrinol. Jap.* **5**, 118–121.

Taylor, T. G. (1969). French molt. *In* "Diseases of Cage and Aviary Birds" (M. L. Petrak, ed.), pp. 237–242. Lea and Febiger, Philadelphia, Pennsylvania.

Thapliyal, J. P., and Saxena, R. N. (1961). Plumage control in Indian Weaver Bird (*Ploceus philippinus*). *Naturwissenschaften* **48**, 741–742.

Thapliyal, J. P., and Tewary, P. D. (1961). Plumage in Lal Munia (*Amandava amandava*). *Science* **134**, 738–739.

Thybusch, D. (1965). Jahreszyklus der Nebennieren bei der Sturmmöwe (*Larus canus* L.) unter Berücksichtigung des Funktionszustandes von Schild- und Keimdrüse sowie des Körpergewichtes. *Z. Wiss. Zool.* **173**, 72–89.

Tixier-Vidal, A., Herlant, M., and Benoit, J. (1962). La préhypophyse du Canard Pékin mâle au cours du cycle annuel. *Arch. Biol.* **73**, 317–368.

Tixier-Vidal, A., Follett, B. K., and Farner, D. S. (1968). The anterior pituitary of the Japanese quail, *Coturnix coturnix japonica*. The cytological effects of photoperiodic stimulation. *Z. Zellforsch. Mikrosk. Anat.* **92**, 610–635.

Tollefson, C. I. (1969). Nutrition. *In* "Diseases of Cage and Aviary Birds" (M. L. Petrak, ed.), pp. 143–167. Lea and Febiger, Philadelphia, Pennsylvania.

Tordoff, H. B., and Dawson, W. R. (1965). The influence of daylight on reproductive timing in the Red Crossbill. *Condor* **67**, 416–422.

van der Meulen, J. B. (1939). Hormonal regulation of molt and ovulation. *World's Poultry Congr. Exposition* [*Proc.*], *7th, 1939* pp. 109–112.

van Tienhoven, A. (1968). "Reproductive Physiology of Vertebrates." Saunders, Philadelphia, Pennsylvania.

Vaugien, L. (1948). Mue, activité thyroïdiene et cycle des gonades chez les oiseaux passériformes. *C. R. Acad. Sci.* **226**, 353–354.

Vaugien, L. (1955). Sur les réactions testicularies du jeune Moineau domestique illuminé à diverses époques de la mauvaise saison. *Bull. Biol. Fr. Belg.* **139**, 218–244.

Voitkevich, A. A. (1966). "The Feathers and Plumage of Birds." October House, New York.

von Faber, H. (1967). Die Beziehungen von Kerngrösse und histologischem Bild zur Thyroxinsekretionsrate der Schilddrüse. *Zool. Anz. (Sonderheft)* **30**, 172–175.

Wagner, H. O. (1961). Schilddrüse und Federausfall. *Z. Vergl. Physiol.* **44**, 565–575.

Wagner, H. O. (1962). Der Einfluss endo- und exogener Faktoren auf die Mauser gekäfigter Vögel. *Z. Vergl. Physiol.* **45**, 337–354.

Wagner, H. O., and Müller, C. (1963). Die Mauser und die den Federausfall fördernden und hemmenden Faktoren. *Z. Morphol. Oekol. Tiere* **53**, 107–151.

Wallgren, H. A. (1954). Energy metabolism of two species of the genus *Emberiza* as correlated with distribution and migration. *Acta Zool. Fenn.* **84**, 1–110.

Ward, P., and D'Cruz, D. (1968). Seasonal changes in the thymus gland of a tropical bird. *Ibis* **110**, 203–205.

Warham, J. (1962). The biology of the Giant Petrel, *Macronectes giganteus*. *Auk* **79**, 139–160.

Watson, G. E. (1962). Molt, age determination, and annual cycle in the Cuban Bobwhite. *Wilson Bull.* **74**, 28–42.

Weise, C. M. (1962). Migratory and gonadal responses of birds on long-continued short daylengths. *Auk* **79**, 161–172.

West, G. C. (1960). Seasonal variations in the energy balance of the Tree Sparrow in relation to migration. *Auk* **77**, 306–329.

Williamson, F. S. L. (1956). The molt and testis cycle of the Anna Hummingbird. *Condor* **58**, 342–366.

Wilson, A. C., and Farner, D. S. (1960). The annual cycle of thyroid activity in White-crowned Sparrows of eastern Washington. *Condor* **62**, 414–425.

Witschi, E. (1961). Sex and secondary sexual characters. *In* "Biology and Comparative Physiology of Birds" (A. J. Marshall, ed.), Vol. 2, pp. 115–168. Academic Press, New York.

Wolfson, A. (1959). Ecologic and physiologic factors in the regulation of spring migration and reproductive cycles in birds. *In* "Comparative Endocrinology" (A. Gorbman, ed.), pp. 38–70. Wiley, New York.

Wolfson, A. (1966). Environmental and neuroendocrine regulation of annual gonadal cycles and migratory behavior in birds. *Recent Progr. Horm. Res.* **22,** 177–244.

Zawadowsky, B. M. (1927). Zur Frage der Wechselbeziehungen zwischen Schilddrüse und Geschlechtsdrüsen bei Hühnern. *Wilhelm Roux' Arch. Entwicklungsmech. Organismen* **110,** 149–182.

Zimmerman, J. L. (1966). Effects of extended tropical photoperiod and temperature on the Dickcissel. *Condor* **68,** 377–387.

Chapter 4

THE BLOOD VASCULAR SYSTEM OF BIRDS

David R. Jones[1] *and Kjell Johansen*[2]

[1]This review was written while the author was supported by operating grants from the National Research Council of Canada.

[2]This review was written while the author was supported by Grant HE 12071 from the National Institutes of Health.

I. Introduction

It is unfortunate but not surprising that the circulatory physiology of birds is often interpreted in the light of more sophisticated experiments that have been performed on mammals. The reason for this association, aside from dubious similarities such as possession of "warm blood" and pseudonomenclature as "higher vertebrates," is perhaps the lack of knowledge concerning circulatory dynamics and their regulation in birds.

There can be no doubt that phylogenetically the avian and mammalian stocks have been long separated. The heart of birds retains typical reptilian characters, and the mammals must have branched off and diverged from the common amniote ancestor before the reptilian type of specialization had begun (Goodrich, 1930). Although in both birds and mammals the circulation is completely divided into lung and body circuits, the mode of division of the heart by the interventricular septum is different. However, birds and mammals share a common feature in that the persisting systemic originates from the left side of the ventricle; but this is where the similarity ends, for in birds it is the right systemic arch and in mammals the left. Many other differences in the morphology of the avian and mammalian circulation exist, and time to produce these differences has also been ample for the evolution of many differences in the cardiovascular function and its control. It may be argued, however, that the consequence of evolution is the production of the most efficient system, and that given the same building materials the end product, if not the same, must be very similar.

In a chapter of this length it is impossible to provide a complete account of the avian blood vascular system. Consequently, we will tend to deal in generalizations, while endeavoring to give references to sources where more detailed information can be found. Although much work has been done, the many important unanswered questions have encouraged the luxury of speculation.

II. Avian Hematology

It is not possible to provide in the space available a complete review of this topic. Fuller accounts of avian hematology can be obtained from a number of sources, including Albritton (1952), Olson (1952), Lucas and Jamroz (1961), Sturkie (1965), and Hemm and Carlton (1967).

Blood consists of liquid in which are suspended the formed elements, the former being the plasma, and the most numerous of the latter being the red blood cells or erythrocytes. Together they form the total blood volume which appears to be in the range of 5–13% of the body weight depending on the species, age, sex, and functional status.

A. The Erythrocytes

The red blood cell is oval in shape and nucleated, as in most inframammalian vertebrates. The red cells are generally larger than mammalian red cells, being 10–15 μ along the long axis and 5–8 μ along the short axis (Groebbels, 1932; Magath and Higgins, 1934). The nuclei are large, measuring in ducks from $5.0 \times 1.5\ \mu$ to $7.0 \times 2.5\ \mu$. The total number of erythrocytes is influenced by age, sex, diet, seasonal and other factors, so any range given must vary widely. Among adults a range of 1.89 to 5.0 million/mm^3 would include all animals reported to date, with the majority falling in the more restricted range of 2.5 to 3.5 million/mm^3. Immature animals show more variation, a value of 5.9 million/mm^3 being reported for the Red-winged Blackbird (*Agelaius phoeniceus*) (Ronald *et al.*, 1968). The Ostrich (*Struthio camelus*) has the largest erythrocytes, but also the smallest number per volume of blood (De Villiers, 1938).

Albritton (1952) reports a packed cell volume (hematocrit) of 39.5% as an estimated mean for birds. Sexually immature chickens, both male and female, have hematocrits of about 29%, whereas at maturity the male hematocrit is 16% higher than the female (Newell and Shaffner, 1950). In most other species of birds sexual differences are not that marked. The hematocrits of both male and female pigeons exceed 50% (Kaplan, 1954; Bond and Gilbert, 1958). The ratio of whole body hematocrit, obtained from separate determinations of red blood cell and plasma volumes, to the venous hematocrit is 0.88 in chickens, a value similar to that found in man (R. R. Cohen, 1967). Two species of penguins showed hematocrits of 45–49% (Lenfant *et al.*, 1969).

Mean red cell volume ranges from 100 to 110 μ^3 in chickens with the highest values in males. Androgen reportedly increases the number and volume of red cells in chickens, while estrogens have the opposite effect (Newell and Shaffner, 1950; Sturkie and Newman, 1951; R. R. Burton *et al.*, 1969). Studies on ducks fail to confirm this. Ronald *et al.* (1968) report that mean corpuscular volume in the Red-winged Blackbird decreases markedly in the first few days after hatching, from around 140 μ^3 to 80–100 μ^3.

Methodological difficulties mean that many of the earlier determinations of hemoglobin levels in birds are subject to considerable error. Albritton (1952) reports that hemoglobin concentration in birds ranges from 9 to 21 gm/100 ml blood. Hemoglobin concentration is affected by age, maturity, sex, and seasons. Surendranathan *et al.* (1968) report values for adult ducks of 12–13 gm/100 ml of blood, whereas Bond and Gilbert (1958) have recorded 18 gm/100 ml of blood in diving and dabbling ducks. Values for chickens are lower than in ducks, and range from 9.5 to 13.5 gm/100 ml (Dukes and Schwarte, 1931; Holmes *et al.*, 1933). Male and female pigeons show values in the range of 16–20 gm/100 ml of blood (Rodnan *et al.*, 1957). Penguins similarly show high values, 16–17.2 gm/100 ml of blood (Lenfant *et al.*, 1969). The average amount of hemoglobin contained in each red cell (MCHC) varies with age (Rostorfer and Rigdon, 1947). Adult ducks average 42.6 pg of hemoglobin per erythrocyte and young ducks average 31.7 pg. More recent determinations give somewhat higher values for ducks, of 45 and 51 pg in male and female, respectively (Surendranathan *et al.*, 1968) to 75 pg by Soliman *et al.* (1966). A value of 45.1 pg has been recorded in turkeys by Usami *et al.* (1970). The mean corpuscular hemoglobin concentration (percent) expresses the amount of hemoglobin as a percentage of the volume of the corpuscle. Most birds show values ranging from 30–40 gm% (D. J. Bell *et al.*, 1964; Soliman *et al.*, 1966; Surendranathan *et al.*, 1968; Usami *et al.*, 1970).

The young red cell is called a reticulocyte because, on vital staining with cresyl blue, a network or reticulum is apparent in the cytoplasm [the precursors of the reticulocytes (erythroblasts) also give this staining reaction]. In man, the number of reticulocytes is about 1% of the total number of erythrocytes except during the first few days of life, when the value may be 2–6%. Spector (1956) reports a mean for ducks of 0.2%, whereas Magath and Higgins (1934) report a mean of 20.7% The number of reticulocytes will vary with the rate of erythrocyte production but, nevertheless, the difference between these two figures is hard to explain. The presence of a nucleus does not seem to confer any advantage to avian erythrocytes with regard to life span compared with those of man. Human red cells have a half-life around 60 days, whereas in chickens the total life span is 28–35 days (Hevesy and Ottesen, 1945; Brace and Altland, 1956; Rodnan *et al.*, 1957), 39–42 days for the duck (Brace and Altland, 1956; Rodnan *et al.*, 1957), and 35–45 days for the pigeon (Rodnan *et al.*, 1957). The relatively short erythrocyte life span in birds may be related to their high energy metabolism and body temperature.

Increased production of red cells occurs both in young and adult chickens (and other species) in response to lowered environmental oxygen tensions (R. R. Burton and Smith, 1969). The effect of low ambient oxygen on blood properties during embryological development is variable, although no compensatory increase in hematocrit and hemoglobin seems to occur until 15 days of development (Chiodi, 1963; Jalavisto *et al.*, 1965; R. R. Burton and Smith, 1969). According to Jalavisto *et al.* (1965), mean corpuscular volume and mean corpuscular hemoglobin concentration are unaffected by differences in ambient oxygen concentration.

B. The Leucocytes

The white blood cells are divided into three groups:

1. Granulocytes—characterized by the presence of granules in the cytoplasm and a lobed nucleus. Three cell types can be recognized by the character of their granules:

(a) Heterophils—the granules are rod- or spindle-shaped acidophilic crystalline bodies. Cells are usually round, about 11 μ in diameter, and the nucleus is polymorphic.

(b) Eosinophils – large spherical granules which appear to crown the cell. The cells are reported as being the same size (11 μ) as heterophils although Andrew (1965) gives an average diameter of only 7.3 μ. The nucleus is often bilobed.

(c) Basophils – granules are large, basophilic, and stain heavily. The cells are spherical and about 9 μ in diameter. The nucleus is round or oval in shape.

2. Lymphocytes – round nongranular cells with a large round nucleus that practically fills the cell substance in the smaller cells. In larger cells there is a marked increase in the cytoplasm but little change in size of the nucleus. In ducks average lymphocyte size is 6 μ with a range from 4 to 8.1 μ (Magath and Higgins, 1934).

3. Monocytes – large oval cells (11 × 13.5 μ) with relatively more cytoplasm than the large lymphocytes. The nucleus is pale-staining, round or indented, and is usually eccentric. The cells are often confused with lymphocytes because there are transitional forms between the two.

The total number of leucocytes in adult avian blood is in the range of 15–30 thousand/mm^3 for both males and females (Sturkie, 1965). The cell count increases with age (Hewitt, 1942). The ratio between leucocytes and erythrocytes is 1:43 in nestling Red-winged Blackbirds, 1:284 in immature, and 1:37 in mature individuals (Ronald *et al.*, 1968).

Differential counts, which often suffer from the difficulty of distinguishing eosinophils from heterophils, have established that lymphocytes are the most numerous in the majority of species studied, making up about 50–60% of the total count, compared with 20–30% heterophils (Fig. 1). In the Ring-necked Pheasant (*Phasianus colchicus*) (Lucas and Jamroz, 1961) and Ostrich (De Villiers, 1938) the heterophils are the most numerous of the white blood cells, but this finding has occasionally been reported for the duck (Lucas and Jamroz, 1961), although the majority of measurements for the duck are much lower (Magath and Higgins, 1934; Surendranathan *et al.*, 1968). Monocytes form about 5–10% of the total white cell count. Basophils and eosinophils together are usually less than 5% of the count (Fig. 1). Hewitt (1942) has found considerable variation in white cell count from day to day in ducks; for instance, lymphocyte numbers varied from 11.0 to 75% in ten ducks on seven different days. In view of this type of variation and that reported by other authors it seems virtually impossible to assign normal values for any one type of white cell in a given bird.

There is little information on the functions of leucocytes in birds, and their role has generally been inferred from conditions in mammals. It appears, however, that monocytes may function in fat metabolism, which is of great importance during the long starvation periods in migratory species (George and Berger, 1966). The lymphocytes play a bacteriophage role and influence transport of antibodies in the blood, in addition to being important as centers of fat synthesis in the liver (George and Naik, 1963).

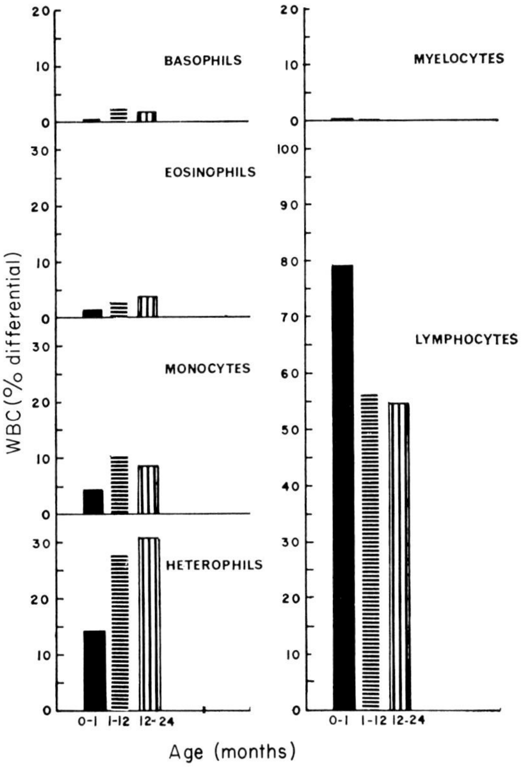

FIG. 1. Differential values of leucocytes in the Red-winged Blackbird (*Agelaius phoeniceus*). (Reproduced by permission of the National Research Council of Canada from the *Canadian Journal of Zoology*, **46**, p. 160, 1968.)

C. THROMBOCYTES

In avian blood the thrombocytes arise from antecedent mononucleated cells, which have a normal blastula stage (Lucas and Jamroz, 1961). It appears that thrombocytes are cells belonging to the erythrocyte series, and Blount (1939) even suggests that a thrombocyte is an erythrocyte in which hemoglobin synthesis has been interrupted. Thrombocytes are of variable size and shape, being typically oval, with a round nucleus and one or more granules present at the poles of the cell in the otherwise clear cytoplasm. The erythrocyte/thrombocyte ratio is about 200. Thrombocytes are essential in blood clotting.

III. Morphology of the Blood Vascular System

A. Heart and Major Blood Vessels

1. Conditions in the Embryo

The heart is a muscular pumping organ and arises as a specialized part of the primary longitudinal ventral vessel. During development the originally straight cylindrical cardiac tube of the embryo becomes differentiated into four primary regions: the posterior sinus venosus, atrium, ventricle, and anterior bulbus cordis (Goodrich, 1930). The walls of the four chambers have characteristic striated muscle cells quite unlike the smooth muscle fibers found in peripheral blood vessels. In no adult craniate is the longitudinal disposition of the four chambers retained, and the heart assumes an S-shaped curve so that the atrium comes to lie dorsally to the ventricle, whose apex points backwards. Following the development of a pulmonary circulation the heart becomes divided, allowing separation of oxygenated and deoxygenated blood. The sinus venosus is much reduced and subdivided, and in many birds the great veins appear to open separately into the right atrium (Röse, 1890; Kern, 1926). The artrial and ventricular cavities are both completely divided in all adult birds to give paired atria and ventricles. The bulbus cordis is obliterated by a caudal continuation of the division of the aorta so that the major vessels arise directly from the outflow tracts of the ventricles themselves.

The heart has a bilateral origin and is first visible in the embryo as two tubes, derived from splanchnic mesoderm, lying on either side of the developing foregut. Following foregut fusion the two heart tubes also fuse, although this fusion may be prevented either mechanically (Gräper, 1907) or chemically (De Haan, 1958a,b) in which case each tube forms a small beating heart (*cardia bifida*). The fused tube has two layers, the outer epimyocardium, which gives rise to the muscles and outer covering, and the inner endocardium.

The first spontaneous contractions occur at the stage of nine or ten pairs of somites and before there is autonomic innervation of the heart. It has been shown morphologically (His, 1893; Abel, 1912; Szepsenwol and Bron, 1935) and confirmed experimentally (Le Grande *et al.*, 1966) that autonomic innervation occurs between the third and sixth days of development in the chick. The heart rate of chick embryos rises rapidly after the initiation of contraction, doubling between the second and eighth day of incubation. From 8 days until hatching the rate of beat is fairly constant at 220 beats/minute (Romanoff, 1960). However, heart rates are affected by many factors such as the amount of blood flowing through the heart (Barry, 1941; Alexander and Glaser, 1941), dissolved substances (e.g., epinephrine) in the blood (Matsumori, 1929), oxygen and carbon dioxide concentrations in the plasma surrounding the heart (Paff and Boucek, 1958), and even visible light (Gimeno *et al.*, 1966).

At the 16–17 somite stage a functional, stop-and-go circulation is established. By 72 hours, blood cells pass smoothly through the heart and large vessels, and backflow during diastole is largely prevented by valvelike accumulations of cardiac jelly at points of transition from sinus venosus to atrium and ventricle to bulbus cordis (truncus arteriosus). The competency of these valves has been questioned, since Paff *et al.* (1965) have found that up to the sixth day of incubation (after which true arterial valves develop) diastolic ventricular pressures are relatively high (5–17 mm H_2O). However, pressure tracings obtained by Van Mierop and Bertuch (1967) from 5 day embryos (Fig. 2) testify to the competency of the endocardial tissue masses in preventing reflux of blood into the ventricle during diastole. The configuration of the tracings supports the view of Patten *et al.* (1948) that such valve action is present in even younger embryos.

The heart assumes its characteristic S-shape, even if freed from mechanical pressures of surrounding tissue, at about the 22–29 somite stages in chick, duck, and turkey. The interatrial septum develops across one of the blood streams in the heart (Bremer, 1932), but the interventricular septum appears, after 4 days of development, between the right and left streams of blood flowing through the heart. The cardiac jelly is of such a consistency that it is probably molded to establish the pattern along which septa later develop (Patten *et al.*, 1948). Septa form by proliferation of tissue (trabeculuted muscle fibers) into the cardiac jelly (von Lindes, 1865). The trabeculations may originally be part of the myocardial wall that grows inwards to extend across the ventricular cavity (Rosenquist, 1970) or the fibers may arise from the endocardial layer (Patten *et al.*, 1948). Stéphan (1959) has shown that the development of the heart septa is dependent on the correct morphological relationships of the various parts of the heart and upon normal circulation through the heart.

Apart from the perforations of the interatrial septum, the heart of the 16-day embryo is morphologically similar to that of the adult. At this stage only a small outer ring of the interatrial septum is intact,

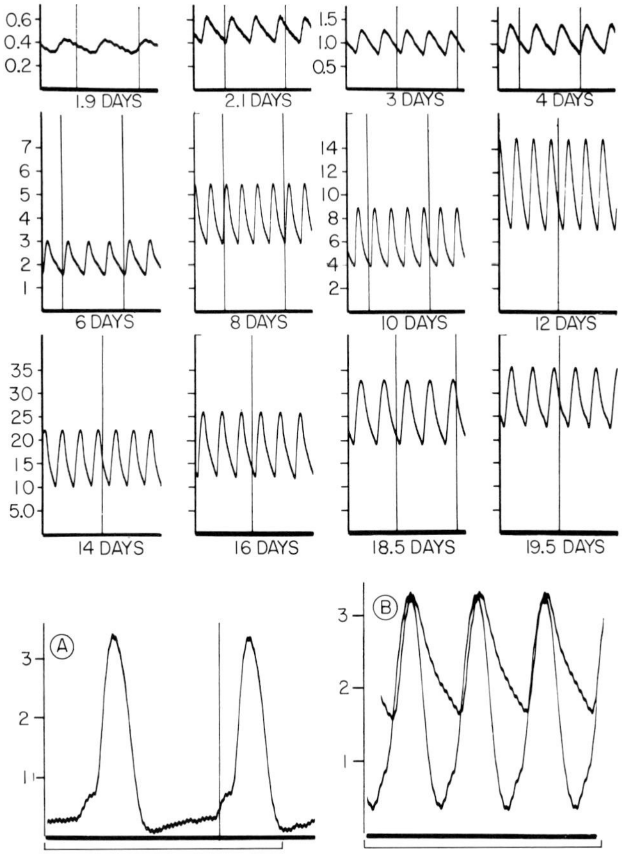

FIG. 2. Upper: arterial pressure tracings at various incubation ages in domestic fowl embryos. Pressure (ordinate) in mm Hg. Time lines, 1 second. Lower: ventricular pressure in the 5½-day embryo after slowing the heart by cooling (A) and superimposed arterial and ventricular pressure tracings of a 6-day embryo (B). Pressure in mm Hg. Horizontal bars equal 1 second. (Van Mierop and Bertuch, 1967, *Amer. J. Physiol.*, by permission.)

the central portion is fenestrated by several large (1 mm diameter) and numerous small holes, but the septum appears functionally closed during atrial systole (Fig. 3) because blood on the right and left sides of the heart is separated (Patten, 1925; Quiring, 1933; White, 1969). Pohlman (1911), however, suggested that there was complete mixing of blood during diastole, but this view has not been supported by Lillie (1908, 1965) and White (1969) on morphological as well as experimental grounds. It appears that oxygenated blood (returning from the chorioallantois) and deoxygenated blood (return-

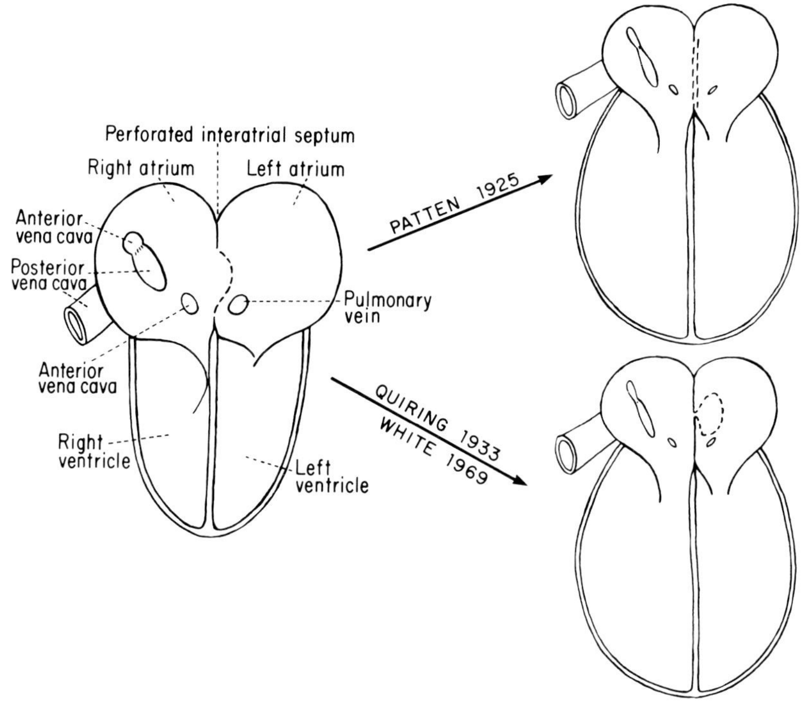

FIG. 3. Possible modes of closure of the interatrial septum in the chick embryo. During atrial systole the fenestrations (broken line) may overlap to occlude the connection between right and left atria (Patten, 1925) or alternatively the central perforated portion of the interatrial septum may be squeezed into the left atrium (Quiring, 1933; White, 1969).

ing from head and body) is kept largely separate within the heart. Oxygenated blood tends to flow via the interatrial septum to the left atrium without complete admixture with deoxygenated blood during passage through the right atrium. A high degree of laminar streaming must be the main reason why the two kinds of blood are kept from mixing completely in the right atrium.

2. Circulation in the Embryo

Cardiac output of chick embryos has been assessed by several authors, with values varying from 0.12 ml/minute at 5 days, 4.8 ml/minute at 12 days, to 6.0–10.4 ml/minute at 16 days of incubation (Hughes, 1949; White, 1969). Arterial blood pressures are low early in incubation but increase to about 20–25 mm Hg at the sixteenth day (Hill and Azuma, 1927; Hughes, 1942). Some authors report a leveling off of arterial blood pressures after this time, but Van Mierop and Bertuch (1967) found a linear relationship between systolic pressure and embryonic weight from 4 days of incubation, with pressures reaching 36/22 mm Hg at the twentieth day. There is a marked fall in pressure, both systolic and diastolic, after the chick has "pipped," which marks the start of breathing and termination of the chorioallantois as a gas exchanger (Van Mierop and Bertuch, 1967). From data of arterial pressure and heart rate, Hughes (1949) estimates that the work done by the chick heart is 3.5 ergs/second at 4 days, increasing to 69 ergs/second at 7 days. The cardiac output is distributed as follows: 36% to head, neck thorax, and heart, 24% to lungs, posterior body, and yolk sac, and 40% to the chorioallantois (White, 1969).

3. *Conditions in the Adult Bird*

In adult birds there is considerable variation in the mode of entry of the large veins into the right atrium. According to Gasch (1888) the sinus venosus does not exist as a separate entity in such birds as *Columba, Ardea, Grus, Otis,* and *Tetrao.* Blood returning from the anterior portion of the body is delivered into the right atrium by the two anterior venae cavae and blood from the viscera and hind regions of the body is delivered by the single posterior vena cava. In other birds, however, the retention of the sinus venosus is indicated by the fact that the posterior vena cava shares a common inflow orifice with either the right anterior vena cava (*Coturnix*) or both anterior venae cavae (*Struthio*). The coronary vein usually opens directly into the left anterior vena cava and may have two or three orifices. In the chicken the posterior vena cava and the right anterior vena cava open together into the sinus venosus, whose slitlike entrance into the right atrium is demarcated by a pair of sinu-atrial valves (Fig. 4). The sinus septum separates the mouths of the left anterior vena cava and coronary vein from the common orifice of the other veins. The pulmonary veins open either separately or together into the left atrium.

The atria are thin-walled but muscular structures acting as storage compartments for blood returning to the heart from the systemic and pulmonary vascular beds. The right atrium is significantly larger than the left. A prominent feature of the avian heart is the presence of dense muscular cords in the dorsal walls of the atria. The median dorsal muscular arch lies above the line of attachment of the interatrial septum (Quiring, 1933). Across the roof of the right and left atria, running transversely, are the lateral muscular arches, which branch into smaller muscular ridges, the *musculi pectinati,* which, in turn, fuse with a muscular band forming the inferior limits of the atria. Contraction tends to obliterate the atrial cavities although pressure changes within the atria are small, as is to be expected in such a thin-walled structure, reaching only 2–3 mm Hg during atrial contraction.

With one, or both, of the sinu-atrial valves lacking in many birds, the importance of atrial contraction for ventricular filling is most likely modest, which is also the case for mammals, since a requisite for an effective atrial contraction is the presence of sinu-atrial valves.

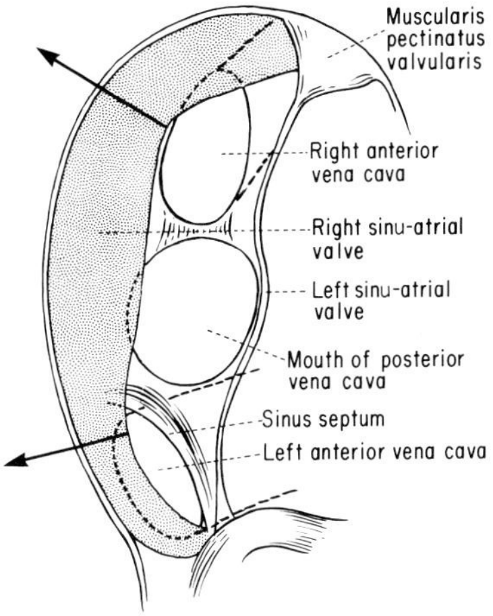

FIG. 4. Diagrammatic representation of the entrance of the caval veins into the right atrium of the chicken. The right sinu-atrial valve is reflected in the direction of the arrows (after Simons, 1960).

The left atrio-ventricular valve of the avian heart is structurally similar to the corresponding mammalian valve, being formed of two cusps, one lateral and one central, although in some species it may be tricuspid (Fig. 5). During ventricular contraction the patency of the valve is supported by fibrous *chordae tendinae* extending from the lower surfaces of the valve cusps to the three papillary muscles of the left ventricle. The right atrio-ventricular valve is very characteristic in birds and consists of a single lateral leaf in the form of a thick muscular flap, made up of atrial and ventricular muscles on its respective sides, between which is an epicardial connective tissue layer. This region is rich in nerve fibers and also has embedded in its fibrous tissue a sheet of Purkinje fibers. The right valve originates at the right side of the base of the pulmonary artery, where it is held by a trabecular muscle (Fig. 5). From this point the valve extends up and backward around the right ventricular wall to the junction of the latter with the interventricular septum. Occasionally, the valve is provided with a membranous part as in the kiwi (*Apteryx*) and the duck. When the right ventricle contracts, the free edge of the valve leaf is forced upward against the bulging interventricular septum and closes the slitlike atrio-ventricular ostium. During diastole the valve drops downward into the ventricle and thus opens the atrio-ventricular ostium.

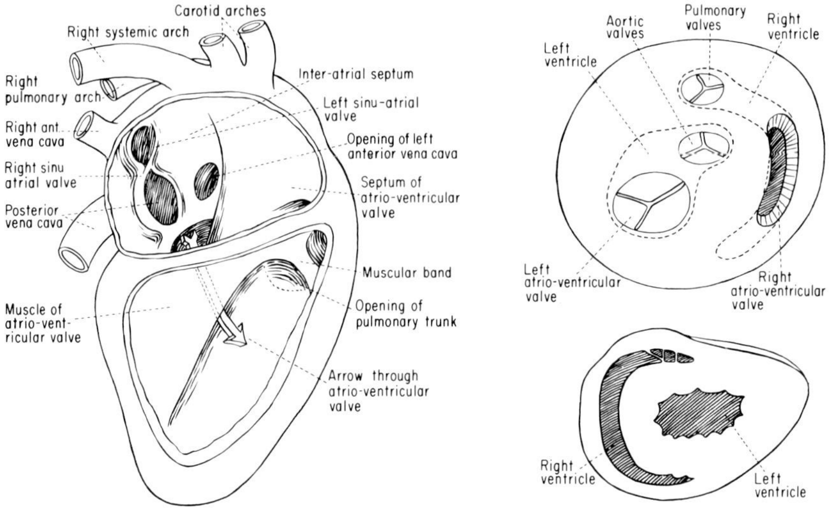

FIG. 5. Left: diagram of the heart of the Mute Swan, *Cygnus olor* (after Goodrich, 1930). Upper right: schematic drawing of the bird heart at the level of and parallel to the atrio-ventricular junction (after Simons, 1960). The right atrio-ventricular orifice is darkly shaded, and the lighter-shaded portion depicts the extent of the right atrio-ventricular valve. Lower right: transverse section through the ventricles of the European Crane, *Grus grus* (after Goodrich, 1930).

The ventricles are very unevenly developed; the wall of the left being some 2–3 times thicker and more muscular than that of the right, which partially surrounds it (Fig. 5). The thickness of the ventricles reflects the pressures that are generated within them. The pressure within the ventricles is raised by the development of tension in the ventricular muscle. It follows from the Law of Laplace that since the radii of curvature will be smaller at the apex than in a relatively flat portion of the ventricle, greater tension must be produced in the walls of the flat portions and the wall will be thicker. The pressures generated by the left ventricle are some 4–5 times as great as those generated by the right ventricle (Fig. 6) and, therefore, the difference in the products of thickness and curvature for the walls of the right and left ventricles will be of this order.

Histologically, the structure of the atria corresponds to that of the ventricular wall. The epicardium of loose connective tissue and elastic fibers is bordered by a simple epithelial layer and is richly innervated by both vagal and sympathetic fibers. Although ganglia of various sizes occur along vessels in the wall, sensory receptor structures, which are common in mammalian atria, have not been reported in avian hearts. The atrial myocardium consists of striated

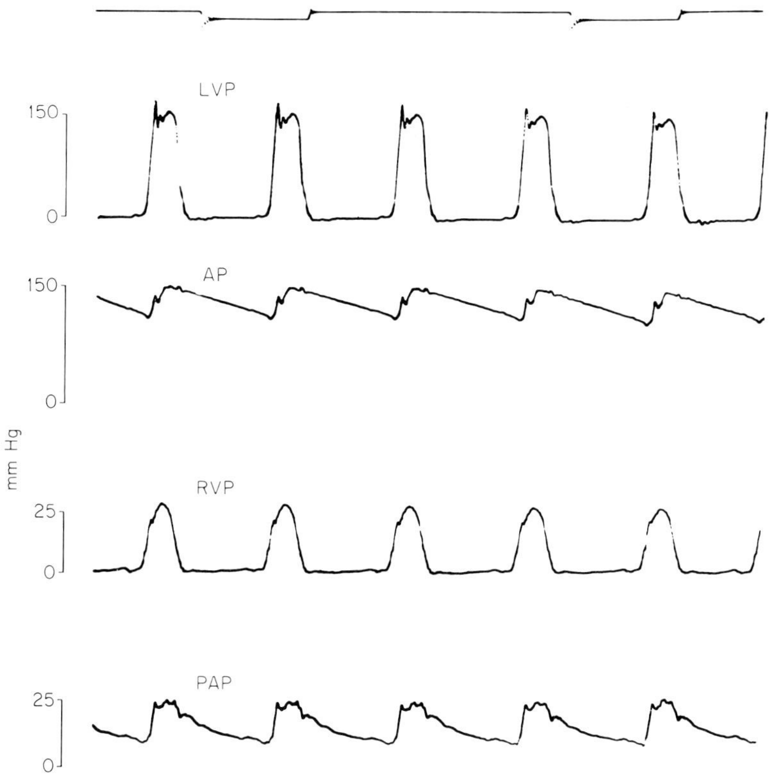

FIG. 6. Pressures in the central arterial system of the domestic duck. Traces, from top to bottom; time (seconds); left ventricular pressure (LVP); aortic pressure (AP); right ventricular pressure (RVP); pulmonary arterial pressure (PAP). (Jones and Langille, 1972.)

muscle fibers on a collagenous basal tissue, myofibrils being transversely striated with granulated sarcoplasm between them. In both atria the myocardium is richly innervated. The endocardium is essentially a thin connective tissue layer that resembles the epicardium. In the sparrow and stork the atrial septum is thinner than in other birds studied and is essentially similar to the endocardium supported by a basal layer of myocardial tissue of varying thickness.

The ventricles, although differing greatly in thickness, are histologically similar. The thick epicardium has a fat layer on its inner border with the myocardium. Proceeding outward there is a collagenous connective tissue layer and a thin elastic layer, terminating at the ventricular surface in a simple epithelium. A nerve plexus innervates the epicardium, but receptor structures again appear to be absent. The apparent absence of receptor structures in the walls of the avian heart is puzzling, for it means that no information about the functional state of the atria and ventricles in birds can be relayed to the central nervous system (Ábrahám, 1969). The ventricular myocardium consists of striated muscle fibers and is less richly innervated than the atrial myocardium. The endocardium consists of a thin connective tissue layer. The ventricular septum is made up of endocardium facing the lumen of the ventricles while its middle portion consists of muscle fibers. The ventricular septum contains Purkinje fibers on both sides.

Avian ventricular muscle fibers, except for those specialized for conduction, have no transverse tubules (T-tubules) and are 5–10 times smaller in diameter than mammalian muscle fibers (Sommer and Steere, 1969) and consequently occur in larger numbers (Hirakow, 1970). The small size of the muscle fibers may be related to the absence of T-tubules. Dense bodies are found in both atrial and ventricular muscle fibers, and these specific granules have been suggested to play some role in the sensitivity of cardiac muscle to acetylcholine, which would explain why avian ventricles exhibit a negative inotropic response to vagal stimulation, for in mammals such granules are only found in the atrial fibers (Sommer and Johnson, 1969). The sarcoplasmic reticulum is well developed and exhibits prominent couplings (Sommer and Johnson, 1969; Hirakow, 1970). The sarcoplasmic tubules of the muscle fibers contain dense granules, which have been suggested to play a role in calcium regulation, their presence being a consequence of the high frequency of contraction. Slautterback (1965) suggests there is a correlation between heart rate, mitochondrial size, and complexity of the mitochondrial cristae structure. Certainly there are large numbers of mitochondria in avian heart muscle (Hirakow, 1970), and in the hummingbird the amounts of mitochondrial and contractile components are practically equivalent (Didio, 1967). Didio (1967) found no glycogen granules in the Swallow-tailed Hummingbird (*Eupetomena m. macroura*) heart, which may correlate with the high metabolic rate; however, the author makes no mention of whether the hearts were taken from active or resting birds.

Hirakow (1970) has pointed out that the essential component dominating the characteristics of the myofiber is the contractile material or myofibril. The development of intercalated discs, mitochondria, and sacroplasmic reticula, as well as a T-system (transverse tubules), in cardiac muscle cells appears to be related to the degree of development of the myofibrils. This concept is illustrated in Fig. 7.

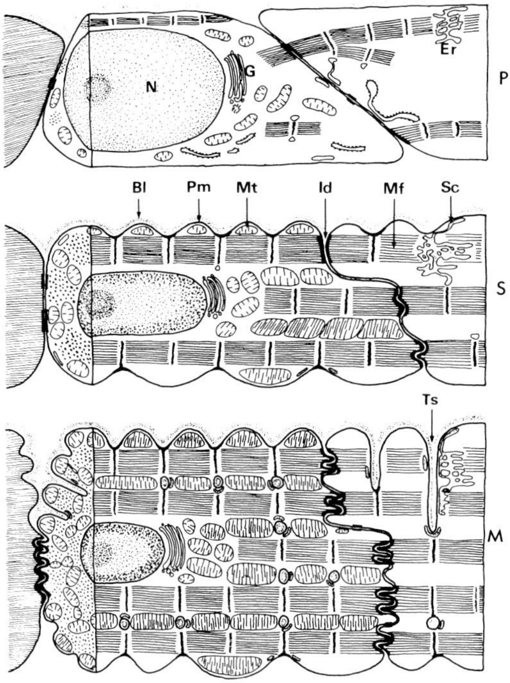

FIG. 7. Schematic drawing to illustrate three fundamental types of vertebrate myocardial cells ultrastructurally defined. P = primitive or embryonic type, S = sauropsidan type, M = mammalian (ventricular) type. Mammalian atrial myofiber may belong to S type. At the left side of each fiber, cross sections of two apposing cells are shown. In the right fiber segment, drawn as a longitudinal section, mitochondria are not included in order to depict the tubular structure more clearly. Granulations such as glycogen and free ribosomes are also excluded. Basic structural components are labeled as follows: Bl = basement lamina, Er = sarcoplasmic or endoplasmic reticulum, G = Golgi complex, Id = intercalated disc, Mf = myofibril, Mt = mitochondria, N = nucleus, Pm = plasma membrane, Sc = special segment of sacroplasmic reticulum in close association with plasma membrane, Ts = transverse tubular system. (Hirakow, 1970, *Amer. J. Cardiol.*, by permission.)

Many birds have extremely large hearts, for instance that of the Scintillant Hummingbird (*Selasphorus scintilla*) may take up 2.4% of the total body weight (Hartman, 1955). While this is an exceptionally high value, it is apparent that avian heart weights, as a percentage of body weight, are greater than those of mammals of similar size. Brush (1966) reports that the logarithm of heart weight is a linear function of the logarithm of body weight in birds greater than 100 gm (slope of regression curve = 1) whereas, below this body weight, the slope is 0.6. The usefulness of a large heart in birds may have adaptive significance as a means of quickly increasing stroke volume by utilization of the systolic reserve volume. Furthermore, a large heart is required to shorten less than a smaller one for ejection of a given volume, but a relative increase in heart-muscle tension (and therefore internal cardiac pressures) prevails in the larger heart. This in turn correlates with higher arterial pressures, which in birds generally exceed those in mammals of comparable size. A large heart size requiring a reduced degree of muscle shortening may also offer some advantages for adequate diastolic filling at the generally high heart rate prevailing in birds. A careful study of heart size, and pressure and volume changes of the contracting avian heart should prove to be rewarding.

Embryonic heart weight increases as the blood volume increases during development (Fig. 8). In fact, it is the plasma volume that

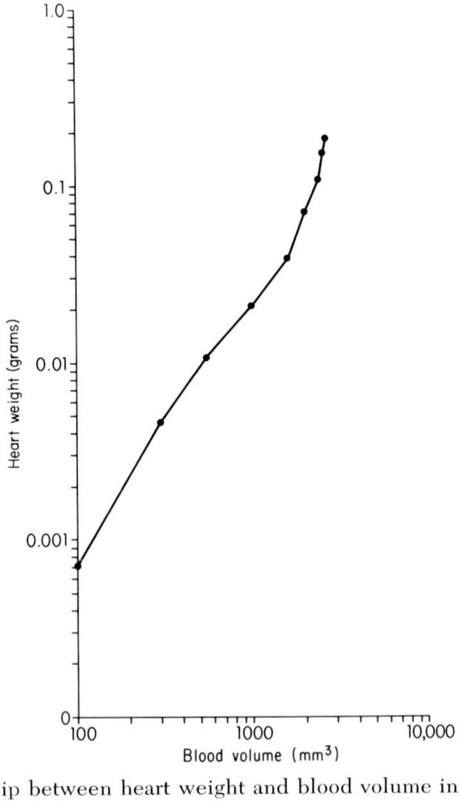

FIG. 8. Relationship between heart weight and blood volume in the chick embryo. (Data from Romanoff, 1960.)

changes during the period between 9 and 18 days development whereas the red cell mass remains constant as related to embryonic weight (Barnes and Jensen, 1959). Burke (1966) has pointed out that the blood oxygen capacity per unit weight decreases as the body weight increases. This has recently been confirmed for the chicken embryo by Tasawa (1971). Spencer (1966) has shown for three genera of porpoises that the relative heart weight (heart weight/body weight) is directly proportional to the blood oxygen capacity. This observation served as a stimulus to define a relationship between relative heart weight and blood volume (Spencer, 1967). The relationship between relative heart weight and blood volume using data for adult birds has not yet been explored. From data presently in the literature it can be deduced that the largest relative heart weights accompany the largest blood volumes, but there are too few points for strong generalizations.

Blood is ejected from the right ventricle into the pulmonary artery, which arises from the right anterior region of the ventricle at the level of the atrio-ventricular border. Three semilunar valves prevent regurgitation from the pulmonary artery to the right ventricle. From the left ventricle blood is pumped into the systemic arch (right aortic arch) and three semilunar valves again prevent reflux of blood from the aorta into the left ventricle during diastole (Fig. 5).

The coronary arteries, supplying the nutritional and respiratory circulation to the heart muscle, arise from the systemic arterial outflow trunk close to the base of the heart. Most birds have two main coronary arteries but some may have three or four (Petren, 1926). The right coronary artery in chickens supplies the ventral side of the heart; the left, which arises from the dorsal side of the aorta, supplies the dorsal side of the heart. The ends and branches of the coronary arteries anastomose freely and many branches are located deep in the myocardium. The coronary arteries take a relatively short course through the thicker wall of the left ventricle, whereas they follow a pattern more parallel to the surface in the thinner walls of the right ventricle and atria and sink into the depth of the myocardium more gradually (Davies, 1930; Fig. 9). There are four major cardiac veins in the fowl heart, lying just beneath the epicardium, with multiple anastomoses among them (Fig. 10). The four major cardiac venous complexes are the great cardiac vein, middle cardiac vein, left circumflex vein, and a small cardiac venous complex. This situation appears fairly typical of birds in general (Neugebauer, 1845). Quiring (1933), McKibben and Christensen (1964), and Lindsay (1967) found no coronary sinus in the heart of chicken, contrary to findings of Kaupp (1918) for this species, and of Neugebauer (1845) for other birds. The terminations of the

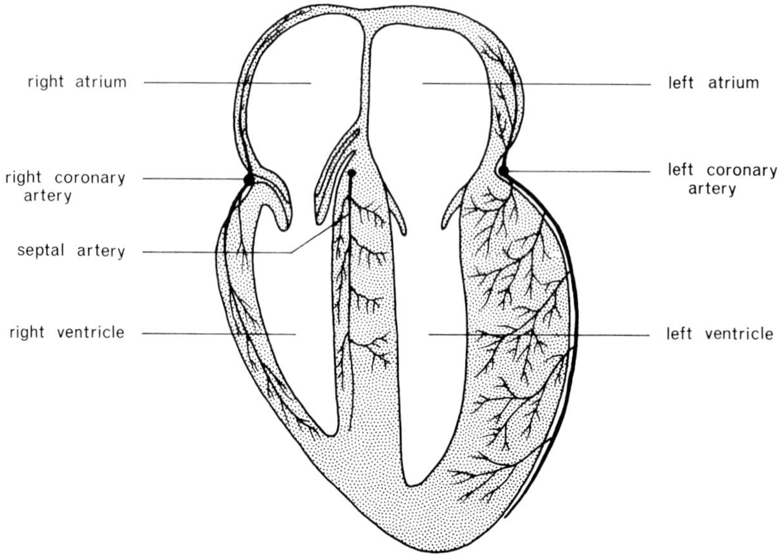

FIG. 9. Diagram of the pattern of the coronary arteries of the heart of the domestic chicken (after Davies, 1930).

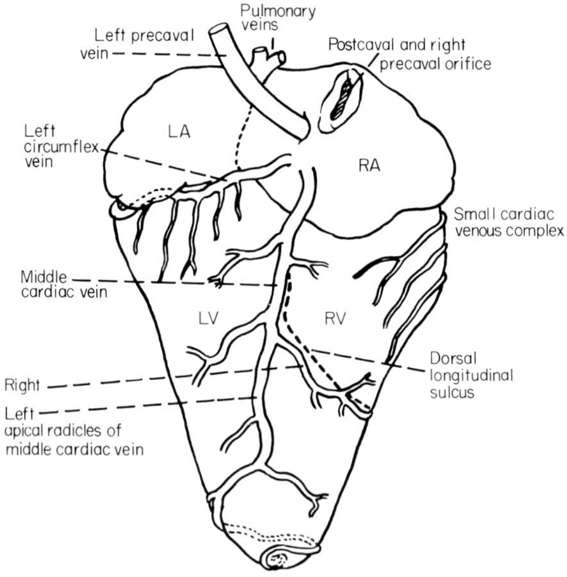

FIG. 10. Dorsal aspect of unopened heart of the domestic chicken illustrating typical pattern of middle cardiac vein and terminal course of left circumflex vein (after Lindsay, 1967).

cardiac veins in the right atrium is thus a matter of some dispute in the chicken. The intra-cardiac openings may be separate from one another and the orifice of the left anterior vena cava (Lindsay, 1967), whereas Neugebauer (1845) and Kaupp (1918) consider that the great and middle cardiac veins open by a common orifice into the left anterior vena cava. Quiring (1933) claims that in the adult fowl the right sinu-atrial valve, at its lower extremity, acts as the coronary valve.

B. The Arterial System

The vascular system undergoes progressive modification during embryonic life. The aorta arises from paired primordia (dorsal aorta), which fuse to form the single definitive aorta. A short ventral aorta also forms during embryogeny as a projection from the heart and becomes connected with the dorsal aorta by six pairs of aortic arches. Not all aortic arches are present at one time because they appear in succession and some are only transitory. The very transitory fifth pair of aortic arches make their appearance last. The whole circulatory system develops from cells that have undergone an orderly migration during gastrulation and the stages that immediately follow it. In fact, cells that give rise to the aortic arches differentiate from a plexus that also gives rise to the anterior cardinal veins (Sabin, 1917), and both sets of vessels combine to attach the ventrally looping cardiac tube to a more stable mediodorsal structure, the foregut (Rosenquist, 1970).

The first and second aortic arches disappear completely but the carotid arteries take origin from the third arch. The left aortic arch (IVth arch) becomes obliterated and disappears, whereas the right arch persists as the permanent arch of the aorta. The Vth arch is temporary and disappears, whereas the VIth arch gives rise to the pulmonary artery, losing its connection with the aorta, when the ductus arteriosus is obliterated soon after hatching (Hughes, 1943; Romanoff, 1960; Harms, 1967). The retention or disappearance of the respective aortic arches seems to be dependent upon the existence of the dilating force exerted by the blood stream. For instance, ligation of the right aortic arch causes the left to develop as in mammals (Stéphan, 1949).

All the arteries of the head or neck are branches of the carotid arteries and are essentially derivatives of the embryonic dorsal and ventral cephalic aortae. The fate of the anterior dorsal aorta is not identical in all species of birds, for there is some variation in adult configuration of the common carotid arteries. The most frequent arrangement is for the common carotids to converge or run parallel in a ventrally placed groove on the cervical vertebrae (Garrod, 1873; Fig. 11A). Consequently, by being close to the axis of rotation of the neck, the carotid arteries are relatively unaffected by neck movement.

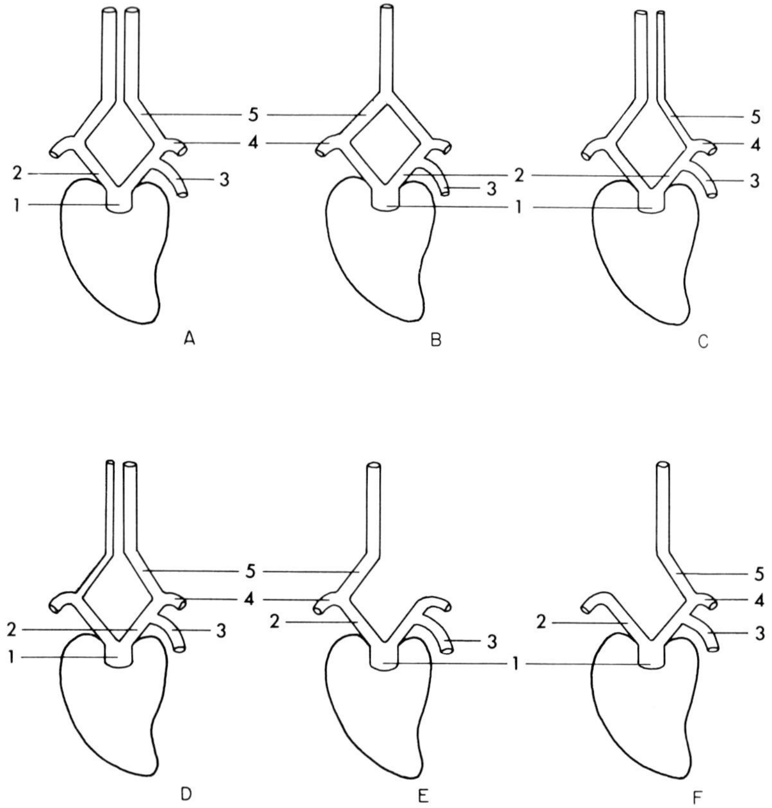

FIG. 11. Dorsal views of the main arteries of birds in the region of the heart: A, most prevalent avian type; B, type found in *Botaurus* and *Butorides*; C, type found in *Phoenicopterus;* D, type found in *Cacatua sulphurea;* E, type found in Passeres; F, type found in two species of *Eupodotis*. (1) aortic root; (2) innominate artery; (3) fourth right aortic arch; (4) subclavian artery; (5) carotid artery (after Glenny, 1940).

However, other configurations may be found (Glenny, 1940, 1943, 1955; Fig. 11). These include a single carotid artery formed from the union of right and left arteries (Fig. 11B), two carotids of unequal size, either the right or the left being smaller (Fig. 11C and D), a single persistent right or left vessel (Fig. 11E and F), or nonparallel carotid arteries.

The internal carotid leaves the common carotid and ascends the intermediate length of the neck in the cervical carotid canal. Leaving the canal below the skull base the internal carotid is located dorsal to the cervical viscera. The internal carotid gives off the stapedial artery upon entering the cranial carotid canal. The cerebral carotid artery, which supplies blood to the choroid plexuses and cerebral hemispheres, is the internal carotid distal to the origin of the stapedial artery. The palatine and sphenomaxillary arteries are branches of the intrasphenoid part of the cerebral carotid. The intrasphenoid segment of the cerebral carotid arteries is of special interest because most birds display some form of intercarotid anastomosis in which both carotids are placed in open communication (Baumel and Gerchman, 1968).

The intercarotid anastomosis appears to be of three main types (Fig. 12) and has been found in all species of birds investigated except

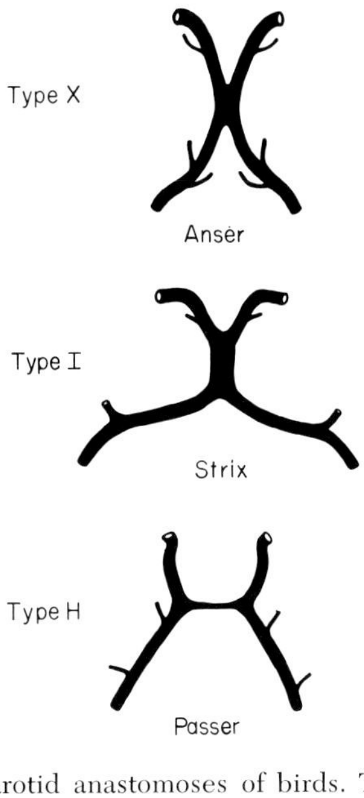

FIG. 12. Types of intercarotid anastomoses of birds. The H-type anastomosis is a transverse vessel of some length; in the X-type the carotids communicate by a side-to-side anastomosis; in the I-type anastomosis the two carotids merge into a single vessel of considerable length. (Baumel and Gerchman, 1968, *Amer. J. Anat.*, by permission.)

those of the suborder Tyranni (Baumel and Gerchman, 1968). There seems to be no correlation between the types of intercarotid anastomosis and the bicarotid or unicarotid condition seen in different birds. Birds do not generally possess a cerebral arterial circle of Willis comparable to that of mammals, since the basilar artery is usually formed unilaterally as the continuation of either left or right caudal rami of the cerebral carotids and not by union of both caudal rami of the cerebral carotid (Fig. 13). Furthermore, a communicans anterior between the anterior cerebral arteries is rare (Baumel, 1967). The intercarotid anastomosis obviously serves as an effective substitute and, since the connection is relatively large in many cases, it must represent a more effective collateral circulation than the mammalian arterial circle. However, blood may also reach the brain by means of extensive cerebral-extracranial anastomoses (Richards and Sykes, 1967). It is unlikely that much blood will reach the brain by way of the vertebral arteries since these unite with the occipitals and have only tenuous connections with the encephalic system (Fig. 13; Richards and Sykes, 1967).

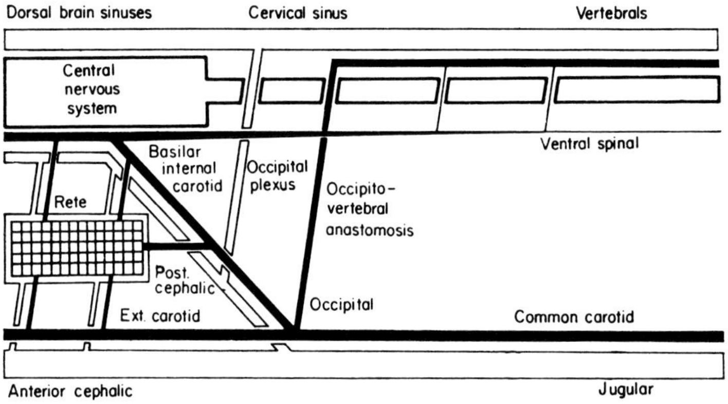

FIG. 13. Diagram illustrating the arrangement of the principal cephalic and cervical blood vessels in the domestic fowl. Dark solid lines, arteries; fine open lines, veins. (Richards and Sykes, 1967, *Comp. Biochem. Physiol.*, by permission.)

The hypophyseal arteries leave the internal (cerebral) carotids after the intercarotid anastomosis, and these small arteries divide repeatedly to form the primary capillary plexus of the median eminence of the neurohypophysis. The capillaries of the primary plexus fuse in the midline of the median eminence to give a series of portal vessels that pass ventrally to the pars distalis of the pituitary (Fig. 14). The primary capillary plexus on the median eminence is organized similarly to that of mammals, consisting of superficial vessels anastomosing in a caudal direction with posthypophyseal and deep vessels, the latter being less well developed than in mammals since they are restricted to the periphery of the superficial vessels. The deep vessels comprise capillary loops and a network of vessels which anastomose with

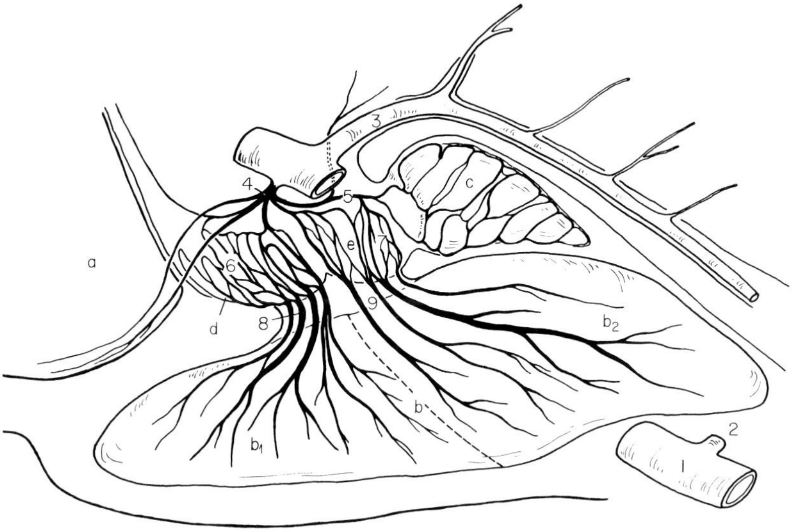

FIG. 14. Diagram of the median eminence and hypophysis and their blood supply (lateral aspect) in the White-crowned Sparrow, *Zonotrichia leucophrys gambelii*. a, optic chiasma; b, pars distalis, b_1, its cephalic lobe; b_2, its caudal lobe; c, neural lobe; d, anterior division of median eminence; e, posterior division of median eminence; 1, internal carotid artery; 2, inter-carotid anastomosis; 3, posterior ramus; 4, infundibular artery; 5, neural lobe artery; 6, anterior primary capillary plexus; 7, posterior primary capillary plexus; 8, anterior group of portal vessels; 9, posterior group of portal vessels. (Vitums *et al.*, 1964, *Z. Zellforsch. Mikrosk. Anat.*, by permission.)

those of the hypothalamus. The deep vessels also establish connections between the blood of the primary capillary plexus and the lumen of the third ventricle. There are no capillaries in the pars distalis, but only wide and irregular sinusoids that surround the glandular cells. The sinusoids drain principally into the paired carotid veins, which leave distally to join the internal jugular veins.

In many birds the hypophyseal portal system is relatively simple (Wingstrand, 1951; Assenmacher, 1952; Hasegawa, 1956) but Vitums *et al.* (1964) describe a more complex arrangement in the White-crowned Sparrow (*Zonotrichia leucophrys*). This arrangement of the portal vessels has also been demonstrated in the Japanese Quail (Sharp and Follett, 1969) and in twelve species of five orders of birds by Dominic and Singh (1969). In all these species there is a distinct point-to-point distribution of portal vessels from anterior and posterior divisions of the median eminence to the cephalic or caudal lobes of the par distalis, respectively (Fig. 14). Consequently, one region of the median eminence alone must control one region of the pars distalis.

The occurrence of anterior and posterior groups of hypophyseal portal vessels is regarded as typically avian by Duvernoy *et al.* (1969).

The endothelial cells of the portal vessels are invested by a definitive basement membrane and by cytoplasm of pericytes, which are oriented spirally to the longitudinal axes of the vessels. The pericytes may play a role in the mechanical support of the vascular wall and also may have a contractile function that might regulate blood flow (Mikami *et al.*, 1970). Furthermore, the endothelial cells themselves may play a part in blood-flow regulation since they often protrude into the vascular lumen giving the appearance of valve-like structures (Mikami *et al.*, 1970).

Although in the foregoing account the main cervical carotid has been referred to as a "common" carotid, this must not be taken to imply that it represents fusion of both the dorsal and ventral carotids. The common carotid is "common" only in the sense that it supplies blood to both internal and external carotids. The ventral carotid disappears after 7.5 days of development, being apparently unable to keep pace with the elongation of the region in which it lies (Fig. 15).

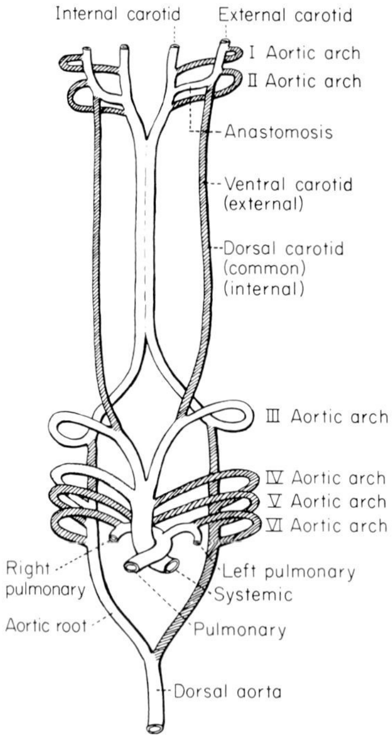

FIG. 15. Modification of the carotid arteries in birds correlated with elongation of the neck. The cross-hatched vessels are not present in adult birds (after Goodrich, 1930).

The external carotid joins the internal carotid and in the 12-day embryo the branches of the former have practically attained their definitive arrangement. A major branch is the occipital artery, which gives off distally a lateral branch to the neck muscles and a branch that is the anterior root of the vertebral artery, which arises from the common carotid (Fig. 13).

Essentially, the systemic and carotid arteries leave the heart together. The common vessel branches, as it leaves the heart, into right and left innominate (brachiocephalic) arteries and the aorta. Each of the former divides into two vessels, the subclavian (secondary) and common carotid. It is around this junction that the carotid body is located, particularly when the subclavian takes as its origin the original root of the external (ventral) carotid. An additional landmark to the position of the carotid body is the junction of the common carotid and vertebral arteries, although this has been disputed (Chowdhary, 1953). Muratori (1932, 1934) emphasizes the intimate relationship that often exists between the carotid body and the wall of the common carotid artery in this region and, bearing in mind the latter's innervation, claims that the region must be the homologue of the carotid sinus of mammals. The nerve fibers supplying this region of the common carotid are derived principally from the cells of the adjacent ganglion nodosum of the vagus (Fig. 16). Terni (1931) suggests that the glossopharyngeal nerve, through the ganglion nodosum, may contribute to the innervation. Chowdhary (1953) describes the sympathetic innervation as by the precarotid sympathetic trunk.

The carotid bodies are small round or ovoid structures about the same size as those of mammals of comparable size (Adams, 1958); e.g., hen, $0.64 \times 0.48 \times 0.36$ mm (Kose, 1907); rat, $0.58 \times 0.37 \times 0.23$ mm (Sato, 1932). They are situated on each side, at the base of the neck, but carotid body tissue has also been found in other areas of this region (Jones and Purves, 1970a). In this respect, Watzka (1943) states that the carotid bodies are paired on each side, whereas Chowdhary (1953) never found more than one on each side in the fowl. In the region of the carotid body are found the pharyngeal derivatives thymus, thyroid, parathyroids, and ultimobranchial body. The carotid body is most intimately related to the parathyroids (Chowdhary, 1953) but may also come into contact with the ultimobranchial body.

The blood supply to the carotid body comes usually from the common carotid artery, either by an artery to the organ or, more often, by one which also supplies the adjacent pharyngeal derivatives, such as the ultimobranchial body (Kose, 1907). The venous drainage is by several veins, some of which proceed from the superior pole to the anterior vena cava; other veins join the parathyroid veins and those from the ultimobranchial body (Adams, 1958).

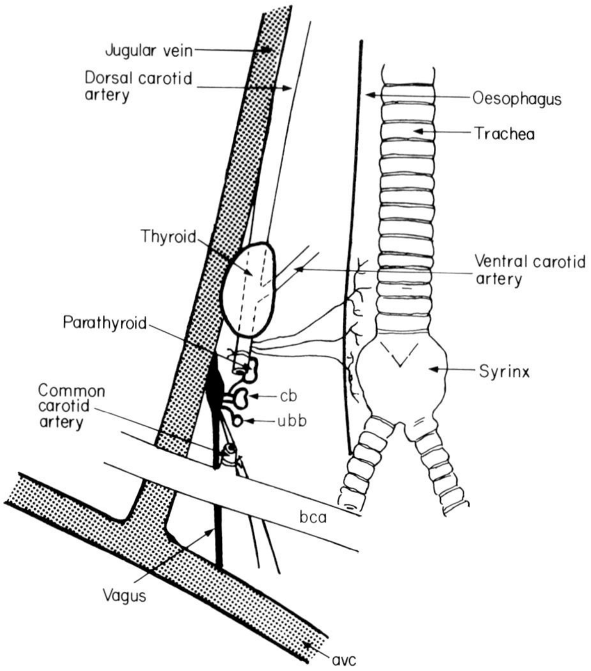

FIG. 16. Diagram showing the structures at the origin of the right common carotid artery in the domestic duck, ventral view. A segment of the carotid artery has been removed to show the dorsally related carotid body (cb) and ultimobranchial body (ubb) and nodose ganglion of the vagus nerve. bca, brachiocephalic artery; avc, anterior vena cava. [Jones and Purves, 1970a, *J. Physiol. (London)*, by permission.]

The carotid body is a second and/or third arch derivative (Chowdhary, 1953; Murillo-Ferrol, 1967). According to Murillo-Ferrol (1967) the first morphological bud appears after 5-days' incubation in chickens and consists of a cellular accumulation that migrates from the residual epiblastic placodes of the second and third cervical clefts towards the contour of the third aortic arch (Fig. 17). From the eighth day of incubation it occupies the situation that it will maintain during adult life.

The nerve supply to the carotid body is not conclusively known. Nearly all authors agree that it receives at least one, but perhaps two or three branches from the vagus ganglion (Fig. 16). However, since it is a second or third arch structure, one might reasonably expect to find it innervated by the glossopharyngeal nerve. Terni (1931) claims innervation of the carotid body by fibers of the glossopharyngeal through the precarotid trunk, but Jones and Purves (1970a) were unable to confirm this. The other possibility is that fibers from the IXth nerve run in the vagus, but Jones and Purves (1970a) could elicit no physiological effects related to carotid-body function following section of all the major anastomoses of the IXth and Xth nerves and the precarotid trunk. Furthermore, both Terni (1927) and Murillo-Ferrol (1967) claim that some fibers may also come from the recurrent laryngeal nerve. This nerve connection was not confirmed by Jones and Purves (1970a) in the duck.

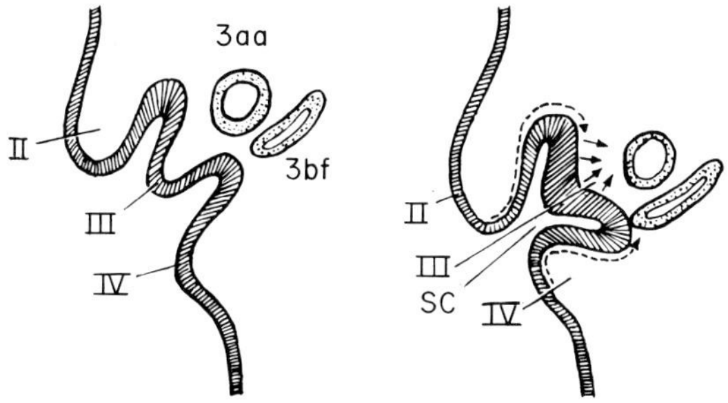

FIG. 17. Diagram to show the evolution of the sinus cervicalis between embryonic stages 24 (left) and 26 (right) in the domestic fowl. The growth of the second branchial arch (II) leaves at the bottom the placodal thickenings of the second and third cervical clefts. The continuous arrows indicate the cellular migration toward the third aortic arch (3aa) to form the carotid body; the dotted arrows indicate the course of the pycnotic current which must destroy the epithelium which covers the cavity of the sinus cervicalis (sc). 3bf = third pharyngeal pouch. (Murillo-Ferrol, N.L. The development of the carotid body in *Gallus domesticus*. *Acta Anat.* **68,** 102–126 (1967), Fig. 7, by permission.]

The microscopic anatomy of the carotid body is similar to that of mammals, as indeed are its functions in regard to respiratory reflex responses to changes in partial pressures of oxygen and carbon dioxide. The carotid body is parenchymatous and is surrounded by a dense connective tissue capsule. The arteries and nerves break up into fine and sometimes tortuous branches, the arteries giving rise to the lobular arterioles, and the heavy myelinated nerve fibers giving rise to an intralobular plexus of non-myelinated fibers ramifying among epitheloid cells. Both Type I and Type II cells are present (de Koch, 1958). The former form the bulk of the carotid body, being large epitheloid cells with a large spherical vesicular nucleus containing a marked nucleolus. These cells are argyrophil and nonsyncytial. It has com-

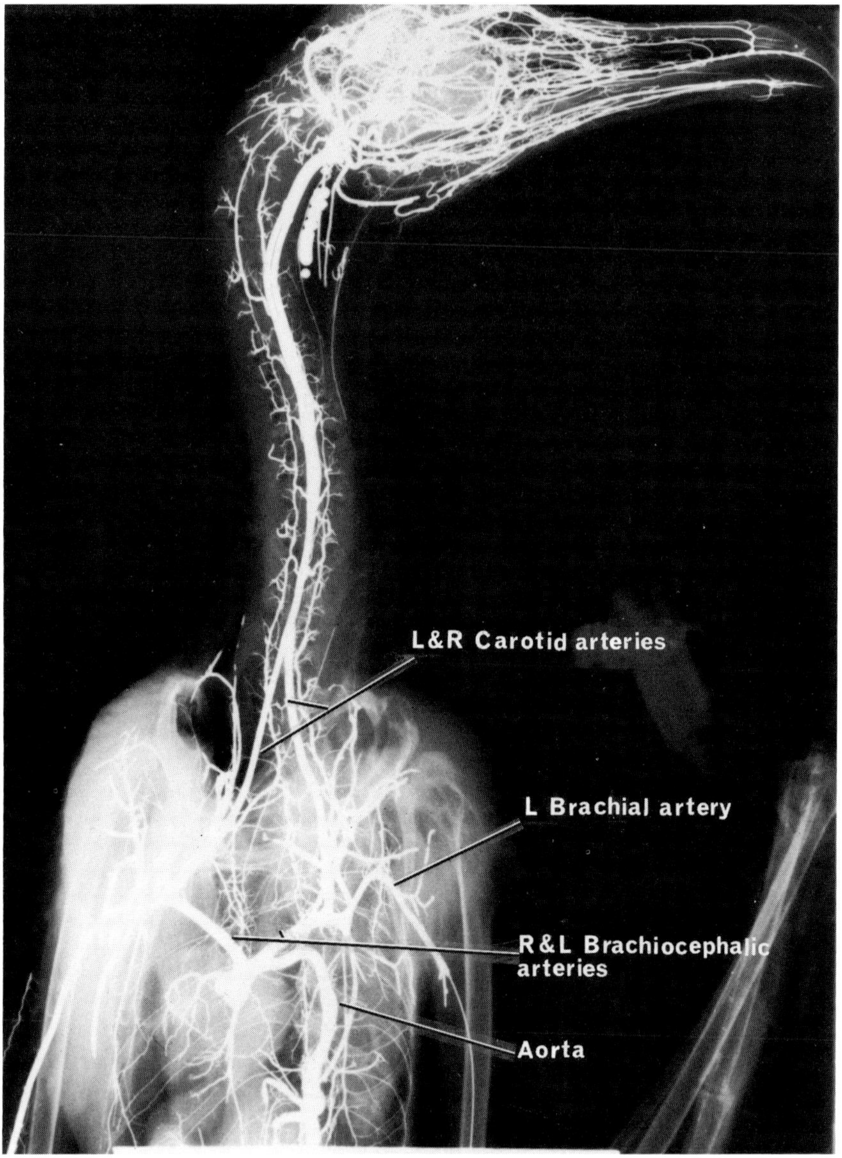

FIG. 18. X-ray photograph showing the major arteries in the head, neck, and chest of a gull (*Larus*). The arteries were injected with mercury.

monly been supposed that these cells are the actual chemoreceptive sites. In addition to these cells, a second less numerous cell type is present (Type II). Type II cells are smaller epitheloid cells with a denser nongranular nucleus. It seems probable that the Type II cells surround those of Type I whether they occur singly or in groups. Both Type I and II cells are innervated.

In the adult bird, blood is supplied to the wing by the brachial artery (Fig. 18). During development the primitive wing bud is supplied by the subclavian artery. The subclavian artery first develops as a branch of the aorta, but in all adults is replaced by a secondary artery that arises from the third arch and is eventually brought to wing level by backward migration of the aortic arches. The other division of the subclavian artery is the pectoral artery, which carries blood to the large pectoral muscle mass. In the adult bird only the right aortic arch persists, although in some species a remnant of the left aortic arch may remain as a solid core of cells (Glenny, 1943) and in a few species, e.g., the Belted Kingfisher *(Ceryle alcyon)*, the left arch may remain patent and functional, though it loses its connection with the root of the aorta (Fig. 19) (Glenny, 1940).

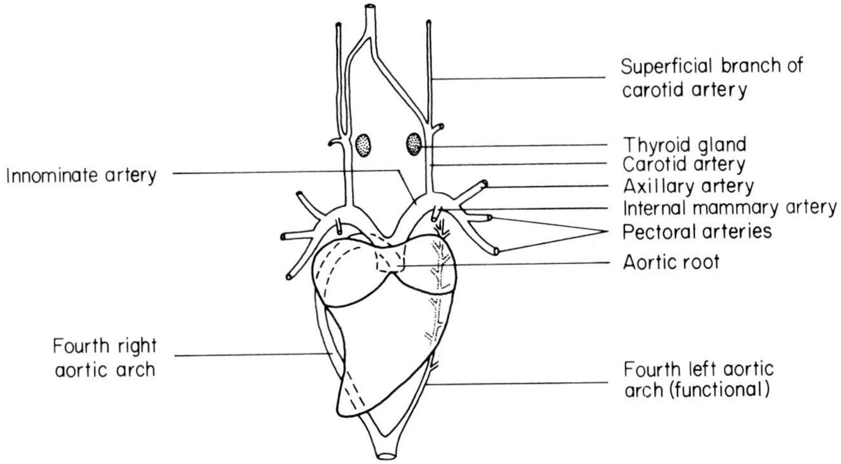

FIG. 19. Ventral view of the main arteries in the region of the heart in the Belted Kingfisher (*Ceryle alcyon*) (after Glenny, 1940).

From the area of the right bronchus the aorta runs caudally ventral to the vertebral column, giving off paired intercostal and lumbar arteries corresponding to the intervals between vertebrae. Blood is supplied to organs within the abdomen and to the legs by the following vessels:

1. Celiac artery	Liver, spleen, glandular stomach, gizzard, intestine, pancreas
2. Cranial mesenteric artery	Most of intestine, pancreas
3. Renal arteries	Kidneys (anterior portion)
4. Femoral arteries	Legs
5. Sciatic arteries	Middle and posterior portions of kidney and legs
6. Caudal mesenteric artery	Rectum and cloaca
7. Hypogastric (iliac) arteries	Walls of pelvis
8. Caudal artery	Tail-termination of aorta

In effect there are three pairs of renal arteries, one from the aorta and two from the sciatic arteries (Siller and Hindle, 1969). Barkow (1829) reported this arrangement in 21 species of birds except for *Ardea cinerea,* in which one pair arises as branches of the femoral artery. The sciatic artery is the major vessel supplying the leg. At the level of the knee it meets and joins the femoral artery to form the popliteal artery. The artery passing into the lower leg divides to form the anterior and posterior tibial arteries. In the tarsal or axillary region of many birds there is an arterio-venous network of vessels referred to as a *rete mirabile* (particularly prominent in wading and aquatic birds). This structure serves as a heat exchanger, since warm arterial blood is brought into close proximity to venous blood that has traversed the foot and is therefore colder (Kahl, 1963; Steen and Steen, 1965). Heat is transferred from arterial to venous blood, thereby reducing heat loss to the environment through peripheral thinner sections like the web of the foot.

Other *retia mirabilia* occur in birds, a notable one being the *rete mirabile ophthalmicum,* which is pear-shaped in the chicken and located in the temporal depression on the left side of the skull (Figs. 13 and 20). Blood probably enters the *rete* along its posterior margin and is collected in vessels that continue anteriorly. In the chicken and pigeon the external ophthalmic artery is the main contributing vessel, but in the Common Ground-Hornbill (*Bucorvus leadbeateri*) the *rete* is formed from branches of the external carotid artery (Wingstrand, 1951; Richards, 1967; Lucas, 1970). A venous *rete* lies directly medial to the arterial *rete* and some intermingling of arterial and venous vessels exists (Fig. 13) (Richards, 1968).

The cranial mesenteric artery is particularly interesting in that a discrete coat of longitudinal muscle is present external to the circular medial muscle (Ball *et al.*, 1963). Bolton (1968a,b) has demonstrated that in the chicken the longitudinal muscle is innervated by excitatory cholinergic and by inhibitory adrenergic nerves, is spontaneously active, and responds to neurohumoral stimuli by large changes in length. C. Bell (1969) points out that it therefore resembles gastrointestinal smooth muscle rather than vascular smooth muscle. The circular muscle, however, resembles vascular smooth muscle and is innervated by adrenergic vasoconstrictor nerves (C. Bell, 1969).

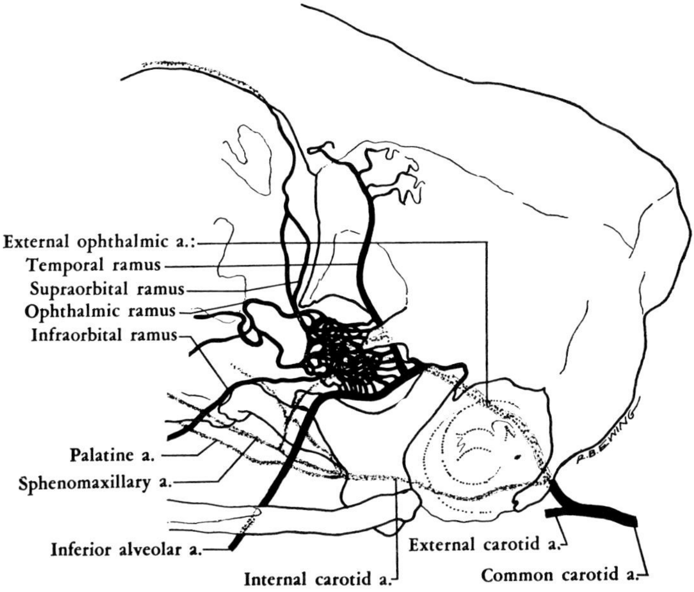

FIG. 20. Arterial rete mirabile ophthalmicum of the chicken. Located in the temporal depression on the left side of the skull. It is assumed that blood enters the arterial rete along its posterior (right) margin and is recollected in vessels that continue anteriorly, but no one has as yet experimentally demonstrated the direction of flow. (Lucas, 1970, *Fed. Proc., Fed. Amer. Soc. Exp. Biol.*, by permission.)

Contraction of the longitudinal muscle has little effect on the resting perfusion pressure in the isolated artery but potentiates responses of perfusion pressure to vasoconstrictor stimuli directly affecting the circular muscle. C. Bell (1969) suggests that the function of the longitudinal muscle is to modulate the response of the artery to vasomotor stimuli and, in view of the fact that the cranial mesenteric artery constitutes the blood supply of virtually all of the intestine, its involvement in restriction of visceral blood flow during sudden stress is indicated.

Nerve plexuses of both sympathetic and myelinated fibers invest the arteries, being located in the *tunica adventitia* and *tunica media*. Ábrahám (1969) reports that there are no nerves of any kind in the *tunica intima.* Bennett and Malmfors (1970) state that the adrenergic innervation of both the heart and arteries of birds is comparable to that of mammals. In fowl, at least, the *tunica adventitia* is very thick and contains smooth muscle fibers. The *tunica media* has a thick elastic layer on its outer border (Kaupp, 1918). The major arteries including the aorta down to its junction with the celiac artery are composed mainly of elastic tissue in the fowl (Fox, 1939). The number of elastic fibers increases during the first 90 days of life and then decreases nonlinearly up to 36 months (Fahr, 1935; Buddecke, 1958; Martins *et al.*, 1970). In the thoracic aorta the amount of collagen increases linearly with age, while muscular tissue stops increasing after about 6–8 months. The cranial mesenteric artery possesses a well-developed outer longitudinal muscle coat (Ball *et al.*, 1963; Bolton, 1968a; Bennett and Malmfors, 1970).

The composition of the vascular wall may show adaptive change to differences in behavior. For instance, the overall size of the iliac artery and vein is much greater in domestic cocks than in cocks trained for fighting, whereas the *tunica adventitia* and *media* are much thicker in the gamecock than the domestic bird (Steeves and Siegel, 1968). The latter probably serves as a protection against aortic rupture in gamecocks during fighting when both heart rate and blood pressure will be high.

C. The Venous System

As with arteries, the veins embryologically first form in a pattern typical of all vertebrates. Formation of cardinal veins in the chick embryo follows shortly after the aortae start to differentiate. With the appearance of ducts of Cuvier at the 14-somite stage the cardinal vein can be differentiated into anterior and posterior divisions. The posterior cardinal veins and ducts of Cuvier do not persist in their original form. The posterior cardinals are gradually transformed into a complex system of vessels that finally fuse in the midline to form the posterior vena cava (postcaval vein). The ducts of Cuvier are eventually incorporated into the anterior venae cavae (precaval veins). The anterior cardinal veins, which elongate as the heart shifts backwards, become the jugular veins.

In the adult bird, on each side, veins issuing from the cranial cavity and orbital veins join to form the jugular vein, which, after receiving a ventral external jugular, joins with the more caudad subclavian vein to become the anterior vena cava (precaval vein). Both the *vena capitis medialis* and *vena capitis lateralis* contribute to the adult jugular vein in all craniates except birds, where that part of the jugular vein derived from the *vena capitis lateralis* is replaced by a secondary outer vein. The jugular veins are unevenly developed on the right and left sides of the neck, the right one usually being the larger. In the anterior region of the neck just behind the angle of the jaw there is an interjugular anastomosis. This provides a bypass system for blood from one jugular vein into the other should one become compressed during neck movement.

Blood from the wing and shoulder region drains into the brachial vein and its branches. The brachial vein unites with the pectoral vein receiving blood from the major and minor flight muscles to form the subclavian vein, a branch of which passes forward to drain the sternum and coracoids. The subclavian vein joins with the jugular to form the anterior vena cava. The subclavian vein first appears in the embryo as a branch of the posterior cardinal vein.

Blood from the tail region is drained by the caudal vein while the femoral and sciatic veins drain blood from the legs. The caudal vein unites with the coccygeo-mesenteric and posterior mesenteric vein (draining the posterior part of the rectum) and immediately divides into a pair of hypogastrics which run into the parenchyma of the kidneys on either side (Fig. 21). Before entering the kidney, the hypogastric vein receives the sacral vein (internal iliac vein) which conducts blood from the pelvic cavity. After traversing the parenchyma

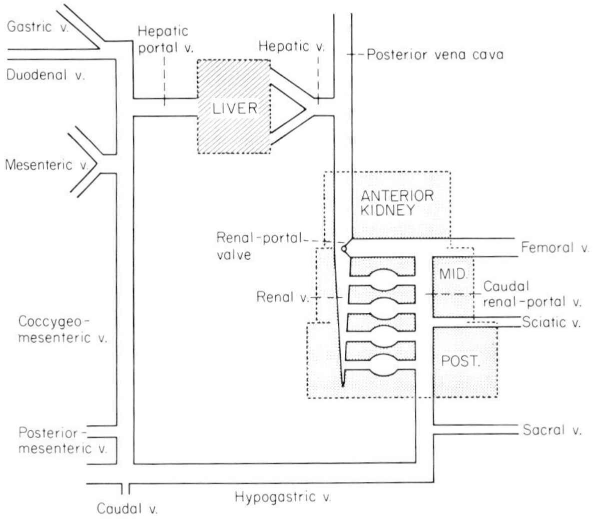

FIG. 21. Schematic diagram of the venous system of the avian body cavity and sacral region.

of the kidneys, the hypogastrics emerge from the anterior ends between the gonads and unite to form the posterior vena cava. The hypogastric vein on its course through the kidney receives the sciatic vein at the level of the middle and posterior kidney lobes and the femoral vein between the anterior and middle kidney lobes. The hypogastric vein on its passage through the kidney is often referred to as the caudal renal portal vein and it maintains uniform diameter and may act as a major shunt of the middle and caudal lobes of the kidney. A smaller cranial renal portal vein connects with a vertebral venous sinus and the anterior vena cava (Akester, 1967). In the anterior region of the kidney there is a large vein lying over its surface, the renal vein, which drains the kidney and unites with the hypogastric vein at the anterior end of the kidney.

An alternative pathway for blood flow from the hind regions of the body is provided by the coccygeo-mesenteric vein (Fig. 21). This vein is unique to birds and arises near the bifurcation of the hypogastrics. Passing forward in the rectal mesentery it receives a number of branches from the posterior end of the small intestine, rectum, and cloaca. Anteriorly, it connects with a portal system of the liver.

The coccygeo-mesenteric vein is very large, out of all proportion to the area of hind gut it serves and the mesentery in which it lies (Akester, 1967), and provides a direct link between the hepatic and renal systems. The question is, does it provide a link by which potentially renal blood could be diverted to the liver or potentially hepatic portal blood could be diverted to the kidneys? Purton (1970), employing a radio-opaque injection technique, has shown that potentially hepatic portal blood may in fact be distributed three ways, if injections are made into the mesenteric vein. These are (1) entire flow to all lobes of liver, (2) half of flow to kidney by way of coccygeo-mesenteric vein and half to the liver, and (3) entire flow to kidney.

Two hepatic veins leave the liver, the larger one being the right one, which receives blood from the mesenteric, pancreatic, and coccygeo-mesenteric veins. The left hepatic vein is supplied by the gastric vein. The hepatic veins join the posterior vena cava. In the embryo the left hepatic vein also receives the umbilical vein (the right disappears early in development) carrying oxygen-rich blood from the chorioallantois. The entrance of the umbilical vein into the hepatic vein is guarded by a netlike structure of strands of connective tissue; at the junctions of the strands swellings appear (White, 1969). It seems that a possible function of these strands is to proliferate and thereby cut off the large orifice of the umbilical vein at the time of hatching, when its main function is lost. It persists in the adult as a small vein draining blood from the ventral body wall (White, 1969).

The veins in the region of the heart are frequently invested with cardiac muscle. Smooth muscle in the great veins is richly innervated with adrenergic nerve fibers, much more so than in mammals, which suggests that the veins in birds may play a more active role in the regional distribution of blood and in venous return (Bennett and Malmfors, 1970). In the anterior venae cavae the nerve fibers run parallel to the muscle fibers in the circular and longitudinal coats and, therefore, form an angular meshwork (Bennett and Malmfors, 1970). The density of adrenergic innervation of the veins draining the gut is much less than that of the corresponding arteries. Unmyelinated nerves are associated with the larger intrahepatic venous and arterial vessels, and large bundles of myelinated nerves occur in the connective tissue sheath of the hepatic portal tract. The vessels of the portal tract receive both adrenergic and cholinergic nerves (Purton, 1970).

D. The Pulmonary Vascular Bed

Parts of the pulmonary arteries are present before the actual pulmonary arches form in the embryo of the chick. However, the two link up and become patent at the 60-hour stage in domestic fowl, when the arteries can be traced to the lung buds where they form anastomoses with the pulmonary veins. By the end of the seventh day, a septum continuous with the interventricular septum has divided the truncus arteriosus into aortic and pulmonary trunks in such a way that the latter opens from the right ventricle. The pulmonary trunks do not stop at the pulmonary arteries but continue as two large ducti arteriosi that join the systemic aorta in a characteristic manner (Fig. 22). The ducti arteriosi provide a pulmonary bypass channel for the nonfunctional lungs. The ducti arteriosi join the systemic arch ventrally and on its left side, while the celiac and omphalomesenteric arteries arise from the right dorso-lateral side of the newly formed aorta. Assuming some degree of laminar flow, it is possible that the celiac and omphalomesenteric arteries receive some blood directly from the left ventricle by way of the systemic arch (Fig. 22; White, 1969). Studies by Grodzínski (1963) on flow pattern in the chick embryo have demonstrated distinctly laminar flow in the vessels he studied. Furthermore, Reynolds numbers calculated for the major arterial vessels are low (White, 1969). The critical Reynolds number for blood, marking transition from laminar to turbulent flow, appears to be between 1000–2000 (McDonald, 1960). White (1969) calculated the following Reynolds numbers for the chick embryo: right and left ductus arteriosus = 37; the aortic arch = 23; dorsal aorta = 172. Consequently, laminar flow patterns can be expected to exist both in the upstream and downstream parts of the junction. Furthermore, this junction is unusual in that vessels of low Reynolds number combine to form a vessel of higher Reynolds number, whereas the reverse case is the more general situation.

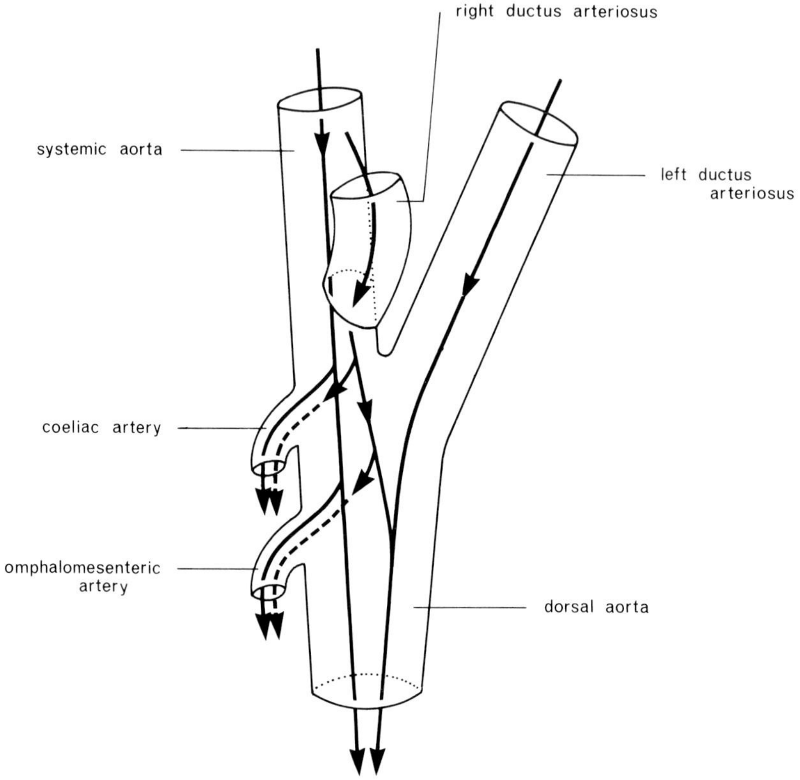

FIG. 22. Junction of the ducti arteriosi and systemic aorta in the chick embryo, showing the possible flow pattern (after White, 1969).

The entire ductus arteriosus becomes a solid core of tissue shortly after the onset of pulmonary respiration. According to Hughes (1943), contraction of the lumen and the resulting obliteration of the ductus begins proximally and progresses distally. Harms (1967) reports a sphincterlike structure formed by the muscle cells of the media at the proximal end of the ductus arteriosus in the chicken. During hatching the ducti arteriosi are closed by contraction of the musculature of the media. The process of occlusion is suggested to be homologous with that in reptiles and mammals. No nerves or presence of cholinesterases have been detected in the wall of the ducti arteriosi, nor is there any smooth muscle response from the ductus in response to application of acetylcholine or L-norepinephrine. It is assumed, in accordance with conditions in mammals, that changes in O_2 tension provide the stimulus that induces the smooth muscle cells to contract and cause the permanent obliteration of the ducti arteriosi. After obliteration of the ducti arteriosi and initiation of pulmonary breathing, all blood passes through the now-enlarged pulmonary arteries, which, after traversing the capillary network of the lung, is collected in the pulmonary veins. The opening of the common pulmonary vein into the sinus venosus is located to the left of the opening of the sinus venosus into the primitive atrium of the heart. The interatrial septum is formed between the sinoatrial opening and the mouth of the pulmonary vein, so that the latter comes to empty into the left atrium.

In the neopulmo the small arteries and veins that supply a respiratory unit run approximately parallel with the parabronchi, whereas in the adult, the arteries tend to cross the parabronchi to supply only restricted regions of the long parabronchi. The blood supply to the capillaries, which surround the air capillaries, comes from smaller vessels that branch off the main arterial trunks (Zeuthen, 1942). The pulmonary artery is much thinner walled than systemic arteries and, as in the mammal, has an outer longitudinal muscle coat. The pulmonary artery is not as densely innervated with adrenergic endings as the pulmonary vein (Bennett and Malmfors, 1970).

E. The Lymphatic System

The lymphatic vessels serve to transport the bulk of the capillary filtrate back into the blood vascular system. The fluid contained in the lymphatic vessels is referred to as lymph. Lymph from the entire body is returned to the blood in the anterior venae cavae. The smallest lymphatic vessels are about the size of capillaries and end blindly, but the largest lymphatic vessels are smaller than the major blood vessels.

The cervical lymphatic vessels are paired trunks and become continuous caudally with the thoracic ducts at the level of the anterior venae cavae (Sala, 1900). The thoracic ducts are the largest lymphatic vessels, and where they join the anterior venae cavae a valve prevents entrance of blood into the lymph channels. The thoracic ducts receive lymphatic vessels draining lymph from the head, neck, lungs, and wings. Caudally, the thoracic ducts are continuous with a similar pair of large para-aortic lymphatic trunks that extend to the legs and collect lymph, through a network of lymph vessels, from the caudal region

(Sala, 1900). At the level of the celiac artery a lymphatic plexus is present.

Lymph glands, which produce the lymphocytes, are present in only a few birds. When present, there are usually two pairs or, less commonly, three pairs. Lymph glands seem to be restricted to certain species of aquatic and wading birds (Jolly, 1910). Jolly (1910) reported three pairs of lymph glands in the goose and swan. These are (1) cervical glands, derived from the cervical lymph sac; (2) lumbar glands, on the para-aortic lymphatic trunks near the femoral artery; and (3) thoracic glands close to the cervical glands. Nodules of lymphatic tissue are scattered throughout various organs of birds, and this probably accounts for the scarcity of definite glandular structures in most avian species.

A few species of adult birds also posses pulsating lymph hearts. The walls of these small oval organs contain muscle, and contractions assist the flow of lymph. The presence of a pair of posterior lymph hearts has been recorded in such diverse groups of birds as the ostrich and swan. Lymph hearts are present during the embryonic life of many birds.

IV. The Electrophysiology of the Heart

In the absence of extrinsic influences the heart rate is set by the spontaneous discharge of a group of specialized muscle cells (pacemaker) in the sinu-atrial node. Cells that are able to function as pacemakers show a characteristic slow depolarization during diastole, the steepness of the depolarization being related to the degree of automaticity inherent in the cell (the fastest cells to depolarize drive the slower), whereas cells not spontaneously active show a steady membrane potential during diastole (Fig. 23). The electrical impulse initiated in the sinu-atrial node travels across the heart. The velocity of conduction of the impulse is about 0.4 to 0.5 m/second for both atria and ventricles.

In the adult heart the propagation of the exciting impulse is carried out by both specialized fibers (Purkinje fibers) and ordinary muscle fibers. The exciting impulse spreads from the sinu-atrial node and is carried by a specialized system of Purkinje fibers through to the ordinary atrial muscle (conduction may be along pectinate muscles; Fig. 23). In turn, the impulse is relayed to the atrio-ventricular node located in the interatrial septum. At the atrio-ventricular node a conduction delay occurs, with conduction velocities slowing to about 0.003 to 0.005 m/second. From the atrio-ventricular node the impulse is rapidly transmitted to the ventricular muscle cells through the specialized conducting system. The Purkinje fiber system of the avian heart is extremely extensive, much more so than in the mammalian heart, and the fibers of the avian atrio-ventricular bundle of conducting tissue lack a fibrous sheath. Both of these factors may be correlated with the rapid heart rate of birds (Davies, 1930).

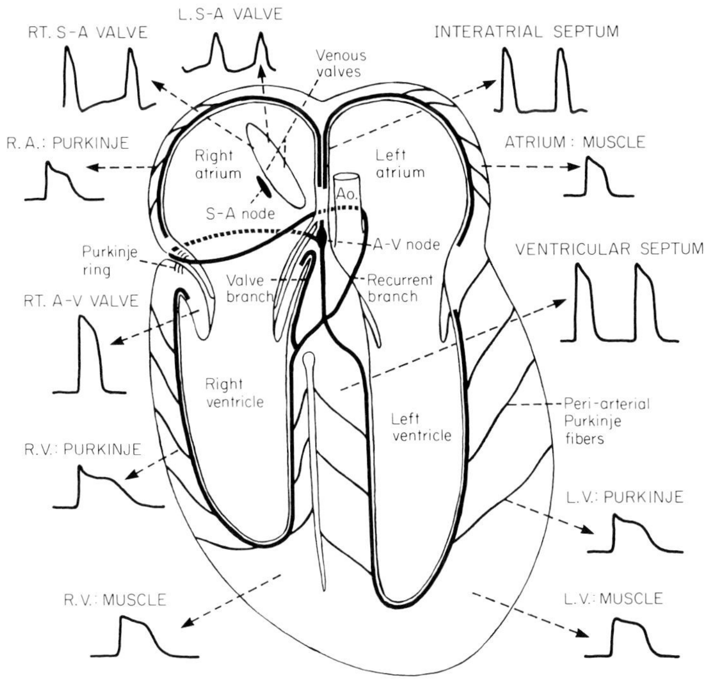

FIG. 23. The Purkinje system of the bird heart (after Davies, 1930) and transmembrane action potentials from chicken and turkey (after Moore, 1965).

The Purkinje fibers form a network on the endocardial surface of both atria and ventricles (Sommer and Johnson, 1969) but some bundles, tracts, or cells may also be found in the myocardium of the left atrium, atrial septum, and ventricles, running mainly along the arteries (Abrahám, 1969). The stimulus-conducting system is poorly supplied with nerves and appears to be independent of the nervous system. Purkinje fibers consist of uniformly thick brick-shaped muscle fibers, containing few myofibrils, comprising as a rule two closely situated nuclei and conjugating with similar muscle elements in various ways (Sommer and Johnson, 1969; Ábrahám, 1969).

The cardiac muscle cells of the chicken are discrete entities but they behave electrophysiologically as a syncytium. This important property could result from the agency of a chemical transmitter or by a particularly low electrical resistance in the parts of the cell membranes where cell apposition is very close (Van Breemen, 1953). Junctional complexes like intercalated discs are relatively common in the chicken myocardium and occur at right angles to the long axis of the myofibrils (Sommer and Johnson, 1969). However, connections along the longitudinal axis (nexuses), joining cells laterally, are few (Sommer and Johnson, 1969).

The Purkinje fibers follow the coronary arteries and, therefore, take a relatively short course through the thick left myocardium. This probably accounts for the rapidity of arrival of the wave of excitation at a given point on the surface of the left ventricular wall in the avian heart (Lewis, 1915). The sequence of depolarization is, according to Kisch (1951), right ventricle apex, right ventricle base, left ventricle base, left ventricle apex. Moore (1965) has mapped epicardial activation and suggests that in the turkey the apical third of the right ventricular epicardium is activated earliest, the upper basilar third is intermediate, and the pulmonary infundibulum region is the last region activated in the whole heart. The anterior one-third of the septal region and the middle region of the left ventricle are activated before the basilar regions, the whole left ventricular epicardium being activated in 12.5 msecond. Lewis (1915) and Mangold (1919) suggest somewhat different sequences. Kisch's suggestion that the conductive system must stimulate the heart muscle only at the places of direct contact of its terminal fibers with heart muscle and not along its entire course receives support from his own work (Kisch, 1951) showing that sub-endocardial muscle is activated about 0.02 second later than the earliest activated sub-epicardial muscle, which in turn suggests short cuts of the conductive system to sub-epicardial muscle. Davies (1930), however, attributes the absence of a fibrous sheath around the avian atrio-ventricular bundle as a requisite of a high heart rate since it allows early and widespread propagation of the impulse from the atria along the bundle to all parts of the ventricles. These suggestions concerning the conduction network are mutually exclusive, and clarification of the problem must await further experimental work.

An indication of the reptilian ancestry of birds can be surmised from the conducting system of the heart. The birds, unlike mammals, possess an atrio-ventricular ring of Purkinje fibers on the right side of the heart which runs up and around the right atrio-ventricular valve (Fig. 24). Davies (1930) suggested that the right valve actively contracts early in ventricular systole so that no reflux of blood occurs before its closure.

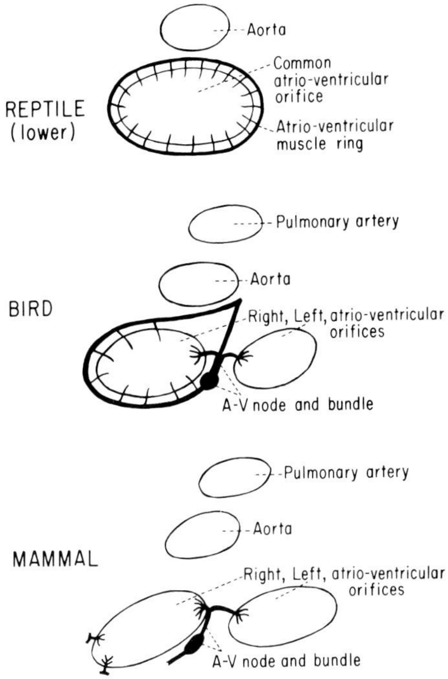

FIG. 24. Diagram of specialized atrio–ventricular connections in lower reptiles, birds, and mammals (after Davies, 1930).

More recently, the conducting systems of the bird heart have been investigated by recording transmembrane potentials from cells in the heart of the chicken and turkey (Moore, 1965, 1967). In the chicken the pacemaker is located in the region where the left sinu-atrial valve and the sinus venosus join together. Action potentials recorded from this junction show diastolic depolarization (prepotentials) with a slow transition to the ascending phase of the actual action potential (Fig. 23), in contrast to the relatively more rapid rise of the action potential recorded in the right sinu-atrial valve itself, indicating that these cells are triggered by other pacemaker cells. The duration of action potentials recorded from ventricular muscle cells is longer than those of atrial muscle cells (Fig. 23). Purkinje fibers display a prominent sharp peak of their action potential, which is followed by a distinct plateau, a feature not recorded from atrial or ventricular muscle cells. The duration of depolarization is also much longer in Purkinje fibers although diastolic depolarization (prepotentials) has not been recorded from avian Purkinje fibers. The longer duration of the Purkinje action potentials as compared with ventricular-muscle action potentials would tend to prevent extrasystole and possible fibrillation by assuring a concerted depolarization of the ventricular muscle (Moore, 1965).

The integrated propagation of the action potential through the heart is expressed by the electrocardiogram. There are three main components of the avain electrocardiogram: a wave associated with atrial systole, a wave associated with ventricular systole, and a wave associated with ventricular repolarization. The time between the atrial and ventricular waves represents the conduction delay at the atrio-ventricular node, and the duration of the ventricular wave indicates the time of ventricular activation and repolarization. These waves, following mammalian terminology, are often referred to as the P, QRS, and T waves, respectively. This is convenient shorthand but has its dangers. For instance, Kisch (1951) reported the presence of P, QRS, and T waves, whereas Mangold (1919) reported that the electrocardiogram of birds has no R component, but instead a deep S wave. Sturkie (1965) reported the presence of P, S, and T waves, a small R, but no Q wave. Interpretation of electrocardiograms is generally difficult since the shape of the waves obviously varies with the position of the recording electrodes and their relationship to the electrical axes of the heart (Hamlin *et al.*, 1969). Sturkie (1965) gives a fairly concise account of electrocardiography in birds, and interested readers are referred to this source.

During embryonic development, long before the specialized conducting systems of the heart come into existence, the electrocardiogram shows great similarity with its definitive adult configuration (Romanoff, 1960). It has been found that the adult type of the electrocardiogram is approximated by the beginning or end of the third day of development in the domestic fowl.

The electrophysiology of the embryonic heart is similar to that of the adult (compare Figs. 23 and 25; Lieberman and Paes de Carvalho, 1965b). Figure 25 also illustrates conduction delay within the embryonic heart. Since the P–R interval is established in the early electrocardiogram (3 to 4 days), before the development of the ventricular node, then conduction delay must be caused by some other agency. By microelectrode studies Lieberman and Paes de Carvalho (1965a,b) have elegantly shown in domestic fowl that almost all the atrio-ventricular delay is localized in a narrow band of tissue that extends for approximately 0.2 mm along the entire atrio-ventricular ring. The conduction velocities in the tissue are about 0.003 to 0.005 m/second, compared with 0.4 to 0.5 m/second for the atria and ventricles.

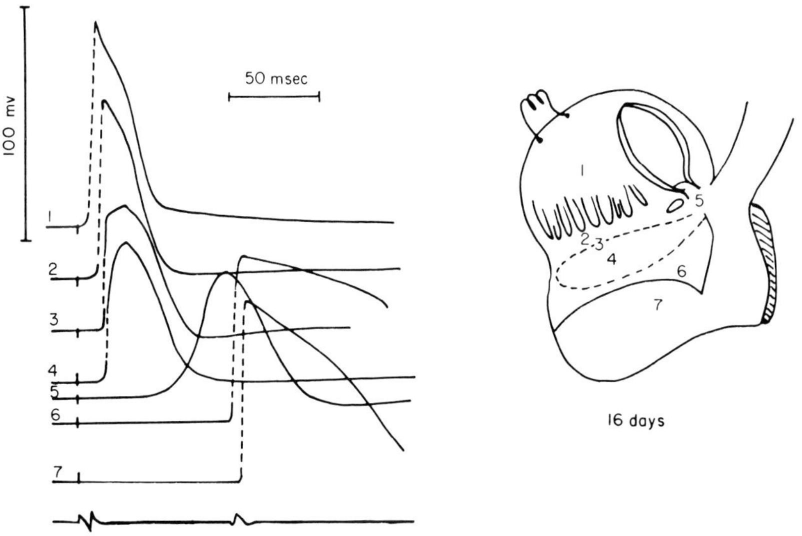

FIG. 25. Tracings of transmembrane action potentials and bipolar electrogram (bottom trace) recorded from the heart of a 16-day-old chicken embryo. (Lieberman and Paes de Carvalho, 1965b, *J. Gen. Physiol.*, by permission.)

Patten (1956) suggests that the atrio-ventricular node starts its development as a counterpart of the left sinus horn, but atrio-ventricular delay occurs before migration of the left sinus horn. However, Lieberman and Paes de Carvalho (1965a,b) suggest that since the atrio-ventricular ring tissue is responsible for delaying atrio-ventricular transmission, then the adult atrio-ventricular node may be a remnant of the embryonic atrio-ventricular ring.

V. General Hemodynamics

The pioneering mathematical analyses of hemodynamics by Womersley (1957) and McDonald (1960) have aroused great interest in modelling the blood-vascular system, but unfortunately little work has been done with birds in this respect. Taylor (1964) considered that a simple Windkessel model may be usefully applied to the arterial system of birds (Fig. 26). The Windkessel theory (Frank, 1899) attempts to establish a relationship between the stroke volume of the heart, the arterial pressure pulse, and the elastic properties of the aorta. The Windkessel represents a lumped system in which pulse-wave velocity is infinite and pressure or flow changes occur simultaneously throughout the system. Some of the consequences and conclusions from examination of such a model have been elegantly described by Taylor (1964), and it will be interesting to bear in mind some of these conclusions when reviewing avian cardiovascular dynamics.

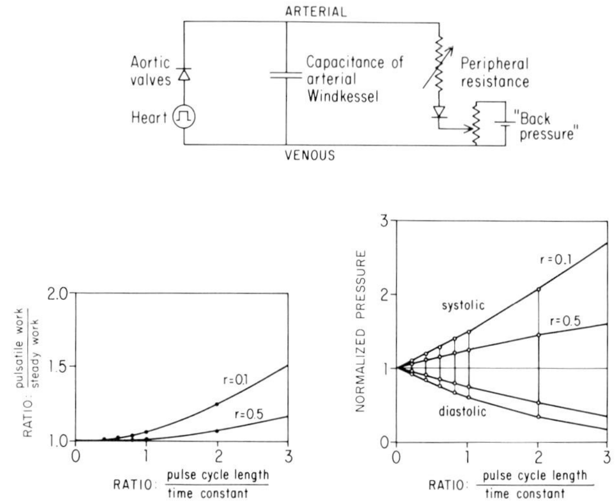

FIG. 26. Upper: a simple Windkessel model of the avian cardiovascular system. Lower left: ratio of pulsatile work to steady work for flow into the Windkessel at two values of r (ratio of the duration of the systolic ejection period to the duration of the cardiac cycle). Lower right: systolic, diastolic, mean, and pulse pressures in the Windkessel at two values of r. The systolic and diastolic pressures were normalized by dividing by the mean pressure (after Taylor, 1964).

A. The Cardiac Output

The cardiac output is the product of the heart rate and stroke volume for a single ventricle. Since the circulation is completely divided, the same amount of blood per unit time will flow, in the steady state, into both the pulmonary and systemic circulations. In other words, the total output of the heart will be twice the cardiac output.

By any standards the heart rate of birds is high, and as a consequence the pulse cycle is short. However, it would appear that resting avian heart rates are lower than those of mammals of comparable body weight; for mammals, the heart rate (beats/minute) = $1355W^{-0.25}$, whereas for birds the heart rate = $763W^{-0.23}$, when body weight (W) is given in grams (Calder, 1968). Normal values for resting birds (e.g., duck, goose, turkey, pigeon, and chicken) are in the range of 150–350 beats/minute, although these rates increase considerably during excitement of stress (Hamlin and Kondrich, 1969). For instance, maximum heart rate recorded by Lasiewski *et al.* (1967) for the Giant Hummingbird *(Patagona gigas)* was 1020 beats/minute.

In view of the high heart rates and large heart size of many birds it can be predicted that cardiac output will be high. Unfortunately, measurements of stroke volume are few and mostly indirect, the majority being done by means of the indicator dilution technique. Stroke volumes for birds, on a per kilogram basis, seem to lie in the range of 0.5 to 3 ml. Cardiac output for the chicken has been reported from 128 to 270 ml/kg·minute (Sapirstein and Hartman, 1959; Sturkie and Vogel, 1959; Butler, 1967a; Piiper *et al.*, 1970). A large variation exists in reported values for cardiac output in ducks. Sturkie (1966) records values in the range of 260 ml/kg·minute, whereas Folkow *et al.* (1967) report values of 560 ml/kg·minute. Jones and Holeton (1972a,b), using both direct and indirect methods, found values in the range of 200–400 ml/kg·minute in resting ducks. The only other available measurements of cardiac output in birds are those for the turkey by Speckmann and Ringer (1963) and Hamlin and Kondrich (1969). The former authors report values of about 111 ml/kg·minute, whereas the latter recorded values of 200 ml/kg·minute.

From available data it appears that the cardiac output of ducks is considerably higher than that of chickens and turkeys. The high cardiac outputs that Folkow *et al.* (1967) have recorded are accompanied by oxygen extraction by the tissues of 3 to 5 ml of oxygen per 100 ml of arterial blood. Consequently, these authors attributed the high cardiac outputs to the high oxygen consumption of ducks, although Butler (1970) has shown that oxygen uptake of ducks is only 80% of the value recorded for chickens of similar weight. Comparisons of cardiac output values in birds can be done only if the animals are in similar thermal states, since marked increases in nonnutritious blood flow to poorly insulated extremities, like the legs, accompany a positive heat load (Djojosugito *et al.*, 1969).

Since blood volumes are seldom more than 10% of the total body weight, the high cardiac outputs must result in short circulation times. Circulation times have been measured in ducks and chickens by the injection of tracers, e.g., acetylcholine (Rodbard and Fink, 1948), heat (Johansen and Aakhus, 1963), and dye (Jones and Holeton, 1972a). For the chicken, injection of acetylcholine into the femoral vein causes a drop in blood pressure after 2–8 seconds, whereas circulation time from the right atrium to the base of the aorta is 2–3 seconds in ducks

(Johansen and Aakhus, 1963). Circulation time (minimal) from the jugular vein to the sciatic artery is about 4–5 seconds in chickens (Henry and Fedde, 1970). Jones and Holeton (1972a) have recorded mean transit times for the pulmonary vascular bed of 7.5 seconds in Muscovy Ducks (*Cairina moschata*), and 4.3 seconds in the White Pekin duck. Henry and Fedde (1970) have recorded average minimal circulation times for the chicken pulmonary vascular bed of 0.6 second, the fastest times for transit from the pulmonary artery to the left ventricle being 0.31 second. Johansen and Aakhus (1963) also deduced circulation times from angio-cardiographic studies of ducks. Contrast first appears in the pulmonary artery one second after injection in the right atrium, representing the duration of three heart beats. The pulmonary capillary phase becomes discernible about 2 seconds after right atrial definition, and within 3 seconds (7–9 heart beats) contrast appears in the aorta.

The cardiac output is distributed to the various organs within the body. In the chicken the percentages of the total cardiac output flowing to the various organs is 4.9% heart, 15.2% kidney, 6.7% liver, 8.6% gut, 1.6% gizzard, 0.4% spleen, 0.4% adrenal, 0.035% thyroid, and 1.08% to the pectoral muscles (Sapirstein and Hartman, 1959). Calculated on a tissue-weight basis, the pectoral muscles receive only one-third of the amount of blood flowing to the legs. The large flow to the legs may be related to heat-dissipation requirements, since naked legs constitute an important radiator surface in birds (see later discussion). Butler and Jones (1971) directly recorded blood flows in the sciatic and carotid arteries of the duck. Total flow to the legs was approximately twice that to the head, and the two flows probably account for 10–20% of the total cardiac output. Total coronary flow in the chicken is about 50% larger than that of a small mammal (Sapirstein and Hartman, 1959).

Subsequent to ventricular filling the heart initially contracts isovolumetrically. No external work is done during this phase, but potential energy is accumulated and stored in the ventricle. The ventricular pressure rises quite sharply until it exceeds the pressure existing in the central arteries. In the chicken this pressure is around 125 mm Hg on the systemic side of the circulation and 12 mm Hg on the pulmonary side (Rodbard *et al.*, 1949). The valves on the outflow tract are then forced open and blood is ejected into the circulation. Ventricular pressures continue to rise, albeit more slowly, and eventually reach peak values around 145 mm Hg for the left ventricle and 27 mm Hg for the right ventricle (Bredeck, 1960). During the ejection phase physical work is done, after which the ventricles relax and ventricular pressures fall sharply. The valves on the outflow tracts close, in response to a reversal of the pressure gradient from the ventricles to the outflow arteries, and reflux of blood is prevented. Diastolic pressures are reported to be about 0 mm Hg for the left ventricle and -0.3 to -2 mm Hg in the right ventricle in chickens (Bredeck, 1960). The subatmospheric pressure prevailing in the right ventricle during diastole may assist ventricular filling by aspirating blood contained in the right atrium and pulmonary veins. A suctional attraction *(vis a tergo)* for blood returning to the heart has been reported for most classes of vertebrates, although the underlying mechanisms vary greatly (Johansen, 1965). The lower diastolic pressure in the right ventricle compared with the left may suggest a greater elastic recoil of the right compared with the left during diastole.

The work done by the ventricles in ejecting blood from the heart may be estimated by the product of the mean arterial pressure and stroke volume. A rigorous estimate of stroke-work requires integration of the pressure–volume curve for one cardiac cycle. Nevertheless, calculation of stroke-work for the left ventricle of mammals using mean pressure and stroke volume are only about 3% less than values obtained by integration (Topham, 1969). That this approximation will hold for the right ventricle is more questionable.

The power output of the ventricles is the rate at which work is done. Using data from chickens previously cited the following values are obtained:

$$
\begin{aligned}
\text{Power output,}\quad \text{left ventricle} &= (135 \times 1330)\ \text{dynes/cm}^2 \times 200\ \text{ml/minute}\\
&= 35.9 \times 10^6\ \text{ergs/minute}\\
\text{right ventricle} &= (18 \times 1330)\ \text{dynes/cm}^2 \times 200\ \text{ml/minute}\\
&= \underline{4.79 \times 10^6\ \text{ergs/minute}}\\
\text{Total power of heart} &= 40.69 \times 10^6\ \text{ergs/minute}
\end{aligned}
$$

The power output of the left ventricle is about 7.5 times that of the right for the chicken, but in the turkey (Hamlin and Kondrich, 1969), where mean arterial blood pressure is 200 mm Hg, the power output of the left ventricle is ten times that of the right. In fact, right ventricular power outputs are more or less the same in both chickens and turkeys. However, production of this work by the heart also involves the cost of maintaining the tension of the heart muscle even when no physical work is done. The energy cost of tension maintenance is far greater than that of the external work, and consequently the efficiency of the heart in moving blood is low. Assuming an efficiency of 10%, the total oxygen uptake of the chicken heart can be calculated, assuming that the utilization of 1 ml of oxygen releases the calorific equivalent of 2.2×10^8 ergs.

$$\text{Oxygen uptake} = \frac{10 \times \text{power output (ergs/minute)}}{2.2 \times 10^8 \text{ (ergs)}} = 1.86 \text{ ml } O_2\text{/minute}$$

Butler (1967a,b) recorded the oxygen uptake of chicken as 26 ml/kg·minute, so the oxygen cost to the metabolism of cardiac activity is 7% of total oxygen consumption.

The systolic ejection period in birds seems to be quite long, occupying in ducks 25 to 33% of the resting cardiac cycle (Johansen and Aakhus, 1963). As heart rate increases it seems unlikely that the ejection phase will be much shortened, so that ejection will tend to occupy a progressively greater proportion of cardiac cycle. In terms of the model shown in Fig. 26, this has important consequences. Using this model, Taylor (1964) has calculated the ratio of the quantity of work per cycle required to maintain the circulation by intermittent input (pulsatile work) to that quantity required if the flow were continuous (steady work). Figure 26 shows the behavior of this ratio and it can be seen that the efficiency of the system decreases as the ratio of the duration of the pulse cycle to the time constant (product of resistance and capacitance) of the system increases. Furthermore, energy cost also becomes larger at constant cycle length as the ratio of the duration of the systolic ejection period to the duration of the cycle decreases (r). Consequently, it is advantageous to have a high heart rate, with systole occupying a large fraction of the cycle. This state of affairs is most likely to be obtained during periods of maximum load on the heart, so that efficiency must decrease somewhat in the resting animal. However, as Taylor (1964) points out, a somewhat inefficient system at rest is a small price to pay for high efficiency at peak demand.

B. Resistance and Capacitance of the Circulation

Peripheral resistance, which is derived from steady-flow conditions, reflects the viscous resistance to blood flow through the circulation. The resistance of a vascular bed is the ratio of the pressure gradient to the flow. The anatomical site of most of the flow resistance lies in the arterioles and capillaries on the systemic side of the circulation. However, even the largest vessels offer some resistance to blood flow. The total peripheral resistance (TPR) is the resistance offered to the total output of the heart by all of the peripheral vascular beds.

The total peripheral resistance is usually calculated by dividing the mean cardiac output (milliliters/minute) into the mean arterial pressure (mm Hg) minus the mean central venous pressure (mm Hg). The latter quantity is small in birds and is usually taken as zero except in diving birds in which the venous pressure may increase during submergence to 15–20 mm Hg. Consequently,

$$\text{TPR} = \frac{\text{Mean arterial pressure (mm Hg)}}{\text{Cardiac output (ml/minute)}}$$

The units of resistance calculated by this equation are peripheral resistance units (PRU). The absolute units of resistance are dynes-second/cm^5. One PRU is equal to 79,600 dynes-second/cm^5. In order that comparisons can be made between animals of different sizes both cardiac output is expressed on a body-weight basis (kg). On this basis the TPR of ducks is in the range of 0.567–0.62 PRU (Sturkie, 1966) and is about half the value for chicken (TPR = 0.88–1.23 PRU; Sturkie and Vogel, 1959), whereas the TPR of the turkey has been variously reported between 1 and 2.28 PRU (Speckmann and Ringer, 1963; Hamlin and Kondrich, 1969). The total peripheral resistance of the pulmonary vascular bed is about 1/7th to 1/10th of that of the systemic vascular bed.

The peripheral resistance of the various vascular beds within the body is not constant, but changes under conditions of activity or stress. For instance, in the resting duck the resistance of the sciatic vascular bed is about half that of the carotid vascular bed, whereas during a dive the resistance of the carotid vascular bed hardly changes while the resistance of the sciatic bed increases by eight times (Butler and Jones, 1971).

Changes in resistance of a vascular bed can be related to the degree of vasomotor tone if these changes are evaluated at constant pressure or flow. For instance, if blood pressure rises and peripheral resistance decreases then it is difficult to decide whether it is passive or active vasodilatation. Since the transmural pressure (pressure across the wall of the blood vessel) has increased, the peripheral vessels may have become bigger due to their elastic properties (passive vasodilatation). However, in actual cases most changes in peripheral resistance are marked enough to rule out passive changes and, furthermore, in ducks increases in peripheral resistance occur even when blood pressure increases during a dive.

The elasticity of the minor and major vessels in the arterial circulation contribute to the capacitance of the arterial system. As a rule the veins are much more distensible than the arteries, but measurements of volume distensibility of the various parts of the avian circulatory system are few. C. S. Roy (1880) studied the relation between internal pressure and volume capacity of blood vessels of various animals, and concluded that the aortic walls are most distensible at pressures corresponding to the normal blood pressure of the animal and that at higher pressures vascular distensibility is impaired. Speckmann and Ringer (1966) have determined pressure–volume curves for the thoracic and abdominal aorta of the turkey (Fig. 27). The calculated values for 2 cm lengths of thoracic and abdominal aorta were 17.2 ± 0.99 μl/mm Hg and 0.64 ± 0.03 μl/mm Hg, respectively, over the physiological range of blood pressures. In other words, the thoracic aorta is relatively compliant, whereas the abdominal aorta is relatively noncompliant. Consequently, the vessels arising from the thoracic aorta are somewhat rigid structures, which has a marked effect on the wave velocities within the circulation (see later).

Speckmann and Ringer (1964, 1966) working with turkeys concluded that the elastic properties of the wall of the thoracic aorta are determined predominantly by the content of muscle, elastin, and collagen at pressures below, within, and above the physiological range. The low distensibility of the abdominal aorta may mean that the amount of collagen in its walls is greater than that of the thoracic aorta.

An important advantage of an elastic peripheral vascular system as opposed to one composed of rigid tubes is that blood pumped into the central arteries during systole will continue to flow out during diastole. Consequently, the intermittent injection of blood into the central arteries is smoothed to give a fairly steady flow in the peripheral arteries. The whole arterial tree participates in this "Windkessel" or pressure-chamber effect, and not just the aorta. In birds the arterial system is relatively stiff, but this may represent a compromise situation related to the very high heart rates in birds, so that the net result of smoothing flow in the periphery is still achieved. Also, a low arterial capacitance may present an advantage in birds in that slight changes in working pressure will not cause large volumes of blood to shift from the venous side of the circulation.

The product of resistance and capacitance of the vascular system expresses the time constant of the system. Taylor (1964) concludes that a system able to adjust to changing requirements for flow should have a short time constant. Undoubtedly, in the life span of many birds, changes in the time constant of the vascular system occur provoked either by changes in resistance or capacitance. Atherosclerosis, which appears to be a characteristic common in most domestic breeds of bird, may do both although Grollman *et al.* (1963) attribute atherosclerosis and hypertension in the fowl to independent processes.

C. The Blood Pressure

When the aortic valves open, the left ventricle is connected to the arterial tree and in essence the whole systolic portion of the pulse in the major arteries is determined by the cardiac ejection curve. Since

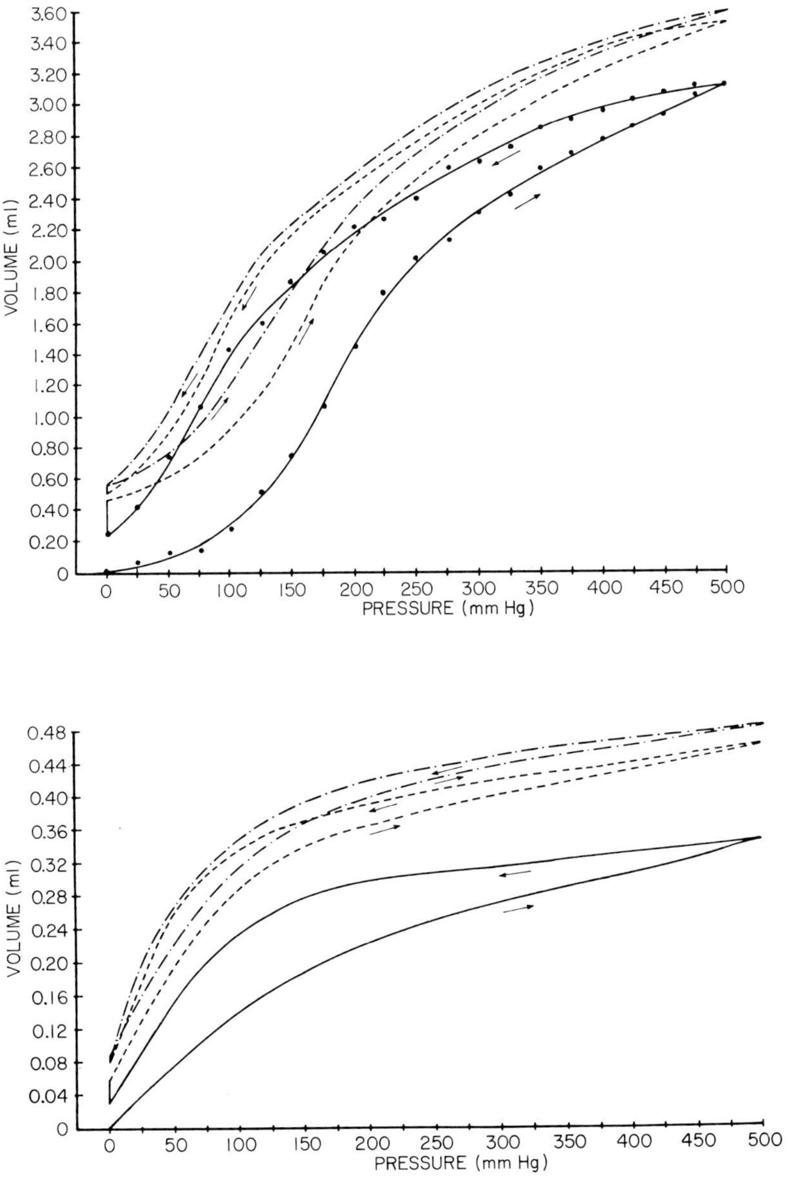

FIG. 27. Upper: three consecutive average volume–pressure curves of a thoracic aorta from a domestic turkey. Lower: three consecutive average volume–pressure curves of an abdominal aorta from a domestic turkey.

The sequence of inflation and deflation is shown by the arrows. Key: first loop (——); second loop (– – –); third loop (– · –). (Speckmann and Ringer, 1966, *Can. J. Physiol. Pharmacol.*, **44,** p. 903, reproduced by permission of the National Research Council of Canada.)

the blood occupying the vascular system has inertia, the major determinants of the cardiac ejection curve are the force of contraction of the heart, the amount of blood forced into the arteries, and their elasticity. It is important to realize that considerable energy is needed not only to set blood in motion, but also to halt the movement of blood already in motion. Following the attainment of peak systolic pressure the central arterial blood pressure declines quite sharply until its fall is interrupted by the incisura, which marks closure of the aortic valves and disconnection of the ventricle from the central arterial system. Arterial blood pressure then declines more gradually until the start of the next ventricular ejection. The minimum pressure reached is the diastolic pressure, and the difference between that and the peak systolic pressure is the pulse pressure.

After the incisura, marking the closure of the aortic valves, the fall in arterial pressure depends on the time constant of the arterial system. The longer the interval between heart beats the lower will be the minimum diastolic level. Taylor (1964) uses the ratio r to describe the duration of the systolic ejection period in relation to the duration of the total cardiac cycle, and finds that for any given value of r, the longer the cycle is in relation to the time constant of the system, the greater will be the fall in diastolic pressure (Fig. 26). If the time constant of the system is fixed, then maintenance of a relatively high diastolic pressure and a low pulse pressure will be achieved with a fast heart rate and with systole occupying a large fraction of the cardiac cycle. Both of these conditions appear to be well met in many birds. For instance, chickens have a very high heart rate (320 beats/minute) and the pulse pressure is only 32 mm Hg (systolic, 157 mm Hg), whereas ducks with about half the heart rate of chickens have a pulse pressure of 56 mm Hg (systolic, 208 mm Hg) (Butler, 1970).

The systemic blood pressure of adult birds is high. Systolic levels may reach 300–400 mm Hg in turkeys (Ringer and Rood, 1959; Speckmann and Ringer, 1963). In both turkeys and chickens, significant sexual differences in blood pressure occur (Sturkie, 1965). Woodbury and Hamilton (1937) and Ringer (1968) have recorded blood pressures for several species of adult birds, and their results substantiate the above generalization. The arterial blood pressure of the turkey is among the highest of all vertebrates. It shows a characteristic increase during the growth period of the bird (Ringer and Rood, 1959; Krista *et al.*, 1965). The high blood pressure and the tendency to formation of atheromatous plaques in both chickens and turkeys (Carnaghan, 1955; Gibson and de Gruchy, 1955; Grollman *et al.*, 1963; Ball, 1964; Krista *et al.*, 1970) are probably the most important underlying causes of death by aortic rupture. Peak systolic pressures in the pulmonary circuit are about one-fifth to one-tenth of those existing in the systemic circuit in chickens, turkeys, and ducks (Butler, 1967a; Hamlin *et al.*, 1969; Jones and Langille, 1972).

A conspicuous feature in the pattern of pressure pulsation found in the mammalian systemic arterial system is a pronounced increase in peak systolic pressure in peripheral compared with central regions ("peaking" of the pressure pulse). In birds the pattern is entirely different, and the arterial system appears to behave as a classical Windkessel since the waveform is virtually unchanged from the central to peripheral arteries (Fig. 28). Consequently, the harmonic components of the pressure pulsation show hardly any evidence of either amplification or reflection of waves (Fig. 28).

The kinetic energy of flow contributes to the total fluid energy in any part of the circulation where blood is in motion. An example shows that under normal conditions the kinetic energy contribution to the systemic arterial energy is less than 913 ergs/ml (equivalent to a pressure of 0.67 mm Hg) if the mean aortic flow velocity is 41.6 cm/second and the density of blood is 1.056 gm/ml (Wirth, 1931). Mean aortic flow velocity quoted above was calculated for a 2 kg duck, assuming a cardiac output of 11.67 ml/second and an aortic root internal diameter of 0.6 cm (Jones, 1970). The example shows quite a small value for the kinetic energy contribution, but during systolic ejection the velocity may be three times the mean, and the kinetic energy value will be nine time as great or 6 mm Hg. Consequently, experimentally measured values of pressure in the ventricle and aortic root may differ as much as this value. Therefore, in the light of these calculations, it would be foolhardy to neglect the kinetic energy contribution to the total fluid energy in either the systemic or pulmonary circulations. Furthermore, when cardiac output increases either during exercise or hypoxia, the kinetic energy term will become correspondingly more important.

In terms of the model shown in Fig. 26, the velocity of the pressure wave should be infinite. This is obviously not so in any real system and the question arises as to whether the velocity of the pressure wave is fast enough so that pressures within all parts of the arterial system can be regarded as changing more or less simultaneously. Pulse-wave velocity (C) can be calculated from measurements of pressure–volume distensibility using the equation

$$C = \sqrt{\frac{V\ \delta P}{\rho\ \delta v}}$$

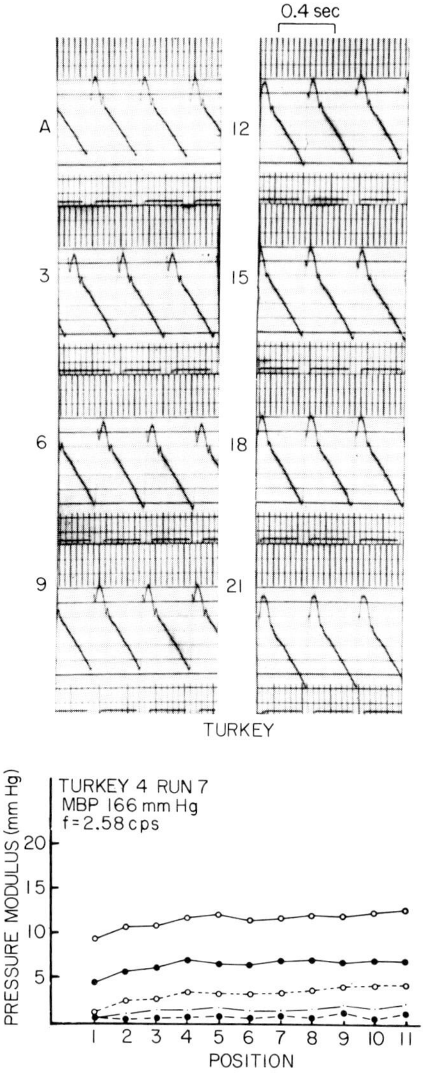

FIG. 28. Upper: Pressure pulses in the aorta and inferior gluteal artery of the domestic turkey, obtained by drawing back a fine nylon catheter from the aortic arch (A) and making recordings at 3 cm intervals. Note that the wave form is almost the same in all positions. Lower: Harmonic analysis of the pressure pulses in the aorta of the domestic turkey. The harmonic components are presented as follows; Top line, fundamental, second, third, fourth, and bottom line, fifth. The interval between positions is 3 cm. [From Taylor, 1964, in "Pulsatile Blood Flow" (E. O. Attinger, ed.). Used with permission of McGraw-Hill Book Company.]

where V is the initial volume, δv and δP are volume and pressure increments, respectively, and ρ is the density of the fluid. It may be calculated from the data of Speckmann and Ringer (1966) that the wave velocity is about 4–6 m/second in the thoracic aorta of the turkey, compared with 30–35 m/second in the abdominal aorta. Although the first value is within the range of velocities quoted for the thoracic aorta of mammals, the value for the abdominal aorta far exceeds that recorded for the dog (8–10 m/second; McDonald, 1960). Consequently, throughout most of the vascular system the pulse-wave velocity appears to be sufficiently high to allow analysis by means of a simple Windkessel model.

D. Blood Flow

The function of the arterial Windkessel is to convert the highly pulsatile flow from the heart into a more or less steady flow at the capillaries. Obviously, the input to both the pulmonary and systemic circulations is highly pulsatile, but the former, a short distance (3–6 cm in the domestic duck) beyond the heart exhibits a steady flow component (Fig. 29b) much different from conditions in the systemic arterial system (Fig. 29a). In the aorta flow increases rapidly during the early part of systolic ejection. After peak blood velocity has been reached the velocity declines (although somewhat slower than it rose), and subsequently a short period of backflow occurs corresponding in time to the closure of the aortic valves. Whereas in the pulmonary artery the closure of the pulmonary valves is marked by an inflexion on the descending limb of the trace, flow continues to decline until interrupted by the next systolic ejection often before it reaches zero flow (Fig. 29b).

Flow at the proximal end of the carotid artery is markedly pulsatile (Fig. 29c), although there is no period of flow reversal associated with closure of the aortic valves. In the sciatic artery, however, there is a considerable diastolic flow component with superimposed flow pulsations generated by systolic ejection (Fig. 29c). One assumes that the closer one gets to the periphery the smoother flow will be, although this must remain conjectural at the moment.

Blood flow in most vessels is laminar, meaning that there is a gradient in velocity across the blood vessel, with the highest velocity being along the central axis and falling to zero at the wall. As has already been noted, resistance to flow of blood is evinced as the energy that is dissipated by friction between the adjacent laminae of blood (except during turbulent flow when the random movement of fluid elements

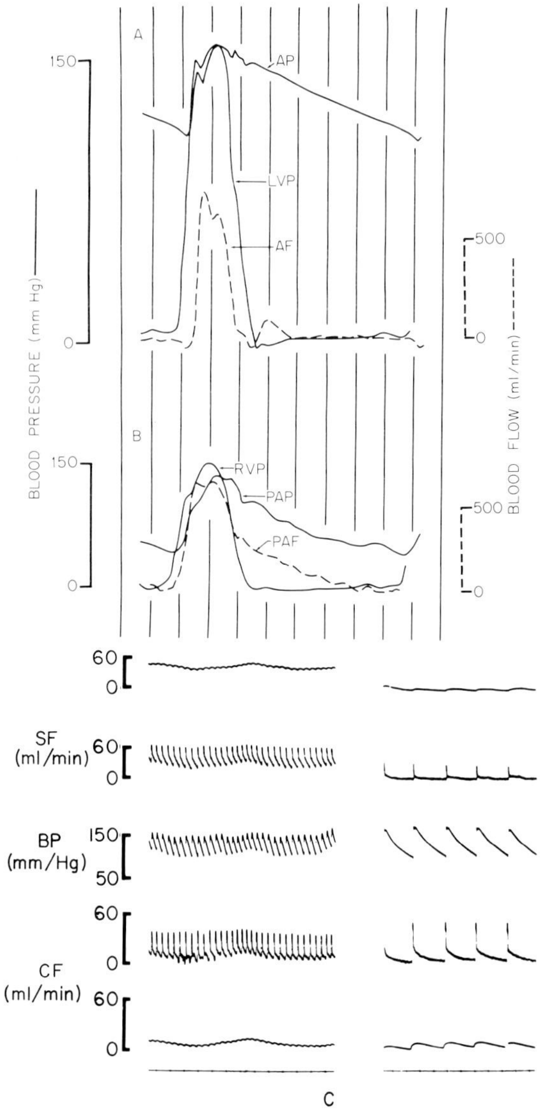

FIG. 29. (A and B) Superimposed pressure (——) and flow (— —) pulses recorded from the heart and aortic (A) and pulmonary (B) circulation of the domestic duck. The vertical lines link coincident points in time and are 0.05 second apart. LVP, left ventricular pressure; AP, aortic pressure; AF, aortic flow; RVP, right ventricular pressure; PAP, pulmonary artery pressure; PAF, pulmonary arterial flow (Jones and Langille, 1972). (c) Traces showing arterial blood pressure (BP) and sciatic (SF) and carotid (CF) artery blood flow before (left) and during (right) a normal dive of 2-minute duration. In each series the traces from top to bottom are mean blood flow through sciatic artery; pulsatile blood flow through sciatic artery; arterial blood pressure; pulsatile blood flow through common carotid artery; mean blood flow through common carotid artery; time marker in seconds. [Butler and Jones, 1971, *J. Physiol. (London)*, by permission.]

and friction between the blood and wall of the vessel becomes important). The factors that influence streamlined or laminar blood flow are described by Poiseuille's equation, which emphasizes the importance of viscosity and the radius of the tube. A reduction in radius to 1/2 its original value reduces flow to 1/16th of what it was for the same pressure drop.

The relative viscosity of avian blood has not been subjected to much study (absolute viscosity is of little practical importance; relative viscosity expresses the value of the viscosity of the fluid compared with water). Vogel (1961) reported the relative viscosity of chicken blood at 42°C to be 3.08 for female and 3.61 for male. The hematocrit is the most important determinant of the relative viscosity, and is considerably higher in male than female chickens (Lucas and Jamroz, 1961). Plasma proteins also have some effect, and in female chickens the viscosity of plasma (1.51) is substantially higher than in males (1.42), which possess lower levels of plasma proteins (Vogel, 1961). Unfortunately, increase in temperature lowers the viscosity of blood more than the viscosity of water, so that recordings of relative viscosity made at 14°–20°C are likely to be somewhat in error considering that normal avian body temperatures are above 40°C. This may explain why Spector (1956) reported higher values, in the range of 4–4.5 for the duck and goose, than were recorded by Vogel (1961) in male chickens.

The kinematic viscosity denotes the tendency of the fluid to become turbulent, and is given by the ratio viscosity/density. The units of kinematic viscosity are "stokes." The kinematic viscosity of whole blood in male chickens is around 3.5 centistokes, whereas plasma has a kinematic viscosity of 1.4 centistokes. This means that plasma is more liable to turbulence than is whole blood, since the higher the kinematic viscosity the less likelihood of turbulence.

During turbulent flow, velocities at right angles to the axis of the tube are abundant and the flow profile is more or less flat, unlike the parabolic flow profile recorded during laminar flow. The likelihood of turbulence occurring in the circulation depends on a number of factors that were empirically derived by Reynolds. Reynolds showed that the critical velocity of flow for turbulence to occur depended on the viscosity and density of the fluid and on the radius of the tube. For a 2 kg duck with an aortic root radius of 0.3 cm and cardiac output of 700 ml/minute the critical velocity would be about 114 cm/second, if the critical Reynolds number for the circulation is 1000 (McDonald, 1960). Peak flow rates in the thoracic aorta and brachiocephalic arteries do not reach these values in the duck (Jones and Langille,

1972), but they may be briefly reached in the proximal aorta (before it divides) at peak ejection.

E. The Relationship Between Pressure and Flow

Application of Poiseuille's equation to the avian circulation will allow estimation of the pressure gradients for the observed flows. However, the pulsatile nature of flow makes a direct application of this equation questionable; due to the inertia of the blood and the high oscillation frequencies, flow amplitude is no longer simply related to pressure gradient. The deviation from Poiseuille's equation and the extent of the phase lag is determined by a nondimensional constant α, where

$$\alpha = R\sqrt{\frac{2\pi n f \rho}{\mu}}$$

where R is the radius of the tube, μ and ρ are the viscosity and density, respectively, of blood, f is the pulse frequency in cycles/second, and n is the order of the harmonic component. When α is 0.5 or less for the fundamental frequency, the phase lag is negligible and the flow conforms approximately with the Poiseuille equation. Calculations for the aorta of the duck (using data that have previously been given and assuming a cycle rate of 3/second) yield a value of α as of about 6.0–7.0 for the fundamental frequency. This may be compared with 1.38–3.5 for the rat, 8.27–10.68 for the dog, and 13.5–16.7 for man, all referring to the root of the aorta (McDonald, 1960). At a value of 6–7 for α in the duck, the estimation of flow by the Poiseuille equation is not reliable, and there appears to be no convenient method for computing the pressure gradient from flow data or vice versa.

Few measured values of the pressure gradient within the central arteries have been reported. Taylor's (1964) recordings show that pressure gradients from the thoracic to the abdominal aorta (21 cm) of the turkey are small, whereas Speckmann and Ringer (1963), in the same species, report a mean difference in pressure between the carotid and tibial arteries of 10 mm Hg. Speckmann and Ringer (1963) suggest that this pressure drop may be caused by the presence of atherosclerotic plaques in the abdominal aorta.

The model of the avian circulation suggests that peak flow at a given point will appear before peak pressure, since we are dealing with a capacitive and resistive network. That this holds for the pressure–flow relationships in the sciatic artery of the duck has been shown by Butler and Jones (1971). Pressure and flow pulses were analyzed from the instant of peak flow to the start of the next cardiac contraction. The pressure and flows were expressed as percentages of maximum pressure and flow recorded in the large number of pulses that were analyzed to construct a single curve. Since peak flow is established before peak pressure, the pressure–flow curves for the sciatic vascular bed are initially concave to the flow axis (Fig. 30). The falling slope of the run-off curve gives an indication of the resistance of the sciatic vascular bed. The slope increases about four fold during 1 minute of submergence. During submergence, zero flow occurred at 56% of the peak pressure and this was approximately twice as high as the intercept obtained by extrapolation of the slope for the surfaced animal to the ordinate, which indicates that there is a yield pressure before measurable flow can commence.

Owing to the viscous properties of blood a small head of pressure is required for commencement of flow in small vessels, but this cannot account for the elevation in yield pressure during diving. A number of factors may contribute to the high yield pressure, such as an increase in venous pressure (Johansen and Aakhus, 1963; Folkow *et al.*, 1967) and any fall in mean systemic pressure that would, either singly or together with other factors, tend to decrease the pressure difference for flow across the capillary bed. Another factor may be that the arterioles become unstable at fairly high pressures and exhibit critical closure when subjected to increased vasomotor tone (A. C. Burton, 1965). Regardless of cause, the effect is real and is represented in the model (Fig. 26) by the variable back voltage ("back pressure") which can be applied to the system.

Blockade of sympathetic α-receptors by means of drugs (see later) abolishes the intense vasoconstriction in the sciatic vascular bed during diving and the yield pressure hardly changes (Fig. 30). This shows that the vasoconstrictor response during diving is mediated by stimulation of adrenergic α-receptors and supports the suggestion that it is active vasomotor tone that contributes to the elevation of the yield pressure (Butler and Jones, 1971).

The technique of using what may be called "run-off" curves to describe the pressure–flow relationships is limited by the fact that it neglects the early period of ventricular systole. There is, in fact, a difference in the slopes of the rising (systolic period) and falling (diastolic period) phases of a complete pressure–flow loop which indicates that pressure–flow relationships may differ in respect of rising and falling pressures; i.e., in the vascular bed vessel recruitment and decrement may be uneven.

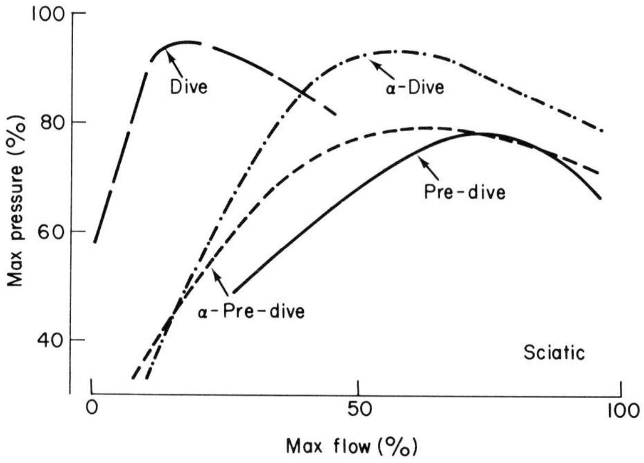

FIG. 30. Relationship between pressure and flow in the sciatic vascular bed in normal and α-blocked ducks. Average values from analysis of 10 to 20 cardiac cycles from the instant of peak flow to the start of the next cardiac contraction. Each pair of pulses was broken down into 15 to 20 points and values were measured at these points. The pressures and flows were expressed as a percentage of the maximum pressure and flow attained in the cardiac cycles analyzed, and averaged graphically. Pressure and flow pulses were analyzed before and after 1 minute submersion in both normal and α-blocked ducks. [Butler and Jones, 1971, *J. Physiol. (London)*, by permission.]

VI. Regulation of the Cardiovascular System

Birds successfully adjust to severe environmental stresses such as temperature, humidity, and altitude. Their behavior patterns may include week-long sustained activity during migratory flight or high tolerance to prolonged asphyxia during diving. Their general metabolism and body temperature exceed those of other vertebrates. These successes are all dependent upon or linked with an efficient cardiovascular transport function. Yet, the level of understanding of cardiovascular control in birds is much inferior to that in mammals.

It is essential that cardiovascular functions are precisely regulated in order to promote and maintain homeostasis. Homeostatic mechanisms dependent upon the cardiovascular system are involved in control of a large number of variables, such as the partial pressures of oxygen and carbon dioxide in the blood, cardiac output, arterial blood pressure, heart rate, distribution of blood flow, temperature regulation, and glomerular filtration. Feedback stimuli, important to all these and other functions, can hence be surmised to alter cardiovascular performance directly or reflexly. Also, higher central nervous centers may overrule local and autonomic adjustments of the circulation to ensure maintenance of an adequate blood supply to tissues particularly sensitive to oxygen lack, such as the brain and heart.

A. AUTOREGULATION OF BLOOD FLOW AND REACTIVE HYPEREMIA

It is well established from studies of mammals that if blood flow to a muscular vascular bed is occluded for a period of time, a much higher level of blood flow will prevail when the occlusion is released. Available evidence accredits this response largely to local and intrinsic factors such as accumulation of metabolic products like CO_2, lactic acid, and H^+ ions in the tissues of the blood vessel walls, with only minor importance attached to neurogenic vasomotor mechanisms. Figure 31 demonstrates blood flow change in the sciatic artery of a

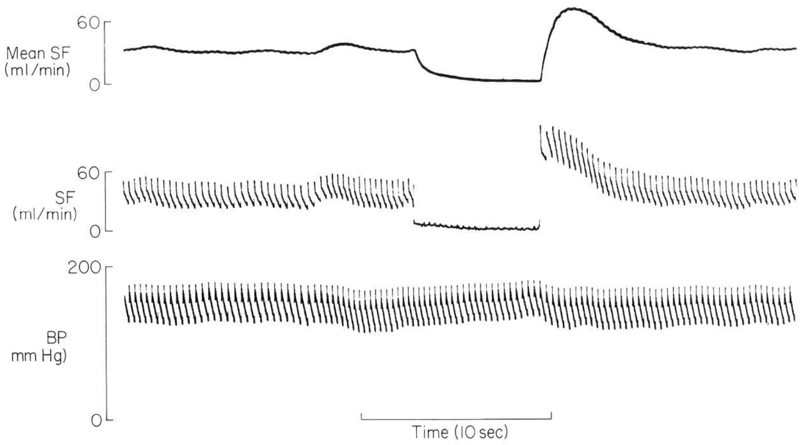

FIG. 31. Change in blood flow in the sciatic artery of a domestic duck following a short period of arterial occlusion (marked by zero flow). Traces, top to bottom: mean sciatic flow; pulsatile sciatic flow and arterial blood pressure (Butler and Jones, 1969).

duck after arterial occlusion. It is apparent on many occasions that the reactive hyperemia is rather weak (Butler and Jones, 1969) compared with that of mammals. This may be a consequence of the high blood flow that normally goes to these areas (Folkow *et al.*, 1967) in that higher demand for oxygen is satisfied by a higher oxygen utilization by the tissues instead of greatly increased flow. In a recent study of blood-flow regulation to the feet of Giant Fulmars, *Macronectes giganteus,* Johansen and Millard (1972) observed a marked reactive hyperemia following transient occlusion of the arterial supply of the foot, even when foot temperature was close to 0°C. This finding indicates that accumulation of metabolites could not have instigated a reactive flow increase, and a neurogenic control component is suggested.

The comprehensive studies of Folkow *et al.* (1966) on regulation of blood flow in skeletal musculature of ducks demonstrate that neurogenic vasoconstriction completely overrides the dilatory effect of local metabolites in ducks, unlike conditions in cats and turkeys where the local regulatory factors predominate and supress the neurogenic vasoconstriction. The authors relate this marked difference to a much higher capillary density in duck muscles and to a denser (ground plexus) network of adrenergic nerve terminals surrounding the arterial musculature in ducks compared with other species. Figure 32 sum-

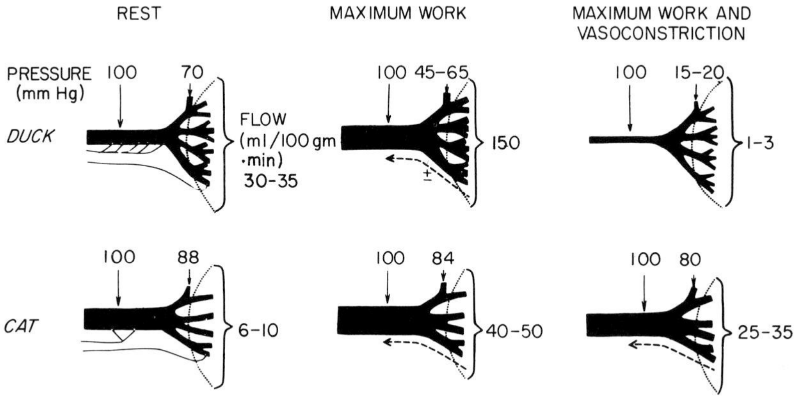

FIG. 32. Diagrams illustrating the differences in innervation and geometry of extra- and intramuscular arteries of the duck and cat, the characteristic pressure gradients along the large, extramuscular arteries, and the rates of blood flow in "rest," maximal work, and maximal work plus intense vasoconstriction. The large artery of the duck is much more densely innervated; further, there are more small arteries per unit muscle (indicated by the dotted line) than in the cat. Values for the "small artery" pressures and for blood flows are approximate for the different situations. The arrow indicates ascending dilatation; in the case of the duck during maximum work, ± means that the ascending dilatation is not a constant feature. (Folkow *et al.*, 1966, *Acta Physiol. Scand.*, by permission.)

marizes their findings schematically by comparison of innervation and geometry of the vascular supply to skeletal muscles in the duck and the cat. The authors discuss the usefulness of the generalized control of muscle blood flow in the duck in relation to diving. It seems reasonable to assume that release of the powerful sympathetic discharge to the peripheral vascular bed of a diving bird (see below) will cause a quicker onset of hyperemia in the richly innervated vascular muscle than could be effected by direct local action of metabolites.

B. Central Nervous Regulation of Cardiovascular Function

There is at present little experimental evidence on the importance of medullary centers in the regulation of cardiovascular performance in birds, nor is there much information available on how higher hypothalamic and cortical centers influence the cardiovascular system.

Dijk (1932), using somewhat crude techniques of stimulation, obtained evidence that the thalamus influences the level of blood pressure in birds. Feigl and Folkow (1963) studied heart rate and systemic arterial blood pressure during stimulation of two specific areas in the mesencephalon and diencephalon of ducks. Stimulation of an area near the midline in the mesencephalon evoked a marked rise in arterial blood pressure, a reduction in heart rate, and an intense vasoconstriction in skeletal musculature. The stimulation also elicited a prompt apnea, and if artificial respiration was administered during the stimulation the cardiovascular responses were much less pronounced or not at all discernible. The authors point out the similarity between the cardiovascular changes associated with diving and with stimulation of this area in the mesencephalon. Stimulation of an area in the diencephalon, again close to a midsagittal plane, evoked an intense vasodilator response in skeletal muscle. The response was interpreted as resulting from activation of cholinergic vasodilator fibers, since the effect was selectively blocked by atropine (Feigl and Folkow, 1963).

D. H. Cohen *et al.* (1970), using a series of retrograde degeneration experiments, found that cells connected with vagal cardioinhibitory fibers in the pigeon are located in the ipsilateral dorsal motor nucleus of the medulla and appear to have their greatest density in the ventral portion of the nucleus approximately 0.5–1.0 mm rostral to the obex. Electrical stimulation of the lateral dorsal motor nucleus produced short-latency bradycardia and hypotension, as did stimulation of the solitary tract and commisural nucleus of Cajal. D. H. Cohen and Schnall (1970) concluded that the response obtained in the latter pair of positions was due to stimulation of afferent fibers. Apnea almost invariably accompanied the bradycardia except when the stimulation was applied to the dorsal motor nucleus. The bradycardia was abolished by bilateral vagotomy and greatly reduced if the ipsilateral vagus was sectioned. β-Sympathetic blockade had no effect on the bradycardia. D. H. Cohen and Schnall (1970) found that stimulation of the dorsal motor nucleus gave results similar to that of stimulation of the peripheral end of the cut vagus and unlike that obtained from stimulation of the cut central end in that the latter gave apnea, a moderate fall in blood pressure, and little change in heart rate. This finding was not surprising in view of the fact that section of the ipsilateral vagus virtually eliminates the response to stimulation of the dorsal motor nucleus. About 70% of stimulations of the dorsal medulla produced tachycardia and hypertension, which were unaffected by β-adrenergic blockade (D. H. Cohen and Schnall, 1970).

Using techniques of microdissection, electrical stimulation, and retrograde degeneration, Macdonald and Cohen (1970) have found in the pigeon that cardioaccelerator preganglionic fibers (sympathetic) probably arise from the most caudal cervical segment of the spinal cord (14), definitely from the upper two thoracic segments (15 and 16), and occasionally from the midthoracic segment (17). The cells representing the origin of the sympathetic preganglionic fibers are localized to a well-defined column dorsal to the central canal. This had previously been suggested by Terni (1923) on the basis of normal adult and embryonic material, and this chain of cells is referred to as the column of Terni. The preganglionic axons enter the sympathetic chain to synapse on postganglionic neurones located in the caudal three cervical ganglia (12–14). Postganglionic cardioaccelerator nerves emerging from these ganglia then anastomose to form a thoracic cardiac nerve. It is of interest to note that in the pigeon stimulation of the right cardiac nerve consistently provokes increases in heart rate, whereas this is not true of the left (Fig. 33) (MacDonald and Cohen, 1970).

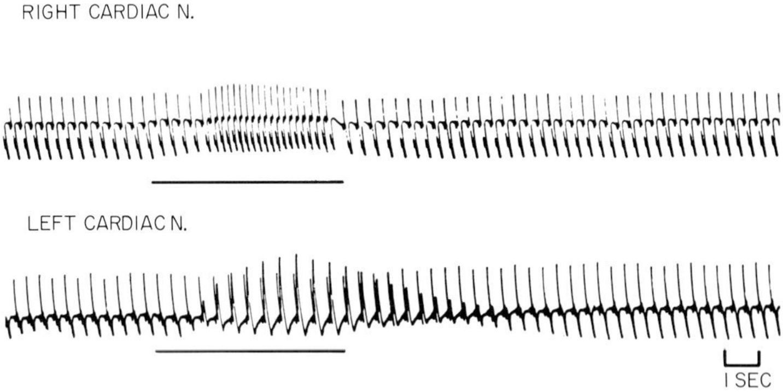

FIG. 33. Representative electrocardiograms of the heart rate response to stimulation of the right and left cardiac nerves in the pigeon. Note the marked cardioacceleration with stimulation of the right in contrast to the left cardiac nerve. Note also the altered T wave response to stimulation of the left cardiac nerve. Rectangular cathodal pulses of 0.5 msec duration, 50 cycles/second, and 25 μA were used. Stimulus duration is indicated by the solid bar below each record. (Macdonald and Cohen, 1970, *J. Comp. Neurol.*, by permission.)

Durkovic and Cohen (1969) investigated the effects of rostral midbrain lesions on conditioning of heart and respiratory rates in pigeons. Conditioning of heart and respiratory rates was established by pairing light and foot-shock. It was established for normal responses of heart and respiratory rates that the integrity of one or more pathways located in the ventromedial tegmentum was required. Deficits in the responses caused by lesions in this area may result from interruption of a descending hypothalamo-tegmental projection.

Other evidence for the participation of the central nervous system in cardiovascular regulation has been obtained by stimulating the central ends of cut peripheral nerves and recording the effects on the blood-vascular system. Central stimulation of the cut left vagus, with the right intact, caused a decrease in blood pressure and bradycardia in the duck. The decrease in blood pressure was still obtained after section of the right vagus, indicating that centrally mediated peripheral mechanisms were involved (Johansen and Reite, 1964). The reverse picture is obtained by stimulating the central end of the right vagus, with the left vagus intact. The tachycardia persists after section of the left vagus, indicating a centrally induced increase in sympathetic tone. Unfortunately, such stimulation is not very specific since numerous centripetal pathways from different organs converge in the vagus nerve (Johansen and Reite, 1964).

C. Peripheral Nervous System and Cardiovascular Control

The nerve transmitter substances acetylcholine and norepinephrine, released on stimulation of the parasympathetic and sympathetic nerves, have profound effects on both the heart and peripheral circulation. The parasympathetic and some sympathetic nerves are cholinergic, and their action is typically depressor centrally and vasodilator peripherally, whereas adrenergic nerves (sympathetic only) are powerful central stimulants and vasoconstrictors peripherally. Many drugs mimic the actions of these transmitter substances and have often proved more convenient for investigation of the role of the peripheral nervous system in cardiovascular control than has nerve section or stimulation. Similarly, several pharmacological agents have been shown to exert specific blocking effects of the cholinergic as well as the several types of adrenergic effector terminals.

In the normal animal, heart rate is set by the interplay of parasympathetic and sympathetic influences on the pacemaker. Stimulation of the parasympathetic or vagus nerve typically reduces heart rate, whereas sympathetic stimulation increases it. Similar antagonistic effects of the double autonomic innervation are apparent in other parts of the circulatory system.

1. The Adrenergic Nerves

The cardiac sympathetic nerve in the chicken arises from the first thoracic ganglion between the first and second rib close to the most caudal branch of the brachial plexus (Tummons and Sturkie, 1969), whereas in the pigeon the last cervical ganglion appears to be the major contributor of sympathetic postganglionic fibers (Macdonald and Cohen, 1970). In the chicken (Fig. 34) the cardiac sympathetic

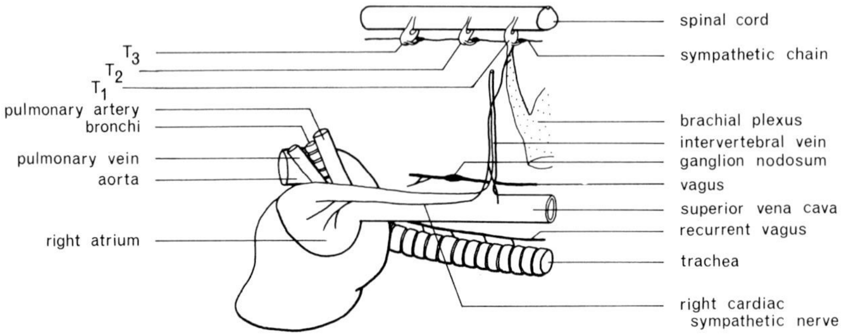

FIG. 34. Diagram of the right cardiac sympathetic nerve in the chicken. T_1–T_3 are thoracic ganglia (after Tummons and Sturkie, 1969).

nerve runs parallel with a small vertebral vein, turning towards the heart close to the junction of this vein with the anterior vena cava. The nerve runs along the anterior vena cava to the heart where it joins with fibers from the vagi to form several cardiac plexuses (Ssinelnikow, 1928; Tummons and Sturkie, 1969). The most densely innervated regions of the heart are the sinu-atrial node, the right atrium, and the atrio-ventricular node (Bennett and Malmfors, 1970). Bolton (1967), by means of field stimulation that excites intramural cardiac nerves, has obtained evidence that the ventricles of the fowl and pigeon, like those of mammals, contain adrenergic nerves.

Tummons and Sturkie (1968) were the first to show directly that the cardio-accelerator nerves of chickens, ducks, and pigeons exerted a marked effect upon heart rate. Stimulation of the right cardiac nerve of chickens that had been bilaterally vagotomized increased heart rate by 48%, whereas stimulation of the left nerve caused an increase of 32%. However, since vagal restraint had been eliminated, these experiments provided no idea of the degree to which heart rate is normally influenced by accelerator and vagal tone.

More recent experiments by the same authors (Tummons and Sturkie, 1969) have demonstrated for the chicken that at the basal resting heart rate of 280–290 beats/minute there is a significant amount of sympathetic and parasympathetic tone that appears to be balanced, since elimination of either system produces equal and opposite changes in heart rate. On the basis of pharmacological experiments Johansen and Reite (1964) had earlier emphasized the delicate balance of sympathetic and parasympathetic tonus to the resting birds' heart. Changes of heart rate induced by excitement appear to result from a decrease in vagal tone and an increase in sympathetic tone. D. H. Cohen and Pitts (1968) concluded that conditioned cardio-acceleration is mediated entirely by the extrinsic cardiac nerves, the principal contributor to the magnitude of the response being the cardiac sympathetic nerves, whereas the shortest latency component of the response is mediated by the vagi. A point of interest is that the rate of the completely denervated chicken heart (280–290 beats/minute) is similar to that in the resting intact birds. Tummons and Sturkie (1969) claim that this indicates that the intrinsic heart rate is determined or regulated by factors such as body temperature, hormones, and metabolic state.

A major difference in the effect of reserpine pretreatment on birds and mammals has been reported by Bolton (1967). Reserpine pretreatment, which depletes stored catecholamines, scarcely alters the effects of sympathetic nerve stimulation on chick ventricular myocardium, whereas it abolishes the effects of stimulating the sympathetic nerves to rabbit or kitten atria (Huković, 1959; Trendelenburg, 1965; Blinks, 1965). This may indicate that the catecholamine stores of the chick heart are very large. However, in intact chickens, 80–90% of cardiac catecholamines are depleted 24 hours after treatment with reserpine, whereas it takes 2 weeks to reach this level after cardiac sympathectomy (Lin *et al.*, 1970).

The detection and elicitation of a response to epinephrine and related compounds is achieved by specialized cell components known as adrenergic receptors and these have been classified as α and β-receptors on the basis of the potency order of a series of related amines (Ahlquist, 1948). In ducks, the α-receptors are associated with vasoconstriction, whereas it has been shown for chickens, ducks, gulls, and pigeons that the β-receptors are associated with adrenergic-positive cardiac inotropic (force) and chronotropic (rate) effects and may also be involved in vasodilatation. The β-receptors can be functionally differentiated into two subgroups, β-1 and β-2, based upon the responses obtained with structurally varied sympathomimetic amines (Lands *et al.*, 1969). Although the properties of the adrenoreceptors in the chick heart suggest that they be classed as β-receptors, Bolton and Bowman (1969) found that norepinephrine was more potent than epinephrine in producing responses, a situation found in some mammalian hearts.

Most of our knowledge on the role of adrenergic nerves in circulatory control has been indirectly obtained by studying the effects of drugs that block the responses stated to be associated with α- or β-receptors. In ducks, injection of epinephrine tartrate typically causes a significant reduction in sciatic artery blood flow and a rise in blood pressure which is usually associated with bradycardia. After adrenergic α-receptor blockage, blood pressure is depressed and the response to epinephrine injection is prevented. This illustrated the importance of α-receptors in the maintenance of peripheral vasomotor tone in ducks. The effects of stimulating intramural cardiac adrenergic nerves are abolished by β-receptor blocking drugs and by adrenergic neuron blocking drugs (Bolton, 1967). Theoretically, the pressure response to epinephrine should be reversed after α-receptor blockade, but this has proved difficult to achieve in the chicken (Dale, 1906; Gibbs, 1928; Vanderbrook and Vos, 1940; Thompson and Coon, 1948; Harvey *et al.*, 1954; Eble, 1963; Bolton and Bowman, 1969). Bolton and Bowman (1969) conclude that the fowl possesses β-receptors in all segments of its vascular system, but when these are stimulated by epinephrine the effects on blood pressure, at least in anaesthetized animals, are masked until α-receptor blockade is well advanced or complete.

Direct evidence of the role of sympathetic nervous systems in peripheral circulatory control in birds has been obtained by Folkow *et al.* (1966). Muscle blood flow in ducks could be almost stopped during reflex vasoconstriction induced by inhalation of 20% CO_2. The large extramuscular arteries constrict far more in ducks than in turkeys or cats, and fail to show ascending dilatation due to accumulation of metabolites. In fact, the proportion of resistance residing in the large artery supplying skeletal muscle of ducks is 30% at rest and 20% during reflex vasoconstriction. These figures compare with 12 and 4.5% for the cat under similar conditions. The femoral artery has a far richer supply of adrenergic nerve terminals in ducks than in cats and turkeys. Figure 32 illustrates these differences between the duck and cat under three conditions. However, it should be pointed out that the resistance of the "large artery" is in series with the resistance of the "small arteries" that contribute 70 to 80% of the peripheral resistance during rest or vasoconstriction. Since the large artery is in series with the small, the resistance of the vascular bed will be given by the value of the greatest resistance under consideration, and this is unlikely to be the large artery (Vonruden *et al.*, 1964; Weale, 1964).

β-receptors can be stimulated by injection of isoprenaline. In ducks this drug has no effect on blood flow or pressure in the sciatic and carotid arteries but causes a significant tachycardia (Butler and Jones, 1971). After β-blockade the tachycardia caused by injection of isoprenaline is prevented. The heart rate and output before and after β-blockade give an idea of the degree of sympathetic tone present to the heart of the animal. In ducks the evidence for its presence or absence is contradictory. Butler and Jones (1968, 1971) and Folkow *et al.* (1967) found no significant effect of β-blockade at rest on any of the cardiovascular parameters they measured (e.g., heart rate, output, blood pressure, and flow in the sciatic and carotid arteries) although earlier work by Johansen and Reite (1964) on vagotomized ducks indicated a high level of sympathetic activity, since after injection of a β-blocking drug the heart rate slowed conspicuously. However, Butler and Jones (1968, 1971) found that the post-dive tachycardia in the domestic duck was unaffected by β-blockade, whereas Folkow *et al.* (1967) report that it was considerably reduced. The latter authors also found that the cardiac output fails to increase during recovery from a dive after β-blockade, a finding corroborated by flow recordings of Butler and Jones (1971). In the β-blocked duck, increases in flow in the sciatic and carotid arteries during recovery from a dive are considerably slower than in the normal animal. However, the possibility must be borne in mind that peripheral β-receptors, active in causing vasodilatation, may also have been blocked.

In the chicken, there seems to be considerable cardiac sympathetic tone in the resting animal, and this has been confirmed by use of β-blocking drugs. Butler (1967b) reported that the heart rate of the chicken decreased by 25% following β-blockade, a value within the range cited by Tummons and Sturkie (1969) following surgical sympathectomy.

2. *The Cholinergic Nerves*

The vagus nerve emerges from the cranium separate from the glossopharyngeal, but there is an anastomosis between them near their point of origin. A side branch of the vagus also connects with the precarotid, pretracheal branch of cranial nerve IX. The vagus continues posteriorly down the neck beside the anterior cervical artery and jugular vein, and eventually swells to form the nodose ganglion. Branches are given off from here to innervate the carotid body and surrounding structures (Fig. 35). From the lower pole of the nodose ganglion one or two branches arise and, after running for a variable distance in the vagal sheath, separate and run medially towards the aortic arch, at least one branch supplying the wall of the aorta (Terni, 1931; Nonidez, 1935). Anterior to the heart the recurrent nerve is given off, and more caudal the vagus divides into viscero-cardiac and pulmonary branches (Fig. 35).

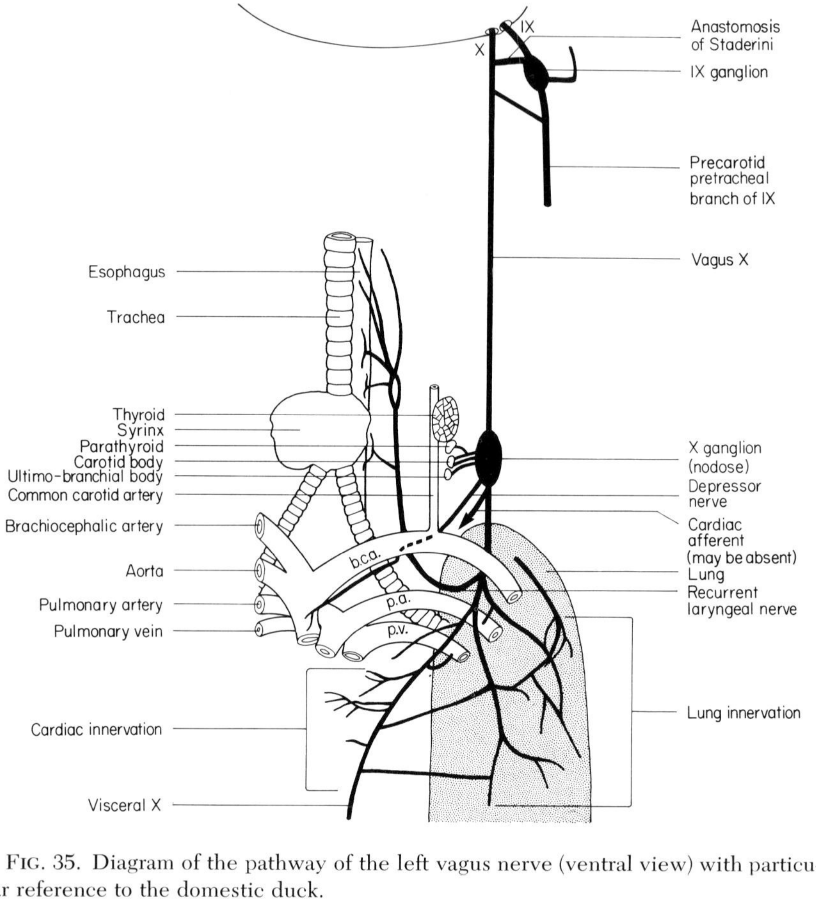

FIG. 35. Diagram of the pathway of the left vagus nerve (ventral view) with particular reference to the domestic duck.

With the exception of recent studies on the regulation of cutaneous blood flow to the webbed feet on some birds (Millard and Johansen, 1972), the effects of cholinergic nerves in the regulation of cardiovascular function in birds have largely been confined to studies on the central circulation. In many birds the vagi exert a powerful tonic inhibiting influence on heart rate, as is shown by the conspicuous increases in heart rate following bilateral vagotomy or injection of atropine. Several authors have reported that one vagus dominates the other in the normal animal. Dominance of the right over the left vagus has been claimed for both the chicken and duck (Johansen and Reite, 1964; Sturkie, 1965). However, Butler and Jones (1968) found, by cold-blocking experiments, that only one vagus is active in cardiac chronotropic control at any one time, and that it could be either the left or right vagus. Cold-block of the active vagus produces an acceleration in heart rate of some 200% if the other vagus is intact. Blockade of the inactive nerve trunk has no significant effect on heart rate. It was found in 16 ducks that the right vagus was active in 12 and the left in the others. However, in one animal the left vagus was initially active, with the right one inactive; 1 hour later the right was active and the left inactive. This switch of activity indicates a certain degree of lability in the avian central nervous system, and also that the right and left cardiac plexuses must both innervate the pacemaker area.

There seems to be considerable species variation in the amount of vagal restraint of the heart. Stübel (1910) attempted to relate these species differences to the activity pattern of the animals in normal life. Species with a large heart in relation to body size (like pigeons, most carnivorous birds, gulls, and ducks) have a marked vagal tonus to the heart, while for cursorial species like chickens the tonic vagal activity is considerably smaller. At least for chickens and ducks this generalization appears to be true, since bilateral vagotomy of ducks increases heart rate by 65–200% (Johansen and Reite, 1964; Butler and Jones, 1968), whereas the same procedure in chickens increases the heart rate by only 20–36% (Butler, 1967a,b; Tummons and Sturkie, 1969).

Stimulation of cut peripheral ends of the vagus nerves produces bradycardia and, depending on the species, cardiac arrest (Knoll, 1897; Thébault, 1898; Dogiel and Archengelsky, 1906; Batelli and Stern, 1908; Jurgens, 1909; Johansen and Reite, 1964; Jones and Purves, 1970a). This effect can be mimicked by injection of acetylcholine, which, depending on the dose level, causes hypotension and bradycardia. Of particular interest in the response to acetylcholine is the tachycardia that follows the transient bradycardia. Durfee (1963) has shown that hypotension induced by drugs is usually followed by tachycardia, a result which suggests that baroreceptors are operative in the chicken. This tachycardia is abolished by vagotomy. Similarly, the bradycardia accompanying drug-induced hypertension is abolished by bilateral vagotomy (Sturkie, 1965).

The effect of acetylcholine, either released by the nerves or introduced by injection, can be blocked with atropine. Following atropinization, the heart rate is usually elevated above normal levels, since the heart has been released from vagal restraint. However, this is not implying that the tachycardia is solely adrenergic. Folkow *et al.* (1967) obtained a substantial increase in heart rate following atropinization of β-blocked ducks. Consequently, some of the increase in heart rate may be due either to a central stimulant action of atropine, which excites ganglionic structures in the myocardium via impulses traveling in the vagus nerves (Donald *et al.*, 1967), or to an intrinsic heart rate that is much higher than the normal basal heart rate [a situation unlike that found in the chicken (Tummons and Sturkie, 1969)].

The contractility of the ventricles may also be affected by vagal activity. Folkow and Yonce (1967) recorded a negative inotropic effect of the vagus in the duck with the same characteristic frequency-response relationship as the negative chronotropic response. They found that contractile force could be reduced to less than half of control during vagal stimulation. The ventricles of both chicken and pigeon also contain cholinergic nerves, the negative inotropic effects of their stimulation being abolished by atropine and hyoscine (Bolton, 1967).

Few authors have commented on the effects of vagotomy on systemic arterial blood pressure in birds. Stübel (1910) recorded a marked increase in blood pressure after bilateral vagotomy, and attributed this to the tachycardia and an active depressor tonus in the vagus acting on the vasomotor center. However, Johansen and Reite (1964) found that bilateral vagotomy caused either an increase or decrease in blood pressure depending on whether the left or right vagus, respectively, was sectioned last. Butler (1967b) reported an initial increase in blood pressure of the chicken after bilateral vagotomy, but in the long term the blood pressure returned to values close to those found in the intact bird.

Vasodilatation in skeletal musculature by activation of cholinergic dilator fibers selectively blocked by atropine has been reported for the duck by Feigl and Folkow (1963); the same type of response also exists in skin, as recently shown by Johansen and Millard (1972) and Millard and Johansen (1972).

D. Baroreceptor Reflexes

The evidence regarding the role of baroreceptors in the avian circulation is very confusing. The best evidence is that drug-induced hypertension is usually accompanied by bradycardia (Durfee, 1963), whereas hypotension is usually followed by tachycardia (Durfee, 1963). On the other hand, Harvey *et al.* (1954) concluded that significant drug-induced pressor reflexes are difficult to obtain from the chicken.

The location of baroreceptors in the circulation of birds might be expected to follow the vertebrate scheme of reflexogenic zones in the carotid arteries and aortic arch. Pressoreceptors can be shown histologically in the adventitia of the aortic arch, innervated by the vagus (Fig. 36) (Heymans and Neil, 1958; Ábrahám, 1969). However, typical specialized sensory nerve endings appear to be absent from nerve branches that terminate in the carotid area of birds, and the fibers terminate in a plexus (Ábrahám, 1969). Jones (1969) recorded afferent vagal activity from ducks associated with various phases of the cardiac cycle. Mechano-receptors are definitely located in the wall of the arch of the ascending aorta and in the pulmonary artery (Jones, 1969) (see Fig. 46). Birds, consequently, seem to be endowed with receptors sensitive to intravascular pressure changes, but the question of their normal importance remains unanswered.

Van der Linden (1934) and Ara (1934) clamped the common carotid artery rostral to the site of the carotid body in chickens and obtained a reflex hypertension that seemed less pronounced than in mammals. Similarly, Muratori (1934) claimed to have elicited alterations in systemic blood pressure by stimulation of this region. Rapid blood loss, with a consequent fall in arterial pressure, results in tachycardia in ducks (Djojosugito *et al.*, 1968). Durfee (1964) was, however, unable to find any pressure reflexogenic areas associated with the carotid arteries in the chicken. Similarly, McGinnis and Ringer (1966) found no significant changes in blood pressure or heart rate in chickens when the blood pressure in one carotid arch was selectively reduced to 25% of normal. The same result was found after contralateral vagotomy, thereby eliminating innervation of the other carotid baroreceptor area. There may be several reasons for this type of deviation apart from species differences. The response may be destroyed by anesthetics. The diving reflexes in ducks, for instance, are prone to disappear during anesthesia. Another possibility is that in birds the aortic arch baroreceptors may effectively buffer or override any stimuli from the carotid baroreceptors.

It has been postulated that cephalic baroreceptors exist in the chicken (Rodbard and Saiki, 1952) but McGinnis and Ringer (1967) could find no evidence in support of this suggestion. They attribute changes in systemic blood pressure following arterial occlusion to cerebral ischemia, assuming that arterial baroreceptors would be activated before mean blood pressure to the head had decreased to a level of 45 to 53 mm Hg.

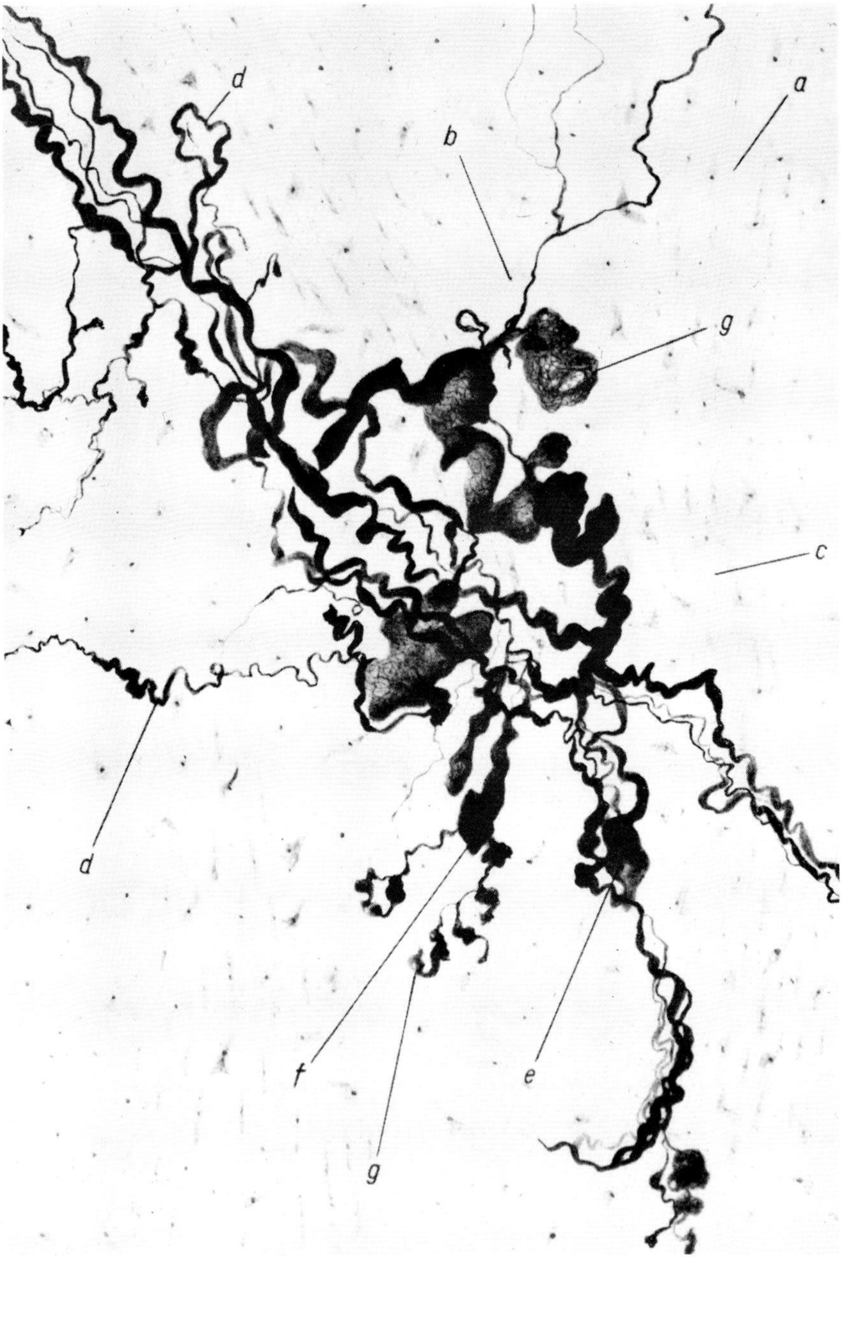
d
a
b
g
c
d
f
e
g

E. Cerebral Ischemic Reflexes

Cerebral ischemia can be produced by clamping both carotid and occipital arteries in the chicken (Richards and Sykes, 1967). In the anesthetized animal cerebral ischemia results in a rise in blood pressure by 100%, whereas heart rate increases by only 13% (Richards and Sykes, 1967). This response seems independent of either baro- or chemoreceptors, since it results whether the carotid arteries are clamped anteriorly or posteriorly of the carotid reflexogenic zone (Jung, 1934; Richards and Sykes, 1967). This work confirms earlier observations of Van der Linden (1934), except that in Van der Linden's experiments systemic hypertension and, therefore, cerebral ischemia occurred after occlusion of both common carotid arteries.

However, Kovách *et al.* (1969) were unable to obtain any pressor response from occlusion of both carotid arteries in the hen, but elicited a marked hypertension in the case of the pigeon and crow (*Corvus*). It may be that differences in cerebral blood supply were responsible for this result, since Richards and Sykes (1967) failed to achieve systemic hypertension following occlusion of both carotids in the fowl. Certainly, in the pigeon and perhaps also the crow the rise in blood pressure following bilateral carotid occlusion appears to be enhanced if the blood pressure is low (Kovách *et al.*, 1969).

An interesting feature of the response to cerebral ischemia in the fowl is that, in the unanesthetized bird, heart rate falls following occlusion of both carotid and occipital arteries, whereas in the anesthetized fowl the heart rate increases. In fact, in respect to changes in heart rate, the unanesthetized fowl shows the same response as is evoked by stagnant hypoxia (produced by clamping the jugular veins) in the anesthetized bird (Richards and Sykes, 1967).

F. Blood Volume Regulation

Several species of birds are able to tolerate blood loss much better than mammals (Kovách and Szász, 1968; Kovách *et al.*, 1969). For instance, most pigeons survive blood loss of 8% of the body weight during prolonged hemorrhage. After hemorrhage of about 50–70% of the total blood volume, blood pressure and cardiac output fall; but within 30–240 minutes a partial recovery occurs. Blood pressure re-

Fig. 36. Pressoreceptors in the adventitia of the aortic arch of the chicken (a) Connective tissue bundle; (b) nucleus of connective tissue cell; (c) smooth muscle tissue; (d) nerve fibre; (e) varix; (f) intermediary plate; (g) end plate, Bielschowsky-Ábrahám's method. Microscopic magnification: 16,000 ×. (Ábrahám, 1969, Pergamon Press, by permission.)

turns to about 80%, and cardiac output to 55% of control values. Peripheral resistance increases in response to hemorrhage (Kovách and Szász, 1968), although treatment with dibenamine has no effect on survival (Kovách *et al.*, 1969). The significance of this latter result in this context is unclear since dibenamine fails to block the vasoconstriction induced by epinephrine and noreprinephrine in the pigeon.

There seems to be a marked difference between pigeons and mammals in the rate of fluid replacement and regulation of blood volume after hemorrhage. Kovách and Balint (1969) have shown that these differences become apparent only during prolonged bleeding because hemodilution continues in the pigeon through the period of blood loss, whereas in the rat no further hemodilution occurs after about 15–20 minutes of bleeding. Hemodilution is achieved by the inflow in isotonic fluid with a low protein content.

The restoration of blood volume most probably results from absorption of tissue fluid across the capillary walls during a reduced capillary pressure. This fall in capillary pressure could be brought about by an increase in the ratio between pre- and postcapillary resistances (Öberg, 1963, 1964) and the resulting changes in arterial and venous pressures. Resistance changes across the capillaries are by far the most important factor in rapid restoration of blood volume in ducks. Blockade of α-adrenergic receptors eliminates vasoconstriction in the skeletal muscle, which forms the major reserve of tissue fluid, and leads to a greatly retarded restoration of blood volume (Djojosugito *et al.*, 1968). At the same time the precapillary sphincters become secondarily relaxed, increasing the capillary surface areas available for absorption. Djojosugito *et al.* (1968) attribute the difference in ability to restore blood volume after hemorrhage in ducks and cats to a very pronounced reflex vasoconstriction in duck skeletal musculature and to a capillary surface area in ducks three to five times that in the cat, a condition that increases the rate of absorption of fluid into the vascular system (Folkow *et al.*, 1966).

VII. Cardiovascular Adjustments to (Habitual) Diving and Alterations in Ambient Gas Composition

A. Cardiovascular Responses to Natural Diving and Experimental Submergence

The tolerance to asphyxia among vertebrates is highest in the naturally diving forms. The physiological mechanisms responsible for the exceptional ability of diving animals to withstand oxygen lack that is fatal to nondiving vertebrates seem to differ only by degree and not in quality from the typical response pattern of vertebrates to asphyxia. For this reason the habitually diving animal has been a favorite subject for physiological studies. General reviews treating diving birds as well as mammals include those of Irving (1934, 1939), Scholander (1940, 1962, 1964), Andersen (1963, 1966), and Elsner and Scholander (1965). Brauer (1965, 1966) has provided two excellent tabular summaries of physiological adjustments to asphyxia and submersion in vertebrates.

The most conspicuous physiological responses to submersion of a diving animal are apnea and the prompt onset of marked cardiovascular changes. These include prominent bradycardia, redistribution of the regional blood flow in the face of relatively modest changes in systemic arterial blood pressure, and a large decrease in cardiac output. These changes serve to economize available oxygen stores. They are generally reflex in nature, and have come to be described by such terms as the diving reflex, diving response, or diving syndrome.

Other factors, such as the level of overall metabolic rate (Pickwell, 1968) and adaptive changes in enzyme kinetics (Blix *et al.*, 1970) to the reduced O_2 availability, are essential features in the tolerance of diving animals to asphyxia. Pickwell (1968) was able to show that total energy metabolism of ducks during prolonged submergence could be reduced as much as 90% from predive levels.

The cardiovascular system contributes to tolerance by rationing available oxygen stores through preferential perfusion of tissues particularly susceptible to damage from oxygen deprivation. The oxygen stores themselves are adaptively increased in diving forms by increases in pulmonary vital capacity, blood volume, blood oxygen capacity, buffering capacity, and other respiratory qualities of blood.

In this chapter emphasis will be given to a review of physiological factors and mechanisms involved in the elicitation of the cardiovascular adjustments to diving and to a description of the hemodynamic consequences and physiological significance of the adjustments for overall diving performance.

1. *Elicitation of Cardiovascular Adjustments to Submergence*

The profound bradycardia associated with diving of ducks was first described around the turn of the century by Richet (1899), Huxley (1913a,b,c), and Lombroso (1913). It was established early that an intact vagus was essential for the development of the bradycardia. More recently, it has been shown that the cardiovascular diving reflexes are independent of higher cerebral activity, since decerebrate ducks show the reflexes as well as intact animals (Andersen, 1963). Central integration of the reflexes is therefore surmised to be medullary. Feigl and Folkow (1963) and Folkow and Rubinstein (1965) have reported that the autonomic reflex adjustments elicited by submersion can be mimicked and reinforced by stimulation of discrete hypothalamic, mesencephalic areas. It is also well established that the response can be modified or conditioned by higher central nervous influence. Other sensory inputs may thus alert the animal, making it aware that diving is imminent and allowing it to anticipate the need for the cardiovascular adjustments.

With regard to the peripheral sensory mechanisms eliciting the diving reflexes, immersion of the head in water seems to be a triggering factor. It has also been claimed, however, that apnea, which is essential for the cardiovascular changes, can be precipitated by the postural changes a duck goes through when preparing for a dive. This alleged role of vestibular functions for the elicitation of bradycardia has been discredited by recent workers (Andersen, 1963; Butler and Jones, 1968). Koppányi and Dooley (1928) have claimed that insufficient ventilation and hypercapnia are responsible for the bradycardia during regular submersion and postural apnea. Recent workers seem to have established beyond doubt that the diving bradycardia is a consequence of apnea (Reite *et al.*, 1963; Butler and Jones, 1968; Cohn *et al.*, 1968). When a duck is provided with a tracheal cannula allowing continued voluntary breathing during water immersion, it becomes apparent that bradycardia is observed only when immersion evokes a concurrent diminution or cessation of breathing (Butler and Jones, 1968). Other workers contend that specific water-immersion receptors exist that trigger a cardiac deceleration when the beak tip touches water or the nares are artificially wetted (Andersen, 1963; Feigl and Folkow, 1963).

Conclusive proof exists that artificial ventilation with air during submergence partially or completely reverses the diving bradycardia (Andersen, 1963; Fiegl and Folkow, 1963; Reite *et al.*, 1963; Cohn *et al.*, 1968), but Andersen (1963) reports that complete recovery with normal post-dive tachycardia occurs only if tidal volumes much larger than normal are administered. In fact, the average value of tidal volume that occurs on surfacing is two to three times normal, and in this light Andersen's result is not surprising (Butler and Jones, 1968). Both Butler and Jones (1968) and Cohn *et al.* (1968) have found that birds would breathe spontaneously, through a tracheal cannula, after submersion of the nares and in these cases neither bradycardia nor other cardiovascular variations are recorded. This finding is at variance with

Andersen's result (1963) showing that this occurred only after trigeminal nerve section, and that the ophthalmic branch of the trigeminal nerve was the most important limb of the afferent nervous pathway.

Despite the conflict of evidence, it seems reasonable to presume that the pronounced circulatory adjustment is a direct consequence of the apnea, and that in the absence of apnea water immersion alone has little or no effect on the response. The importance of water immersion is in eliciting the apnea. The question is: where are the water receptors which, by their discharge, can successfully inhibit the respiratory center? Andersen (1963) claims that the receptors are located in the beak and nares and are innervated by the trigeminal nerve. But their action cannot be invariant or predominant, since animals continue to breathe through a tracheal cannula with the nares underwater (Butler and Jones, 1968; Cohn *et al.*, 1968). Butler and Jones (1968) recorded apnea only when water made contact with the glottis or some other area within the internal respiratory passages, thus confirming the earlier results of Huxley (1913a). Jones and Purves (1970a) dissected free and stimulated the central end of the glottal branch of the IXth cranial nerve on both sides and showed that stimulation caused apnea and bradycradia, the latter being as great as that observed in normal ducks during submergence (Fig. 37). It may be that under

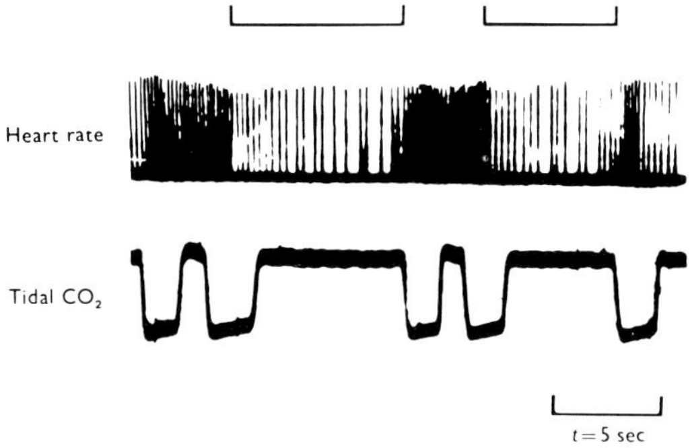

FIG. 37. The effect of stimulating the glottal branches on each side of the IXth cranial nerves, exposed under local anesthesia in a 9-week-old domestic duck, upon heart rate (upper trace) and tidal CO_2 (lower trace). The nerves were stimulated for the periods indicated by the solid black lines at 8 V, 8/second and 100 μsec duration. The CO_2 trace is intended to indicate the phase of respiration only, inspiration upwards, $t = 5$ seconds. [Jones and Purves, 1970a, *J. Physiol. (London)*, by permission.]

natural conditions stimulation of glottal receptors by water contributes substantially to any reflex inhibition of respiration initiated by receptors in the nares.

Bradycardia develops after apnea in the duck (Fig. 38), and the time course of this cardiovascular event is worthy of special consideration. There is an initial decrease of heart rate (after a latent period of about 6 seconds) which occurs rapidly with a half-time of 8 seconds, followed

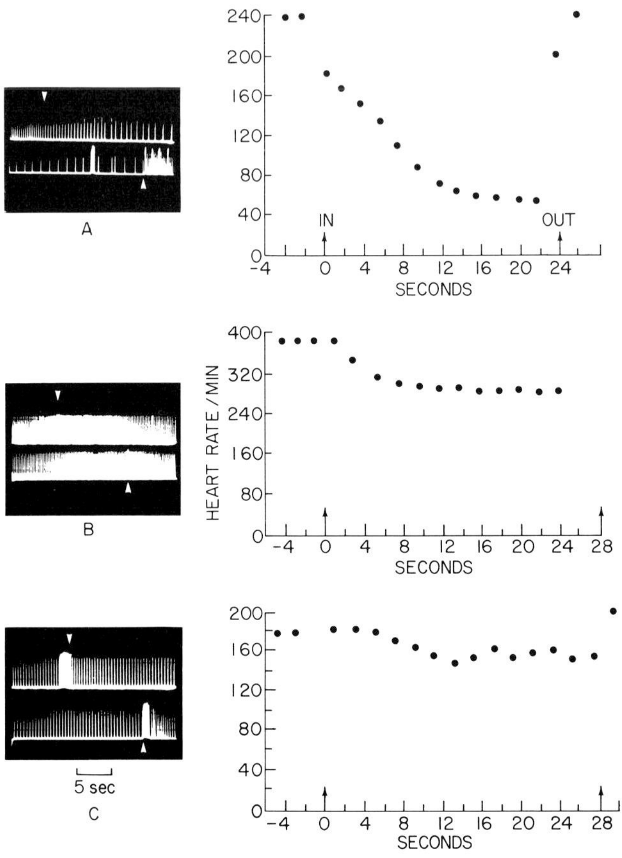

FIG. 38. The effect of submergence of the head in water upon heart rate in a 6-week-old Khaki Campbell duck. On the left, the oscilloscope record, the upper and lower traces being continuous and the period of submergence being indicated by the white arrowheads. The solid vertical patches on the oscilloscope traces represent short bursts of tachycardia associated with struggles. On the right, a plot of heart frequency per minute against time. A, control; unanesthetized duck with all nerves intact. B, 1 hour following operation in which the carotid bodies were denervated. C, 6 days later, unanesthetized duck. [Jones and Purves, 1970a, *J. Physiol. (London)*, by permission.]

by a slow component with a half time of 94 seconds (Jones and Purves, 1970a; Fig. 39). This sequence has been corroborated by many authors.

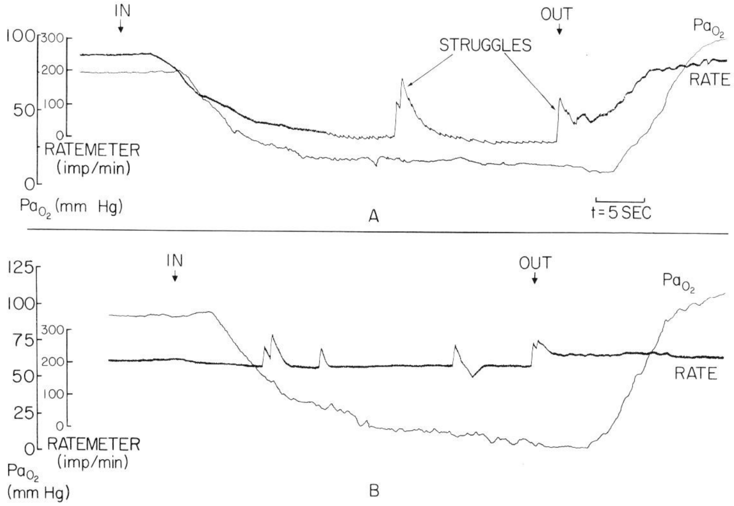

FIG. 39. Changes in heart beat frequency (rate-meter output, beats/minute) and in brachiocephalic artery oxygen tension (mm Hg) during a period of submergence of the head in water (indicated by the IN–OUT arrows) in a 6-week-old Khaki Campbell unanesthetized duck. A, control with all nerves intact; B, the same duck 3 weeks later, after denervation of the carotid bodies. [Jones and Purves, 1970a, *J. Physiol. (London)*, by permission.]

Following the latent period the heart rate invariably starts to decline before any fall in arterial O_2 tension, regardless of whether the duck is normal or has its carotid bodies denervated (Fig. 39). Consequently, the initial decline in frequency must be independent of carotid chemo- or baroreceptor activity. It may be that afferent activity from receptors in the lungs has a stimulatory effect on the heart when the duck is breathing normally, as has been reported for mammals by Anrep *et al.* (1936), so the lack of this stimulation during apnea may contribute to the initial fall in heart rate.

Several investigators have found that hypercapnia and hypoxia tend to potentiate the cardiovascular effects of submergence. Feigl and Folkow (1963) found hypercapnia to be more important than hypoxia, whereas Cohn *et al.* (1968) found that hypoxia accelerated the events and hyperoxia delayed them. This work indicates that chemoreceptors are likely to be involved in development and maintenance of the physiological responses to submersion. The proposition that carotid body chemoreceptors participate in the initiation of the cardiovascular responses to submersion is not without support from other sources. Daly and Scott (1958, 1962) have shown that the primary effect of stimulating carotid body chemoreceptors in the dog is bradycardia and vasoconstriction in muscle, skin, and intestine, but that this is obscured in spontaneously breathing animals by the reflex effects of the hyperpnea which develops. Further, Comroe and Mortimer (1964) have shown that selective stimulation of the carotid body chemoreceptors in the dog during apnea causes bradycardia and hypertension.

Hollenberg and Uvnäs (1963) divided the tissues surrounding the carotid bodies in the duck between ligatures, thereby eliminating the carotid body from any participation in the diving response. During submergence of these denervated ducks there was no bradycardia, and blood pressure rose markedly. Jones and Purves (1970a) found two major effects on the circulation after denervation of the carotid bodies in ducks. First, the profound bradycardia characteristic of intact ducks was not observed in animals with denervated carotid bodies, and heart rate fell by only about 10% during submergence (Fig. 38b,c). Second, the "biphasic" fall in arterial O_2 tension characteristic of intact animals (Hollenberg and Uvnäs, 1963; Jones and Purves, 1970a; Butler and Jones, 1971) was not so marked, and a more steady and rapid decline in Pa_{O_2} attended prolonged submergence (Fig. 39). This indicated that the oxygen-conserving mechanisms obtained from the bradycardia and the regional redistribution of the circulating blood were deranged. A pattern of change in arterial blood gas composition similar to that seen in denervated ducks was observed in ducks given an α-receptor blocking agent (Butler and Jones, 1971). These observations suggested the importance of carotid body chemoreceptors to both the central and peripheral cardiovascular adjustments. However, Holm and Sørensen (1970) suggested that the absence of diving bradycardia in the denervated ducks may indicate only a new circulatory state caused by the baroreceptor denervation that inhibited the diving response. The main evidence for this conclusion rested on the fact that stimulation of peripheral chemoreceptors by injection of cyanide and nicotine into the aorta of anesthetized ducks elicited hyperventilation and tachycardia, and not bradycardia.

The possibility remains open that baroreceptor activity determines in part the central and peripheral cardiovascular adjustments to diving (Andersen, 1966). Recent evidence indicates that if peripheral vasoconstriction during a dive is prevented by administration of an α-receptor blocking agent (dibenyline), blood pressure falls as heart rate declines. In all animals investigated in this manner, the bradycardia "breaks" after some time and heart rate and blood pressure return to control levels despite the continued submersion of the animal (Butler and Jones, 1971). However, since dibenyline also prevents neuronal or muscle uptake of catecholamines and thereby inactivation, the breakthrough may be due to gradually increasing circulatory catecholamine levels causing cardiac β-receptor stimulation. The response to diving after injection of atropine perhaps provides more convincing evidence for the involvement of baroreceptors in the response of the normal animal. Atropine not only affects cardiac chronotropic activity but also abolishes a negative inotropic response mediated via the vagus (Folkow and Yonce, 1967). However, in the face of an eightfold increase in vascular resistance, as exhibited by the normal duck during a dive, pressure should increase proportionately. However, mean blood pressure rises by only 50%, so there must be an overriding of vasoconstrictor activity which is probably mediated through arterial baroreceptors.

The efferent side of the cardiovascular diving reflex is mediated via parasympathetic and sympathetic pathways. Injection of atropine or bilateral vagotomy obliterates the cardiac chronotropic response to submergence (Butler and Jones, 1968). So, the heart must be exposed to greatly increased vagal activity which affects not only its rate but also its contractility (Folkow *et al.*, 1967; Folkow and Yonce, 1967). In addition, evidence of an inhibition of sympathetic tone to the heart during submersion has been provided by Folkow *et al.* (1967) but has not been confirmed by other investigators (Butler and Jones, 1968, 1971). Since α-receptor blockade prevents effective redistribution of blood flow during submersion it is reasonable to conclude that an intense sympathetic discharge to the various vascular beds attends a normal dive (Kobinger and Oda, 1969). Folkow *et al.* (1967) point out that an opposite change in sympathetic activity to the heart and blood vessels, such as demonstrated by their data, is contrary to the general view that sympathetic discharge must always be generalized. In fact, the changes in sympathetic activity during submersion proposed by Folkow *et al.* (1967) would indicate that the sympathetic nervous system is capable of highly differentiated action.

When a duck surfaces, the cardiovascular adjustments established during the dive are completely reversed at the time breathing is resumed, but not before (Butler and Jones, 1968, 1971). From the onset of breathing a pronounced tachycardia is established apparently independent of changes in arterial blood gas composition (Jones and Purves, 1970a). Restoration of normal blood gas tensions with concomitant changes in chemoreceptor drive does not appear to be important for the heart rate changes. The balance of evidence suggests that afferent impulses from pulmonary stretch receptors and other parts of the respiratory system, triggered when breathing starts, are involved (Butler and Jones, 1968). Other receptors such as intrapulmonary receptors sensitive to changes in carbon dioxide in the lung may be involved (Fedde and Petersen, 1970). Andersen (1963) suggests that an increase in venous return, which probably takes place on emersion, is more important for abolition of diving bradycardia. Termination of a dive causes a marked decrease in vagal activity and a likely increase in sympathetic discharge to the heart combined with a reduction in sympathetic activity at the periphery. Since there is a rise in blood pressure at the end of the dive it appears that the central and peripheral components are not in phase but that the former occur faster than the latter (Butler and Jones, 1971). Owing to the fact that venous pressure is high at the end of the dive (Johansen and Aakhus, 1963; Folkow *et al.*, 1967) the abolition of vagal restraint on the heart leads to a prompt increase in cardiac output. However, Butler and Jones (1968) could find no evidence for the participation of the sympathetic nervous system in the cardiac chronotropic response to submergence and recovery. β-Blockade has no effect on the post-dive tachycardia, and a tachycardia of similar proportions can be achieved by cold-block of the "active" vagus even during a dive. In other words, it is envisaged that during recovery vagal tone is reduced to a subnormal level that provokes the tachycardia. These findings are in agreement with those of Kobinger and Oda (1969), who have investigated the effects of bretylium and guanethidine on the recovery of heart rate at the termination of submergence.

Most, if not all, studies on cardiovascular adjustments to diving or submersion in birds have been done on animals subjected to various degrees of restraint. Such forced dives may distort the cardiovascular responses considerably from those occurring in a freely diving, unrestrained animal. In a recent study on two species of penguins, *Pygoscelis adeliae* and *Pygoscelis papua*, radiotelemetry was utilized to transmit cardiovascular information from freely swimming animals (Millard *et al.*, 1972). After implantation of Doppler shift flowmeter transducers on carotid and femoral arteries and a chronic catheter in a systemic artery for blood pressure measurements and blood sampling, the birds were allowed to swim and dive freely towing a lightweight Styrofoam float carrying the miniaturized FM radiotelemetry equipment. When diving voluntarily, penguins practiced a "porpoising"-like swimming and diving. A rapid pace of forward, underwater swimming was interrupted at variable intervals of 20 seconds to 120 seconds by a flight-like motion at the surface of the water, allowing split-second breathing before plunging back and continuing the same underwater swimming. During these voluntary dives, bradycardia was always evident, although in general to a lesser degree than when restrained birds were submerged in a tub. Frequently, an anticipatory increase in heart rate preceded emersion and breathing, particularly towards the end of the longer breath-holding periods (Fig. 40).

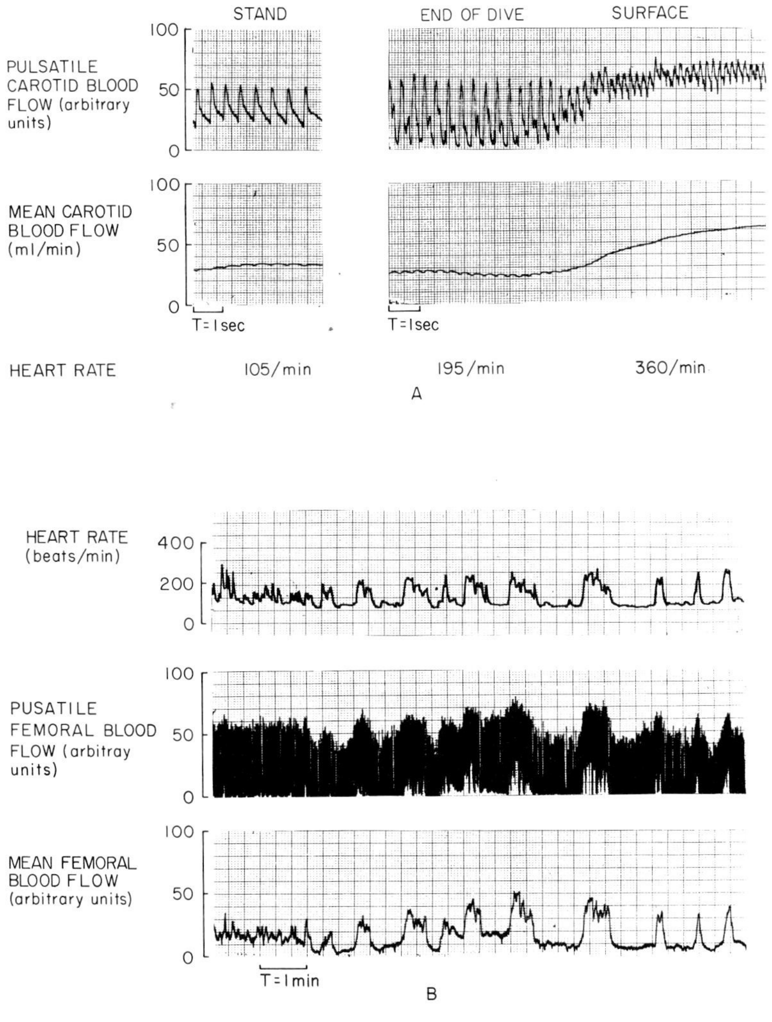

FIG. 40. Cardiovascular adjustments in the Adélie Penguin (*Pygoscelis adeliae*) during voluntary, unrestrained diving. Carotid blood flow (A) and fermoral blood flow (B) obtained by FM telemetry. Diving periods during femoral flow are designated by bars (Millard and Johansen, 1972).

Femoral arterial flow was reduced by no more than 50% of the values obtained during breathing. The flow did not fall to zero between beats as was observed during forced dives. It should be noted that the legs of swimming penguins are not actively used in underwater propulsion. Carotid flow of birds in water in a surfaced position was higher than values obtained when quiet on land. During dives, the carotid flow decreased although proportionately not as much as femoral flow and never below the values recorded when the birds were on land. The reduction in arterial O_2 tension and concurrent increase in arterial pH were far more pronounced during forced dives than during voluntary diving. Figure 40 shows tracings of carotid and femoral flow changes during voluntary dives transmitted by FM telemetry. These free-dive studies on penguins suggest that the quality of the diving responses are in accord with the responses to forced diving but the magnitude of the latter indicates the need for caution when discussing the physiological implications of the responses.

2. *Homeostatic Significance of the Cardiovascular Adjustments to Diving*

The principal advantage occurring from the cardiovascular adjustment during diving is the preferential perfusion of selective vascular beds most susceptible to anoxic damage. These include the brain, heart, sensory organs, and some endocrine glands. The profound redistribution of blood flow, with almost complete curtailment of blood flow to skeletal muscle, skin, gastrointestinal organs, and kidneys, was first documented for skeletal muscle by means of the thermostrohmuhr technique (Irving, 1938). A shut-down of muscular blood flow has also been demonstrated with plethysmographic techniques (Johansen and Krog, 1959; Folkow *et al.*, 1966), by a drop-flow technique (Djojosugito *et al.*, 1969), and by fractional distribution of $Rb^{86}Cl$ (Johansen, 1964). A reduction in blood flow to other vascular beds such as skin and gastrointestinal organs in submerged ducks has also been demonstrated using similar techniques (Hollenberg and Uvnäs, 1963; Johansen, 1964).

Inference about the flow distribution can be gained from the technique using fractional distribution of $^{86}RbCl$. Figure 41 summarizes the relative difference in blood flow distribution to various organs of ducks expressed as the percentage of ^{86}Rb activity deposited. Note that the activity to the heart, eye, and adrenal glands is actually higher than pre-dive values. This must not be taken to show an increase in oxygen availability to these organs, since arterial O_2 saturation steadily declines during a dive. One potentially important finding emerging from this study and parallel studies on blood flow distribution during arousal from hibernation in mammals (Johansen, 1961) concerns the possibility that the increased sympathetic vasoconstrictor tonus may be segmentally oriented, since the tissues and organs posterior to the heart show a profound vasoconstriction, while tissues anterior to the heart appear to show an increased blood flow. This idea has never been contested or examined with the use of other techniques.

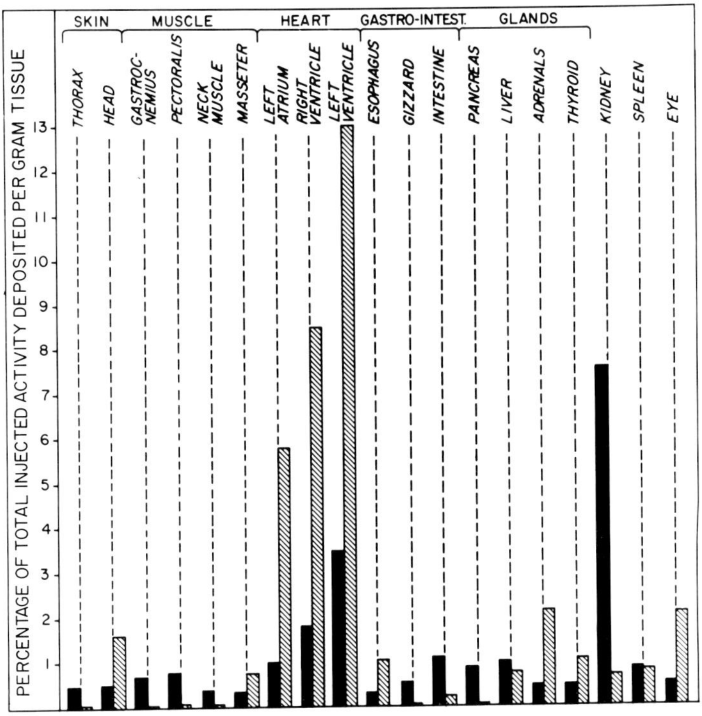

FIG. 41. Percentage of total injected activity of ^{86}Rb deposited per gram tissue in various organs of domestic ducks. The black bars indicate values during normal breathing, the striped bars values obtained during submergence. The graph represents average values from eight ducks. (Johansen, 1964, *Acta Physiol. Scand.*, by permission.)

The interesting study on muscle and cutaneous blood flow adjustments to submersion in ducks by Djojosugito *et al.* (1969) merits considerable attention. While there is a massive muscle vasoconstriction with virtual cessation of blood flow, the vasoconstrictor fibers to the prominent arterio-venous anastomoses in the skin of the web are not activated. The authors contend that this is to allow a significantly improved utilization of the venous oxygen stores during diving by bringing the large peripheral venous volumes into the circulation and thus making O_2 available for the selectively perfused organs such as the brain and heart. They argue that without such a prominent nonnutritive shunt flow the peripheral venous depots might become stagnant. However, measurement of blood flow in the sciatic artery during a dive shows it to be less than 4 ml/minute in ducks of 2–3 kg body weight, so it seems unlikely that flow of this order will contribute much to prevention of stagnation on the venous side of the system (Butler and Jones, 1971).

There is a scarcity of information about cardiac output during diving in birds. Using thermodilution and dye dilution techniques, Folkow *et al.* (1967) have recorded up to a twentyfold reduction in heart rate and cardiac output during submersion. The reduction in cardiac output is mainly parallel to the decline in heart rate, but not entirely since a 10–20% decline in stroke volume occurs. However, direct recordings of blood flow in the pulmonary artery have shown, on the average, no change in stroke volume during diving (Jones and Holeton, 1972b; Fig. 42). In any event these findings merit attention in relation to

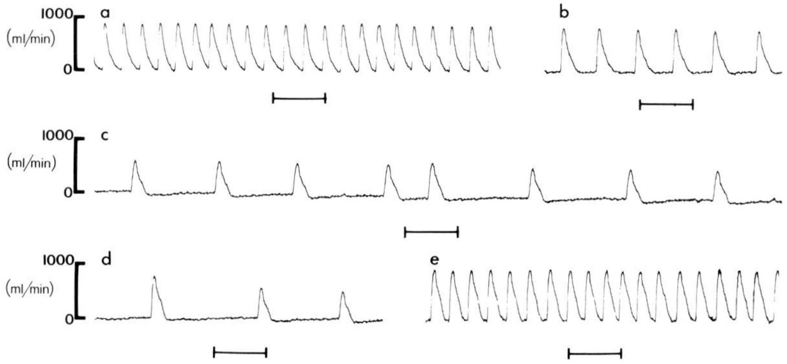

FIG. 42. Blood flow in the pulmonary artery before, during, and after submergence in a domestic duck. a, pre-dive. b, dive – 30 seconds. c, dive – 90 seconds. d, dive – 120 seconds. e, recovery – 20 seconds. The marker under each trace indicates one second (Jones and Holeton, 1972b).

data showing an increase in central venous pressure (Johansen and Aakhus, 1963; Folkow *et al.*, 1967) and end-diastolic size of the heart (Aakhus and Johansen, 1964) in diving ducks. It therefore seems safe to assume that in some ducks there is a decrease in myocardial contractility during submergence. Folkow and Yonce (1967) have shown that the contractility of the heart ventricles in the diving duck is markedly depressed by an increased vagal tonus, which also forms the efferent pathway for the cardiac chronotropic response (Butler and

Jones, 1968). The reduced cardiac output during diving affords some metabolic saving by reducing the work done by the heart.

An increased resistance to pulmonary outflow was surmised by Eliassen (1960) from pressure profiles recorded in the pulmonary artery of submerged ducks. Angiocardiography of ducks during submersion (Aakhus and Johansen, 1964) similarly revealed that the caliber of the pulmonary arteries was much reduced during diving. The conjectured increase in pulmonary vascular resistance may be related to an increase in intrathoracic pressure during the dive or may be caused by a direct effect of the low alveolar oxygen tension on the pulmonary vascular smooth muscle, as described for mammals in general (Duke, 1957). The presence of a decreased transmural pressure in the pulmonary vascular bed will reduce the pulmonary blood volume during submersion. This fact in connection with a marked reduction in pulmonary flow may offer a rational adjustment to the reduced oxygen availability in the lung during prolonged asphyxia.

B. Cardiovascular Responses to Changes in Environmental Gas Composition

1. *Hypoxia*

Exposure of chickens to hypoxia causes an initial increase in heart rate and a decrease in blood pressure (Durfee and Sturkie, 1963). Butler (1967b) and Richards and Sykes (1967) found that the tachycardia was transient, being followed by bradycardia when Pa_{O_2} falls to around 40–50 mm Hg. Both authors reported a fall in mean blood pressure, although in one case not until bradycardia was evident (Butler, 1967b). The tachycardia is prevented by bilateral vagotomy and adrenergic β-receptor blockade, but not the later bradycardia and fall in blood pressure, which appear to be due to a direct effect of hypoxia. Butler (1967b) reported a continued increase in respiratory frequency, whereas Richards and Sykes (1967) found, after an initial increase, a marked decline which was accompanied by acidosis. However, Ray and Fedde (1969) using a unidirectional artificial respirator held inspired carbon dioxide tension constant and obtained hypotension and tachycardia in chickens exposed to inspired oxygen tensions of 90 mm Hg or less. Some of the variability in the above results may be due to the fact that the experimental techniques used may or may not entail change in arterial carbon dioxide tensions, which may have pronounced cardiovascular effects.

The response of chickens to hypoxia is not mirrored by that of other birds, such as the pigeon and duck. These birds show only slight hypotension at reduced Pa_{O_2} (Pa_{O_2} < 50 mm Hg) and maintain heart rate at its control value until Pa_{O_2} is about 35 mm Hg, below which tachycardia occurs (Butler, 1970). Jones and Purves (1970b) have found that when Pa_{CO_2} is maintained at control levels in ducks, the heart rate responses at a Pa_{O_2} of 37–45 mm Hg are variable but never more than 14% changed from control values.

Cardiac output increases in the chicken during hypoxia due to increases in both heart rate and stroke volume. However, there is little change in total sciatic artery flow, but a marked change in the nature of the flow pulse, which occasionally reaches zero flow (Fig. 43). β-Adrenergic blockade reduces the cardiac output during hypoxia, most likely and largely due to the fact that tachycardia is eliminated.

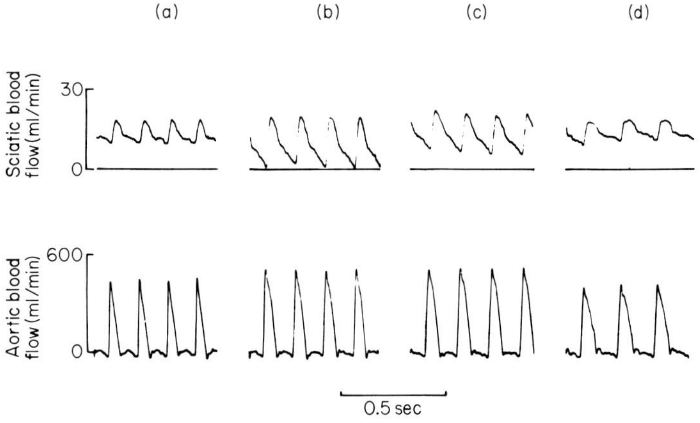

FIG. 43. Intact cockerel, 996 gm. Effect of hypoxia and hypercapnic hypoxia on blood flow in the descending aorta and sciatic artery. Animal breathing: (a) air, (b) 10% O_2 for 5 minutes, (c) 10% O_2 for 10 minutes, (d) 10% O_2 + 5% CO_2 for 2.5 minutes (Butler, 1967a).

Pulmonary arterial pressure increases during hypoxia due to the increase in cardiac output, pulmonary vasoconstriction, or both. However, the rise in pulmonary arterial pressure due to breathing 10% O_2 in N_2 is prevented by α-blockade, indicating the importance of vasoconstriction to the response (Butler, 1967a,b).

The effects of chronic hypoxia on the cardiovascular system of birds have been studied by exposing the animals to simulated or actual altitudes. Acclimation to hypoxia causes many changes in cardiovascular parameters, such as increases in hemoglobin concentration, total circulating erythrocyte volume, total blood volume, and cardiac ventricular hypertrophy and chronic pulmonary arterial hypertension (Viault, 1892; Jolyet and Sellier, 1895; Sellier, 1896; Schuman and Rosenquist, 1897; Kocian, 1936; Vezzani, 1939; Rostorfer and Rigdon, 1945; Smith and Abbott, 1961; Jalavisto *et al.*, 1965; Dunson, 1965; R. R. Cohen, 1965, 1969a,b; Rosse and Waldmann, 1966; R. R. Burton and Smith, 1967; R. R. Burton *et al.*, 1968). In the chicken, the hypoxia at 3800 m altitude (barometric pressure 470 mm Hg) causes a 47% reduction in arterial oxygen tension with no significant changes in arterial carbon dioxide tension and pH.

Altland (1961) reported a greater altitude tolerance in pigeons than in chickens, and Rostorfer and Rigdon (1946, 1947) showed that ducks are more tolerant to simulated high altitudes than are chickens. It is interesting to note that newly hatched chicks (less than 8 days old) are 85% more tolerant to hypoxia than adults (R. R. Burton and Carlisle, 1969). The polycythemia exhibited by chronically hypoxic chickens is also present in ducks and pigeons (R. R. Cohen, 1969a; McGrath, 1970). Erythrocyte counts increase rapidly during the first 20 days of hypoxia in pintail ducks. Venous hematocrit rises by 30%, total erythrocyte volume by 68%, and total blood volume by 23% during simulated altitude of 7000 m (310 mm Hg barometric pressure).

Ventricular hypertrophy has been noted in chickens as a result of chronic hypoxia and has been related to the birds' hematocrit, presumably as an adaptation to increased viscosity of the blood (R. R. Burton and Smith, 1967). McGrath (1970) similarly observed right ventricular hypertrophy in response to chronic hypoxia in pigeons. In the chicken, R. R. Burton and Smith (1967) argue that right ventricular hypertrophy, unlike left, is independent of the hematocrit and more likely due to chronic pulmonary arterial hypertension exhibited by birds at 3800 m elevation. Pulmonary arterial blood pressures are about twice those at sea level and have a good correlation to the relative right ventricular mass of the individual birds (R. R. Burton *et al.*, 1968). In the female chicken, each 1 mm Hg increase in the mean pulmonary arterial pressure results in a 41 mg increase in right ventricular mass. Venous hematocrit and blood volume elevated by acclimation to hypoxia are readjusted to normal values within 20 days after the birds return to normal environmental oxygen tensions (R. R. Burton *et al.*, 1968; R. R. Cohen, 1969b).

2. *Hypercapnia*

Cardiovascular responses to hypercapnia seem to vary depending on the condition of the animal. Anesthetized chickens (Richards and Sykes, 1967; Ray and Fedde, 1969). show an increase in heart rate in response to levels of 1–20% CO_2 in inspired air, whereas Butler (1967a,b), working with unanesthetized chickens, reports a bradycardia and fall in cardiac output when CO_2 in inhaled air is 5%. Ray and Fedde (1969) could find little effect of hypercapnia on blood pressure. The bradycardia reported by Butler (1967a,b) was not abolished by either β-adrenergic blockade or vagotomy, indicating a direct effect of carbon dioxide on the pacemaker of the heart. Hypercapnia (5% CO_2 in inspired air) has little effect on pulmonary arterial pressure (Butler, 1967a,b). Bradycardia during hypercapnia (arterial CO_2 tension $>$ 65 mm Hg) has also been recorded in ducks (Jones and Purves, 1970b).

3. *Combined Hypoxia and Hypercapnia*

Breathing hypoxic-hypercapnic gas mixtures causes varying degrees of bradycardia in the chicken; the reduction in heart rate varies from 10 to 30% (Richards and Sykes, 1967; Scheid and Piiper, 1970). A blood pressure increase has been reported by Butler (1967a), whereas Richards and Sykes (1967) and Scheid and Piiper (1970) found a reduced blood pressure. In Butler's (1967a) experiments the increase in pressure was abolished by α-adrenergic receptor blockade suggesting that, in unanesthetized birds, carbon dioxide increases sympathetic constrictor activity. A point requiring further investigation is that bradycardia caused by hypoxic hypercapnia is also abolished by β-receptor blockade (Butler, 1967a). Hypercapnia together with hypoxia causes a reduction of the elevation in pulmonary artery pressure caused by hypoxia alone.

It seems pertinent to differentiate between the effects of asphyxia and those caused by submergence. During asphyxia, induced by blocking or clamping the trachea, birds are seldom quiet and often attempt respiratory movements. During submergence birds are generally quiet and do not, for obvious reasons, attempt to breathe.

In the anesthetized chicken, 1 minute of tracheal occlusion causes an initial tachycardia followed by a slight bradycardia. During tachycardia, blood pressure falls by about 36% but later increases so that when bradycardia is most pronounced blood pressure is about 21% above control values (Richards and Sykes, 1967). Asphyxia in chickens also causes a doubling of pulmonary arterial pressure (R. R. Burton *et al.*, 1968). In ducks, tracheal occlusion results in a reduction in heart rate that is not as pronounced as that occurring during submergence (Andersen, 1963; Feigl and Folkow, 1963). A stimulating effect of the attempted breathing movements on heart rate is a likely explanation for this difference (Butler and Jones, 1968).

VIII. Cardiovascular Performance During Flight

Flight is the typical form of avian locomotion, but due to technical difficulties in measurement there is as yet little information on the cardiovascular adjustments to flight. Oxygen consumption increases during flight by about 5 times over resting values for the Budgerigar *(Melopsittacus undulatus)* (Tucker, 1968b) and 12–14 times resting for the pigeon [Tucker (1968b) from data of LeFebvre (1964) and Lasiewski and Dawson (1967)], Evening Grosbeak *(Coccothraustes vespertinus)*, and Black Duck *(Anas rubripes)* (Berger *et al.*, 1970b). Since the amount of oxygen consumed is transported from the lungs to the tissues by the cardiovascular system we may conveniently discuss adjustments to flight by using the Fick equation for O_2 transport:

$$V_{O_2} = \text{heart rate} \times \text{stroke volume} \times \text{A–V } O_2 \text{ difference}$$

To meet an increased oxygen demand a bird can increase one or more of the right-hand factors in the above equation. On the basis of the few recorded values and some assumptions, the relative importance of the various factors contributing to oxygen delivery is discussed below.

A. Heart Rate

From recordings of blood pressure, Eliassen (1963) concluded that there was no change in heart rate in the Great Black-backed Gull (*Larus marinus*) during rest and in flight covering about 100 m distance. However, O. Z. Roy and Hart (1966) and Hart and O. Z. Roy (1966) reported increases in heart rate in the pigeon of just over four times the resting level during short flights. Heart rate also showed a relatively slow post-flight recovery, which may be indicative of an oxygen debt incurred during the exercise. From an investigation of 12 species of birds during rest, in flight, or immediately post-flight in comparison with data from other authors, Berger *et al.* (1970b) established that the relationship between heart rate and body weight was $HR = 12.4W^{-0.209}$ in resting birds, and $HR = 25.1W^{-0.157}$ in flying birds (where HR = heart rate in beats/second, W = body weight in grams). The exponents are significantly different between rest and flight and indicate that the increase in heart rate during flight is less in smaller birds (about 2 times resting) than in larger birds (about three to four times resting) (Berger *et al.*, 1970b). Consequently, heart rate can increase between two to four times depending on the species.

B. Stroke Volume

The stroke volume of humans and dogs may double during exercise (Astrand *et al.*, 1964; Ceretelli *et al.*, 1964). Few measurements of stroke volume in birds have been made, particularly under conditions when changes could be expected. Folkow *et al.* (1967) have reported an increase in stroke volume of about 1.4 times during the immediate post-dive recovery in the duck. If stroke volume can increase under these conditions, it would seem reasonable to assume increases of the same order during vigorous exercise. Millard *et al.* (1972), studying freely exercising penguins (walking, running, skidding) on land by radiotelemetry of femoral blood flow and systemic arterial blood pressure, recorded heart rate increases from 90 to 180/minute upon transition from rest to vigorous running. Femoral blood flow rose concomittantly to four times the resting value (Fig. 44). Mean sys-

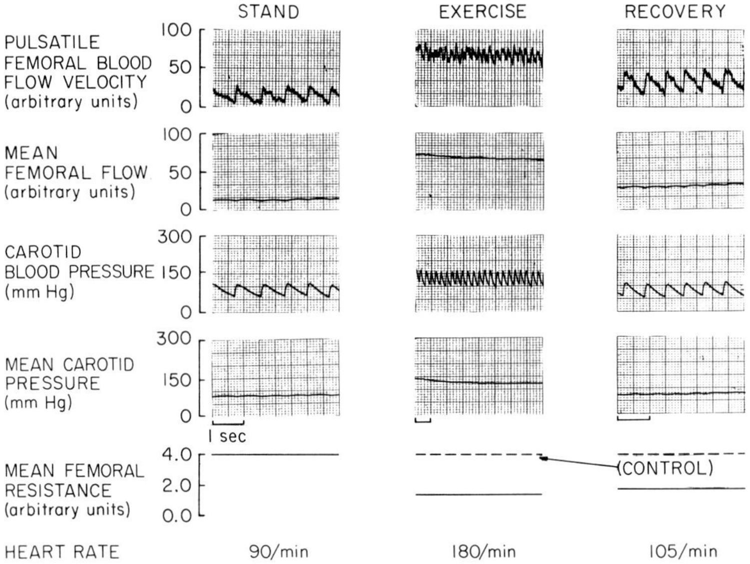

FIG. 44. Alterations of femoral blood flow, arterial blood pressure, and heart rate in an unrestrained Gentoo Penguin (*Pygoscelis papua*) in response to exercise on land. Data were obtained by FM radiotelemetry (Millard and Johansen, 1972).

temic arterial blood pressure increased from 80 to 125 mm Hg, while femoral resistance decreased by 70%. Treadmill exercise, to which the penguins adjusted easily, evoked far smaller responses, with the femoral blood flow increase less than twice resting levels.

C. Arterial–Venous Oxygen Difference

Folkow *et al.* (1967) reported the arterio-venous oxygen (A–V O_2) difference in ducks as about 3–5 vol %. Since duck blood has an oxygen capacity of at least 15 vol % (Wastl and Leiner, 1931), ducks should have a large venous oxygen reservoir for utilization during exercise. However, Wastl and Leiner (1931) reported duck A–V O_2 difference as high as 9.5 vol %, which would reduce the venous oxygen stores markedly. Jones and Holeton (1972a) measured arterial and venous oxygen tensions in ducks and obtained [using the oxygen dissociation curve of Andersen and Løvø (1967) and data of Wastl and Leiner (1931)] an A–V O_2 difference of about 6.5 vol %. Obviously, the functional significance of a venous oxygen reservoir depends not only on the A–V O_2 difference but also on the oxygen capacity and general respiratory properties of the blood. For instance, Lenfant *et al.* (1969) point out that flying birds need to facilitate oxygen supply to the tissues, and for this reason can be expected to have blood properties that display a low affinity for oxygen and a large Bohr effect.

In ducks during hypocapnic hypoxia (Jones and Holeton, 1972a), venous oxygen tension fell to around 30 mm Hg, a value which might be easily expected during exercise. If arterial oxygen tension remained unchanged, an A–V O_2 difference of about 12 vol % would obtain, or nearly double that in the resting animal. Other flying birds appear to have large venous oxygen reserves; for instance, the oxygen capacity of pigeon blood is 20 vol %, while the A–V O_2 difference in resting pigeons is only about half this value.

It is of considerable interest that Folkow *et al.* (1967) have commented on the exceptionally high blood flow to skeletal musculature in resting ducks compared with cats, inferring from visual observation that in ducks at rest the venous blood has a very high oxygen saturation. Consequently, ducks have a very large venous oxygen reserve on hand for swiftly changing needs in oxygen transport. Another significant consequence of the high venous O_2 tensions in birds at rest would be steep gradients in O_2 tension between the tissue capillaries and the cells and (or) high O_2 tensions in tissues, both of which may be of great importance to birds prepared for flight. The generally low affinity of avian hemoglobin for O_2 would tend to amplify this factor still further.

D. Oxygen Uptake (A × B × C)

The maximum potential for compensatory increase in oxygen transport is about twelvefold above the resting level (heart rate increases four times, stroke volume increases one and a half times, A–V O_2 difference increases two times). However, for small birds that exhibit increases in heart rate of only about two times, the product of stroke volume and the A–V O_2 difference must increase by six times above the resting value if an increase in oxygen transport of twelve times the resting value is to be achieved. Tucker (1968b) reports increases in oxygen uptake of only five times in flying Budgerigars weighing 35 gm. This increase in oxygen uptake could easily be met within the general scheme of compensatory capacity for O_2 delivery outlined above.

If, in order to increase O_2 transport, the flying bird depends in part on an expansion of the A–V O_2 difference during flight, then the ratio between ventilation ($\dot{V}_L$) and perfusion ($\dot{Q}_b$) of the lungs cannot remain constant. Since the available evidence suggests that expired oxygen tensions do not change during exercise, the increase in oxygen consumption must be matched by a similar increase in ventilation (Berger *et al.*, 1970b). Consequently, if oxygen consumption and ventilation increase twelve times, then a $\dot{V}_L:\dot{Q}_b$ ratio of 1 at rest will increase to 2 during flight, with a doubling of the A–V O_2 difference.

An additional burden on oxygen transport during flight exists when birds are flying at very high altitudes. There are numerous reports of birds flying at altitudes of 6000 m or more (Aymar, 1935; Meinertzhagen, 1955; Lack, 1960; Nisbet, 1963; Manville, 1963). A record altitude of 7940 m was reported by Hunt (1954). At an altitude of 8000 m a bird must cope with an oxygen availability that is only slightly more than 1/3 that at sea level. Tucker (1968a) has investigated this problem by exposing House Sparrows *(Passer domesticus)* to a simulated altitude of 6100 m and an environmental temperature of 5°C. The latter tended to raise metabolic rates above the basal level. At 6100 m, oxygen uptake in sparrows at rest was 2.2 times basal, and both heart and breathing frequencies were higher than at sea level. Despite an increase in ventilation of 77%, oxygen extraction increased from 7 to 12% of the effective ventilation. Carbon dioxide and oxygen tensions in the lungs fell by 12 and 72 mm Hg, respectively, from the normal sea-level values. At this oxygen tension the blood was only 24% saturated with oxygen. However, the birds remained normally active, unlike mice, which under the same conditions died after prolonged exposure. Tucker (1968a) has calculated for a hypothetical sparrow flying at 6100 m, with an oxygen demand of 8 times basal, that effective ventilation will be 11.6 liters/kg·minute and, if venous blood contains no oxygen, cardiac output will be 2.78 liters/kg·minute; this gives a ventilation/perfusion ratio of about 4:1. House Sparrows can fly for 2 to 5 m and can gain height at the simulated altitudes used by Tucker (1968a).

In any event, flight at high altitudes must place extraordinary demands on the cardiovascular and respiratory systems, although Dolnik (1969) has reported that the increase in oxygen uptake of migrating birds is much lower than for individuals that only indulge in bursts of activity. Ignoring prolonged anaerobiosis as a contributor to the energy balance during sustained flight at high altitudes, it would appear that high flying birds possess a considerable resistance to tissue anoxia, which is also reflected in the ability to retain control of cardiovascular functions at low Pa_{O_2} (Butler, 1970). The respiratory properties of blood, such as the low affinity for oxygen and the pronounced Bohr shift seen in most birds, are advantageous to O_2 transport since the disproportionate increase in ventilation must bring the bird into respiratory alkalosis (Jones and Holeton, 1972a).

A recent study of cardiorespiratory responses to exercise in penguins (Millard *et al.*, 1972) revealed a consistent expansion of the A–V P_{O_2} difference, sometimes corresponding to more than a threefold increase in O_2 utilization. Notably, this change resulted not only from a reduction of P_{O_2} of mixed venous blood but also from a consistent increase in the arterial P_{O_2}. Surprisingly, resting arterial P_{O_2} was quite low, often corresponding to arterial saturations between 70 and 80%. It can be calculated from the data of Millard *et al.* (1972) that the Gentoo Penguin, *Pygoscelis papua*, could change its perfusion requirement, $\dot{Q}_b/\dot{V}_{O_2}$, from about 18 ml blood/ml O_2 during rest to about 5 ml blood/ml O_2 after 12 minutes of walking on a treadmill at 2 miles/hour.

IX. The Cardiovascular System and Gas Exchange

Much discussion has been given to the point that unequal distribution of ventilation and perfusion in various parts of the mammalian lung will lead to regional inequalities in gas exchange (Rahn and Farhi, 1964; West, 1970). To achieve maximum efficiency in gas exchange the ventilation/perfusion ratio must be uniformly distributed and at an optimum value throughout the lung. Owing to the configuration of the gas-exchanging surfaces of the avian lung the adverse effects of such factors as gravity (West, 1970) can be eliminated, although an unchanging ventilation/perfusion ratio throughout the lung is probably not present (Zeuthen, 1942; Scheid and Piiper, 1969, 1970). Even if the ventilation/perfusion ratio is uniform, the problem remains to estimate its optimal value, both from the point of view of gas tensions within the lung and overall transport performance (Rahn and Farhi, 1962).

A general method of assessing the overall ventilation/perfusion ratio for 95% oxygen saturation of arterial blood has been presented by Jones *et al.* (1970). The analysis relates equations for oxygen uptake (from the air being ventilated and by the blood being perfused through the lungs) by means of a factor ΔP_A, which describes the "resistance" of the whole system to gas exchange in terms of millimeters of mercury. The factor ΔP_A includes the effects of air and blood shunts as well as any temporal or spatial diffusion barriers within the lung. The results of such an analysis extended to the duck are illustrated in Fig. 45. The horizontal bars represent the ventilation/perfusion ratio necessary to maintain a given arterial oxygen tension. For 95% saturation of arterial blood the required ventilation/perfusion ratio is 1.

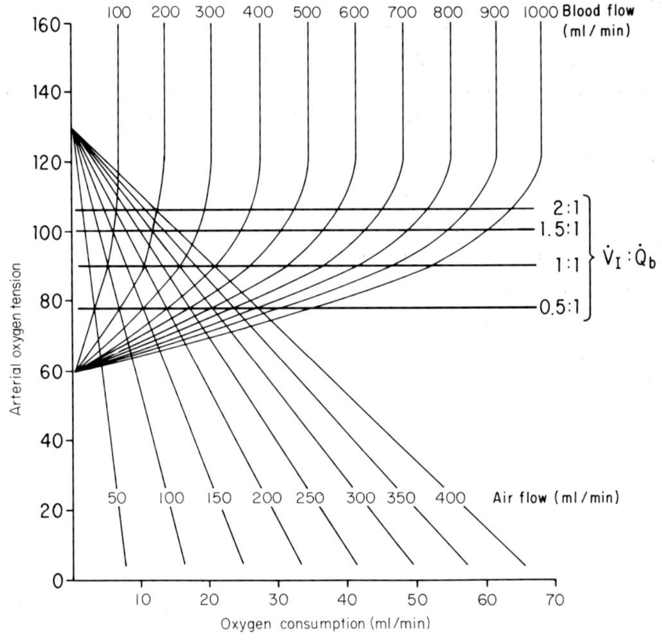

FIG. 45. A graphical analysis of gas exchange in the domestic duck. The horizontal bars indicate the overall ventilation : perfusion ratio for a given arterial oxygen tension. A ratio of unity gives an arterial oxygen tension of 90 mm Hg, a level commonly recorded in the duck. Based on data of Jones and Holeton (1972a).

Few studies have been directed to recording ventilation/perfusion ratios, although many recordings of ventilation and some of cardiac output alone have been made (Sapirstein and Hartman, 1959; Sturkie and Vogel, 1959; Sturkie, 1965; King and Payne, 1964; Butler, 1967a,

1970; Folkow *et al.*, 1967). Piiper *et al.* (1970) and Jones and Holeton (1972a) have measured ventilation/perfusion ratios in the chicken and duck, respectively. It can be calculated from the mean values of Piiper *et al.* (1970) that the ventilation/perfusion ratio for chickens is 1.78. Taking approximate mean values from the literature, the ventilation/perfusion ratio, corrected for difference of body weight, would come out closer to 1 for chickens. However, in the duck, Jones and Holeton (1972a) have recorded ventilation/perfusion ratios in the range predicted by the analysis. In resting birds, the total lung ventilation is less than perfusion, with an average value for the ventilation/perfusion ratios around 0.85. Ratios of this order might be expected from data in the literature, since the cardiac output of resting ducks has been variously reported between 300 and 600 ml/kg·minute (Sturkie, 1966; Folkow *et al.*, 1967), whereas ventilation volume is in the range of 3–400 ml/kg·minute (Butler and Jones, 1968; Jones and Holeton, 1972a). The ventilation/perfusion ratio established experimentally for ducks is in the range of optimal ventilation/perfusion ratio in man at rest, although the latter is more strictly defined and refers to the relationship at the gas exchanging surface (the alveoli). During hypocapnic hypoxia the ventilation/perfusion ratio of ducks increases, and this might be predicted as the appropriate response, since the oxygen dissociation curve is shifting to the left and a given reduction in the oxygen content of the air will not provoke the same reduction in oxygen content of the blood.

In the analysis, the magnitude of ΔPA is obviously important, since changes in air and (or) blood flow are related to ΔPA in a complex manner. The time that blood resides in the lung capillaries will depend on the cardiac output and on the total volume of blood that can be contained in the capillaries. In adult chickens the amount of blood in the lungs is between 4 and 6 ml/kg (R. R. Burton and Smith, 1968), so at normal cardiac output residence time in the capillaries is very short, of the order of 1–3 seconds. In mammals, red blood cells remain in contact with the alveolar surface for about 0.75 second, while only 0.3 second is required for complete equilibration of oxygen between alveolar gas and blood (Roughton, 1945, Staub *et al.*, 1962; Comroe, 1965). If capillary volume and oxygen content of inspired air are constant, then an increase in cardiac output will decrease the time blood resides in the lung and tend to decrease arterial oxygen saturation. In this respect Henry and Fedde (1970) occasionally recorded pulmonary transit times as short as 0.31 second in chickens, and claim that, owing to the even shorter time blood is in the capillaries themselves, complete equilibration of gas pressures between air and blood capillaries may not occur. This observation is supported by the fact that arterial oxygen tensions recorded from birds are often low (Morgan and Chichester, 1935; Sykes, 1960; Ray, 1966; Millard *et al.*, 1972). The contention that ventilation/perfusion equality is controlled by homeostatic mechanisms receives some support from the coupling between the respiratory and circulatory functions that has often been reported in birds. Sturkie (1965) reports that oscillations in blood pressure associated with breathing in birds appear to be caused by oscillations of intrathoracic pressure. The blood pressure is highest when the intrathoracic pressure is lowest (e.g., at peak inspiration). Jones and Holeton (1972b) recorded flow in the pulmonary artery and found in some ducks that it increased as inspiration proceeded, reaching a maximum at peak inspiration. If this is also true of the systemic circulation, then it may be the cause of the elevation in systemic blood pressure.

A more easily observed functional coupling is between the heart rate and breathing rate. This relationship usually takes the form of an increase in heart rate during inspiration, and is referred to as sinus arrhythmia. Sinus arrhythmia is particularly noticeable in ducks on emergence from a dive (Andersen, 1963; Reite *et al.*, 1963). Butler and Jones (1968) also recorded a close relationship between heart rate and respiratory frequency over extended periods of time. If respiratory frequency was low, then heart rate was also low. A ratio between heart rate and respiratory frequency of about 10–12:1 was commonly found in ducks.

In flying birds the heart rate is usually related to the wing beat frequency, 1:1 or 1:2 (Berger *et al.*, 1970a,b; Aulie, 1971). Also, the respiratory patterns appear to be coordinated with wing motion (Hart and Roy, 1966; Berger *et al.*, 1970a), although only in the pigeon and Common Crow *(Corvus brachyrhynchos)* is the relationship 1:1. Synchrony may be important from the point of view of gas exchange, but in finches (*Sporophila* sp.) the heart rate is exactly half the wingbeat frequency and the pectoral muscles perform two twitch contractions every time the heart is in diastole (Aulie, 1971). For this reason, Aulie (1971) has proposed that the pectoral muscles in small birds may work like a "venous pump," enhancing venous return by their contraction when the heart is in diastole. In view of the high cardiac outputs that have been calculated to occur during flight, considerable advantage may come from such a mechanism for improved venous return.

McCready *et al.* (1966) have investigated the neural origin of the relationship between ventilation rate and heart rate in dogs, cats,

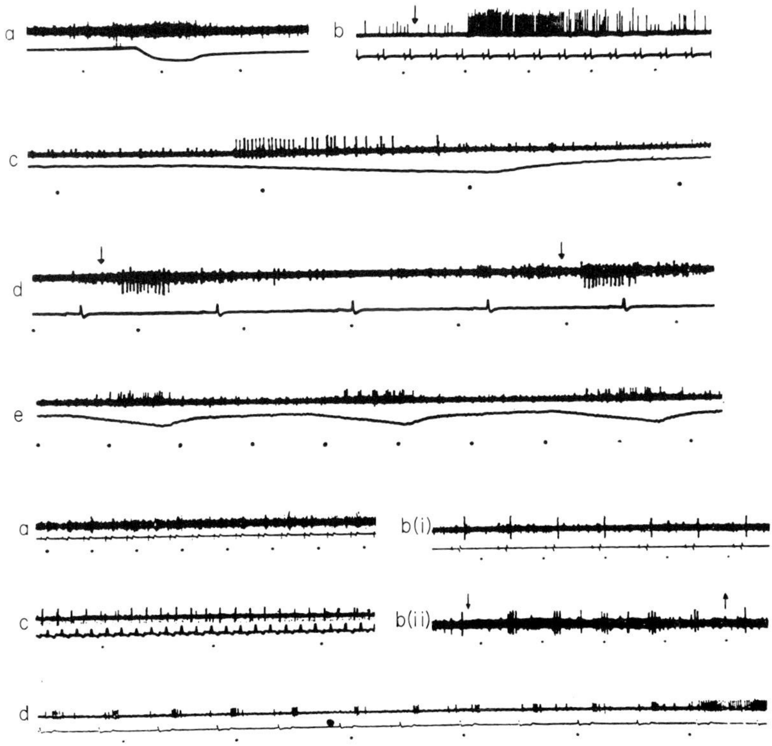

FIG. 46. Upper: Afferent nervous activity associated with the respiratory cycle recorded from the vagi. a. Domestic duck. Upper trace, mass discharge from a vagal slip. Middle trace, displacement of sternum (down on trace = inspiration). b. Muscovy Duck (*Cairina moschata*). Upper trace, afferent activity during maintained inflation of the lung (retouched). Arrow marks start of inflation. Middle trace EKG. c. Muscovy Duck. Upper trace, afferent activity during normal inspiration. Middle trace, displacement of sternum (down on trace = inspiration). d. Khaki Campbell duck. Upper trace afferent activity during rhythmic artificial ventilation. Arrows mark start of inflation. Middle trace EKG. e. Muscovy Duck. Upper trace, afferent activity during normal inspiration. Middle trace, displacement of sternum (down on trace = inspiration). In all records the bottom trace is a 1-second time mark.

Lower: Afferent nervous activity associated with the cardiac cycle recorded from the vagi. a. Muscovy Duck. Upper trace, mass discharge associated with the cardiac cycle. Middle trace, EKG. b. Muscovy Duck. (i) Upper trace, single discharge occurring soon after the QRS complex. Middle trace, EKG. (ii) Effect of locating a probe in the region of the pulmonary artery. Probe positioned at the downward-pointing arrow and removed at the upward-pointing arrow. c. Khaki Campbell duck. Upper trace, afferent discharge associated with the P wave of the EKG. Middle trace, EKG. d. Muscovy Duck. Upper trace, afferent discharge associated with the repolarization wave of the EKG. The discharge continues during inspiration, marked by discharge similar to that shown in Fig. 1c. Middle trace EKG. In all records the bottom trace is a 1-second time marker. (Jones, 1969, *Comp. Biochem. Physiol.*, by permission.)

and chickens. Their studies suggest that lung inflation stimulates stretch receptors in the lungs and thorax which send impulses to the cardiac regulatory center via the vagi and, also, that radiation of impulses from the respiratory to the cardiac regulatory center occurs. These events will in turn enhance sympathetic outflow to the heart and bring about a central suppression of vagal tone, thereby matching an increased breathing rate with an increased heart rate. However, recent evidence suggests that neural and metabolic factors may also be involved, since the type of coupling described by Butler and Jones (1968) for ducks is masked if arterial carbon dioxide tension is controlled at a fixed value and not allowed to vary with changes in breathing frequency (Jones and Purves, 1970b).

On the grounds that the avian lung has been described as relatively nonexpansive (Fedde *et al.*, 1963) it may be questioned whether stretch receptors exist in the lungs. However, both King *et al.* (1968) and Jones (1969) have recorded afferent activity from the vagus nerve associated with phases of the respiratory cycle (Fig. 46), although in neither case was the source of this activity precisely identified.

X. The Role of the Cardiovascular System in Excretion and Osmoregulation

The kidney is not always the most important organ of excretion, and it now appears that only in mammals does the kidney function as the major organ of osmoregulation. In birds for instance, and especially in marine birds that are frequently subjected to salt loading, the salt-secreting nasal glands produce hypertonic solutions of sodium chloride, the concentration of which is often several times as high as the maximum urine concentration (see Chapter 9).

The renal arteries provide the kidney with a high–pressure blood supply that is essential to glomerular filtration (Fig. 47). However, there has been much controversy about whether birds receive a low-pressure blood supply by way of a renal portal system. The renal portal system might presumably play an important role for reabsorption and secretion in the kidney. A renal portal system certainly is present in fish, amphibia, and reptiles, although it is lost in the adult mammal. Bradley and Grahame (1960), on the basis of arterial casts and injection methods, suggest the presence of a renal portal system because of the greater caliber and different character of the efferent venous vessels as compared with the afferent vessels. Spanner (1925) has shown in the adult bird that there is a valve that prevents direct flow of blood from the external iliac and caudal renal veins into the posterior vena

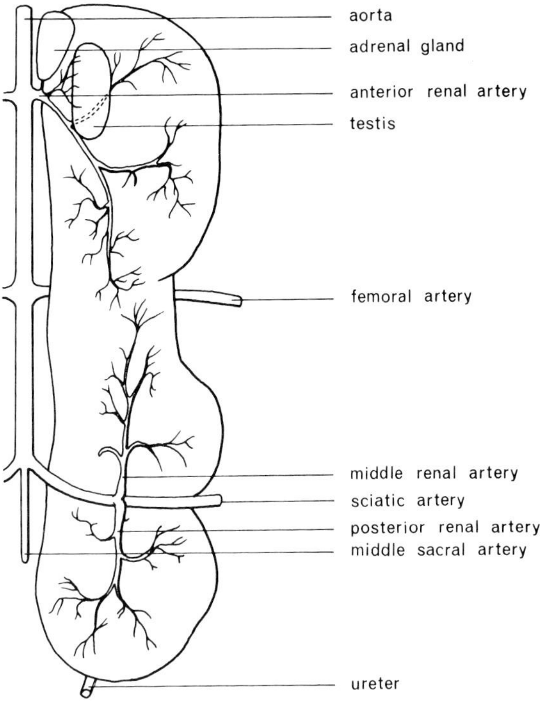

FIG. 47. Diagram of the arterial supply to the kidney of the domestic fowl. There are three pairs of renal arteries, the anterior arising directly from the aorta, the middle and posterior ones from the sciatic artery. The femoral artery gives off no supply vessels to the kidney. The ureters are supplied on each side by branches from all three renal arteries (after Siller and Hindle, 1969).

cava (Fig. 48). The valve contains both circular and radial muscle fibers and is of variable structure, either a simple diaphragm with a central aperture, funnel-shaped, or sieve-like with numerous small apertures (Simons, 1960). When this valve is shut a renal portal system is established, since the blood then must enter the capillary system of the kidneys before being transported by the renal veins into the posterior vena cava.

The renal portal valve is unique in that it may be the only intravascular structure in nonmammalian vertebrates, with the exception of hepatic vein sphincters in elasmobranchs (Johansen and Hanson, 1967), to contain smooth muscle and a nerve supply. Both circumstantial and direct histochemical evidence has been provided for both cholinergic and adrenergic innervation of the valve (Rennick and Gandia, 1954; Gilbert, 1961; Akester and Mann, 1969; Doležel and

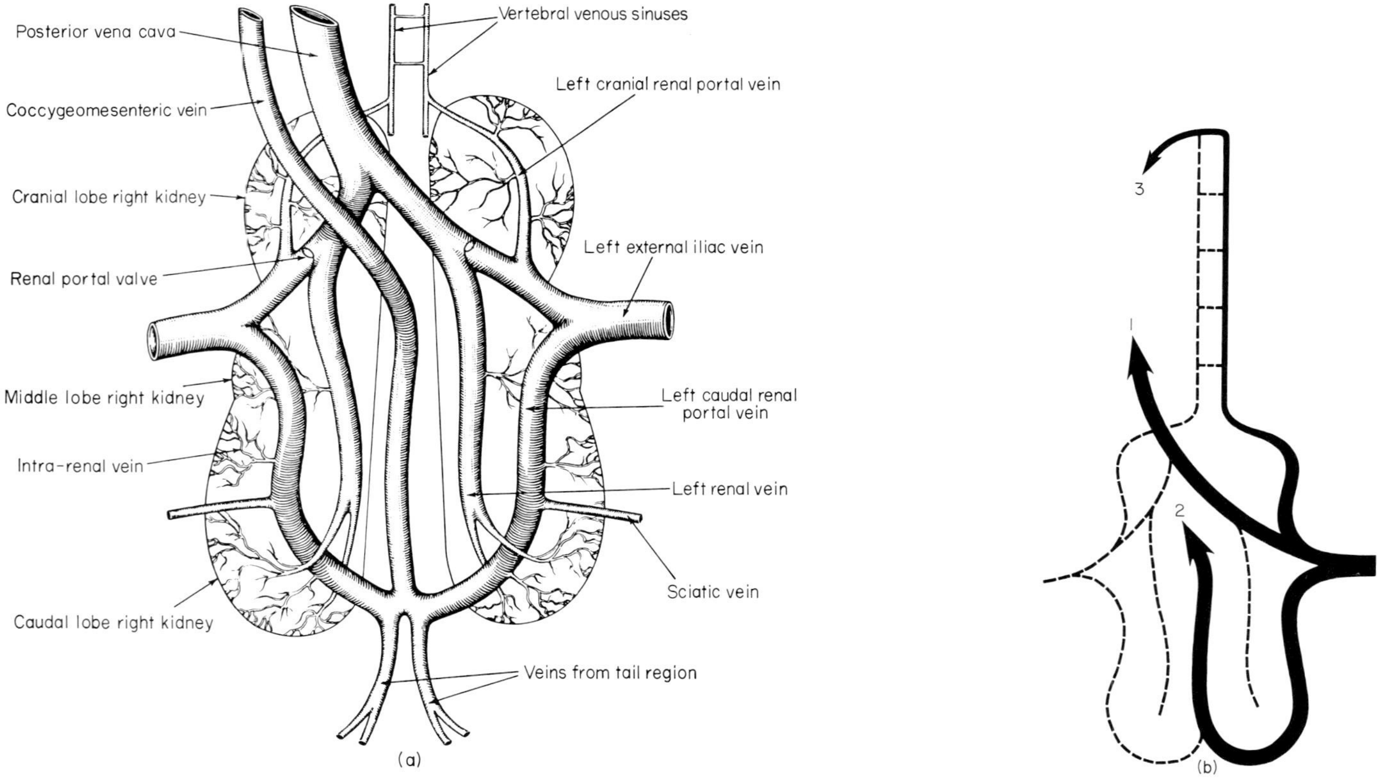

FIG. 48. (a) The principal veins associated with the kidneys in the domestic fowl (ventro-dorsal view). (b) The three renal portal shunts (ventro-dorsal view). 1, Renal portal valve and posterior vena cava. 2, Caudal renal portal vein and coccygco-mesenteric vein. 3, Cranial renal portal vein and vertebral venous sinuses. Shunts 1 and 2 have been shown to function individually and all three have been shown to function collectively, to provide a complete bypass of kidney tissue. However, more commonly they function as partial shunts in which part of the renal portal flow bypasses the kidney while the remainder enters it. Only the shunts of the left kidney have been marked. (Akester, 1967, *J. Anat.*, by permission.)

Žlábek, 1969; Bennett and Malmfors, 1970). Akester and Mann (1969) have shown that the distribution of adrenergic and cholinergic fibers is identical, and suggest that the two types of fibers run together in the same axon bundles with the same Schwann cells, thus providing a very precise and effective control system.

Sperber (1946, 1948) demonstrated from injection experiments that the renal portal veins provided a functional afferent inflow to the kidneys. A substance (phenol red) injected into a leg vein was detectable in the renal tubules of the same side before being detected in the ureter of the opposite side. In other words, venous blood had reached the corresponding kidney before entering the general circulation.

The venous system of the visceral cavity can be represented as in Fig. 49. Since the liver and kidneys are perfused in parallel by venous blood, the blood flow through the liver and kidneys will be reciprocally related to their resistances (Fig. 49, R_1 R_2). The valve at the end of

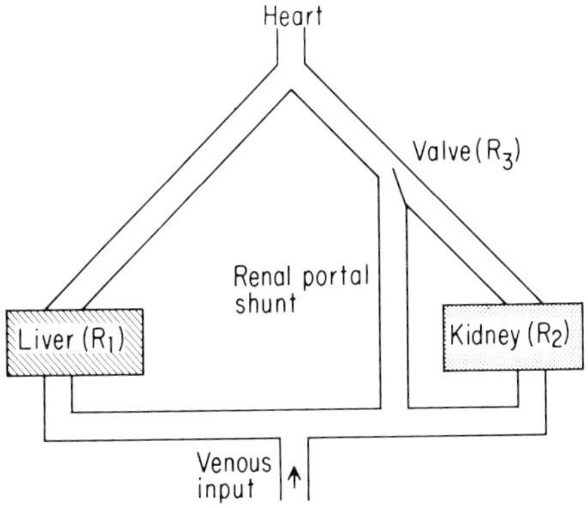

FIG. 49. Simplified schematic diagram of the renal portal shunt in a bird. R_1, R_2, and R_3 represent the resistance to blood flow of the liver, kidneys, and renal portal valve, respectively.

the renal portal shunt presents another resistance (Fig. 49, R_3) to blood flow. When this valve is shut, blood will flow through the liver and kidneys in inverse relationship to their resistances. However, when the valve opens and its resistance is reduced, then blood will start to flow along the renal portal shunt. If the resistance (Fig. 49, R_3) is very low, virtually all the blood will pass along the shunt, thus effectively bypassing both the liver and kidney capillary circulations. Rennick and Gandia (1954) have studied the pharmacology of the valve both *in vitro* and *in vivo*. They believe that the valve opens under conditions of high sympathetic discharge, which might ensure adequate venous return under conditions of high cardiac output.

Akester (1967) has, in fact, shown that three renal portal shunts operate, by which all or part of the renal portal flow can bypass the kidney tissue. These are (1) via the renal portal valve and posterior vena cava, (2) via the caudal renal portal vein and the coccygeo-mesenteric vein, and (3) via the cranial renal portal vein and vertebral venous sinuses (Fig. 48). Furthermore, Akester (1967) has demonstrated that blood flows in opposite directions along the coccygeo-mesentric vein on different occasions, and finds circumstantial evidence for powerful vasomotor activity at the point of origin of the two renal portal veins and at the points where branches leave these veins to enter kidney tissue.

XI. The Role of Peripheral Circulation in Temperature Regulation

The high metabolic rate and body temperature of birds require precise mechanisms for dissipation and conservation of heat. Even without sweat glands and with a well-insulated body significant amounts of heat can escape; but heat loss undoubtedly occurs most readily by evaporative cooling from the respiratory tract and by dissipation from the naked feet. Measurements on herons and gulls indicate that at low ambient air temperatures less than 10% of the total heat production is dissipated through the legs, whereas at air temperatures of 35°C almost the entire metabolic heat production can be lost through the legs. With legs immersed in water, heat loss is increased more than four times compared with feet in air at the same temperature (Steen and Steen, 1965). The same authors note that panting or gular fluttering at high ambient temperatures promptly decreases or stops when the naked feet are immersed in water. The importance of the legs in temperature regulation was also emphasized by Kahl (1963), working with American Wood-Storks (*Mycteria americana*). These birds showed a rapidly increasing body temperature if the legs were artificially insulated. At high air temperatures, the storks increase evaporative heat loss by depositing their fluid excreta on their naked legs. Heat loss, whether by evaporative cooling from the respiratory tract, from wetted naked extremities, or by convection or conduction from the legs, must depend on heat being transported by the vascular system to the evaporative or dissipating surfaces by the blood flow.

On this basis, it can be surmised that blood circulation through avian legs and feet is highly adapted for passage of a variable, well-controlled blood flow in relation to temperature regulation. Birds display some notable structural specializations in the vascular anatomy of their legs. Hyrtl (1863), more than 100 years ago, described the presence of rete mirabile or vascular bundles of small arteries and veins running tightly packed in a countercurrent arrangement in the legs of several birds, particularly in aquatic and wading birds. These structures are also found in the extremities of some mammals, where they are surmised to play a role in thermoregulation by allowing the cooler venous blood returning from the periphery of an extremity to be warmed by the counter flowing warm arterial blood, with the net result of a markedly reduced heat loss. Experimental work on the retia mirabilia of mammals has proven their effectiveness in such countercurrent heat exchange (Scholander and Krog, 1957; Scholander, 1958).

However, when in a positive heat load with a need for effective dissipation of heat, countercurrent heat exchange in retia mirabilia would offer no advantage. On this basis it has been suggested that at low ambient temperature, when heat conservation is called for, an effective countercurrent exchange takes place in the vascular exchangers, whereas at high ambient temperature or with the bird in a positive heat load from exercise, the markedly increased leg blood flow is allowed to bypass the rete and to exchange heat by dissipation at the outer surfaces of the legs (Steen and Steen, 1965). There has been little experimental work to contest this hypothesis. Johansen and Tønnesen (1969) measured blood flow to the legs of a gull using clearance of krypton-85 from epicutaneously labeled skin. At ambient temperatures between 4° and 27°C, blood flow in the web varied between zero and 18 ml/100 gm web per minute. Calculation of heat loss at high ambient temperature on the basis of blood flow and blood temperature gave values corresponding closely with those directly measured by Steen and Steen (1965); but similar calculations at low ambient temperature gave heat loss values several times those directly measured. The authors interpreted this as a result of a countercurrent heat exchange, reducing the heat loss at low temperature. The authors also point out that if diffusion of ^{85}Kr between blood and tissues did not reach equilibrium, then the method is ill-suited for quantitative measurements of blood flow. Such lack of diffusion equilibrium would certainly exist if the vascular system of the leg contains direct arteriovenous anastomoses. This has indeed been reported to be the case in many birds and Schumacher (1916) noted that numerous anastomoses are found in the toes and webs of birds, particularly on the plantar aspects of the feet. Anastomoses have also been found in the metatarsal and tarsal segments of the feet, but none in the feathered portion of the skin. Whereas the retia mirabilia have been reported in only a few birds, the profuse presence of A–V anastomoses appears common to all birds studied (Grant and Bland, 1931). Grant (1930), in his careful studies of peripheral vascular reactions to cold, particularly noticed many A–V anastomoses in the ears of the rabbit and proposed that they play an important role in temperature regulation by allowing large increases in peripheral blood flow otherwise impossible if the flow change had to pass through the regular nutritional channels (capillaries) of the peripheral vascular beds. Similar conclusions about the importance of A–V anastomoses have been advanced for the vascular response of the human finger to cold (Lewis, 1930). In a series of recent papers (Millard and Johansen, 1972; Johansen and Millard, 1972) the peripheral circulation and its control in several species of Antarctic marine birds have been studied by direct measurement of blood flow to the legs using indwelling flowmeters. In addition, changes in blood flow distribution in response to changing ambient conditions, to nerve stimulations, and to locally applied autonomic agents were evaluated by intracutaneous and intravascular temperature measurements and by local A–V O_2 differences. Antarctic birds, like all polar homeotherms, are so well insulated that effective heat dissipation during and following exercise poses as great or greater a problem than heat conservation in the severe cold that these animals encounter.

In the Giant Fulmar, *Macronectes giganteus*, the blood flow to one leg with the bird at rest in a thermoneutral condition ranges between 5 and 15 ml/100 gm tissue per minute. Maximum recorded blood flow with the bird in a positive heat load could exceed 100 ml/100 gm per minute, or about 50 ml/minute to one foot. At maximum blood flow, the intracutaneous and subcutaneous temperatures were less than 1°C below central core temperature. The heat capacity of the thin web was remarkably low, since both intravascular as well as subcutaneous peripheral temperatures showed pulsatile variations in phase and frequency with the directly measured flow. A–V O_2 differences in the web revealed that less than 1/20 of the arterial O_2 content is consumed by the foot in a thermoneutral condition, giving a $\dot{Q}/\dot{V}_{O_2}$ value of about 100 ml/ml O_2. At maximum blood flow the A–V O_2 difference became immeasurable, with a $\dot{Q}/\dot{V}_{O_2}$ approaching infinity, and practically all blood flow shunted in non-nutritional channels.

When the leg of a bird in a thermoneutral state was immersed in iced brine (−1°C), an immediate and pronounced increase in blood flow occurred in the leg, commonly amounting to about 2.5 times the flow prevailing before immersion (Fig. 50). This cold-flush response appears not to have been reported for vertebrates previously, but re-

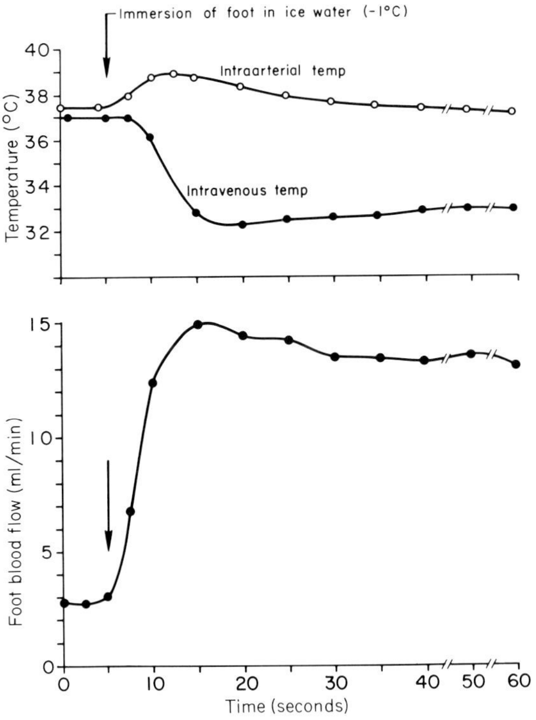

FIG. 50. Upper: intraarterial and intravenous temperatures following immersion of the foot of the Giant Fulmar, *Macronectes giganteus*, in iced water (−1°C). Lower: immersion of the foot in iced brine (−1°C) causes prompt increase in blood flow to the foot. In both graphs the foot was immersed at the arrow (after Johansen and Millard, 1972).

sembles the cold-induced vasodilatation observed in mammals, which follows the cold vasoconstriction initially elicited by immersion in ice water (Lewis, 1930). The cold-flush may be regarded as protective against tissue freezing and damage. It results in web temperatures remaining above water temperature after prolonged exposure (several hours) without causing any drop in deep central temperatures. Measurement and intravenous temperatures in the leg of the Giant Fulmar revealed some striking gradients, which could amount to as much as 25°C between shank and web along less than 15 cm (Fig. 51). Venous temperatures were much lower than arterial temperatures at the same site, suggesting little effective countercurrent exchange (Fig. 50). Yet, the more central intravascular temperatures in the tarsal region of the foot changed much less than the peripheral. When the intravascular thermistor catheters were

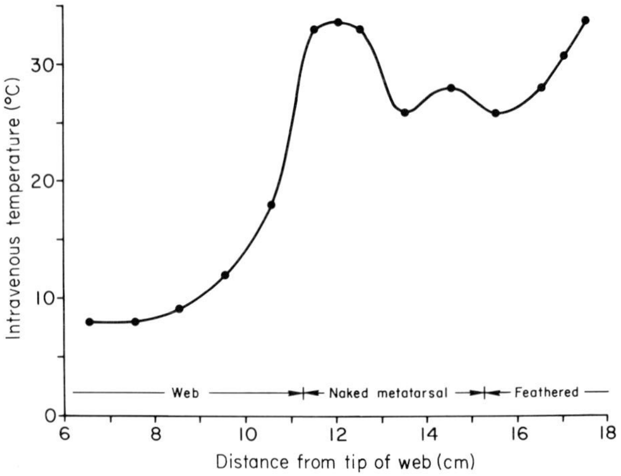

FIG. 51. Intravenous temperature gradient in the foot of the Giant Fulmar, *Macronectes giganteus*, during immersion of the foot in iced brine (from Johansen and Millard, 1972).

withdrawn successive lengths, it became apparent that the strikingly large gradients in venous temperatures were not linear but could even make sudden reversals to higher blood temperatures at peripheral sites (Fig. 51). This fact as well as other evidence (Johansen and Millard, 1972) led to the conclusion that A–V shunting of blood in the leg is the principal mechanism for increasing the total flow as well as for rewarming the cold venous blood returning from the web of the foot. Marked variations in the blood temperature as well as in the directly recorded flow occurred. Increased flow was also associated with marked increases in venous pressure in the foot. It is of considerable interest to compare this situation in the avian foot with recent studies on the importance of A–V shunt flow in mammalian peripheral tissues (Golenhofen, 1968). Plenk and Püschmann (1971) recently discovered discrete sphincter muscles and a valve mechanism where A–V anastomoses branch off from skin arteries on the plantar surface of the rat foot.

The experiments by Millard and Johansen (1972) and Johansen and Millard (1972) revealed that the conspicuous changes of blood flow to the leg of the birds studied are governed by a dual neurogenic mechanism. At low and moderate web temperature with the core temperature in its lower range, cutting the foot nerves most commonly caused an increase in blood flow to the foot, indicating release from vasoconstrictor tonus. On the other hand, at high web temperatures and high blood flow, cutting the foot nerves brought about a decrease in blood flow, suggesting the release from an active vasodilator mechanism. Stimulation of peripheral cut ends of the nerves gave vasodilatation and (or) vasoconstriction, depending on the nerves selected (Fig. 52). The responses could be mimicked by autonomic agents injected intraarterially in the foot. Atropine blocked vasodilatation from both peripheral nerve stimulation and acetylcholine injection. Adrenergic α-receptors were activated by phenylephrine, causing immediate vasoconstriction, while the β-adrenergic agent isoprenaline caused no observable change in blood flow in the foot. Millard

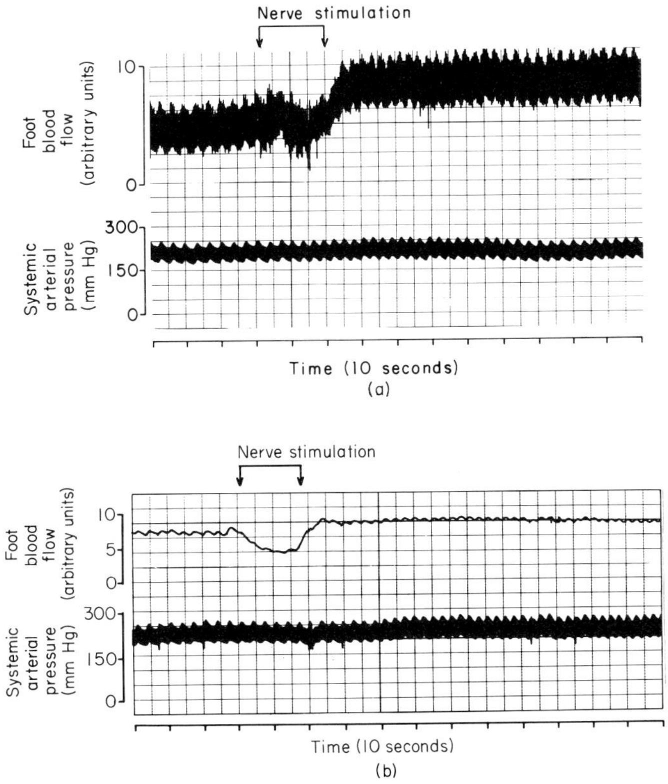

FIG. 52. (a) Marked vasodilatation and increase in foot blood flow resulting from stimulation of a cut peripheral nerve to the foot of the Giant Fulmar, *Macronectes giganteus*. (b) Transient decrease in foot blood flow from stimulation of cut peripheral end of foot nerve.

and Johansen (1972) concluded that the conspicuous changes of blood flow in the foot of the Giant Fulmar are largely effected by changes in A–V shunt flow and controlled by a synergistic, neurogenic mechanism involving both a cholinergic vasodilator and an adrenergic vasoconstrictor component.

REFERENCES

Aakhus, T., and Johansen, K. (1964). Angiocardiography of the duck during submersion asphyxia. *Acta Physiol. Scand.* **63,** 10.

Abel, W. (1912) Further observations on the development of the sympathetic nervous system in the chick. *J. Anat. Physiol. Norm. Pathol. Homme Anim.* **47,** 35.

Ábrahám, A. (1969). "Microscopic Innervation of the Heart and Blood Vessels in Vertebrates Including Man." Pergamon, Oxford.

Adams, W. R. (1958). "Comparative Morphology of the Carotid Body and Carotid Sinus." Thomas, Springfield, Illinois.

Ahlquist, R. P. (1948). A study of adrenotrophic receptors. *Amer. J. Physiol.* **153,** 586.

Akester, A. R. (1967). Renal portal shunts in the kidney of the domestic fowl. *J. Anat.* **101,** 569.

Akester, A. R., and Mann, S. P. (1969). Adrenergic and cholinergic innervation of the renal portal valve in the domestic fowl. *J. Anat.* **104,** 241.

Albritton, E. C. (1952). "Standard Values in Blood." Saunders, Philadelphia, Pennsylvania.

Alexander, R. S., and Glaser, O. (1941). Progressive acceleration in embryonic hearts. *J. Exp. Zool.* **87,** 17.

Altland, P. D. (1961). Altitude tolerance of chickens and pigeons. *J. Appl. Physiol.* **16,** 141.

Andersen, H. T. (1963). "The Elicitation of Physiological Responses to Diving." Norwegian Research Council for Science and Humanities, Universitetsforlaget.

Andersen, H. T. (1966). Physiological adaptations in diving vertebrates. *Physiol. Rev.* **46,** 212.

Andersen, H. T., and Løvø, A. (1967). Indirect estimation of partial pressure of oxygen in arterial blood of diving ducks. *Resp. Physiol.* **2,** 163.

Andrew, W. (1965). "Comparative Hematology." Grune & Stratton, New York.

Anrep, G. V., Pasqual, W., and Rössler, R. (1936). Respiratory variations in heart rate. I. The reflex mechanism of the respiratory arrhythmia. *Proc. Roy. Soc., Ser. B* **119,** 191.

Ara, G. (1934). Il riflesso di pagano-hering nel pollo. *Arch. Fisiol.* **33,** 332.

Assenmacher, I. (1952). La vascularization du complexe hypophysaire chez le canard domestique. *Arch. Anat. Microsc. Morphol. Exp.* **41,** 69.

Åstrand, P., Cuddy, T. E., Saltin, B., and Sternberg, J. (1964). Cardiac output during submaximal and maximal work. *J. Appl. Physiol.* **19,** 268.

Aulie, A. (1971). Co-ordination between the activity of the heart and the flight muscles during flight in small birds. *Comp. Biochem. Physiol.* **38A,** 91.

Aymar, G. C. (1935). "Bird Flight." Dodd, Mead, New York.

Ball, R. A. (1964). A study of vascular diseases in the turkey. Ph.D. Thesis, University of Minnesota, St. Paul.

Ball, R. A., Sautter, J. H., and Katter, M. S. (1963). Morphological characteristics of the anterior mesenteric artery of the fowl. *Anat. Rec.* **146,** 251.

Barkow, H. (1829). Anatomisch-physiologische Untersuchungen, vorzüglich über das Schlagadersystem der Vögel. *Arch. Anat. Physiol., Physiol. Abt.* **12,** 305.

Barnes, A. E., and Jensen, W. N. (1959). Blood volume and red cell concentration in the normal chick embryo. *Amer. J. Physiol.* **197,** 403.

Barry, A. (1941). The effect of exsanguination on the heart of the embryonic chick. *J. Exp. Zool.* **88,** 1.

Batelli, E., and Stern, L. (1908). Excitabilité du nerf vague chez le canard. *C. R. Soc. Biol.* **2,** 505.

Baumel, J. (1967). The characteristic asymmetrical distribution of the posterior cerebral artery of birds. *Acta Anat.* **67,** 523.

Baumel, J., and Gerchman, L. (1968). The avian intercarotid anastomosis and its homologue in other vertebrates. *Amer. J. Anat.* **122,** 1.

Bell, C. (1969). Indirect cholinergic vasomotor control of intestinal blood flow in the domestic chicken. *J. Physiol. (London)* **205,** 317.

Bell, D. J., Bird, T. P., and McIndoe, W. M. (1964). Changes in erythrocyte levels and mean corpuscular haemoglobin concentration in hens during the laying cycle. *Comp. Biochem. Physiol.* **14,** 83.

Bennett, T., and Malmfors, T. (1970). The adrenergic nervous system of the domestic fowl. *Z. Zellforsch. Mikrosk. Anat.* **106,** 22.

Berger, M., Hart, J. S., and Roy, O. Z. (1970a). The co-ordination between respiration and wing beats in birds. *Z. Vergl. Physiol.* **66,** 190.

Berger, M., Hart, J. S., and Roy, O. Z. (1970b). Respiration, oxygen consumption and heart rate in some birds during rest and flight. *Z. Vergl. Physiol.* **66,** 201.

Blinks, J. R. (1965). Field stimulation as a means of effecting the graded release of autonomic transmitters in isolated heart muscle. *J. Pharmacol. Exp. Ther.* **151,** 221.

Blix, A. S., Berg, T., and Fyhn, H. J. (1970). Lactic acid dehydrogenase isoenzymes in a diving mammal, the common seal *Phoca vitulina vitulina. Int. Biochem.* **1,** 292.

Blount, W. P. (1939). Thrombocyte formation in the domestic hen. *Vet. J.* **95,** 195.

Bolton, T. B. (1967). Intramural nerves in the ventricular myocardium of the domestic fowl and other animals. *Brit. J. Pharmacol. Chemother.* **31,** 253.

Bolton, T. B. (1968a). Studies on the longitudinal muscle of the anterior mesenteric artery of the domestic fowl. *J. Physiol. (London)* **196,** 273.

Bolton, T. B. (1968b). Electrical and mechanical activity of the longitudinal muscle of the anterior mesenteric artery of the domestic fowl. *J. Physiol. (London)* **196,** 283.

Bolton, T. B., and Bowman, W. C. (1969). Adrenoreceptors in the cardiovascular system of the domestic fowl. *Eur. J. Pharmacol.* **5,** 121.

Bond, C. F., and Gilbert, P. W. (1958). Comparative study of blood volume in representative aquatic and non-aquatic birds. *Amer. J. Physiol.* **194,** 519.

Brace, K., and Altland, P. D. (1956). Life span of the duck and chicken erythrocyte as determined with C^{14}. *Proc. Soc. Exp. Biol. Med.* **92,** 615.

Bradley, O. L., and Grahame, T. (1960). "The Structure of the Fowl." Oliver & Boyd, Edinburgh.

Brauer, R. W. (1965). Circulatory and metabolic responses to diving and to apnea. *Proc. 2nd Annu. Conf. Biol. Sonar Diving Mammals* pp. 63–71.

Brauer, R. W. (1966). Circulatory and metabolic responses to diving and apnea: Vertebrates. *In* "Environmental Biology" (P. L. Altman and D. S. Dittmer, eds.), pp. 344–349. Fed. Amer. Soc. Exp. Biol., Bethesda, Maryland.

Bredeck, H. E. (1960). Intraventricular blood pressure in chickens. *Amer. J. Physiol.* **198,** 153.

Bremer, J. L. (1932). The presence and influence of two spiral streams in the heart of the chick embryo. *Amer. J. Anat.* **49,** 409.

Brush, A. H. (1966). Avian heart size and cardiovascular performance. *Auk* **83,** 266.

Buddecke, E. (1958). Untersuchungen zur Chemie der Arterienwand. III. Veränderung und Beeinflussung des Aortenbindegewebe bei tierexperimenteller Arteriosklerose. *Hoppe-Seyler's Z. Physiol. Chem.* **310,** 199.

Burke, J. D. (1966). Vertebrate blood oxygen capacity and body weight. *Nature (London)* **212,** 46.

Burton, A. C. (1965). "Physiology and Biophysics of the Circulation." Yearbook Publ., Chicago, Illinois.

Burton, R. R., and Carlisle, J. C. (1969). Acute hypoxia tolerance of the chick. *Poultry Sci.* **48,** 1265.

Burton, R. R., and Smith, A. H. (1967). The effect of polycythemia and chronic hypoxia on heart mass in the chicken. *J. Appl. Physiol.* **22,** 782.

Burton, R. R., and Smith, A. H. (1968). Blood and air volumes in the avian lung. *Poultry Sci.* **47,** 85.

Burton, R. R., and Smith, A. H. (1969). Induction of cardiac hypertrophy and polycythemia in the developing chick at high altitude. *Fed. Proc., Fed. Amer. Soc. Exp. Biol.* **28,** 1170.

Burton, R. R., Besch, E. L., and Smith, A. H. (1968). Effect of chronic hypoxia on the pulmonary arterial blood pressure of the chicken. *Amer. J. Physiol.* **214,** 1438.

Burton, R. R., Smith, A. H., Carlisle, J. C., and Sluka, S. J. (1969). Role of hematocrit, heart mass, and high altitude exposure in acute hypoxia tolerance. *J. Appl. Physiol.* **27,** 49.

Butler, P. J. (1967a). Studies on avian cardiovascular and respiratory systems. Ph.D. Thesis, University of East Anglia, Norwich.

Butler, P. J. (1967b). The effect of progressive hypoxia on the respiratory and cardiovascular systems of the chicken. *J. Physiol. (London)* **191,** 309.

Butler, P. J. (1970). The effect of progressive hypoxia on the respiratory and cardiovascular systems of the pigeon and duck. *J. Physiol. (London)* **210,** 527.

Butler, P. J., and Jones, D. R. (1968). Onset of and recovery from diving bradycardia in ducks. *J. Physiol. (London)* **196,** 255.

Butler, P. J., and Jones, D. R. (1969). Unpublished data.

Butler, P. J., and Jones, D. R. (1971). The effect of variations in heart rate and regional distribution of blood flow on the normal pressor response to diving in ducks. *J. Physiol. (London)* **214,** 457.

Calder, W. A. (1968). Respiratory and heart rates of birds at rest. *Condor* **70,** 358–365.

Carnaghan, R. B. A. (1955). Atheroma of the aorta associated with dissecting aneurysms in turkeys. *Vet. Rec.* **67,** 568.

Ceretelli, P., Piiper, J., Mangili, F., Cuttica, F., and Ricci, B. (1964). Circulation in exercising dogs. *J. Appl. Physiol.* **19,** 29.

Chiodi, H. (1963). Action of high altitude chronic hypoxia on organisms during various growth stages. *Perspect. Biol.* pp. 336–367.

Chowdhary, D. S. (1953). A comparative study of the carotid sinus of vertebrates. II. The carotid body and "carotid sinus" of the fowl (*Gallus domesticus*). Ph.D. Thesis, University of Edinburgh, Edinburgh.

Cohen, D. H., and Pitts, L. H. (1968). Vagal and sympathetic components of conditioned cardio-acceleration in the pigeon. *Brain Res.* **9,** 15.

Cohen, D. H., and Schnall, A. M. (1970). Medullary cells of origin of vagal cardioinhibitory fibers in the pigeon. II. Electrical stimulation of the dorsal motor nucleus. *J. Comp. Neurol.* **140,** 321.

Cohen, D. H., Schnall, A. M., Macdonald, R. L., and Pitts, L. H. (1970). Medullary cells of origin of vagal cardioinhibitory fibers in the pigeon. I. Anatomical studies of peripheral vagus nerve and the dorsal motor nucleus. *J. Comp. Neurol.* **140,** 299.

Cohen, R. R. (1965). Avian adjustment of blood oxygen capacity in and after chronic hypoxic stress as seen in the Pintail duck (*Anas acuta*). *Amer. Zool.* **5,** 207.

Cohen, R. R. (1967). An estimation of percentage trapped plasma in normal chicken microhematocrits, using Cr^{51}. *Poultry Sci.* **46,** 219.

Cohen, R. R. (1969a). Total and relative erythrocyte levels of Pintail ducks (*Anas acuta*) in chronic decompression hypoxia. *Physiol. Zool.* **42,** 108.

Cohen, R. R. (1969b). Recovery of erythrocyte levels following chronic decompression hypoxia in Pintail ducks (*Anas acuta*). *Physiol. Zool.* **42,** 120.

Cohn, J. E., Krog, J., and Shannon, R. (1968). Cardiopulmonary responses to head immersion in domestic geese. *J. Appl. Physiol.* **25,** 36.

Comroe, J. H., Jr. (1965). "Physiology of Respiration." Yearbook Publ., Chicago, Illinois.

Comroe, J. H., Jr., and Mortimer, L. (1964). The respiratory and cardiovascular responses of temporally separated aortic and carotid bodies to cyanide, nicotine, phenyldiguanide, and serotonin. *J. Pharmacol. Exp. Ther.* **146,** 33.

Dale, H. H. (1906). On some physiological actions of ergot. *J. Physiol. (London)* **34,** 163.

Daly, M. De B., and Scott, M. J. (1958). The effects of stimulation of the carotid body chemoreceptors on heart rate in the dog. *J. Physiol. (London)* **144,** 148.

Daly, M. De B., and Scott, M. J. (1962). An analysis of the primary cardiovascular reflex effects of stimulation of the carotid body chemoreceptors in the dog. *J. Physiol. (London)* **162,** 555.

Davies, F. (1930). The conducting system of the bird's heart. *J. Anat.* **64,** 9.

De Haan, R. L. (1958a). Cell migration and morphogenetic movements. *In* "The Chemical Basis of Development" (W. D. McElroy and B. Glass, eds.), pp. 339–374. Johns Hopkins Press, Baltimore, Maryland.

De Haan, R. L. (1958b). Modification of cell-migration patterns in the early chick embryo. *Proc. Nat. Acad. Sci. U.S.* **44,** 32.

de Koch, L. L. (1958). On the carotid body of certain birds. *Acta Anat.* **35,** 161.

De Villiers, O. T. (1938). The blood of the Ostrich. *Onderstepoort J. Vet. Sci. Anim. Ind.* **2,** 419.

Didio, L. J. A. (1967). Myocardial ultrastructure and electrocardiograms of the hummingbird under normal and experimental conditions. *Anat. Rec.* **159,** 335.

Dijk, J. A. (1932). *Arch. Neer. Physiol.* **17,** 495 (cited from Prosser *et al.*, 1950).

Djojosugito, A. M., Folkow, B., and Kovách, A. G. B. (1968). The mechanisms behind the rapid blood volume restoration after hemorrhage in birds. *Acta Physiol. Scand.* **74,** 114.

Djojosugito, A. M., Folkow, B., and Yonce, L. R. (1969). Neurogenic adjustments of muscle blood flow, cutaneous A-V shunt flow and venous tone during "diving" in ducks. *Acta Physiol. Scand.* **75,** 377.

Dogiel, J., and Archengelsky, K. (1906). Der Bewegungshemmende und der motorische Nervenapparat des Herzens. *Arch. Gesamte Physiol. Menschen Tiere* **113,** 1.

Doležel, S., and Žlábek, K. (1969). Über einen monaminergen Mechanismus in Nierenpfortadersystem der Vögel. *Z. Zellforsch. Mikrosk. Anat.* **100,** 527.

Dolnik, V. R. (1969). Bioenergetics of the flying bird. *Zh. Obshch. Biol.* **30,** 273.

Dominic, C. J., and Singh, R. M. (1969). Anterior and posterior groups of portal vessels in the avian pituitary. *Gen. Comp. Endocrinol.* **13,** 22.

Donald, D. E., Samuelhoff, S. L., and Ferguson, D. (1967). Mechanisms of tachycardia caused by atropine in conscious dogs. *Amer. J. Physiol.* **212,** 901.

Duke, H. N. (1957). Observations on the effects of hypoxia on the pulmonary vascular bed. *J. Physiol. (London)* **135,** 45.

Dukes, H. H., and Schwarte, L. H. (1931). The haemoglobin content of the blood of fowls. *Amer. J. Physiol.* **96,** 89.

Dunson, W. A. (1965). Adaptation of heart and lung weight to high altitude in the robin. *Condor* **67,** 215.

Durfee, W. K. (1963). Cardiovascular reflex mechanism in the fowl. Thesis, Rutgers University, New Brunswick, New Jersey.

Durfee, W. K. (1964). Cardiovascular reflex mechanisms in the fowl. *Diss. Abstr.* **24,** 2966.

Durfee, W. K., and Sturkie, P. D. (1963). Some cardiovascular responses to anoxia in the fowl. *Fed. Proc., Fed. Amer. Soc. Exp. Biol.* **22,** 182.

Durkovic, R. G., and Cohen, D. H. (1969). Effects of rostral midbrain lesions on conditioning of heart- and respiratory-rate responses in pigeons. *J. Comp. Physiol. Psychol.* **68,** 184.

Duvernoy, H., Gainet, F., and Koritké, J. G. (1969). Sur la vascularization de l'hypophyse des oiseaux. *J. Neuro-Visc. Relat.* **31,** 109.

Eble, J. N. (1963). An analysis of the depressor and pressor effects of tryptamine in the chicken. *J. Pharmacol. Exp. Ther.* **139,** 31.

Eliassen, E. (1960). Cardiovascular responses to submersion asphyxia in avian divers. *Arbok Univ. Bergen, Mat. Natur. Ser.* **2,** 1.

Eliassen, E. (1963). Preliminary results from new methods of investigating the physiology of birds in flight. *Ibis* **105,** 234.

Elsner, R. W., and Scholander, P. F. (1965). Circulatory adaptations to diving in animals and man. *In* "Physiology of Breath-hold Diving and the Ama of Japan" (H. Rahn, ed.), pp. 281–294. Nat. Acad. Sci., Nat. Res. Counc., Washington, D. C.

Fahr, H. O. (1935). Die "Arteriosklerose" beim Haushuhn. Dissertation, Veterinär Medizinischen Fakultät Giessen.

Fedde, M. R., and Petersen, D. F. (1970). Intrapulmonary receptor response to changes in airway-gas composition in *Gallus domesticus*. *J. Physiol. (London)* **209,** 609.

Fedde, M. R., Burger, R. F., and Kitchell, R. L. (1963). Localization of vagal afferents involved in the maintenance of normal avian respiration. *Poultry Sci.* **42,** 1224.

Feigl, E., and Folkow, B. (1963). Cardiovascular responses in "diving" and during brain stimulation in ducks. *Acta Physiol. Scand.* **57,** 99.

Folkow, B., and Rubinstein, E. H. (1965). Effect of brain stimulation on "diving" in ducks. *Hvalrad. Skr. Nor. Vidensk-Akad. Oslo* **48,** 30.

Folkow, B., and Yonce, L. R. (1967). The negative inotropic effect of vagal stimulation on the heart ventricles of the duck. *Acta Physiol. Scand.* **71,** 77.

Folkow, B., Fuxe, K., and Sonnenschein, R. R. (1966). Responses of skeletal musculature and its vasculature during "diving" in the duck: Peculiarities of the adrenergic vascoconstrictor innervation. *Acta Physiol. Scand.* **67,** 327.

Folkow, B., Nilsson, N. J., and Yonce, L. R. (1967). Effects of "diving" on cardiac output in ducks. *Acta Physiol. Scand.* **70,** 347.

Fox, H. (1939). Some comments on arteriosclerosis in wild animals and birds. *Bull. N. Y. Acad. Med.* [2] **15,** 748.

Frank, O. (1899). Grundform des arteriellen Pulses *Z. Biol.* **37,** 483.

Garrod, A. H. (1873). On the carotid arteries of birds. *Proc. Zool. Soc. London* **40,** 457.

Gasch, F. R. (1888). Beiträge zur vergleichenden Anatomie des Herzens der Vögel und Reptilien. *Arch. Naturgesch.* **54,** 119.

George, J. C., and Berger, A. J. (1966). "Avian Myology." Academic Press, New York.

George, J. C., and Naik, D. V. (1963). Haematopoietic nodules as centres of fat synthesis in the liver of migratory starling, *Pastor roseus*. *Quart. J. Microsc. Sci.* **104,** 393.

Gibbs, O. S. (1928). The renal blood flow of the bird. *J. Pharmacol. Exp. Ther.* **34,** 277.

Gibson, E. A., and de Gruchy, P. H. (1955). Aortic rupture in turkeys subsequent to dissecting aneurysm. *Vet. Rec.* **67,** 650.

Gilbert, A. B. (1961). The innervation of the renal portal valve of the domestic fowl. *J. Anat.* **95,** 594.

Gimeno, M. A., Roberts, C. M., and Webb, J. L. (1966). Acceleration of rate of the early chick heart by visible light. *Nature (London)* **214,** 1014.

Glenny, F. H. (1940). A systematic study of the main arteries in the region of the heart-Aves. *Anat. Rec.* **76,** 371.

Glenny, F. H. (1943). A systematic study of the main arteries in the region of the heart. Aves, VI, Trogoniformes, Part I. *Auk* **60,** 235.

Glenny, F. H. (1955). Modifications of pattern in the aortic arch system of birds and their phylogenetic significance. *Proc. U. S. Nat. Mus.* **104,** 525.

Golenhofen, K. (1968). Die Arterio-venose Anastomisis. *In* "Aktuelle Probleme in der Angiologie" (F. Hammersen and D. Gross, eds.), Vol. 2, pp. 67–81. Verlag Hans Huber, Bern.

Goodrich, E. S. (1930). "Studies on the Structure and Development of Vertebrates." Macmillan, New York.

Grant, R. T. (1930). Observations on direct communications between arteries and veins in the rabbit's ear. *Heart* **15,** 281.

Grant, R. T., and Bland, E. F. (1931). Observations on arteriovenous anastomosis in human skin and in the bird's foot with special reference to the reaction to cold. *Heart* **15,** 385.

Gräper, L. (1907). Untersuchungen über die Herzbildung der Vögel. *Arch. Entwicklungsmech. Organismen* **24,** 375.

Grodzínski, Z. (1963). Prad krwi w mlodej tarczy zarodkowej kurczęcia. *Acta Biol. Cracov., Ser. Zool.* **6,** 39.

Groebbels, F. (1932). "Der Vogel," Vol. I. Borntraeger, Berlin.

Grollman, A., Ashworth, C., and Suki, W. (1963). Atherosclerosis in the chicken. *Arch. Pathol.* **75,** 56.

Hamlin, R. L., and Kondrich, R. M. (1969). Hypertension, regulation of heart rate, and possible mechanism contributing to aortic rupture in turkeys. *Fed. Proc., Fed. Amer. Soc. Exp. Biol.* **28,** 402.

Hamlin, R. L., Pipers, F. S., Kondrich, R. M., and Smith, G. R. (1969). QRS component of the orthogonal lead, special magnitude and special velocity of electrocardiograms, and vectorcardiograms of turkeys. *J. Electrocardiol.* **2,** 127.

Harms, D. (1967). Über den Verschluss der Ductus arteriosi von *Gallus domesticus*. *Z. Zellforsch. Mikrosk. Anat.* **81,** 433.

Hart, J. S., and Roy, O. Z. (1966). Respiratory and cardiac responses to flight in pigeons. *Physiol. Zool.* **39,** 291.

Hartman, F. A. (1955). Heart weight in birds. *Condor* **57,** 221.

Harvey, S. C., Copen, E. G., Eskelson, D. W., Graff, S. R., Poulsen, L. D., and Rasmussen, D. L. (1954). Autonomic pharmacology of the chicken with particular reference to adrenergic blockade. *J. Pharmacol. Exp. Ther.* **112,** 8.

Hasegawa, K. (1956). On the vascular supply of hypophysis and of hypothalamus in the domestic fowl. *Fukuoka Acta Med.* **47,** 89.

Hemm, R., and Carlton, W. W. (1967). Review of duck hematology. *Poultry Sci.* **56,** 956.

Henry, J. D., and Fedde, M. R. (1970). Pulmonary circulation time in the chicken. *Poultry Sci.* **49,** 1286.

Hevesy, G., and Ottesen, J. (1945). Life cycle of the red corpuscles of the hen. *Nature (London)* **156**, 534.

Hewitt, R. (1942). Studies in the host parasite relationship in untreated infections with *Plasmodium lophurae* in ducks. *Amer. J. Hyg.* **36**, 6.

Heymans, C., and Neil, E. (1958). "Reflexogenic Areas of the Cardiovascular System." Churchill, London.

Hill, L., and Azuma, Y. (1927). Blood pressure in the two-three day chick embryo. *J. Physiol. (London)* **62**, 27.

Hirakow, R. (1970). Ultrastructural characteristics of the mammalian and sauropsidan heart. *Amer. J. Cardiol.* **25**, 195.

His, W., Jr. (1893). Die Entwicklung des Herznerven system bei Wirbelthieren. *Abh. sächs. Ges. (Akad.) Wiss. Math.-Phys. Kl.* **18**, 1.

Hollenberg, N. K., and Uvnäs, B. (1963). The role of the cardiovascular response in the resistance to asphyxia in avian divers. *Acta Physiol. Scand.* **58**, 150.

Holm, B., and Sørensen, S. C. (1970). Evidence against the role of the peripheral chemoreceptors in the diving bradycardia in ducks. *Acta Physiol. Scand.* **79**, 44A.

Holmes, A. D., Piggot, M. G., and Campbell, P. A. (1933). The haemoglobin content of chicken blood. *J. Biol. Chem.* **103**, 657.

Hughes, A. F. W. (1942). The blood pressure of the chick embryo during development. *J. Exp. Biol.* **19**, 232.

Hughes, A. F. W. (1943). The histogenesis of the arteries of the chick embryo. *J. Anat.* **77**, 266.

Hughes, A. F. W. (1949). The heart output of the chick embryo. *J. Roy. Microsc. Soc.* **69**, 145.

Huković, S. (1959). Isolated rabbit atria with sympathetic nerve supply. *Brit. J. Pharmacol. Chemother.* **14**, 372.

Hunt, J. (1954). "The Conquest of Everest." Dutton, New York.

Huxley, F. M. (1913a). On the reflex nature of apnoea in the duck in diving. I. The reflex nature of submersion apnoea. *Quart. J. Exp. Physiol.* **6**, 147.

Huxley, F. M. (1913b). On the reflex nature of apnoea in the duck in diving. II. Reflex postural apnoea. *Quart. J. Exp. Physiol.* **6**, 159.

Huxley, F. M. (1913c). On the resistance to asphyxia of the duck in diving. *Quart. J. Exp. Physiol.* **6**, 183.

Hyrtl, J. (1863). Neue Wundernetze und Geflechte bei Vögeln und Säugethieren. *Denkschr. Akad. Wiss. Wien* **22**, 113.

Irving, L. (1934). On the ability of warm-blooded animals to survive without breathing. *Sci. Mon.* **38**, 422.

Irving, L. (1938). Changes in the blood flow through the brain and muscles during the arrest of breathing. *Amer. J. Physiol.* **122**, 207.

Irving, L. (1939). Respiration in diving mammals. *Physiol. Rev.* **19**, 112.

Jalavisto, E., Kuorinka, I., and Kyllästinen, M. (1965). Responsiveness of the erythron to variations of oxygen tension in the chick embryo and young chicken. *Acta Physiol. Scand.* **63**, 479.

Johansen, K. (1961). Distribution of blood in the arousing hibernator. *Acta Physiol. Scand.* **52**, 379.

Johansen, K. (1964). Regional distribution of circulating blood during submersion asphyxia in the duck. *Acta Physiol. Scand.* **62**, 1.

Johansen, K. (1965). Cardiovascular dynamics in fishes, amphibians and reptiles. *Ann. N. Y. Acad. Sci.* **127**, 414.

Johansen, K., and Aakhus, T. (1963). Central cardiovascular responses to submersion

asphyxia in the duck. *Amer. J. Physiol.* **205,** 1167.
Johansen, K., and Hanson, D. (1967). Hepatic vein sphincters in elasmobranchs and their significance in controlling hepatic blood flow. *J. Exp. Biol.* **46,** 195.
Johansen, K., and Krog, J. (1959). Peripheral circulatory response to submersion asphyxia in the duck. *Acta Physiol. Scand.* **46,** 194.
Johansen, K., and Millard, R. W. (1972). Peripheral vascular control in the giant petrel, *Macronectes giganteus.* In preparation.
Johansen, K., and Reite, O. B. (1964). Cardiovascular responses to vagal stimulation and cardioaccelerator nerve blockade in birds. *Comp. Biochem. Physiol.* **12,** 479.
Johansen, K., and Tønnesen, K. H. (1969). Blood flow in the interdigital web of sea gull at low temperatures. *Acta Physiol. Scand.* **76,** 21A.
Jolly, J. (1910). Recherches sur les ganglions lymphatiques des oiseaux. *Arch. Anat. Microsc. Morphol. Exp.* **11,** 179.
Jolyet, E., and Sellier, J. (1895). L'hyperglobulie dans l'asphyxie expérimentale. *C. R. Soc. Biol.* **47,** 381.
Jones, D. R. (1969). Avian afferent vagal activity related to respiratory and cardiac cycles. *Comp. Biochem. Physiol.* **28,** 961.
Jones, D. R. (1970). Unpublished data.
Jones, D. R., and Holeton, G. F. (1972a). Cardiovascular and respiratory responses of ducks to progressive hypocapnic hypoxia. *J. Exp. Biol.* **56,** 657.
Jones, D. R., and Holeton, G. F. (1972b). Cardiac output of ducks during diving. *Comp. Biochem. Physiol.* **41A,** 639.
Jones, D. R., and Langille, B. L. (1972). Central cardiovascular dynamics in ducks. *Amer. J. Physiol.* (in press).
Jones, D. R., and Purves, M. J. (1970a). The carotid body in the duck and the consequences of its denervation upon the cardiac responses to immersion. *J. Physiol. (London)* **211,** 279.
Jones, D. R., and Purves, M. J. (1970b). The effect of carotid body denervation upon the respiratory response to hypoxia and hypercapnia in the duck. *J. Physiol. (London)* **211,** 295.
Jones, D. R., Randall, D. J., and Jarman, G. M. (1970). A graphical analysis of oxygen transfer in fish. *Resp. Physiol.* **10,** 285.
Jung, F. (1934). Physiologische Versuche über Pressorezeptoren an der Karotisteilungsstelle bei Vögeln. *Z. Kreislaufforsch.* **26,** 328.
Jurgens, H. (1909). Über die Wirkung des Nervus Vagus auf das Herz der Vögel. *Arch. Gesamte Physiol. Menschen Tiere* **129,** 506.
Kahl, M. P. (1963). Thermoregulation in the Wood Stork with special reference to the role of the legs. *Physiol. Zool.* **36,** 141.
Kaplan, H. M. (1954). Sex differences in the packed cell volume of vertebrate blood. *Science* **120,** 1044.
Kaupp, B. F. (1918). "The Anatomy of the Domestic Fowl." Saunders, Philadelphia, Pennsylvania.
Kern, A. (1926). Das Vögelherz. Untersuchungen an *Gallus domesticus. Gegenbaurs Jahrb.* **56,** 264.
King, A. S., and Payne, D. C. (1964). Normal breathing and the effects of posture in *Gallus domesticus. J. Physiol. (London)* **174,** 340.
King, A. S., Molony, V., McLelland, J., Bowsher, D. R., and Mortimer, M. F. (1968). Afferent respiratory pathways in the avian vagus. *Experientia* **24,** 1017.
Kisch, B. (1951). The electrocardiogram of birds (chicken, duck, pigeon). *Exp. Med. Surg.* **9,** 103.

Knoll, P. (1897). Über die Wirkung des Herzvagus bei Warmblütern. *Arch. Gesamte Physiol. Menschen Tiere* **67,** 587.

Kobinger, W., and Oda, M. (1969). Effects of sympathetic blocking substances on the diving reflex in ducks. *Eur. J. Pharmacol.* **7,** 289.

Kocian, V. (1936). Krevni obraz u ptáků přizměnách barometrického tlaku. *Vest. Cesk. Spolecnosti Zool.* **3,** 37.

Koppányi, T., and Dooley, M. S. (1928). The cause of cardiac slowing accompanying postural apnea in the duck. *Amer. J. Physiol.* **85,** 311.

Kose, W. (1907). Die Paraganglien bei den Vögeln. *Arch. Mikrosk. Anat. Entwicklungsmech.* **69,** 665.

Kovách, A. G. B., and Balint, T. (1969). Comparative study of haemodilation after haemorrhage in the pigeon and the rat. *Acta Physiol. Acad. Sci. Hung.* **35,** 231.

Kovách, A. G. B., and Szász, E. (1968). Survival of pigeon after graded haemorrhage. *Acta Physiol. Acad. Sci. Hung.* **34,** 301.

Kovách, A. G. B., Szász, E., and Pilmayer, N. (1969). The mortality of various avian and mammalian species following blood loss. *Acta Physiol. Acad. Sci. Hung.* **35,** 109.

Krista, L. M., Waibel, P. E., and Burger, R. E. (1965). The influence of dietary alterations, hormones, and blood pressure on the incidence of dissecting aneurysms in the turkey. *Poultry Sci.* **44,** 15.

Krista, L. M., Waibel, P. E., Shoffner, R. N., and Sautter, J. H. (1970). A study of aortic rupture and performance as influenced by selection for hypertension and hypotension in the turkey. *Poultry Sci.* **49,** 405.

Lack, D. (1960). The height of bird migration. *Brit. Birds* **53,** 5.

Lands, A. M., Luduena, F. P., and Buzzo, H. J. (1969). Adrenotrophic B-receptors in the frog and chicken. *Life Sci.* **8,** 373.

Lasiewski, R. C., and Dawson, W. R. (1967). A re-examination of the relation between standard metabolic rate and body weight in birds. *Condor* **69,** 13.

Lasiewski, R. C., Weathers, W. W., and Bernstein, M. H. (1967). Physiological responses of the giant hummingbird, *Patagona gigas. Comp. Biochem. Physiol.* **23,** 797.

Le Febvre, E. A. (1964). The use of D_2O^{18} for measuring energy metabolism in *Columba livia* at rest and in flight. *Auk* **81,** 403.

Le Grande, M. C., Paff, G. H., and Boucek, R. J. (1966). Initiation of vagal control of heart rate in the embryonic chick. *Anat. Rec.* **155,** 163.

Lenfant, C., Kooyman, G. L., Elsner, R., and Drabek, C. M. (1969). Respiratory function of blood of the Adélie penguin, *Pygoscelis adeliae. Amer. J. Physiol.* **216,** 1598.

Lewis, T. (1915). The spread of the excitatory process in the vertebrate heart. V. The bird's heart. *Phil. Trans. Roy. Soc. London, Ser. B* **207,** 298.

Lewis, T. (1930). Observations on the reactions of the vessels of the human skin to cold. *Heart* **15,** 177.

Lieberman, M., and Paes de Carvalho, A. (1965a). The electrophysiological organization of the embryonic chick heart. *J. Gen. Physiol.* **49,** 351.

Lieberman, M., and Paes de Carvalho, A. (1965b). The spread of excitation in the embryonic chick heart. *J. Gen. Physiol.* **49,** 365.

Lillie, F. R. (1908). "Development of the Chick," 1st ed. Holt, New York.

Lillie, F. R. (1965). "Development of the Chick," 3rd ed. Holt, New York.

Lin, Y. C., Sturkie, P. D., and Tummons, J. (1970). Effect of cardiac sympathectomy, reserpine, and environmental temperatures on the catecholamine levels in the chicken heart. *Can. J. Physiol. Pharmacol.* **48,** 182.

Lindsay, F. E. (1967). The cardiac veins of *Gallus domesticus. J. Anat.* **101,** 555.

Lombroso, U. (1913). Über die Reflexhemmung des Herzens während der reflektorischen Atmungshemmung bei verschiedenen Tieren. *Z. Biol.* **61,** 517.

Lucas, A. M. (1970). Avian functional anatomic problems. *Fed. Proc., Fed. Amer. Soc. Exp. Biol.* **29,** 1641.
Lucas, A. M., and Jamroz, C. (1961). "Atlas of Avian Haematology," Agr. Monogr. No. 25. U. S. Dept. of Agriculture, Washington, D. C.
McCready, J. D., Vallbona, C., and Hoff, H. E. (1966). Neural origin of the respiratory-heart rate response. *Amer. J. Physiol.* **211,** 323.
McDonald, D. A. (1960). "Blood Flow in Arteries." Arnold, London.
Macdonald, R. L., and Cohen, D. H. (1970). Cells of origin of sympathetic pre- and postganglionic cardio-acceleratory fibers in the pigeon. *J. Comp. Neurol.* **140,** 343.
McGinnis, C. H., and Ringer, R. K. (1966). Carotid sinus reflex in the chicken. *Poultry Sci.* **45,** 402.
McGinnis, C. H., and Ringer, R. K. (1967). Arterial occlusion and cephalic baroreceptors in the chicken. *Amer. J. Vet. Res.* **28,** 1117.
McGrath, J. J. (1970). Effects of chronic hypoxia on hematocrit ratios and heart size in the pigeon. *Life Sci.* **9,** 451.
McKibben, J. S., and Christensen, G. C. (1964). The venous return from the interventricular septum of the heart. *Amer. J. Vet. Res.* **25,** 512.
Magath, T. B., and Higgins, G. M. (1934). The blood of the normal duck. *Folia Haematol. (Leipzig)* **51,** 230.
Mangold, E. (1919). Electrographische Untersuchungen des Erregungsverlaufes im Vogelherzen. *Pfluegers Arch. Gesamte Physiol. Menschen Tiere* **175,** 328.
Manville, R. H. (1963). Altitude record for Mallard. *Wilson Bull.* **75,** 92.
Martins, L. F., Medeiros, L. F., Ferri, S., and Moraes, N. (1970). Histometric study of age related changes in thoracic aorta of *Gallus gallus. Z. Anat. Entwicklungsgesch.* **131,** 347.
Matsumori, T. (1929). The effect of adrenaline on the heart of the chick embryo. *Endocrinology* **13,** 537.
Meinertzhagen, R. (1955). The speed and altitude of bird flight. *Ibis* **97,** 81.
Mikami, S., Oksche, A., Farner, D. S., and Vitums, A. (1970). Fine structure of the vessels of the hypophyseal portal system of the White-crowned Sparrow, *Zonotrichia leucophrys gambelii. Z. Zellforsch. Mikrosk. Anat.* **106,** 155.
Millard, R. W., and Johansen, K. (1972). Active neurogenic vasodilation in skin. *Physiologist* (in press).
Millard, R. W., Johansen, K., and Milsom, G. (1972). Unpublished data.
Moore, E. N. (1965). Experimental electrophysiological studies on avian hearts. *Ann. N. Y. Acad. Sci.* **127,** Art. 1, 127.
Moore, E. N. (1967). Phylogenetic observations on specialized cardiac tissues. *Bull. N. Y. Acad. Med.* [2] **43,** 1138.
Morgan, V. E., and Chichester, D. F. (1935). Properties of the blood of the domestic fowl. *J. Biol. Chem.* **110,** 285.
Muratori, G. (1932). Contributo all'innervazione del tessuto paragangliare annesso al sistema del vago (glomo carotico, paragangli estravagali ed intravagali) e all'innervazione del seno carotideo. *Anat. Anz.* **75,** 115.
Muratori, G. (1934). Contributo istologiche e sperimentali sull' innervazione della zone arteriosa glomo-carotidea. *Arch. Ital. Anat. Embriol.* **33,** 421.
Murillo-Ferrol, N. L. (1967). The development of the carotid body in *Gallus domesticus. Acta Anat.* **68,** 102.
Neugebauer, L. A. (1845). De venis avium. *Nova Acta Leopold.-Carol.* **13,** 521.
Newell, G. W., and Shaffner, C. S. (1950). Blood volume determinations in chickens. *Poultry Sci.* **29,** 78.

Nisbet, I. C. T. (1963). Measurements with radar of the height of nocturnal migration over Cape Code, Massachusetts. *Bird-Banding* **34,** 57.

Nonidez, J. F. (1935). The presence of depressor nerves in the aorta and carotid of birds. *Anat. Rec.* **62,** 47.

Öberg, B. (1963). Aspects on the reflex control of capillary filtration transfer between blood and interstitial fluid. *Med. Exp.* **9,** 49.

Öberg, B. (1964). Effects of cardiovascular reflexes on net capillary fluid transfer. *Acta Physiol. Scand.* **62,** Suppl. 229.

Olson, C. (1952). Avian haematology. *In* "Diseases of Poultry" (H. E. Biester and L. H. Schwarte, eds.), 3rd ed., pp. 71–91. Iowa State Coll. Press, Ames.

Paff, G. H., and Boucek, R. J. (1958). Effects of different oxygen and carbon dioxide concentrations on the activity of the embryonic chick heart. *Circ. Res.* **6,** 88.

Paff, G. H., Boucek, R. J., and Gutten, G. S. (1965). Ventricular blood pressures and competency of valves in the early embryonic chick heart. *Anat. Rec.* **151,** 119.

Patten, B. M. (1925). The interatrial septum in the chick heart. *Anat. Rec.* **30,** 53.

Patten, B. M. (1956). The development of the sino-ventricular conduction system. *Univ. Mich. Med. Bull.* **22,** 1.

Patten, B. M., Kramer, T. C., and Barry, A. (1948). Valvular action in the embryonic chick heart by localized apposition of endocardial masses. *Anat. Rec.* **102,** 299.

Petren, T. (1926). Die Coronararterien des Vogelherzens. *Gegenbaurs Jahrb.* **56,** 239.

Pickwell, G. V. (1968). Energy metabolism in ducks during submergence asphyxia: Assessment by a direct method. *Comp. Biochem. Physiol.* **27,** 455.

Piiper, J., Drees, F., and Scheid, P. (1970). Gas exchange in the domestic fowl during spontaneous breathing and artificial ventilation. *Resp. Physiol.* **9,** 234.

Plenk, H., Jr., and Püschman, H. (1971). Arterio-venöse Kurzschlüsse mit Klappenmechanismen im Bereich der dorsalen Haut der Rattentunterpfote. *Z. Zellforsch. Mikrosk. Anat.* **118,** 243.

Pohlman, A. G. (1911). The circulation of mixed blood in the embryo mammal and bird and in the adult reptile, amphibian and fish. Cited by White (1969).

Prosser, C. L., Brown, F. A., Bishop, D. E., Jahn, T. L., and Wulff, V. J., eds. (1950) "Comparative Animal Physiology." Saunders, Philadelphia, Pennsylvania.

Purton, M. D. (1970). Blood flow in the liver of the domestic fowl. *J. Anat.* **106,** 189.

Quiring, D. P. (1933). The development of the sinu-atrial region of the chick heart. *J. Morphol.* **55,** 81.

Rahn, H., and Farhi, L. E. (1962). Ventilation-perfusion relationship. *Pulmonary Struct. Funct., Ciba Found. Symp., 1962* p. 139.

Rahn, H., and Farhi, L. E. (1964). Ventilation, perfusion and gas exchange – the $\dot{V}_A/\dot{Q}_B$ concept. *In* "Handbook of Physiology" (Amer. Physiol. Soc., J. Field, ed.), Sect. 3, Vol. I, p. 735. Williams & Wilkins, Baltimore, Maryland.

Ray, P. J. (1966). Physiological effects of variation in levels of respired CO_2 and O_2 in the chicken. M.Sc. Thesis, Kansas State University, Manhattan.

Ray, P. J., and Fedde, M. R. (1969). Responses to alterations in respiratory P_{O_2} and P_{CO_2} in the chicken. *Resp. Physiol.* **6,** 135.

Reite, O. B., Krog, J., and Johansen, K. (1963). Development of bradycardia during submersion of the duck. *Nature (London)* **200,** 684.

Rennick, B. R., and Gandia, H. (1954). Pharmacology of smooth muscle in valve in renal portal circulation of birds. *Proc. Soc. Exp. Biol. Med.* **85,** 234.

Richards, S. A. (1967). Anatomy of the arteries of the head of the domestic fowl. *J. Zool.* **152,** 221.

Richards, S. A. (1968). Vagal control of thermal panting in mammals and birds. *J. Physiol. (London)* **199,** 89.

Richards, S. A., and Sykes, A. H. (1967). The effects of hypoxia, hypercapnia and asphyxia in the domestic fowl (*Gallus domesticus*). *Comp. Biochem. Physiol.* **21**, 691.

Richet, C. (1899). De la résistance des canards à l'asphyxie. *J. Physiol. Pathol. Gen.* **1**, 641.

Ringer, R. K. (1968). Blood pressure of Japanese and Bobwhite quail. *Poultry Sci.* **47**, 1602.

Ringer, R. K., and Rood, K. (1959). Hemodynamic changes associated with aging in the broad-breasted Bronze turkey. *Poultry Sci.* **38**, 395.

Rodbard, S., and Fink, A. (1948). Effects of body temperature changes on the circulation time in the chicken. *Amer. J. Physiol.* **152**, 383.

Rodbard, S., and Saiki, H. (1952). Mechanism of the pressor response to increased intracranial pressure. *Amer. J. Physiol.* **163**, 234.

Rodbard, S., Brown, F., and Katz, L. N. (1949). The pulmonary arterial pressure. *Amer. heart J.* **38**, 863.

Rodnan, G. P., Ebaugh, F., Jr., and Fox, M. R. S. (1957). The life span of the red blood cell and the red blood cell volume in the chicken, pigeon and duck as estimated by the use of $Na_2Cr^{51}O_4$ with observations on red cell turnover rate in the mammal, bird and reptile. *Blood* **12**, 355.

Romanoff, A. L. (1960). "The Avian Embryo." Macmillan, New York.

Ronald, R., Foster, M. E., and Dyer, M. I. (1968). Physical properties of blood in the red-winged blackbird (*Agelaius phoeniceus*). *Can. J. Zool.* **46**, 157.

Röse, C. (1890). Beitrag zur vergleichenden Anatomie des Herzens. *Gegenbaurs Jahrb.* **16**, 27.

Rosenquist, G. (1970). Aortic arches in the chick embryo: Origin of the cells as determined by radioautographic mapping. *Anat. Rec.* **168**, 351.

Rosse, W. F., and Waldmann, T. A. (1966). Factors controlling erythropoiesis in birds. *Blood* **27**, 654.

Rostorfer, H. H., and Rigdon, R. H. (1945). Anoxia in malaria: An experimental study on ducks. *J. Lab. Clin. Med.* **30**, 860.

Rostorfer, H. H., and Rigdon, R. H. (1946). A study of oxygen transport in the blood of young and adult domestic ducks. *Amer. J. Physiol.* **146**, 222.

Rostorfer, H. H., and Rigdon, R. H. (1947). The relation of blood oxygen transport to resistance to anoxia in chicks and ducklings. *Biol. Bull.* **92**, 23.

Roughton, F. J. W. (1945). The average time spent by the blood in the human lung capillary and its relation to the rates of CO uptake and elimination in man. *Amer. J. Physiol.* **143**, 621.

Roy, C. S. (1880). The elastic properties of the arterial wall. *J. Physiol. (London)* **3**, 125.

Roy, O. Z., and Hart, J. S. (1966). A multi-channel transmitter for the physiological study of birds in flight. *Med. Biol. Eng.* **4**, 457.

Sabin, F. R. (1917). Origin and development of the primitive vessels of the chick and of the pig. *Contrib. Embryol. Carnegie Inst.* **6**, 63.

Sala, L. (1900). Sullo suiluppo dei cuori linfatici e dei dotti toracici nell'embrione di pollo. *Ric. Lab. Anat. Norm. Univ. Roma* **7**, 263.

Sapirstein, L. A., and Hartman, F. A. (1959). Cardiac output and its distribution in the chicken. *Amer. J. Physiol.* **196**, 751.

Sato, S. (1932). *Igaku Kenkyu* **6**, 707 (cited by Adams, 1958).

Scheid, P., and Piiper, J. (1969). Volume, ventilation and compliance of the respiratory system in the domestic fowl. *Resp. Physiol.* **6**, 298.

Scheid, P., and Piiper, J. (1970). Analysis of gas exchange in the avian lung: Theory and experiments in the domestic fowl. *Resp. Physiol.* **9**, 246.

Scholander, P. F. (1940). Experimental investigations on the respiratory function in diving mammals and birds. *Hvalrad. Skr. Nor. Vidensk.-Akad. Oslo* **22,** 1.

Scholander, P. F. (1958). Counter current exchange. A principle in Biology. *Hvalrad. Skr. Nor. Vidensk.-Akad. Oslo* **44,** 1–24.

Scholander, P. F. (1962). Physiological adaptation to diving in animals and man. *Harvey Lect.* **57,** 93.

Scholander, P. F. (1964). Animals in aquatic environments: diving mammals and birds. *In* "Handbook of Physiology" (D. B. Dill, E. F. Adolph, and C. G. Wilber, eds.), Sect. 4, pp. 729–739. Amer. Physiol. Soc., Washington, D.C.

Scholander, P. F., and Krog, J. (1957). Counter current heat exchange and vascular bundles in sloths. *J. Appl. Physiol.* **10,** 405.

Schumacher, S. V. (1916). Arterio-venose anastomosis in den Zehen der Vögel. *Arch. Mikrosk. Anat. Entwicklungsmech.* **87,** 309.

Schuman, O., and Rosenquist, E. (1897). Über die Natur der Blutveränderungen im Gebirge. *Arch. Gesamte Physiol. Menschen Tiere* **68,** 55.

Sellier, J. (1896). Influence de la tension de l'oxygène sur l'hématopoiese et les combustions respiratoires. Thesis, University of Bordeaux, France.

Sharp, P. J., and Follett, B. K. (1969). The blood supply to the pituitary and basal hypothalamus in the Japanese quail (*Coturnix coturnix japonica*). *J. Anat.* **104,** 227.

Siller, W. G., and Hindle, R. M. (1969). The arterial blood supply to the kidney of the fowl. *J. Anat.* **104,** 117.

Simons, J. R. (1960). The blood vascular system. *In* "Biology and Comparative Physiology of Birds" (A. J. Marshall, ed.), Vol. 1, pp. 345–362. Academic Press, New York.

Slautterback, D. B. (1965). Mitochondria in cardiac muscle cells of the canary and some other birds. *J. Cell Biol.* **24,** 1.

Smith, A. H., and Abbott, U. K. (1961). Adaptation of the domestic fowl to high altitude. *Poultry Sci.* **40,** 1459.

Soliman, M. K., Elamrousi, S., and Ahmed, A. A. S. (1966). Cytological and biochemical studies on the blood of normal and spirochaete-infected ducks. *Zentralbl. Veterinaermed., Reine B* **13,** 82.

Sommer, J. R., and Johnson, E. A. (1969). Cardiac muscle. A comparative ultrastructural study with special reference to frog and chicken hearts. *Z. Zellforsch. Mikrosk. Anat.* **98,** 437.

Sommer, J. R., and Steere, R. J. (1969). A propos: Transverse tubules in chicken cardiac muscle. *Fed. Proc., Fed. Amer. Soc. Exp. Biol.* **28,** 328.

Spanner, R. (1925). Der Pfortaderkreislauf in der Vogelniere. *Gegenbaurs Jahrb.* **54.**

Speckmann, E. W., and Ringer, R. K. (1963). The cardiac output and carotid and tibial blood pressure of the turkey. *Can. J. Biochem. Physiol.* **41,** 2337.

Speckmann, E. W., and Ringer, R. K. (1964). Static elastic modulus of the turkey aorta. *Can. J. Physiol. Pharmacol.* **42,** 553.

Speckmann, E. W., and Ringer, R. K. (1966). Volume pressure relationships in the turkey aorta. *Can. J. Physiol. Pharmacol.* **44,** 901.

Spector, W. S. (1956). "Handbook of Biological Data." Saunders, Philadelphia, Pennsylvania.

Spencer, R. P. (1966). Relative heart weight in porpoises. *Science* **152,** 230.

Spencer, R. P. (1967). A blood volume heart weight relationship. *J. Theor. Biol.* **17,** 441.

Sperber, I. (1946). A new method for the study of renal tubular excretion in birds. *Nature (London)* **158,** 131.

Sperber, I. (1948). Investigations on the circulatory system of the avian kidney. *Zool. Bidr. Uppsala* **27,** 429.

Ssinelnikow, R. (1928). Die Herznerven der Vögel. *Z. Anat. Entwicklungsgesch.* **86,** 540.

Staub, N. C., Bishop, J. M., and Foster, R. E. (1962). Importance of diffusion and chemical reaction rates in O_2 uptake in the lung. *J. Appl. Physiol.* **17,** 21.

Steen, I., and Steen, J. B. (1965). The importance of the legs in the thermoregulation of birds. *Acta Physiol. Scand.* **63,** 283.

Steeves, H. R., and Siegel, P. B. (1968). Comparative histology of the vascular system in the domestic and game cock. *Experientia* **24,** 937.

Stéphan, F. (1949). Les suppléances obtenues expérimentalement dans le système des arcs aortiques de l'embryon d'oiseau. *C. R. Soc. Anat.* **36,** 647.

Stéphan, F. (1959). Etude expérimentale du cloisonnement du coeur chez l'embryon de poulet, *Gallus gallus. Proc. Int. Congr. Zool., 15th, 1958* p. 615.

Stübel, H. S. (1910). Beiträge zur Kenntnis der Physiologie des Blutkreislaufes der verschiedenen Vogelarten. *Arch. Gesamte Physiol. Menschen Tiere* **135,** 249.

Sturkie, P. D. (1965). "Avian Physiology," 2nd ed. Cornell Univ. Press, Ithaca, New York.

Sturkie, P. D. (1966). Cardiac output in ducks. *Proc. Soc. Exp. Biol. Med.* **123,** 487.

Sturkie, P. D., and Newman, H. J. (1951). Plasma proteins of chickens as influenced by the time of laying, ovulation, number of blood samples taken and plasma volume. *Poultry Sci.* **30,** 240.

Sturkie, P. D., and Vogel, J. A. (1959). Cardiac output, central blood volume, and peripheral resistance in chickens. *Amer. J. Physiol.* **197,** 1165.

Surendranathan, K. P., Nair, S. G., and Simon, K. J. (1968). Haematological constituents of duck. *Indian Vet. J.* **45,** 311.

Sykes, A. H. (1960). A note of the determination of oxygen in the blood of the fowl. *Poultry Sci.* **39,** 16.

Szepsenwol, J., and Bron, A. (1935). Le premier contact du système nerveux vagosympathetique avec l'appareil cardio-vasculaire chez les embryons d'oiseaux. *C. R. Soc. Biol.* **118,** 946.

Tasawa, H. (1971). Measurement of respiratory parameters in blood of chicken embryo. *J. Appl. Physiol.* **30,** 17.

Taylor, M. G. (1964). Wave travel in arteries and the design of the cardiovascular system. *In* "Pulsatile Blood Flow" (E. O. Attinger, ed.), pp. 343–372. McGraw-Hill, New York.

Terni, T. (1923). Ricerche anatomiche sul sistema nervoso autonomo degli uccelli. *Arch. Ital. Anat. Embriol.* **20,** 433.

Terni, T. (1927). Il corpo ultimobrachiale degli uccelli. Ricerche embriologiche anatomiche e histologische sul *Gallus dom. Arch. Ital. Anat. Embriol.* **24,** 407.

Terni, T. (1931). Il Simpatico cervicale degli amnioti: (Ricerche di morfologia comparata). *Z. Anat. Entwicklungsgesch.* **96,** 289.

Thébault, M. V. (1898). Etude des rapports qu'existent entre les systèmes pneumogastrique et sympathique chez les oiseaux. *Ann. Sci. Natur. Zool.* [2] **6,** 142.

Thompson, R. M., and Coon, J. M. (1948). Effect of adrenolytic agents on the response to pressor substances in the domestic fowl. *Fed. Proc., Fed. Amer. Soc. Exp. Biol.* **7,** 259.

Topham, W. S. (1969). Comparison of methods for calculation of left ventricular stroke work. *J. Appl. Physiol.* **27,** 767.

Trendelenburg, U. (1965). The effect of sympathetic nerve stimulation on isolated atria of guinea pigs and rabbits pretreated with reserpine. *J. Pharmacol. Exp. Ther.* **147,** 313.

Tucker, V. A. (1968a). Respiratory physiology of House Sparrows in relation to high altitude flight. *J. Exp. Biol.* **48,** 55.

Tucker, V. A. (1968b). Respiratory exchange and evaporative water loss in the flying Budgerigar. *J. Exp. Biol.* **48,** 67.

Tummons, J., and Sturkie, P. D. (1968). Cardio-accelerator nerve stimulation in birds. *Life Sci.* **7,** 377.

Tummons, J. L., and Sturkie, P. D. (1969). Nervous control of heart rate during excitement in the adult white leghorn cock. *Amer. J. Physiol.* **216,** 1437.

Usami, S., Magazinovic, V., Chien, S., and Gregersen, M. I. (1970). Viscosity of turkey blood: Rheology of nucleated erythrocytes. *Microvas. Res.* **2,** 489.

Van Breemen, V. L. (1953). Intercalated discs in heart muscle studied with the electron microscope. *Anat. Rec.* **117,** 49.

Vanderbrook, M. J., and Vos, B. J. (1940). The pharmacodynamics of the domestic fowl with respect to ergonovine and ergotamine. *Quart. J. Exp. Physiol.* **30,** 173.

Van der Linden, P. (1934). Recherches sur les réflexes sinocarotidiens chez les oiseaux. *Arch. Int. Physiol.* **40,** 59.

Van Mierop, L. H. S., and Bertuch, C. J. (1967). Development of arterial blood pressure in the chick embryo. *Amer. J. Physiol.* **212,** 43.

Vezzani, V. (1939). Influence of a sojourn in the mountains on the blood composition, body development, and egg production of white leghorn pullets. *World's Poultry Congr. Exposition* [*Proc.*], *7th, 1939* p. 117.

Viault, F. (1892). Action physiologique des climats de montagne. *C. R. Acad. Sci.* **114,** 1562.

Vitums, A., Mikami, S., Oksche, A., and Farner, D. S. (1964). Vascularization of the hypothalamo-hypophysial complex in the White-crowned Sparrow, *Zonotrichia leucophrys gambelii. Z. Zellforsch. Mikrosk. Anat.* **64,** 541.

Vogel, J. (1961). Studies on cardiac output in the chicken. Thesis, Rutgers University, New Brunswick, New Jersey.

von Lindes, G. (1865). Ein Beitrag zur Entwicklungsgeschichte des Herzens. Inaugural Dissertation, Dorpat. Heinrich Laakmann.

Vonruden, W. J., Blaisdell, F. W., Hall, A. D., and Thomas, A. N. (1964). Multiple arterial stenoses. *Arch. Surg. (Chicago)* **89,** 307.

Wastl, H., and Leiner, G. (1931). Beobachtungen über die Blutgase bei Vögeln. I. *Pfluegers Arch. Gesamte Physiol. Menschen Tiere* **227,** 368.

Watzka, M. (1943). *In* "Handbuch der mikroskopischen Anatomie des Menschen" (W. von Möllendorff, ed.), Vol. 6, p. 262. Springer-Verlag, Berlin and New York.

Weale, F. E. (1964). Series and parallel resistances in steady bloodflow. *Brit. J. Surg.* **51,** 623.

West, J. B. (1970). "Ventilation/Blood Flow and Gas Exchange," 2nd ed. Blackwell, Oxford.

White, P. T. (1969). Experimental studies on the heart of the chick embryo at two stages of development. Ph.D. Thesis, University of London, London.

Wingstrand, K. G. (1951). "The Structure and Development of the Avian Pituitary." Gleerup, Lund.

Wirth, D. (1931). "Grundlage einer klinischen Haematologie der Haustiere." Urban & Schwarzenberg, Munich.

Womersley, J. R. (1957). "An Elastic Tube Theory of Pulse Transmission and Oscillatory Flow in Mammalian Arteries," Tech. Rep. No. WADC-TR-56-614. Wright Air Force Development Center.

Woodbury, R. A., and Hamilton, W. F. (1937). Blood pressure studies in small animals. *Amer. J. Physiol.* **119,** 663.

Zeuthen, E. (1942). The ventilation of the respiratory tract in birds. *Kgl. Dan. Vidensk. Selsk., Biol. Med.* **17,** 1.

Chapter 5

RESPIRATORY FUNCTION IN BIRDS

Robert C. Lasiewski[1]

[1]Deceased 2 August 1971.

I. Introduction

The avian respiratory system may be the most complicated in the animal kingdom and differs markedly in its fundamental architecture from that of mammals. In mammals, gaseous exchange occurs in clusters of tiny blind sacs (alveoli). In birds, gaseous exchange takes place across shorter diffusion distances and relatively smaller surface areas than in mammals, in minute air capillaries which open into tubular tertiary bronchi. A number of pairs of pulmonary air sacs connect to the relatively small avian lungs, providing a totally different pattern of respiratory air circulation than that found in mammals.

Despite many anatomical and physiological studies over the past century, the physiology of the avian respiratory apparatus is poorly understood. Its fundamental structure was first described by Harvey in 1651. The area of avian respiration has been reviewed at least six times within the past dozen years (Salt and Zeuthen, 1960; J. R. King and Farner, 1964; Salt, 1964; Sturkie, 1966; A. S. King, 1966; A. S. King and Molony, 1971), and there is little point in reviewing the literature once again. The six reviews include over 300 references, yet the available data do not permit resolution of some of the most fundamental aspects of anatomy, of patterns of air flow, of control of ventilation, and so on.

Mammalian respiration has been studied in depth, primarily because of its medical relevance. For example, when the American Physiological Society sponsored a comprehensive review of mammalian respiration, the result was two weighty volumes published in 1964 and 1965, totalling more than 1700 pages with over 700 figures.

The respiratory systems of birds and mammals are able to accommodate the requirements of respiratory gas exchange, evaporative cooling, and vocalization, although these two vertebrate classes employ radically different respiratory mechanisms.

Since the area of avian respiration has been reviewed several times recently, I shall neither include all previous work in my discussion, nor provide a comprehensive survey of the literature. Instead, the present chapter attempts to summarize the more recent developments in the field of avian respiration, to provide recent references as starting points for further work, and to indicate areas where further research is needed.

II. Anatomy

Air sacs are not unique to birds and were probably foreshadowed in the respiratory apparatuses of their reptilian ancestors. The lungs of

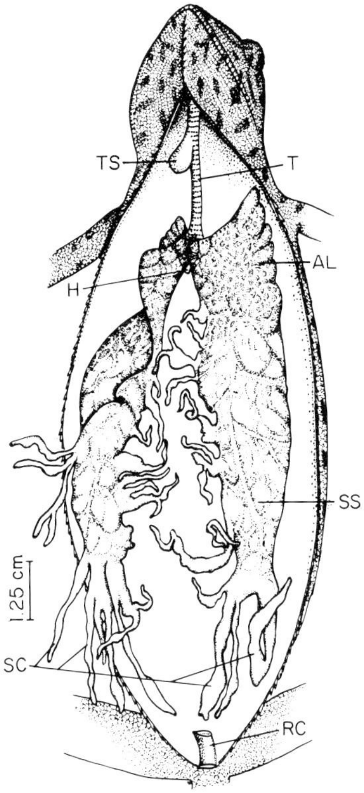

FIG. 1. The lung of *Chamaeleon zeylanicus*, showing the alveolar (AL), semisaccular (SS), and saccular (SC) portions. The tracheal sac (TS), trachea (T), heart (H), and rectum (RC) are also represented. From George and Shah (1965).

some lizards and snakes are differentiated into an anterior alveolar portion and a posterior saccular portion (George and Shah, 1965). The chameleon, *Chamaeleon zeylanicus*, shows a spectacular differentiation of its respiratory system (Fig. 1), which possesses a tracheal sac and extensive lungs consisting of alveolar, semisaccular, and saccular portions. The saccular portions of reptilian lungs are less well vascularized than the alveolar portions, and, although a variety of functions have been attributed to their unusual structure, their respiratory purpose has not been studied. Air sacs are also found in some larger active insects.

The respiratory apparatus of birds consists of relatively small paired lungs, a number (6 to 14, but usually 8 or 9) of voluminous, poorly vascularized pulmonary air sacs, and the trachea and bronchi (Fig. 2). Diverticula from air sacs may penetrate the avian skeleton. The total volume of the respiratory system occupies approximately 5 to 20% (mean value of about 15%) of the total volume of the avian body. The

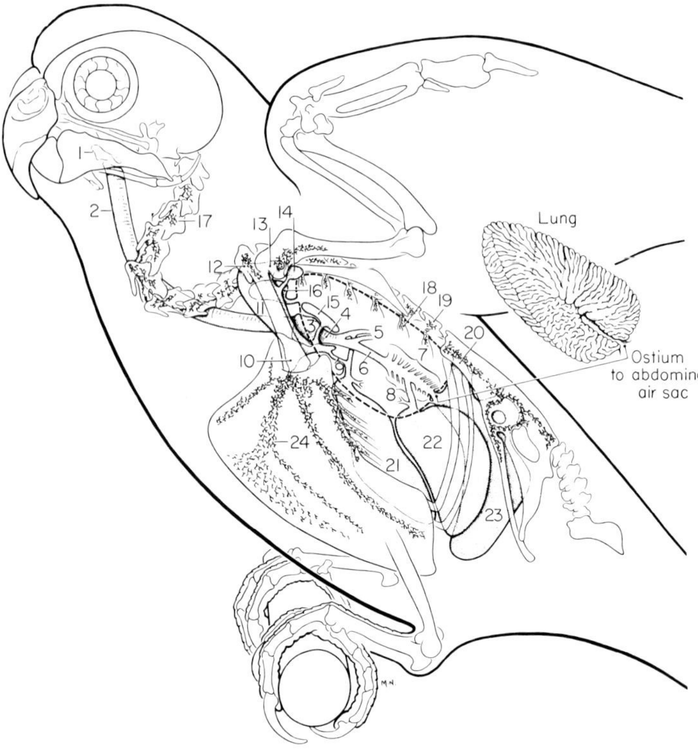

FIG. 2. Lateral view of the respiratory system and skeleton of the Budgerigar, *Melopsittacus undulatus*. The surface view of the lung is displaced to show internal structures. 1, cricoid cartilage of the larynx; 2, trachea; 3, syrinx; 4,6, left primary bronchus; 5, craniomedial secondary bronchi; 7, caudodorsal secondary bronchi; 8, caudoventral secondary bronchi; 9, direct connection of clavicular sac; 10, 11, 13, 14, 15, diverticula of clavicular sac; 12, 17, 18, 19, 20, 24, pneumatic spaces in skeleton; 16, cervical sac; 21, cranial thoracic sac; 22, caudal thoracic sac; 23, abdominal sac. From Evans (1969) with minor changes in terminology to correspond to text.

literature on the anatomy of this system is vast, confusing, and often contradictory, and controversy still exists about the detailed anatomy of some of the commonest avian species. The following description pertains to a "generalized" bird, although the literature makes it abundantly clear that detailed anatomical variations occur between species (A. S. King, 1957). Nevertheless, the general anatomical features appear to be similar in such diverse birds as hummingbirds, domestic fowl, and Ostriches. A. S. King's (1966) authoritative review of the functional anatomy of avian lungs and air sacs is recommended to readers interested in specific details or references. The anatomical terminology used herein follows that of King.

A. Upper Respiratory Tract

In the resting bird, tidal air enters through paired external nares into elongated nasal chambers. These paired chambers are divided into three regions, each containing a concha (turbinal) on its lateral wall (Portmann, 1961), which cleanse, modify, and sample respiratory air. From the nasal chambers air passes through internal nares to the buccal cavity.

From the buccal and anterior pharyngeal cavities, tidal air enters the trachea through the glottis, a slitlike opening surrounded by the larynx. The larynx is a valvular structure that opens and closes the glottis during the respiratory cycle and aids in preventing aspiration of foreign material into the lower respiratory tract. Unlike mammals, the larynx of birds is generally not the site of avian sound production (see Section VI,C).

Air then proceeds through the tubular trachea, the lumen of which is usually circular in cross section and is maintained by a large number of cartilaginous or partially ossified overlapping rings. There is some dispute about the number of tracheal rings, their degree of overlap, and the tracheal musculature in different species (see McLelland, 1965, for discussion and references).

At the caudal end of the trachea, air passes through the syrinx, the major sound-producing organ in most birds. The syrinx is located inside the thoracic cavity at the bifurcation of the trachea into two primary bronchi. Three types of syrinxes are recognized: tracheal, bronchotracheal, and bronchial (Evans, 1969). Two to nine pairs of syringeal muscles have been reported in different species, the superior singers having more complicated musculature (see Chamberlain *et al.*, 1968, for references).

The paired primary bronchi have a cartilaginous framework of incomplete rings and proceed laterally and caudally from the syrinx into the lungs. As the primary bronchi reach the lungs, the cartilaginous framework is reduced or absent medially and ventrally.

B. Lungs and Air Passages

The paired lungs of birds are relatively small and inexpansible and are located in the dorsal portion of the thorax (Fig. 2). They are somewhat flattened, with their dorsolateral surfaces closely apposed to five or six vertebral ribs which furrow their surfaces.

Although the lungs of birds are regularly described as small, there are few quantitative data supporting these statements. That avian lungs constitute a relatively constant proportion of body weight may be seen in Fig. 3 in which the equation relating lung weight to body weight has a slope near unity. Although birds and mammals of similar body weights tend to have comparable lung weights and similar specific gravities, the total volume of the avian lung is less than that of the mammalian lung due to smaller avian lung air volumes (see Section III,A).

The interconnections between the primary, secondary, and tertiary bronchi in the lungs of birds are labyrinthine (Fig. 4) and differ greatly from the treelike branching of the mammalian respiratory passages. The primary bronchi course through the length of the lungs, decreasing in diameter caudally and giving rise to two major groups of secondary bronchi, the anterior craniomedial and posterior caudodorsal series (Fig. 5). These two series have been described in the earlier literature as the ventro- and dorsobronchi, or the ento- and ectobronchi. A. S. King (1966) suggests that two lesser series of secondary bronchi also regularly stem from the primary bronchi, the caudoventral and caudolateral series (Fig. 5), which may have diameters similar to those of tertiary bronchi. Akester (1960) even considers many of these to be tertiary bronchi.

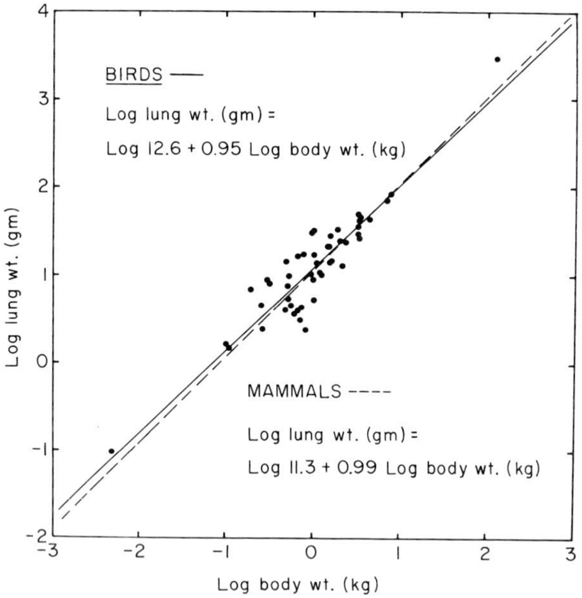

FIG. 3. The relationship between lung weight and body weight in birds, excluding passerines. Data are from Spector (1956) and Dittmer and Grebe (1958), and when their data were clearly duplicated, only one point was used. Solid line represents equation fit to avian data (Lasiewski and Calder, 1971), while dashed line represents equation fit to mammalian data (Stahl, 1967).

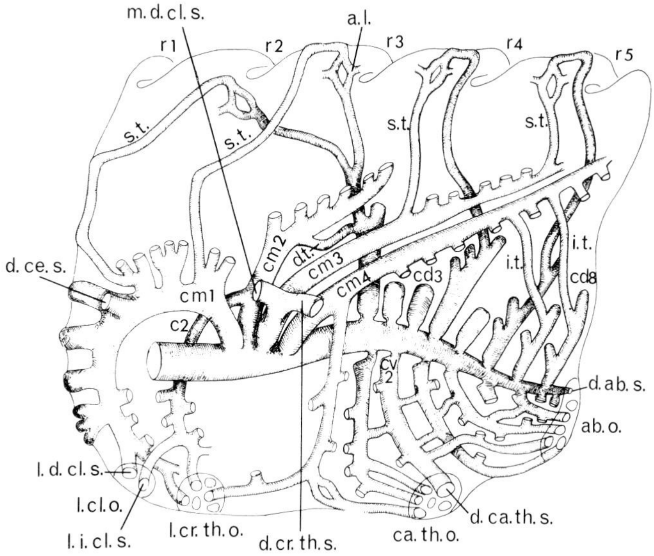

FIG. 4. Medial view of the right lung of the adult domestic chicken. Ostium means the general area of connection between air sac and lung. m.d.cl.s., medial direct connection to clavicular sac; d.cr.th.s., direct connection of cranial thoracic sac; d.ca.th.s., direct connection of caudal thoracic sac; d.ab.s., direct connection of abdominal sac; l.d.cl.s., lateral direct connection of clavicular sac; l.i.cl.s., lateral indirect connection of clavicular sac; d.ce.s., direct connection of cervical sac; ab.o., ostium of abdominal sac; ca.th.o., ostium of caudal thoracic sac; l.cr.th.o., lateral ostium of cranial thoracic sac, comprising indirect connections only; l.cl.o., lateral ostium of clavicular sac; cm, craniomedial secondary bronchi; cd, caudodorsal secondary bronchi; cv, caudoventral secondary bronchi; c2, circumflex branch of second craniomedial secondary bronchus; s.t., superficial tertiary bronchi; i.t., tertiary bronchi of intermediate depth; d.t., a deep tertiary bronchus; a.l., anastomotic line; r1 to r5, five impressions made by ribs. From A. S. King (1966).

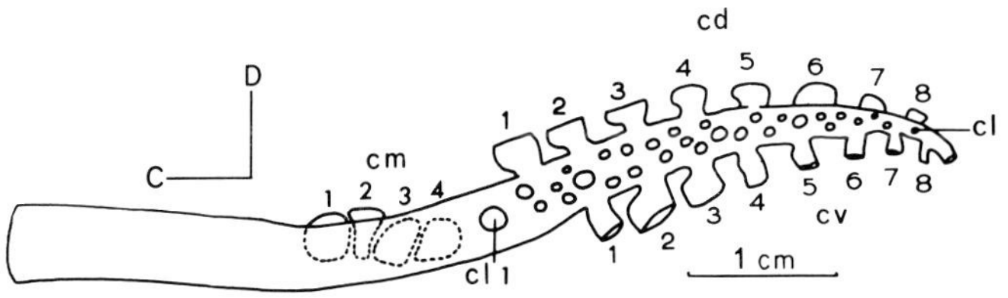

FIG. 5. Lateral view of a typical primary bronchus of adult domestic chicken, left side. cm, craniomedial secondary bronchi; cd, caudodorsal secondary bronchi; cv, caudoventral secondary bronchi; cl, caudolateral secondary bronchi; D, dorsal; C, cranial. (From A. S. King, 1966.)

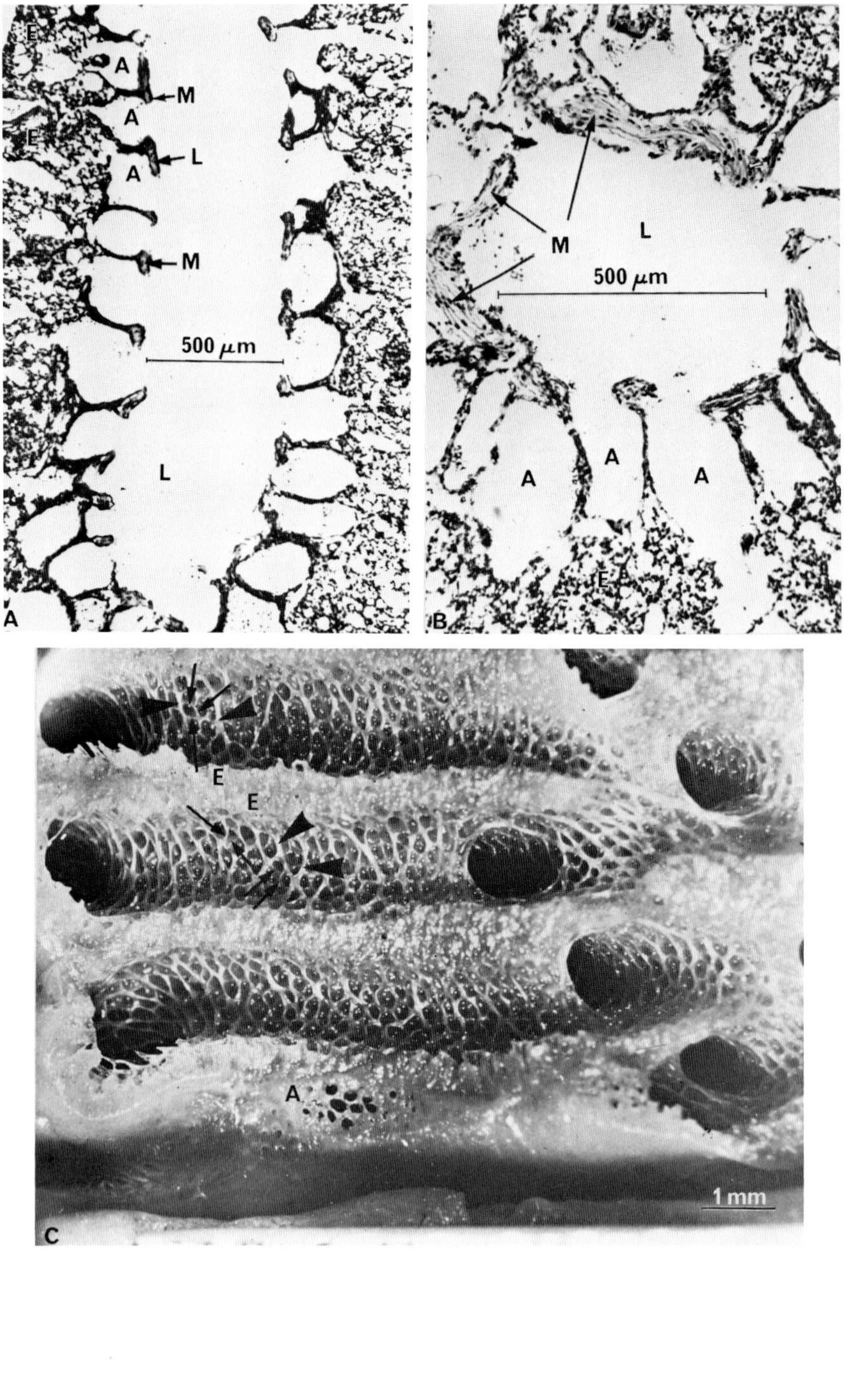
E
A
M
A
E
L
A
M
500 μm
L
A
M
L
500 μm
A
A
A
E
B
E
E
A
1 mm
C

The number of craniomedial secondary bronchi is normally four, while that of the caudodorsal bronchi is usually seven. The numbers and orientation of the secondary bronchi arising from the primary bronchus have been studied thoroughly in surprisingly few species, even though the anatomy of this region may prove to be crucial in understanding the aerodynamic patterns of respiratory flow through the avian lung (see Section III,B).

Numerous tubular tertiary bronchi arise from the secondary bronchi, and along with the functional respiratory exchange regions, the air capillaries (Fig. 6), they form the bulk of the lung. A large proportion of the tertiary bronchi form a series of more or less parallel circuits, which connect the craniomedial and caudodorsal secondary bronchi. Other tertiary bronchi may join the caudoventral and caudolateral secondary bronchi to each other and to the other series of secondary bronchi. In some species the tertiary bronchi anastomose freely. A. S. King and Cowie (1969) estimate that the total number of tertiary bronchi in the lung of the domestic fowl is 300–500. Akester (1960) reports that the duck and the pigeon, which are both strong flying birds, have "at least four times as many tertiary bronchi per unit volume of lung as are present in the relatively flightless fowl." Bretz (1969) estimates that ducks have 1800 tertiary bronchi. In the lungs of the fowl the longest tertiary bronchi are superficial and 3 to 4 cm long, while the shortest bronchi are deep and as short as 1 cm long (see Payne, 1960).

The smooth muscles on the walls of the tertiary bronchi are penetrated by many openings called atria (Fig. 6). In the fowl, atria are typically 0.1 to 0.2 mm in diameter (A. S. King and Cowie, 1969) and lead to the air capillaries, where gaseous exchange occurs. Massive sphincterlike bands of smooth muscle surround the openings of the tertiary bronchi at their points of origin from the secondary bronchi. Other muscle bundles in the tertiary bronchi form large spiral bands

FIG. 6. The fine structure of tertiary bronchi in the domestic chicken. (A) Photomicrograph of tertiary bronchus cut longitudinally, from 5-week-old chicken. L, lumen of tertiary bronchus; E, exchange areas; A, atria; M, arrows to bundles of bronchial muscle covered by simple squamous epithelium. (B) Photomicrograph of tertiary bronchus cut transversely; symbols as in (A). (C) Dissected lung of adult chicken. Three tertiary bronchi have been opened longitudinally, but turn into the depth of field on the left. Five more tertiary bronchi open on the right. Examples of large spiral muscle bands are indicated by four large arrows, while examples of smaller irregular atrial bundles are indicated by five small arrows. The pockets between the bundles of bronchial muscle are atria, a group of which have been cut tangentially at A. The atria lead into the exchange areas, E. Photography by T. S. Fleming, Liverpool University, U. K. (From A. S. King and Cowie, 1969.)

and small oblique bundles (Fig. 6C). This musculature may vary the caliber of the airways. The ultrastructure of the smooth muscle and its innervation in the avian lung have been examined by Cook and King (1970).

Although earlier workers concluded that avian lungs lacked surfactant, more recent studies have confirmed the presence of surface-active phospholipids (Pattle and Hopkinson, 1963; Fujiwara *et al.*, 1970; Clements *et al.*, 1970). Tyler and Pangborn (1964) describe a unique laminated membrane surface in the epithelial cells of the tertiary bronchi and atria, which they suggest is responsible for the surface-tension-reducing properties of avian lung extracts.

C. Air Sacs

The pulmonary air sacs, which arise from the lungs, occupy the major portion of the volume of the respiratory systems of all birds. The air sacs in some species may completely surround the heart, liver, kidneys, testes, ovaries, and intestines (Hamlet and Fisher, 1967). Most birds also have cervicocephalic air sacs, which arise from the nasal or tympanic cavities, or both. Some species may also have pharyngotracheal air sacs, which arise from the pharynx or trachea. The cervicocephalic and pharyngotracheal air sac systems are less well developed than the pulmonary air sacs, and are obscure in function. Hereafter in this chapter, the term "air sacs" will be used to refer to the pulmonary air sacs.

The number of air sacs may vary from 12 in Charadriiformes and Ciconiiformes (perhaps 14 in some Ciconiiformes) to as few as perhaps 6 or 7 in the passeriform House Sparrow, *Passer domesticus* (Wetherbee, 1951; A. S. King, 1966), and 7 in the Common Loon, *Gavia immer* (Gier, 1952) and turkey, *Meleagris gallopavo* (A. S. King and Atherton, 1970). In most birds, the 6 primordial pairs are reduced by fusion to 8 or 9 air sacs. A bird with 9 air sacs would have a single unpaired clavicular sac and 4 paired sacs: the cervical, cranial thoracic, caudal thoracic, and abdominal sacs (see Figs. 2 and 7). Earlier workers used a plethora of terminologies for air sacs. Many of these terms are summarized in Table II of A. S. King's (1966) review.

The air sacs constitute about 80% of the total avian respiratory volume. The cervical air sacs are usually the smallest, while the abdominal sacs are generally the largest, although in hummingbirds and Ostriches the caudal thoracic sacs are the largest (Stanislaus, 1937; Schmidt-Nielsen *et al.*, 1969). The unpaired clavicular sac is generally one of the larger air sacs. The caudal thoracic sac is generally larger than the cranial thoracic sac, although, as A. S. King (1966)

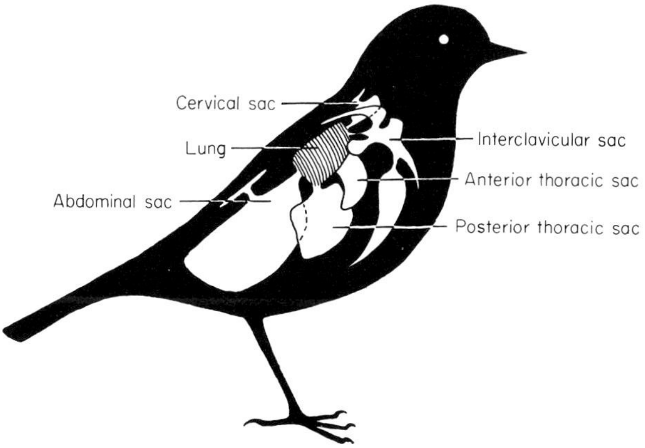

FIG. 7. Diagrammatic representation of the position of lungs and air sacs in a bird's body. (From Salt, 1964.)

points out, the situation is reversed in two of the most common experimental birds, the domestic chicken and pigeon. He states, "It is perhaps unfortunate that these two birds have been used so widely for experimental purposes without being anatomically typical of the class generally." See A. S. King (1966, Table III) for summary of measured volumes of lungs and air sacs.

A variety of diverticula arise from the main chambers of the clavicular, cervical, and abdominal sacs, and these may differ considerably between species. The diverticula may be relatively simple or highly complicated, as in the subcutaneous diverticula of the cervical sac of some Pelecaniformes or the lateral diverticula of the clavicular sac of many avian species (portions of which may penetrate the skeleton).

D. CONNECTIONS OF THE AIR SACS TO AIR PASSAGES IN THE LUNGS

Generally, the individual air sacs have one direct connection to a primary or secondary bronchus and may have several indirect connections (sometimes called recurrent bronchi) to other secondary or tertiary bronchi (see Fig. 4). Because of their connections to the craniomedial series of secondary bronchi as well as their anterior position in the thoracic cavity, the cervical, clavicular, and cranial thoracic sacs are considered anterior sacs. In contrast, the posterior sacs (the abdominal and caudal thoracic air sacs) are directly connected to the caudal ends of the primary bronchi and to the second caudoventral secondary bronchi, respectively.

The following description is based on Payne's (1960) work on the domestic fowl. A. S. King (1966) states that the connections of air sacs to airways seem to be similar in most birds. The cervical sacs have direct connections to the first craniomedial secondary bronchi but have no indirect connections. The clavicular sac has two direct connections to each lung, because of its derivation from two primordial pairs of sacs. Its major connection is to the third craniomedial secondary bronchus, while its other connection joins the sac to the first craniomedial secondary bronchus via a smaller tube. The cranial thoracic sacs connect directly to the third craniomedial secondary bronchi by means of a common canal with the clavicular sac and indirectly to the first, second, and fourth craniomedial secondary bronchi.

The caudal thoracic sacs connect directly with the second caudoventral secondary bronchi and indirectly through several connections to the first caudoventral secondary bronchi. The abdominal sacs connect directly to the caudal end of the primary bronchi. Several indirect connections join this sac to the last several caudoventral secondary bronchi. For further details, see Fig. 4 and A. S. King's (1966) clear comparative presentation of this complicated anatomical material.

E. Pleural Membranes

Two membranous structures, the pulmonary aponeurosis and the oblique septum, are closely associated with the avian respiratory system, and are often referred to as avian diaphragms. However, neither structure is homologous to the mammalian diaphragm, and thus the term avian diaphragm should be discarded.

Although the two pleural membranes are difficult to visualize without dissection, Salt and Zeuthen's description (1960) is helpful. It states that in form the two pleural membranes

> are like two roofs, one above the other, both having the same ridgepole. The ridgepole is the ventral surface of the vertebral column. The dorsal (and flatter) one [the pulmonary aponeurosis] stretches from the dorsal mid-line to the lateral walls of the thorax. It fuses with the abdominal diaphragm posteriorly. The ventral one [the oblique septum] stretches from the same suspension point to the lateral margins of the sternum in the anterior part of the body, where it is fused to the pericardial sac and reaches posteriorly to the synsacrum.

Both pleural membranes contain muscular components, although their function in respiration is unknown. The pulmonary aponeurosis attaches to the lateral walls of the thorax via a fringe of striated muscle. The oblique septum, on the other hand, has smooth muscle on its medial border. The functions of these two membranous structures, intimately associated with the respiratory system, are unknown and thus deserve further study.

F. Related Skeletal Elements

The pneumatized bones of birds have received much attention, although little is known about their respiratory function. The diverticula of some air sacs may penetrate the skeleton in some species (see A. S. King, 1957, 1966). Large strong-flying birds have extensively pneumatized skeletons, while small birds and some nonfliers show comparatively little pneumatization. Some large dinosaurs, however, had pneumatized bones, while crocodiles have parts of their skulls pneumatized by extensions of the cervicocephalic air sac system. Although simple causality may not exist, a variety of functions have been ascribed to the aerated bones of birds, particularly relating to powers of flight, to strength with minimum weight, and to respiration (Bellairs and Jenkin, 1960).

Because of the dorsal location of the lungs in avian thoracic cavities, the respiratory movements of the rib cage appear to act primarily upon the air sacs in effecting tidal air movements. During inspiration, the thoracic cage increases in both dorsoventral and transverse diameters, with the increase being greatest in the caudal portion of the thoracic cage. Kadono and Okada (1962) have studied the respiratory muscles of chickens with the electromyographic technique and have found that the expiratory muscles "acted as vigorously" as the inspiratory muscles.

III. Respiration of Resting Birds

Despite the efforts of many clever research workers, the avian respiratory system has not yielded its secrets easily. It is still difficult to make quantitative statements about the physiology of respiration, even in resting birds, and only recently has a modicum of reliable data on respiratory function in flying and heat-stressed birds become available.

While it might be desirable to confine the following discussion to a typical bird with capacities for flight, such data are not available. Salt and Zeuthen (1960) alluded to the problem of species peculiarities causing confusion in understanding avian respiration, so they attempted to confine their discussion to the domestic pigeon "unless otherwise specified." While I agree with their logic, the available physiological data on pigeons are rarely totally acceptable, and in addition, the anatomy of this species may be atypical (see Section II, C). Much of the existing data on avian respiration are for the domestic fowl, a species that is a poor flier and is hardly a typical bird. I shall draw evidence from a variety of species, realizing full well that species differences in physiology, as in anatomy, do exist. The ideal "generalized" bird for physiological studies should be a large flying bird, the anatomy of which is reasonably representative of the class. Because large birds have slower rates of physiological processes, the rapid response times of modern instrumentation may allow a more accurate dissection of sequential events than has been hitherto possible. Data from nonflying or poor-flying species, however useful, can always be criticized as being atypical of the class.

Standard respiratory physiological symbols and terminology as defined by Comroe *et al.* (1950) will be used in the following discussion. They include the following: f = frequency of respiration as breaths/minute, V_T = tidal volume in milliliters, $\dot{V}_E$ = minute volume in milliliters/minute, $\dot{V}_{O_2}$ = oxygen consumption per unit time, W = body weight in kilograms, and $F_{I_{O_2}}$ and $F_{E_{O_2}}$ = the fraction of oxygen in inspired and expired air respectively. Also, T_a = ambient temperature, T_b = core body temperature, and EWL = evaporative water loss per unit time. All temperatures will be in °C and all gas volumes reduced to standard temperature and pressure.

A. Respiratory Variables and Body Size in Resting Birds

Reliable physiological data on avian respiratory variables are scarce. Furthermore, respiratory rates (and presumably some other respiratory variables) are easily disturbed in birds (Calder, 1968). The incompleteness of the data in the literature prompted Lasiewski and Calder (1971) to survey the available data and to provide a preliminary allometric analysis of the relationships between avian respiratory variables and body weight.

A variety of physiological and morphological attributes are related to the body weight of organisms in a manner that can be described by the power function equation:

$$Y = aW^b \tag{1}$$

where Y is some variable, W is body weight, and a and b are empirically derived constants. This equation can be written in the more convenient logarithmic form:

$$\log Y = \log a + b \log W \tag{2}$$

recognizable as a mathematical expression of a straight line. Given a series of data conforming with these equations, investigators have usually transformed their data to logarithms and then estimated an equation by least-squares regression analysis.[2]

Body weights of living birds span almost five log cycles (about 2 gm to 150,000 gm). Respiratory data for some variables are available for small, medium, and large birds. Lasiewski and Calder (1971) proposed that if a few data points spanning a wide range of body weights are available for a given physiological variable and if the variances of the *Y* values are not large, the "preliminary" regression line fitted to these points may suggest the "true" relationship to a surprising degree. Lasiewski and Calder calculated values for *a* and *b* for a variety of avian respiratory variables, and these are summarized in Table I. Values for *a* and *b* relating mammalian and other avian respiratory variables to body weight are also presented in Table I. Since body weight is in kilograms, the *a* value represents the *Y*-intercept at 1 kg (log 1 = 0). Given relatively similar slopes of the comparable equations for birds and mammals (as demonstrated by the relatively low values, in most cases, for $b_{\text{birds}}/b_{\text{mammals}}$), the ratio $a_{\text{birds}}/a_{\text{mammals}}$ provides a convenient basis for comparing these variables between birds and mammals. Ratios of $a_{\text{birds}}/a_{\text{mammals}}$ near 1.0 (with similar *b* values) indicate fundamental similarities between avian and mammalian parameters.

Examination of the *a*/*a* ratios for birds/mammals or for two variables within the avian data suggests several tentative statements regarding avian and mammalian respiration. The accuracy of the following statements will be greatest, on the average, when body weight is 1 kg. If the $b_{\text{birds}}/b_{\text{mammals}}$ deviates significantly from 0.00, the reliability of the following statements diminishes toward either end of the weight range.

Lung weights of birds and mammals are similar at a given body weight (Fig. 3), although the lungs of birds are more compact (Table I). The specific gravities of avian and mammalian lung tissues are near one, and the relative compactness of avian lungs is due to their small air volumes in comparison to those of mammals (Burton and Smith, 1968). The volumes of the total respiratory systems of birds are several

[2] However, the choice of the appropriate model for calculating the parameter estimates of *a* and *b* will depend upon whether the data conform to the assumptions of least-squares regression theory [i.e., that the deviations between the predicted and measured values of *Y* are normally distributed with a constant variance (homoscedasticity)]. If the data are heteroscedastic (lack of constant variance), appropriate transformation of these data (e.g., logarithmic) will stabilize the variance. See Zar (1968, 1970), Lasiewski and Dawson (1969), and Glass (1969) for discussions of this point in relation to metabolism and body weight.

TABLE I

COMPARISONS OF POWER LAW PARAMETER ESTIMATES RELATING RESPIRATORY VARIABLES TO BODY WEIGHT IN BIRDS AND MAMMALS[a]

Variable	Birds[b] a	Birds[b] b	Mammals[c] a	Mammals[c] b	$\frac{a_{birds}}{a_{mammals}}$	$\frac{b_{birds}}{b_{mammals}}$
Lung wt. (gm)	12.6	0.95	11.3	0.99	1.12	−0.04
Total lung vol. (ml)	29.6	0.94	53.5	1.06[o]	0.55	−0.12
Lung air vol. (ml)	9.9	0.76[e]			0.19	−0.30
Tracheal vol. (ml)	3.70	1.09[f]	0.82	1.18[p]	4.51	−0.09
Total respiratory system vol. (ml)	160.8	0.91	54.4	1.06[q]	2.96	−0.15
Tidal vol. (ml)	13.2	1.08	7.69	1.04	1.72	0.04
Respiratory rate (min^{-1})	17.2	−0.31[g]	53.5	−0.26	0.32	−0.05
Minute vol. (ml/min)	284.0	0.77	379	0.80	0.75	−0.03
Standard metabolism (cm^3 O_2/min)	11.3	0.72[h]	11.6	0.76	0.97	−0.04
Total compliance (ml/cm H_2O)	5.83?	1.04?[e,i,j]	1.56	1.04	3.7	–
CO diffusing capacity (ml/min per mm Hg)	0.55?	1.14?[e,i,k]	0.22	1.14	2.5	–
Body area (m^2)	0.09	0.74[e,l]	0.11	0.65	0.82	0.09
Evaporative water loss (mg/min)	24.2	0.61[m]	38.8	0.88[r]	0.62	−0.27
Thermal conductance, [cm^3 O_2 (gm hr °C)$^{-1}$]	0.025	−0.51[d,n]	0.031	−0.51[s]	0.81	0.00
Heart wt. (gm)	8.2	0.91	5.8	0.98	1.41	−0.07
Cardiac rate (min^{-1})	155.8	−0.23[d]	241	−0.25	0.65	0.02

[a] From Lasiewski and Calder (1971). Statistical fits to the logarithmic form of the power law equation, $\log Y = \log a + b \log W$; where W is in kg; a represents the Y-intercept at 1 kg; b represents the slope.

[b] Avian parameter estimates for a and b for all birds except passerines unless otherwise noted; from Lasiewski and Calder (1971) unless otherwise noted.

[c] Mammalian parameter estimates for a and b from Stahl (1967) unless otherwise noted.

[d] For all birds including passerines.

[e] For domestic chickens only.

[f] Hinds and Calder (1971).

[g] Calder (1968).

[h] Lasiewski and Dawson (1967).

[i] Slope assumed to be the same as in mammals; a obtained by calculation from data point and assumed slope.

[j] Scheid and Piiper (1969).

[k] Piiper *et al.* (1969).

[l] Leighton *et al.* (1966).

Footnotes continued on facing page.

times greater than those of similar-sized mammals, largely because of the voluminous air sac system (Table I). Avian tracheae are longer (2.7×) and have greater diameters (1.2×) than their mammalian counterparts, and the tracheal volumes are, on the average, more than four times greater (Hinds and Calder, 1971).

The lower respiratory frequencies (f) of birds (Fig. 8; Table I) are partially offset by higher avian tidal volumes (V_T). Avian minute volumes ($\dot{V}_E$) appear to be somewhat lower than those of mammals.

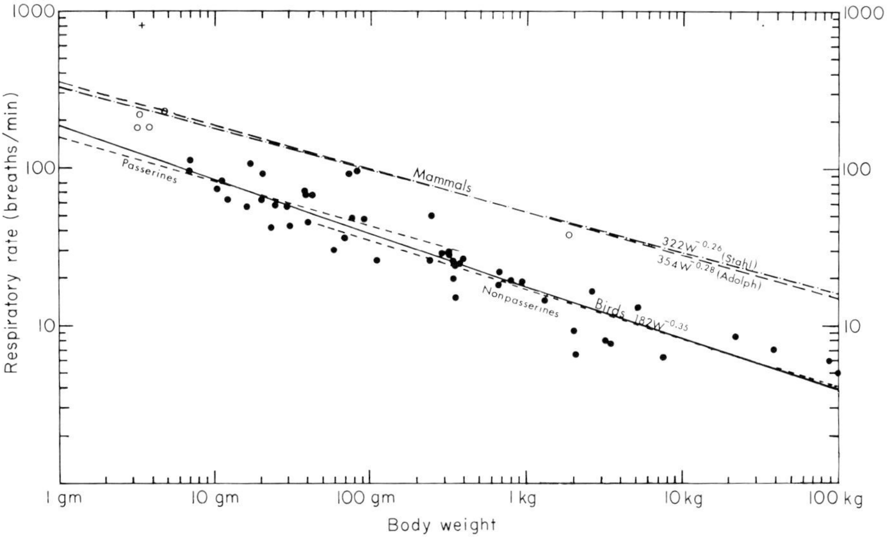

FIG. 8. The relation between respiratory rate and body weight in unrestrained, resting birds (shaded circles). Unshaded circles are for restrained fowl and hummingbirds, and these data were not used in calculation of avian regression lines. The solid line represents a regression equation fitted to the data on resting birds. The upper dashed lines represent equations fitted to mammalian data by Adolph (1949) and Stahl (1967). (From Calder, 1968.)

The values of *a* and *b* summarized in Table I can be used to make predictions about other respiratory variables. For example, steady-state oxygen consumption ($\dot{V}O_2$) must be equivalent to the product of

[m] Crawford and Lasiewski (1968).
[n] Lasiewski *et al.* (1967).
[o] Listed by Stahl (1967) as total lung capacity.
[p] Estimated from Fig. 1 of Tenney and Bartlett (1967).
[q] Total lung capacity plus tracheal volume.
[r] Chew (1965; personal communication, cited in Crawford and Lasiewski, 1968).
[s] Herreid and Kessel (1967).

f, VT, and the difference in oxygen content of the inspired and expired air ($F_{I_{O_2}} - F_{E_{O_2}}$), as shown in the following equation:

$$f \cdot V_T \cdot (F_{I_{O_2}} - F_{E_{O_2}}) = \dot{V}_{O_2} \quad (3)$$

This equation can be rearranged as follows:

$$\dot{V}_{O_2}(f \cdot V_T)^{-1} = (F_{I_{O_2}} - F_{E_{O_2}}) \quad (4)$$

Since the parameter estimates of *a* and *b* for all the variables on the left-hand side of Eq. (4) are available (Table I), they can be used to "predict" the relationship between avian $F_{I_{O_2}} - F_{E_{O_2}}$ and body weight:

$$11.3W^{0.72}(17.2W^{-0.31} \cdot 13.2W^{1.08})^{-1} = 0.05W^{-0.05} \quad (5)$$

Equation (5) "predicts" that the oxygen removal from tidal air in birds is 5% of the total air volume and is relatively independent of body weight.

One intriguing implication of Lasiewski and Calder's (1971) analysis is that avian VT is equivalent to lung air volume plus tracheal volume. Their analysis suggests that an amount of air equivalent to VT minus tracheal dead space passes over the respiratory surfaces during each respiratory cycle in resting birds, assuming that the unidirectional air flow hypothesis is correct (see Section III,B). Potentially, then, lung air in birds is completely renewed during each breath. This may contribute to the higher gas exchange efficiency of avian lungs when compared to mammals, since less ventilation is required for achieving a certain arterialization, and with equal ventilation a higher degree of arterialization is reached (Scheid and Piiper, 1970).

The total compliance may be much greater in birds than in mammals, although this statement is based only on a pair of points for the domestic fowl and pigeon (Scheid and Piiper, 1969; Crawford and Kampe, 1971). The posterior air sacs of birds extend well beyond the limits of the thoracic cage, and since total compliance is a reflection of the distensibility of the lungs, air sacs, and body wall, the higher avian total compliance seems reasonable.

The arterial CO_2 tension (Pa_{CO_2}) appears to be independent of *W* in birds (average approximately 28 mm Hg; Frankel, 1965; Chiodi and Terman, 1965; Calder and Schmidt-Nielsen, 1968; Schmidt-Nielsen *et al.*, 1969) and less than 70% of Pa_{CO_2} in mammals (average approximately 41 mm Hg; Dejours, 1966; Tenney and Bartlett, 1967). More extensive washout of CO_2 occurs from the tertiary bronchi exchange areas of birds than from the alveoli of mammals, even though avian $\dot{V}_E$ may be less. This may indicate greater efficiency in the avian flow-through lung. The CO diffusing capacity of the blood of birds (an indication of O_2 diffusibility) is higher than that of mammals (Table I; Piiper *et al.*, 1969).

Heart weights of birds are larger than those of mammals (Table I), but this is somewhat offset by lower avian cardiac rates. Cardiac outputs in birds and mammals appear to be similar (Lasiewski and Calder, 1971).

Despite the incompleteness of the avian respiratory data, the allometric analysis by Lasiewski and Calder permits a series of preliminary quantitative comparisons and conclusions to be drawn and predictions to be made, many of which were previously impossible.

B. Air Movements Within the Avian Respiratory Tract

Few areas related to avian respiration have been subject to as much speculation and surrounded by as much controversy as this one. Sturkie (1966) noted that, "Unfortunately, more heat than light has been generated upon this subject." Salt and Zeuthen (1960) also took note of the confusion and stated that "In view of the meager experimental evidence available, this discussion . . . has long been overdone, and the importance of the problem raised has perhaps been overestimated." Nevertheless, the problem of air movement within the avian respiratory tract is fundamental to our understanding of avian respiration and must be clarified.

The two major hypotheses that have been advanced regarding air movement in the respiratory system of birds propose markedly different air flow patterns in the exchange areas of the tertiary bronchi. The "reciprocal flow hypothesis" proposes that on inspiration, air flows through the lungs to the air sacs, and on expiration it flows from the air sacs through the lungs (Zeuthen, 1942; Salt and Zeuthen, 1960). The "unidirectional flow hypothesis" postulates a more complicated air flow pattern in the respiratory system and a unidirectional air movement in the tertiary bronchi (Hazelhoff, 1951; Biggs and King, 1957; and others). A third hypothesis was proposed by Shepard *et al.* (1959), but as Sturkie (1966) pointed out, their results "are contrary to all other data on gas analyses of air sacs" and "add confusion to an already confused picture." Workers have used a variety of indirect techniques to attempt to gain insight into this question, but only recently have more direct techniques been employed.

Most of the recent data support the unidirectional air flow hypothesis and confute the reciprocal flow hypothesis. Rather than reviewing all of the available data once again (it has been discussed in much detail in earlier reviews), I intend to present some of the more important recent physiological findings.

1. There is a complicated sequence of arrival times of inspired air in air sacs (Cohn *et al.*, 1963; Schmidt-Nielsen *et al.*, 1969). If an

Ostrich (*Struthio camelus*) inhales a single breath of pure oxygen (Fig. 9), some of the oxygen appears in the abdominal and caudal thoracic air sacs by the end of the first inspiration, but there is a delay of at least one full respiratory cycle before any of the inspired oxygen appears in the anterior sacs. The oxygen appearing in the anterior sacs of the Ostrich during the second and third respiratory cycles must be derived from oxygen that has passed through the lungs and posterior sacs previously.

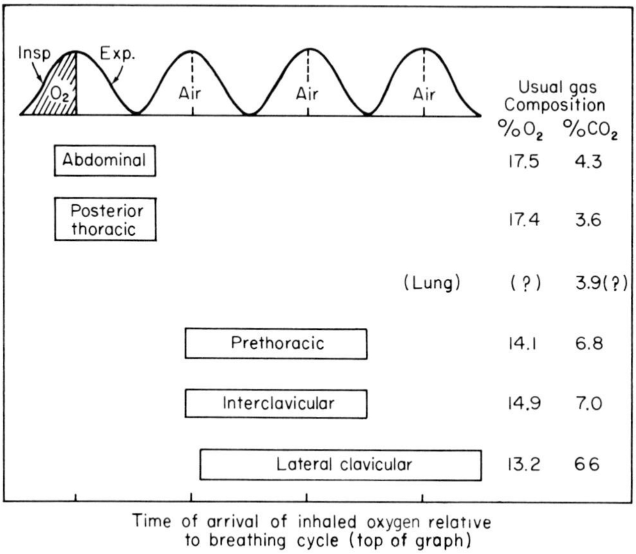

FIG. 9. Summary of arrival times in air sacs of inhaled oxygen, in relation to the respiratory cycle of the Ostrich. Arrival times were recorded with a rapid-responding oxygen electrode. Respiratory cycles are indicated at the top of the graph; the shaded portion of the first cycle denotes a single inspiration of pure oxygen. The left side of each horizontal bar indicates the earliest observed initial arrival of oxygen in that particular sac, while the right side of each bar represents the latest time at which the oxygen concentration in that sac was observed to increase. Resting respiratory rate was about 6 cycles/minute. (From Schmidt-Nielsen *et al.*, 1969.)

2. Bretz (1969) and Bretz and Schmidt-Nielsen (1971) determined air flow directions in the lungs of ducks, using a paired thermistor device. They found similar patterns of air flow in resting, panting, and anesthetized ducks. During inspiration, a large pulse of air moves caudally through the primary bronchus (mesobronchus), and a portion of this enters the caudodorsal bronchi moving toward the tertiary bronchi. At the same time, Bretz recorded a small flow of air moving from the primary bronchus through the craniomedial bronchi toward

the tertiary bronchi. During expiration, a very small pulse of air moves cranially in that portion of the primary bronchus between the caudodorsal and craniomedial secondary bronchi. Simultaneously, a large pulse of air moves from the primary bronchus through the caudodorsal bronchi toward the tertiary bronchi, and a large pulse of air moves from the tertiary bronchi through the craniomedial bronchi towards the primary bronchus. Scheid and Piiper (1970) obtained direct measurements of the pathway of respired gas in duck lungs which are consistent with Bretz and Schmidt-Nielsen's findings.

3. Pressure measurements within the avian respiratory system indicate (Fig. 10) that all air sacs are emptying and filling synchronously

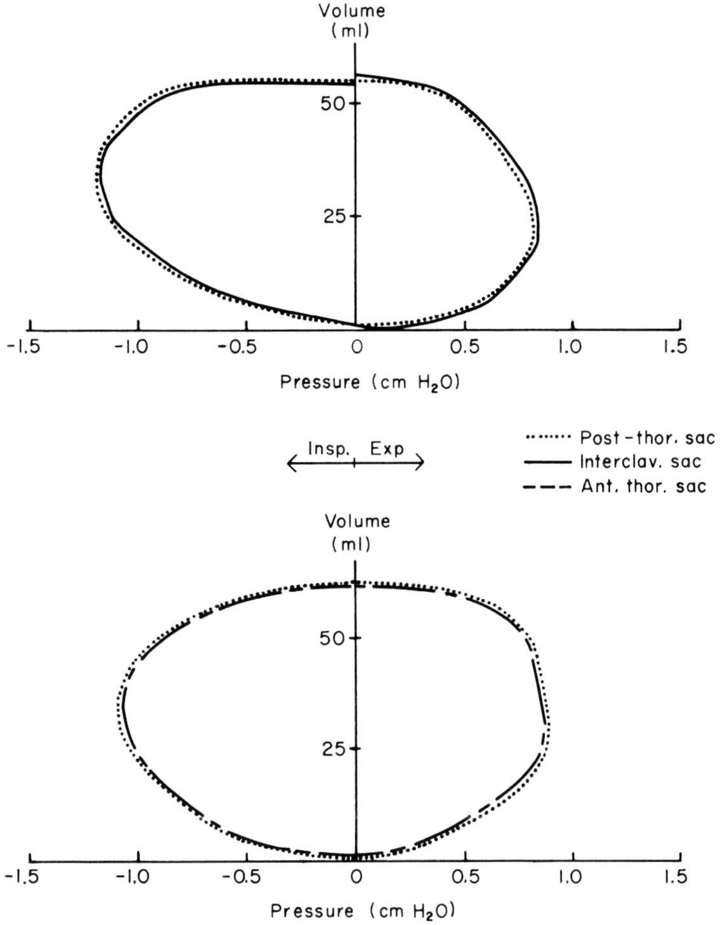

FIG. 10. The relationships between air sac pressures and tidal volumes in the domestic goose. Air sac pressures were recorded simultaneously in two different sacs, along with tidal volumes. (From Cohn and Shannon, 1968.)

in a large variety of species including pigeons, crows, chickens, ducks, geese, and Ostriches (Cohn and Shannon, 1968; Schmidt-Nielsen *et al.*, 1969; and others). Intersac pressure differences are small during any segment of the respiratory cycle.

4. Measurement of CO_2 and O_2 concentrations in air sacs in many species generally shows higher CO_2 and lower O_2 concentrations in the anterior sacs than in the posterior sacs (Table II). Where end tidal O_2 and CO_2 concentrations are available, they tend to be closer to gas concentrations in the anterior sacs.

5. All the air sacs are well ventilated in the resting Ostrich (Schmidt-Nielsen *et al.*, 1969) and in the chicken (Shepard *et al.*, 1959; Fedde,

TABLE II
PERCENTAGES OF O_2 AND CO_2 IN THE AIR SACS AND END-TIDAL AIR

Species (Reference)	Cervical sac		Clavicular sac		Cranial thoracic sac		Caudal thoracic sac		Abdominal sac		End-tidal air	
	O_2	CO_2	O_2	CO_2	O_2	CO_2	O_2	CO_2	O_2	CO_2	O_2	CO_2
Domestic Pigeon												
(Soum, 1896)			12.8	6.4	15.2	5.0					14.7	3.8
(Scharnke, 1934)			15.4	4.5	14.9	5.0			16.3	3.2		
(Scharnke, 1935)			15.8	4.8	15.0	4.1			16.5	3.7		
(Scharnke, 1938)			15.5	4.5	15.0	4.8	15.8	4.2	15.7	3.9		
Domestic Duck												
(Dotterweich, 1933)			11.9	6.5					18.5	2.3	13.6	4.9
(Vos, 1935)			14.2	5.6	14.9	5.1	17.7	2.7	18.1	2.5	14.3	5.1
(Soum, 1896)									17.1	2.7	15.8	3.0
(Bretz, 1969)	12.4	6.4	12.9	6.1	12.2	6.2			15.1	4.9		
(Bretz, 1969[a])	16.3	2.8	14.5	5.5	18.5	2.4			18.5	2.3		
Domestic Goose												
(Cohn and Shannon, 1968)				4.4		3.5		1.4		1.3		4.4
Domestic Chicken												
(Zeuthen, 1942)					15.4	4.8	14.9	5.3	18.8	2.3		
(Graham, 1939)	15.6	3.2	14.6	5.0	16.3	3.2	17.4	3.4	19.0	2.0		
(Makowski, 1938)			15.5	4.5	16.6	3.8	17.0	3.0	18.3	2.0	13.5 (trachea)	6.5
(Shepard *et al.*, 1959)	12.3	5.5	13.4	5.3	14.8	5.4	15.3	4.9	14.5	3.6		
Ostrich												
(Schmidt-Nielsen *et al.*, 1969)	13.2	6.6	14.9	7.0	14.1	6.8	17.4	3.6	17.5	4.3		
(Schmidt-Nielsen *et al.*, 1969[a])				< 1.5			20.0	1.5				

[a]Gas compositions from panting birds.

1971) and are almost certainly so in other avian species. Evidence to the contrary is advanced by the calculations of Zeuthen (1942) on his data and on those of Vos (1935). Both investigators caused birds to inhale a foreign gas (Zeuthen, H_2; Vos, O_2) for several breaths (Zeuthen, two to six; Vos, four or more), and they then determined the concentrations of this gas in the air sacs. From these, Zeuthen calculated partition percentages which seemed to indicate that most of the inhaled gas was in the posterior sacs. He concluded that the anterior sacs were relatively unimportant in resting ventilation. These calculations are misleading, though, since they inherently assume that air passes to all air sacs and that partition percentages after several breaths are merely summations of a series of similar events. This assumption is not valid *if* the unidirectional air flow hypothesis is correct. For example, in the Ostrich discussed above (Section III,B,1), air inspired during the first respiration does not begin to arrive at the anterior sacs until the second respiratory cycle, and this may be the case in other species as well. Had Zeuthen or Vos determined gas percentages in various sacs at the end of one respiratory cycle, they probably would have found no inspired gas going to the anterior sacs. Yet Zeuthen's calculations of partition percentages constitute the primary supporting evidence for the reciprocal flow hypothesis.

6. Arterial P_{CO_2} concentrations for 10 species of birds averaged 28 mm CO_2 (Calder and Schmidt-Nielsen, 1968; Piiper *et al.*, 1970). These arterial P_{CO_2} concentrations are intermediate between the anterior and posterior air sac CO_2 concentrations. Slightly higher resting values for arterial P_{CO_2} were obtained by Chiodi and Terman (1965), Haning and Thompson (1964), and Tucker (1968a).

The majority of these data, as well as most studies examining the distribution of fine particles introduced into the avian respiratory system (Dotterweich, 1930; Vos, 1935; Hazelhoff, 1951; and others), and the clever humeral breathing experiments of Biggs and King (1957), are incompatible with the reciprocal flow hypothesis. They are consistent with the unidirectional flow hypothesis, although not all data can be encompassed by it (Walter, 1934; Shepard *et al.*, 1959; and a few others).

A schematic diagram representing a version of the unidirectional air flow hypothesis (which seems compatible with most of the reliable data) is shown in Fig. 11. The pattern of air flow can be visualized by following a "packet" of air as it enters the avian respiratory system. During the first inspiration (Fig. 11,I), some air passes through the trachea and primary bronchi into the posterior air sacs (P), and some air flows into the tertiary bronchi of the lungs (T) via the caudal group

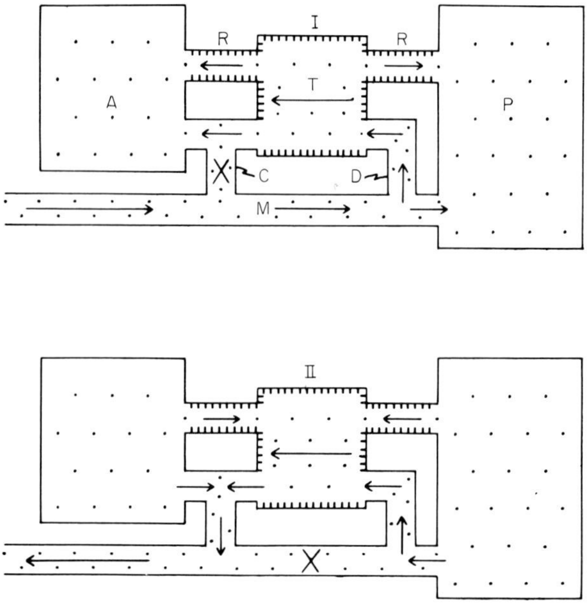

FIG. 11. Schematic representation of the directions of air flow in the avian respiratory system, during both inspiration (I) and expiration (II), according to the unidirectional flow hypothesis. M, primary bronchi; C, craniomedial secondary bronchi; D, caudal group of secondary bronchi; T, tertiary bronchi and respiratory exchange surfaces; R, indirect connections between air sacs and tertiary bronchi; A, anterior air sacs; P, posterior air sacs. The large X signifies area of little or no air flow. (Figure adapted from Bretz and Schmidt-Nielsen, 1971.)

of secondary bronchi (D). Little or no air enters the lungs or anterior sacs via the craniomedial secondary bronchi (C). During the first expiration (Fig. 11,II), air leaving the posterior air sacs passes through the tertiary bronchi via the caudal group of secondary bronchi → tertiary bronchi → craniomedial secondary bronchi circuit (Hazelhoff's d-p-v circuit). At the same time, little or no air passes from the posterior sacs to the outside by the direct route through the primary bronchus. During subsequent respiratory cycles, the air entering the anterior sacs (A) is derived from air that has already been through the tertiary bronchi exchange regions. Air expired from the anterior sac system passes out of the trachea via the craniomedial secondary bronchi. No data exist yet on the pattern of air flow in the indirect connections from air sacs (R), although the air flow directions shown in R in Fig. 11 are most consistent with air flow in other segments of the respiratory system. Air is therefore passing through the tertiary bronchi in the same direction during inspiration and expiration: from the caudal group of secondary bronchi to the craniomedial secondary bronchi. With one-way flow of air through the tertiary bronchi, the possibility exists of countercurrent exchange between the air flowing in the tertiary bronchi and the blood vascular bed surrounding the tertiary bronchi (Schmidt-Nielsen *et al.*, 1969; Bretz, 1969; Bretz and Schmidt-Nielsen, 1971).

It is uncertain at present how the air flow patterns hypothesized are effected. Most workers lean toward aerodynamic explanations based on the dynamics of fluid flow, since no anatomical valves, per se, are evident in the avian lung. Schmidt-Nielsen *et al.* (1969) and Bretz (1969) provide evidence that air flow in the respiratory systems of resting Ostriches and ducks is driven by low pressure differentials, is probably laminar (low Reynolds numbers), and in ducks according to Bretz is of low velocity. There is no direct evidence, except for the existence of the pattern and the lack of anatomical valves, that fluid dynamic air flow does control the pattern of air circulation. Conceivably, the musculature of the pleural membranes (see Section II, E) or the bronchial muscles (see Section II,B; A. S. King and Cowie, 1969) could play important roles in controlling the directions of air flow. More studies are needed in this area and might involve visualization of air flow direction by the use of radiopaque gases and modern cinematographic X-ray technology, or some other fresh, direct techniques.

C. Regulation of Avian Respiration

Our understanding of the mechanisms regulating avian respiration is based partially on avian data, but primarily on inference from data on mammalian respiration. Given the striking basic differences between avian and mammalian respiratory anatomy and patterns of air flow, as well as differences in other respiratory attributes, there seems little reason to expect, *a priori*, that the respiratory control mechanisms are necessarily similar. The mechanisms are not completely understood in birds, then, and more experimentation is required. For example, several of the most recent reviews and studies of the peripheral control of avian respiration arrived at fundamentally different conclusions (Fedde, 1970; Peterson, 1970; Jones, 1969; A. S. King *et al.*, 1968).

1. *Central Nervous System Control Mechanisms*

Our comprehension of central nervous system control of resting respiration in birds was influenced importantly by von Saalfeld's (1936) studies of the effects of transection of the brains of pigeons on respiration. He concluded that the primary respiratory center in birds is located in the medulla oblongata (as it is in mammals) and that complete extirpation of the pigeon cerebrum does not affect respiration (von Saalfeld, 1936; Rogers, 1928). A respiratory center controlling panting in the pigeon was described by von Saalfeld and several subsequent workers as being in the antero-dorsal portion of the midbrain (see Section IV,D). Further studies on the role of the central nervous system in the control of avian respiration are needed.

2. Peripheral Control of Respiration

The existence of a mammalian-type Hering–Breuer reflex based on stretch receptors has not been unequivocally demonstrated in birds, despite a number of recent reports to the contrary (Sinha, 1958; Blankart, 1960; Richards, 1968; Jones, 1969). Although the results from these studies were interpreted in terms of stretch-receptor activity, they also can be interpreted by an alternative hypothesis relating to chemoreceptors rather than mechanoreceptors. Several studies have suggested that the stretch reflex may not be important in maintenance of normal avian respiration (Fedde *et al.*, 1961; Sturkie, 1966; Eaton *et al.*, 1971).

Fedde (1970) suggests that since the lungs of birds are relatively inexpansible, stretch receptors would probably not be activated by distortion. He proposes instead that the apnea produced by inflation of the avian respiratory system results from stimulation of pulmonary chemoreceptors that are sensitive to the low carbon dioxide concentration of the inflating gas. Fedde (1970) and Eaton *et al.* (1971) report that when the inflating gas contains 5 to 8% CO_2, apnea is not produced in the fowl. The existence of avian stretch receptors, then, still appears to be open to question.

The avian respiratory system is remarkably sensitive to levels of CO_2 concentration, particularly in the inhaled air, but perhaps also in the blood. Several authors have demonstrated that ventilation of the avian lung with gas containing low concentrations of CO_2 produces respiratory apnea (Van Matre, 1957; Ray and Fedde, 1969; and others), while inhalation of higher CO_2 concentration stimulates respiratory rate or amplitude, or both. There is, however, considerable variation between species (probably partially due to differences in techniques). Respiration in aquatic birds, for example, may be inhibited by high CO_2 concentrations (Hiestand and Randall, 1941; Zimmer, 1935).

Recent experiments by Peterson and Fedde (1968), Peterson (1970), and Fedde and Peterson (1970) provide convincing evidence of the existence of receptors sensitive to the removal of CO_2 from the venti-

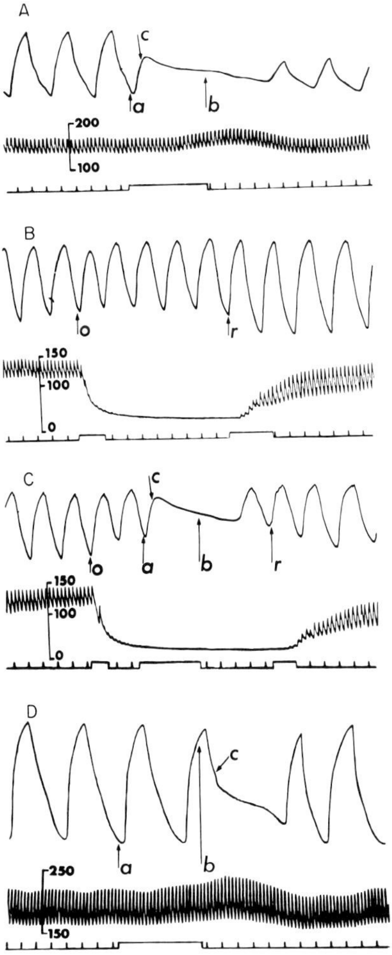

FIG. 12. Respiratory responses of the domestic chicken to elimination of CO_2 from unidirectionally ventilated gas and to occlusion of pulmonary blood flow. Top trace in each record represents respiratory movements (inspiration up, expiration down); middle trace is blood pressure; and lower trace is time line in seconds. (A) Control response to CO_2 elimination. At a, CO_2 was removed from the ventilatory gas; b, CO_2 was restored to the ventilatory gas; and c, the respiratory response began. (B) Control response to brief occlusion of the pulmonary vessels. The right pulmonary vessels were ligated. At o, the left pulmonary artery and vein were occluded, and at *r*, the occlusion was removed. (C) Response to elimination of CO_2 from the respiratory gas following the occlusion of pulmonary circulation. Symbols are the same as in parts A and B. (D) Response to elimination of CO_2 from the ventilatory gas after bilateral thoracic vagotomy. Symbols are the same as in part A. In this case the response occurred in 5.9 seconds after CO_2-free gas reached the trachea. (From Peterson and Fedde, 1968. Copyright 1968 by the American Association for the Advancement of Science.)

lating gas (Fig. 12). Their experiments take advantage of the air sac system and parallel arrangement of tertiary bronchi by opening the abdominal air sacs and applying unidirectional artificial ventilation to the respiratory system with humidified gas of known composition. This technique offers the advantage of almost total elimination of dead-space gas while permitting quick independent control of the air composition of the lungs. Peterson and Fedde's experiments indicate that rapid lowering of airway CO_2 concentrations induces the initiation of apnea within 0.5 to 0.6 second, whether blood is flowing through the lungs or not (Fig. 12A,C). Their experiments present strong evidence for the location of these receptors in the lung. Bilateral vagotomy markedly delays the response time (Fig. 12), but does not eliminate sensitivity to CO_2, which suggests the existence of additional CO_2-sensitive areas in the chicken. Elimination of the rapid response to CO_2 after bilateral vagotomy suggests that the vagus nerves contain afferent fibers innervating the CO_2-sensitive receptor sites of the lung.

Afferent vagal activity synchronous with the respiratory cycle of fowl was documented by A. S. King *et al.* (1968), who recorded single unit discharges in the peripheral stump of the midcervical vagus (Fig. 13). This was the first demonstration of afferent activity in the avian vagus nerve "capable of continuously informing the central nervous system of the state of the respiratory cycle." Peterson (1970) and Fedde (1970) present additional evidence from the vagal afferent fibers of fowl that indicate marked sensitivity of lung receptors to CO_2. The firing rates of the receptors they studied were inversely related to the level of CO_2 ventilated through the lungs (via unidirectional artificial respiration). Their single unit recordings indicate a range of sensitivities of receptors to CO_2. Some units cease firing when CO_2 levels reach 5%, while others persist until CO_2 levels reach 14%. Receptors sensitive to CO_2 concentrations could inform the central nervous system about the gas tension in the lung and thus about the metabolic activities of the bird. Cook and King (1969) provided electron microscopic evidence for neurite-receptor complexes in the epithelium of the primary bronchus of the chicken.

Carotid and aortic bodies have been described in birds, although the physiology of these structures has not been elucidated (Tcheng *et al.*, 1963). Henry and Fedde (1970) studied the minimal pulmonary circulation time in chickens and concluded that aortic and carotid bodies do not appear to be involved in the response described by Peterson and Fedde (see Fig. 12). Additional receptors in the mucosa of the tracheae of chickens have been reported by Graham (1940) and

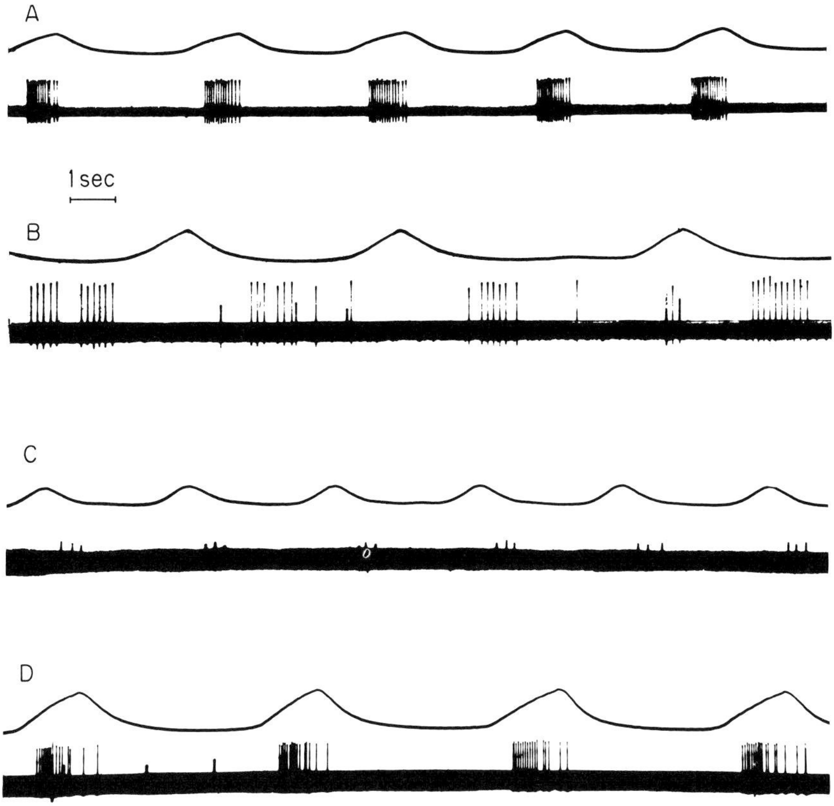

FIG. 13. Recordings from single respiratory units in the peripheral stump of the mid-cervical right vagus nerves of the domestic chicken. The afferent respiratory units are firing (lower traces in A, B, C, and D) in phase with eupneic (resting) breathing. The left vagi are intact. In the upper traces, inspiration is upward. (From A. S. King *et al.*, 1968.)

Hiestand and Randall (1941) as being sensitive to air flow. However, Fedde (1970) proposes that these receptors are not sensitive to air flow per se, but respond to noxious stimuli such as dry cold air or acid vapors. Activity of receptors in the upper respiratory tract of ducks may produce apnea such as that associated with diving (Andersen, 1963; Feigl and Folkow, 1963; Jones, 1969).

Although the existence of a Hering–Breuer reflex based on stretch receptors in birds has not been clearly established, most workers agree that bilateral cervical vagotomy causes a profound decrease in respiratory frequency and an increase in the amplitude of respiration in chickens, ducks, pigeons, and presumably in other birds (Fedde *et al.*, 1963; Fedde and Burger, 1963; Richards, 1969; and others). Bilateral thoracic vagotomy that does not destroy the innervation of the carotid and aortic bodies alters respiration similarly to cervical vagotomy, suggesting again that the receptors important in control of normal respiration are located in the lungs.

D. Evaporative Water Loss in Resting Birds

Until recently, most workers assumed that respiratory evaporation was the predominant avenue of avian evaporative water loss (EWL).

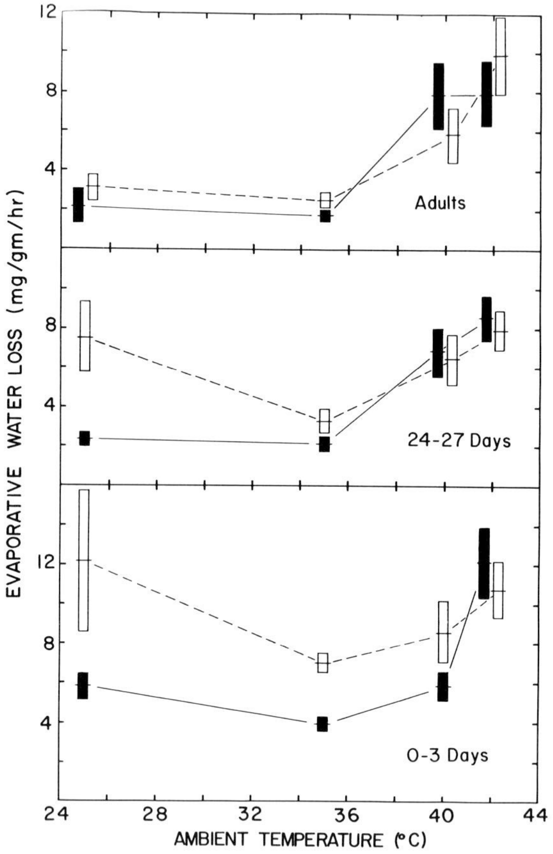

FIG. 14. The relationships of weight-specific respiratory and cutaneous evaporation to ambient temperature in three age groups of Painted Quail, *Coturnix chinensis*. Horizontal lines represent mean values, and rectangles enclose two standard errors on each side of the mean. Shaded rectangles connected by solid lines represent respiratory evaporation; unshaded rectangles connected by dashed lines represent cutaneous evaporation. (From Bernstein, 1970.)

The assumption seemed reasonable to many, since birds possess a feathered layer of insulation, are not known to have sweat glands, and a few measurements indicate low cutaneous EWL.

Several studies have recently documented the importance of cutaneous EWL as a major component of the total EWL in some species. Bernstein (1969, 1970, 1971a,b) has demonstrated that cutaneous EWL represents a considerable proportion of the total EWL in the Painted Quail, *Coturnix chinensis* (Fig. 14). Furthermore, during heat stress, the cutaneous evaporation of *C. chinensis* increased along with evaporation due to panting and gular flutter. Smith (1969) and Smith and Suthers (1969) demonstrated the importance of cutaneous EWL in resting and heat-stressed pigeons and also showed that it increased during heat stress. The mechanism of this increase in the cutaneous component of EWL is unclear. Lasiewski *et al.* (1971a) and Bernstein (1971b) found considerable cutaneous evaporation in five additional species of small birds (several of them from xeric habitats). The only avian species examined to date that does not lose appreciable water through the skin is the Ostrich. Schmidt-Nielsen *et al.* (1969) found that cutaneous EWL was a small fraction of total EWL in both resting and heat-stressed Ostriches.

The available data on total EWL of birds are rarely totally acceptable or comparable. Although ambient humidity is an important determinant of the rate of evaporation, most studies seldom controlled this variable. Most avian EWL data were obtained by using an open-flow system, in which the importance of air flow rate as a determinant of humidity was overlooked [see Lasiewski *et al.* (1966a,b) for discussions of this problem and potential solutions].

Evaporative water loss in resting birds is directly related to body weight (Fig. 15) in species ranging from hummingbirds to Ostriches (Crawford and Lasiewski, 1968). Larger birds lose less water per unit weight than do smaller species, as shown by the equation that relates EWL to W and which has a slope of less than 1. Total daily water turnover of four species (measured with tritiated water) is approximately four times higher than total EWL of resting birds (unshaded circles in Fig. 15; Ohmart *et al.*, 1970). The difference between total and evaporative avenues of water loss is less than fourfold under natural conditions however, since the data indicated by shaded circles in Fig. 15 do not reflect evaporative levels during activity, while the total water turnover involves active as well as resting levels.

The total amount of water evaporated in birds increases with ambient temperature (T_a) in most species. EWL increases markedly at high T_a's, as birds enhance respiratory water loss by panting and

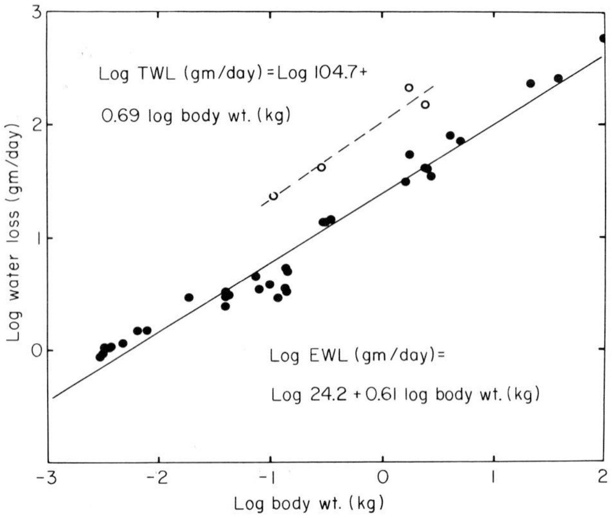

FIG. 15. The relation between evaporative water loss and total water loss and body weight in birds. Shaded circles represent evaporative water loss of resting birds at ambient temperatures within or below their respective zones of thermal neutrality, and the solid line represents the regression line fitted to these data (Crawford and Lasiewski, 1968). The unshaded circles represent total water loss (measured with tritiated water) of unstressed birds, and the dashed line represents the regression line fitted to the data (Ohmart *et al.*, 1970).

gular flutter as well as facilitate cutaneous evaporation. At T_a's below the zone of thermal neutrality, EWL decreases with temperature, despite the higher respiration and metabolism associated with thermoregulation in the cold. This is a reflection of temporal countercurrent heat exchange in the nasal region, which results in a recondensation of some of the water contained in expired air (Schmidt-Nielsen *et al.*, 1970). The large surfaces of the turbinals in the nasal chambers of birds (Section II,A) contribute to the heat exchange between respiratory air and surrounding tissues. The respiratory water recondensation results in savings of heat as well as water.

IV. Respiration of Heat-Stressed Birds

All avian species studied in this respect increase respiratory frequency (f) during heat stress, thereby facilitating evaporative cooling. Avian tidal volumes (VT) may remain constant, increase, or decrease with increased f, depending upon the species and the conditions, although few data of this sort are available. Birds of several orders supplement respiratory evaporation of panting with gular flutter, a rapid vibration of the membranous gular region which is driven by the hyoid apparatus. Dawson and Hudson (1970) and Richards (1970a) have reviewed aspects of avian responses to heat stress.

A. Patterns of Panting and Gular Flutter

Several patterns of avian panting and gular flutter have thus far emerged and are presented schematically in Fig. 16. The relations between f and body temperature (T_b) conform to two general patterns: some species show a gradual increase in f with hyperthermic T_b (Fig. 16,A,C,E), while others exhibit two relatively distinct levels of f (Fig. 16,B,D,F).

Since respiratory patterns during heat stress have been studied in few birds, patterns other than those in Fig. 16 may emerge with more study. Several passeriform birds and frogmouths (Caprimulgiformes:

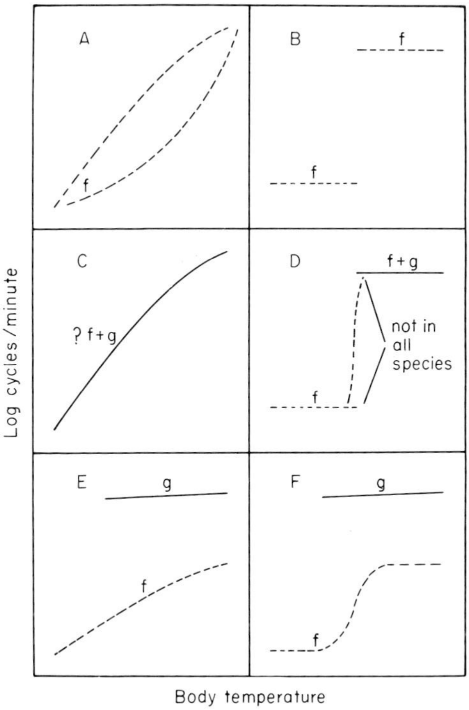

FIG. 16. The relations between breathing rate (f) and gular flutter rate (g) and body temperature in heat-stressed birds. The patterns depicted schematically in this figure have been observed in various species (see text).

Podargidae) gradually increase f with T_b and do not gular flutter (Fig. 16A; Kendeigh, 1944; Salt, 1964; Lasiewski and Bartholomew, 1966; Lustick, 1968; Lasiewski *et al.*, 1970). Ostriches tend to exhibit two distinct levels of f. At rest, breathing rates are 4 to 12 cycles/minute, while panting rates are relatively constant at 40 to 60/minute (Fig. 16B; Crawford and Schmidt-Nielsen, 1967; Schmidt-Nielsen *et al.*, 1969). Bobwhite (*Colinus virginianus*), California Quail (*Lophortyx californicus*), and Gambel's Quail (*L. gambelii*) (Galliformes: Phasianidae) have gular flutter frequencies that increase with T_b and are probably synchronous with f (Fig. 16C; Lasiewski *et al.*, 1966b), although this pattern has not been adequately studied. Cormorants (Pelecaniformes: Phalacrocoracidae), Poorwills (*Phalaenoptilus nuttallii;* Caprimulgiformes: Caprimulgidae), and mousebirds (Coliiformes) exhibit a gular flutter frequency that is relatively independent of T_a and T_b and more rapid than f, which increases gradually with T_b (Fig. 16E, Lasiewski and Bartholomew, 1966; Bartholomew *et al.*, 1968; Lasiewski and Snyder, 1969; Lasiewski, 1969; Bartholomew and Trost, 1970), although at high T_b Poorwills pant and gular flutter synchronously (Lasiewski and Seymour, 1972). Roadrunners (*Geococcyx californianus*), pigeons, doves, owls, and Painted Quail have synchronous gular flutter and panting frequencies, and respiratory rates that show two distinct levels of response, reminiscent of panting in dogs (Figs. 16D and 17; Calder and Schmidt-Nielsen, 1967; Bartholomew *et al.*, 1968; Lasiewski *et al.*, 1971a; Lasiewski and Seymour, 1972). Pelicans exhibit a gular flutter frequency that is relatively independent of T_b, while f increases rapidly over a narrow range of T_b from resting to panting levels (Fig. 16F; Bartholomew *et al.*, 1968).

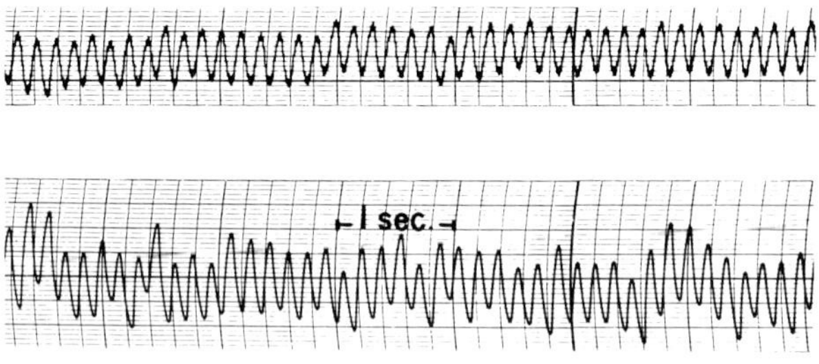

FIG. 17. Synchronous gular flutter (upper trace) and panting respiration (lower trace) in the Roadrunner, *Geococcyx californianus*. The simultaneous records were obtained with impedance transducers (Lasiewski *et al.*, 1971a). See Dawson and Hudson (1970, p. 265) for impedance recordings of asynchronous gular flutter and breathing movements in a young Cattle Egret (*Bubulcus ibis*).

Dehydrated Ostriches exposed to heat stress reduce the amount of water expended for evaporative cooling to below those levels for hydrated birds, which seems due to a decrease in panting respiratory rate with dehydration (Crawford and Schmidt-Nielsen, 1967).

Several authors have suggested that gular flutter and perhaps panting in birds occur at resonant frequencies, a phenomenon similar to that described by Crawford (1962) in panting dogs. Recently, Crawford and Kampe (1971) presented evidence that domestic pigeons pant at the resonant frequency of the respiratory system. Lacey (1965) and Lacey and Burger (1972) tested and rejected the hypothesis that panting rates in chickens are determined by the resonant frequencies of the chest–lung systems, after they found no significant changes in panting rates which changed elastance and inertance of these systems. Lasiewski and Bartholomew (1966) weighted the gular region of a Poorwill and found no change in gular flutter frequency.

B. Sites of Respiratory Evaporative Cooling

Panting causes increased evaporation from the nasal, buccal, and upper pharyngeal regions from the trachea, and (at least in the Ostrich) perhaps from the walls of the air sacs. The temperatures of the evaporating surfaces may be as much as several degrees below ambient or cloacal temperatures (Kallir, 1930; Wilson *et al.*, 1952; Lasiewski and Bartholomew, 1966; Calder and Schmidt-Nielsen, 1968; Lasiewski and Snyder, 1969; Schmidt-Nielsen *et al.*, 1969; Lasiewski *et al.*, 1970), although the surface temperatures alone cannot be used to assess the relative contributions of various regions to total EWL.

Gular flutter permits evaporation from moist surfaces of the upper digestive tract (portions of the pharynx and anterior esophagus) not normally exposed to tidal air and brings additional external air across the buccal region (Fig. 18). In the Roadrunner, panting and gular flutter are synchronized (see Fig. 17), and inspiration and expansion of the bucco-pharyngo-esophageal cavity (a single "gular flutter") occur simultaneously, as do expiration and contraction of the cavity (Lasiewski and Bartholomew, 1971).

C. Enhancement of Ventilation Volumes

Few data on avian respiratory minute volumes ($\dot{V}_E$) are available, and those that are available are necessarily based on techniques which interfere with respiration to some degree. All species studied increase $\dot{V}_E$ ($= f \cdot V_T$) in response to heat stress. However, at least two

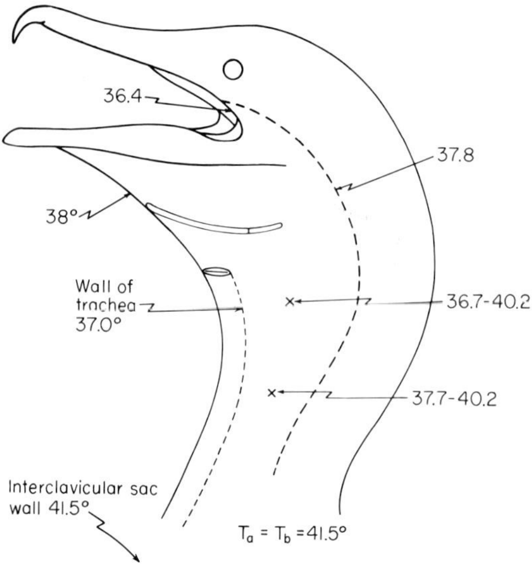

FIG. 18. Representative evaporative surface temperatures in a 6.5-week-old Double-crested Cormorant, *Phalacrocorax auritus*. Ambient temperature was held at body temperature during the measurements. (From Lasiewski and Snyder, 1969.)

patterns of response are evident in recorded VT. In the first pattern, pigeons, chickens, ducks, and Ostriches decrease VT below resting levels during panting at moderate heat loads (although the percentage decrease in VT varies considerably between species (von Saalfeld, 1936; Frankel *et al.*, 1962; Hart and Roy, 1966; Schmidt-Nielsen *et al.*, 1969; Smith, 1969; Bretz, 1969; Berger *et al.*, 1970b), which is similar to the panting response of dogs. In the second pattern, birds appear to maintain relatively constant VT over a wide range of f, as suggested for several passerine species by the calculations of Salt (1964). In these birds, increased $\dot{V}E$ is accomplished by a gradual rise in f with T_b (Fig. 17A). All avian species studied in this respect increase VT during severe heat stress, which in pigeons, chickens, and ducks is accompanied by decreasing f as T_b approaches lethal levels.

The metabolic cost of gular flutter or panting, or both, does not appear to be high (Lasiewski, 1969; Lasiewski and Seymour, 1972), despite a variety of statements to the contrary in the literature. While most workers have been concerned with the evaporative effects of enhanced $\dot{V}E$, Seymour (1972) has proposed that variations in $\dot{V}E$ due to panting may impose significant alterations in convective heat exchange in the respiratory tract. At high or low T_a, these alterations may be considered as the convective "cost" or "advantage" of panting.

The increased $\dot{V}E$ resulting from panting causes O_2 and CO_2 concentrations within the respiratory tract of pigeons, ducks, and Ostriches to approach those of air (Scharnke, 1935; Schmidt-Nielsen *et al.*, 1969; Bretz, 1969). After Ostriches had been exposed to heat stress for several hours, the CO_2 concentration in the clavicular sac (~ 1.5%) was still higher than that in the caudal thoracic sacs (0.5–1.0%), suggesting that a complicated pattern of air flow may still continue to exist in the respiratory tract of panting Ostriches.

Although it seems possible that increased ventilation could be accomplished while bypassing the respiratory exchange surfaces, most avian species studied undergo hypocapnia and alkalosis during long bouts of panting and gular flutter (Linsley and Burger, 1964; Calder and Schmidt-Nielsen, 1967, 1968; Frankel and Frascella, 1968), which is similar to that found in panting dogs. It may be that in some species, panting and gular flutter during moderate heat stress would not cause alkalosis, although this has not been examined. Bretz (1969) found directions of air flow in the lungs of panting and resting ducks to be similar, although this does not appear to be true for the Ostrich. Ostriches pant vigorously for up to 8 hours without becoming alkalotic, suggesting internal regulation of air flow by means of shunting much of the ventilation volume away from the respiratory exchange surfaces (Schmidt-Nielsen *et al.*, 1969). The mechanisms involved are uncertain, although the shunting could be accomplished by the increase in resistance to air flow caused by graded contractions of the massive sphincterlike bands of smooth musculature surrounding the openings of the tertiary bronchi from the secondary bronchi, or by constriction of the bundles of smooth musculature surrounding the walls of the tertiary bronchi and atrial openings, or by both of these mechanisms. (See A. S. King and Cowie, 1969.)

Lasiewski (1969) reports increased CO_2 production and a resultant rise in the apparent respiratory quotient when Poorwills commence active evaporative cooling at high T_a's. This is presumably a reflection of higher ventilation rates and consequent blow-off of CO_2 from the respiratory system, and not a shift in foodstuffs metabolized.

Although gular flutter clearly supplements panting, it has not been practicable to assess the relative contribution of gular flutter per se to the noncutaneous component of total EWL. Lasiewski (1969) examined short-term rates of evaporation in the Poorwill during exposure to heat stress, using a recording electrobalance. Gular flutter in the species may be viewed as an ON–OFF system. When OFF, EWL is relatively constant over the T_a range studied (Fig. 19). When ON, gular flutter frequency is relatively constant (Lasiewski and Bartholomew, 1966), although rates of evaporation increase with the area and ampli-

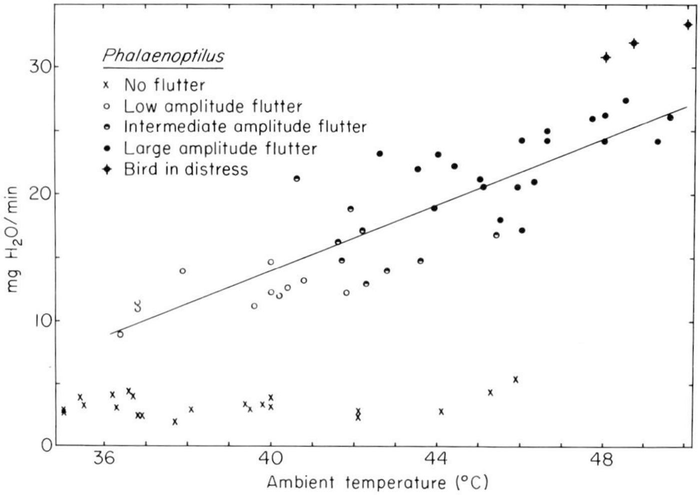

FIG. 19. Relation between short-term rates of evaporative water loss and ambient temperatures in the Poorwill, *Phalaenoptilus nuttallii*. Water vapor pressure was maintained between 10 and 15 mm Hg. Each value represents a stable rate of weight loss, recorded with a Cahn RH Electrobalance, for at least 1 minute during gular flutter or no flutter. (From Lasiewski, 1969.)

tude of gular region moved (Fig. 19). Since f also increases during bouts of gular flutter in the Poorwill, these data do not permit satisfactory separation of instantaneous rates of evaporation due to gular flutter from those due to increased respiratory rates. The data do suggest, however, that gular flutter may contribute more than one-half of the total EWL at T_a above 39.5°.

D. Regulation of Respiration During Heat Stress

The role of the nervous system in the control of respiration of heat-stressed birds has been reviewed by Salt and Zeuthen (1960), Sturkie (1966), and Richards (1968, 1970a) and need not be discussed in detail here. Panting in birds appears to be mediated by a "panting center" in the anterior dorsal midbrain and by neurons in the anterior hypothalamus and preoptic areas (von Saalfeld, 1936; Åkerman *et al.*, 1960; Kanematsu *et al.*, 1967; Lepkovsky *et al.*, 1968; Richards, 1970b). Destruction of the midbrain or hypothalamic area abolishes the panting response. However, panting can be reinstated by injection of lobeline (von Saalfeld, 1936) or by electrical stimulation of the dorsal midbrain (Sinha, 1959). Panting in intact birds can be elicited by stimulation of hypothalamic or preoptic areas (Sinha, 1959; Åkerman *et al.*, 1960; Richards, 1971).

Bilaterally vagotomized pigeons can still pant (von Saalfeld, 1936; Sinha, 1959; Richards, 1968), although Richards reports that vagotomy abolishes rapid respiration in fowl, ducks, and quail. In some vagotomized fowl, afferent vagal stimulation would support panting after ablation of the hypothalamus, but not of the midbrain. The midbrain region thus appears essential for the response.

Rautenberg (1969) provides evidence for the existence in pigeons of thermosensitive structures in the vertebral canal which are "most likely localized within the spinal cord." Selective heating of spinal cords with thermodes evokes panting in pigeons exposed to thermoneutral temperatures, while cooling of the spinal cords of heat-stressed birds lowers respiratory rates from panting to resting levels.

V. Respiration of Flying Birds

The strenuous metabolic demands of sustained avian flight are met by oxidative metabolism, and respiration must therefore provide sufficient gaseous exchange during this activity. Through flight, birds are able to move farther and faster than any other organisms known to fly at altitudes that cause severe respiratory stress to mammals. However, because of the technical difficulties of obtaining data from flying birds, there is little physiological information available on this subject. Modern techniques hold considerable promise for the understanding of the physiology of avian flight, and recent studies now permit more precise statements to be made than were possible a decade ago.

A. Respiratory Exchange

Oxygen consumption ($\dot{V}O_2$) during flight has been determined only for hovering hummingbirds (Pearson, 1950; Lasiewski, 1963), for Budgerigars (*Melopsittacus undulatus*) flying in a wind tunnel (Tucker, 1968b), and for three avian species during short flights (Berger *et al.*, 1970b). A variety of less direct techniques has been used for estimating flight metabolism, although these techniques require more assumptions than do those for $\dot{V}O_2$ [see Tucker (1968b), Farner (1970), and Berger *et al.* (1970b) for recent reviews of the subject]. Although there is considerable spread in proposed levels of flight metabolism, the more reliable estimates indicate a 7- to 15-fold increase over standard metabolic levels. Flight metabolism will be discussed in more detail in other chapters in these volumes.

B. Respiratory Rates, Tidal and Ventilation Volumes During Flight

Although available data are limited and rarely totally acceptable, most birds studied appear to show some degree of coordination between respiration and wing-beat frequency, although coordination is not obligatory. Several earlier studies found in some species no synchronization between f and wing beat (Fraenkel, 1934; Tomlinson, 1963; Tucker, 1968b). Tucker (1968b) suggested that body size might limit the possibility of synchronized f and wing beat, and that below a certain weight birds would not be able to ventilate their lungs efficiently at wing-beat frequency. Recent studies by Berger *et al.* (1970a) on 10 species indicate that respiration of birds in flight is usually coordinated with wing beats. This does not mean that f and wing beat are necessarily synchronous, but that "the transition from inspiration to expiration or vice versa" occurs "preferentially in certain phases of the wing cycle" (Berger *et al.*, 1970a). Pigeons and crows have synchronous wing beats and respiration (1:1), although 11 other types of coordination were found in other species, including coordination in which the ratio between frequencies was not a whole number (e.g., 3.5:1). For example, ducks may have 3, 3.5, 4, 4.5, or 5 wing beats per respiratory cycle. In the 10 species studied, the beginning of inspiration occurs most frequently during or at the end of the wing upstroke, while the beginning of expiration occurs at the end of the downstroke. The data of Berger *et al.* are impressive, although it should be noted that the flights they studied were of short duration and presumably during the take-off period. They found large variability in types of coordination between and within species, some of which could be due to experimental technique.

Tucker (1968b) found no fixed relation between f and wing beat in Budgerigars flying in a wind tunnel. Wing-beat frequencies were constant at 14/second at all flight speeds, whereas f changed with flight speed (Fig. 20). Berger *et al.* (1970a) do not feel, however, that Tucker's data exclude the possibility of coordination of respiration and wing beats in the Budgerigar.

Lung ventilation and $\dot{V}O_2$ appear to increase proportionately in the Budgerigar (Tucker, 1968b) and in Evening Grosbeaks (*Coccothraustes vespertinus*), Ring-billed Gulls (*Larus delawarensis*), and Black Ducks (*Anas rubripes*) (Berger *et al.*, 1970b). For pigeons, however, the available data are inconclusive on this point (Hart and Roy, 1966). At 20°C, $\dot{V}O_2$ of flying Budgerigars increases 4.9-fold over resting levels, while $\dot{V}E$ increases 4.7-fold. Both $\dot{V}O_2$ and f are lowest at a flight speed of 35 km/hour. In all three species studied by Berger *et al.*

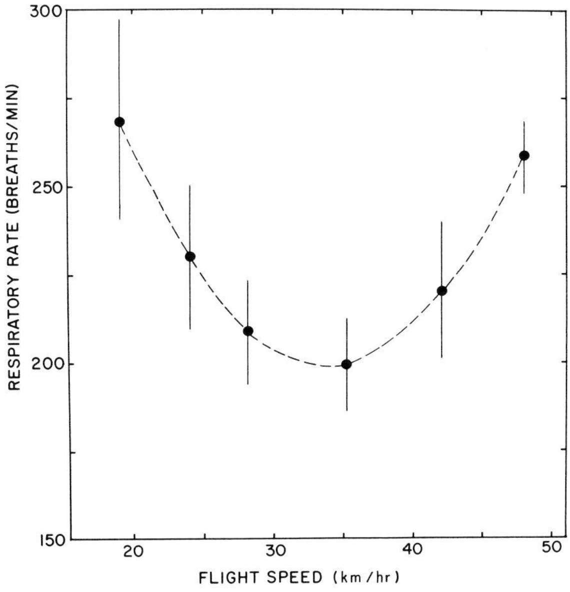

FIG. 20. Respiratory rate of two Budgerigars during level flight in a wind tunnel, at various speeds. Each point is a mean, calculated from five measurements on each bird, and the vertical bars represent two standard errors on either side of the mean. (Redrawn from Tucker, 1968b.)

(1970b), augmentation of $\dot{V}_E$ during flight was effected by increasing V_T as well as f. In contrast, V_T of flying and resting pigeons are similar (~ 5 ml). The increase in flight $\dot{V}_E$ results from increasing f (Hart and Roy, 1966). Immediately after the pigeons land, there is a threefold increase in V_T, a comparable decrease in f, and little change in $\dot{V}_E$.

It seems unlikely that the contractions of the massive flight muscles, the *Mm. pectoralis* and *supracoracoideus,* have no effect on avian respiration in those birds that do not have a 1:1 ratio between respiration and wing beat. Tucker (1968b) suggested that perhaps the externally measured respiratory rate reflects only the slow movements of air in the primary and secondary bronchi of the lungs. The flight muscles may impose a faster rhythm upon internal air movement, enhancing ventilation of air in the tertiary bronchi and air capillaries and some of the air sacs. This seems possible, given the complex anatomy of the avian respiratory system.

C. Evaporative Cooling During Flight

Zeuthen's (1942) calculations of the energetic expenditures and heat dissipation during flight in pigeons (based on aerodynamic engineering concepts) suggest that evaporative water loss is the major avenue of heat loss in flight. Recent data indicate that his estimates of energy metabolism and evaporative water loss during flight are too high. At moderate temperatures, flying pigeons and Budgerigars lose approximately 13 to 22% of their heat production by evaporation (LeFebvre, 1964; Hart and Roy, 1966; Tucker, 1968b), while the remainder is lost via convection and radiation. Flying Budgerigars increase the evaporative component with ambient temperature, and at 36° to 37°C lose 40% of their heat production via evaporation (Tucker, 1968b). At 36° to 37°C, the Budgerigars extend the unfeathered portions of their legs into the slip stream (presumably increasing rates of convective heat loss) and will not fly for the full 20-minute experimental period.

Long-distance migrations of many species provide less direct evidence for lower rates of evaporation during flight than had been earlier predicted, since the birds could not replace the high evaporative water loss by metabolic water. Odum *et al.* (1964) present evidence for the homeostatic maintenance of the nonfat body components during migration. They believe that "fuel, not dehydration, is the limiting factor in long migrations as long as weather conditions remain favorable for sustained flight."

VI. Respiration During Specialized Activities

A. Torpidity

Biologists have been slow to accept the occurrence of avian torpidity (this discussion will be confined to species that are capable of allowing their T_b to drop to within a few degrees of T_a and are capable of physiological arousal). The most thorough evidence for avian torpidity exists for hummingbirds (Apodiformes: Trochilidae), which can but do not necessarily enter torpor nightly (for references, see Lasiewski and Lasiewski, 1967; Hainsworth and Wolf, 1970) and for nightjars (Caprimulgiformes: Caprimulgidae), which may engage in torpor on a daily or more extended basis (see Bartholomew *et al.*, 1962; Peiponen, 1966; Austin and Bradley, 1969; Dawson and Fisher, 1969). Swifts (Apodiformes: Apodidae) may also engage in torpidity (see Stager, 1965, Ramsey, 1970), as do mousebirds (Coliiformes; Bartholomew and Trost, 1970) but have been less well studied. Although there is much circumstantial evidence for the occurrence of torpidity in swallows (Passeriformes: Hirundinidae), physiological evidence is lacking (see McAtee, 1947; Lasiewski and Thompson, 1966; Keskpaik and Lyuleyeva, 1968).

Energy expenditures during torpor are reduced to many-fold below that of homeothermic levels and are inversely related to T_b. Although there are several sets of data on avian respiration and torpor, there is no truly adequate study on this topic for any species. The oxygen consumption of torpid hummingbirds is highly sensitive to temperature ($Q_{10} = 4.1$; Lasiewski, 1963), suggesting that at low T_a, respiration during torpor would be low indeed. Bartholomew *et al.* (1957) and Lasiewski (1964) reported sporadic or irregular breathing in torpid hummingbirds, with long periods of apnea at low temperatures. Peiponen (1966) reported no observable respiratory movements in the European Nightjar *(Caprimulgus europaeus)* while in deep torpor ($T_b = 7.0°C$). The extremely low requirements of gaseous exchange in torpid birds at low T_a's may be provided for through diffusion or through diffusion plus sporadic breathing. Torpidity at intermediate temperatures requires more oxygen. Careful studies of avian respiration during torpor using modern electronic sensors might be profitable.

Arousal from torpor is an active process. During arousal, f increases with T_b and may even surpass the f of the homeothermic bird (Bartholomew *et al.*, 1957; Peiponen, 1966; Austin and Bradley, 1969).

B. Diving

Many aquatic birds do all or most of their feeding in the water. They may travel considerable distances underwater, capture prey or other food items, and remain submerged voluntarily for periods of up to several minutes. Most studies dealing with avian diving adaptations have stressed the importance of cardiovascular function (for references, see Eliassen, 1960; Andersen, 1966; Butler and Jones, 1968). The respiratory correlates of avian diving are not as well understood.

Grebes can increase their specific gravity (presumably by exhaling air from the respiratory system and perhaps by compressing their feathers), enabling them to submerge quietly in the water (Schorger, 1947). Loons and other birds may exhale before diving (Zimmer, 1935; Schorger, 1947). The oxygen stores of diving birds may be adequate to maintain oxidative metabolism during short dives, but they are inadequate to maintain the long forced-submergence periods endured by avian divers. Richet (1899) analyzed the gas composition of the respiratory system of diving ducks and found that oxygen consumption during a dive decreased, a result corroborated several times by later workers (Andersen, 1959; Pickwell, 1968). Furthermore, the oxygen debt that must be repaid after a dive is less than would be expected if oxygen consumption during a dive were at pre-dive levels. Avian divers may make only a few ventilations between each dive (see

Eliassen, 1960). Pickwell (1968) elegantly demonstrated by direct calorimetric techniques that the total energy metabolism of ducks was reduced more than 90% during a prolonged experimental submergence. Alcids and some other birds swim underwater by "flying" with their wings. Presumably, the movements of their pectoral muscles facilitate utilization of the available oxygen in the respiratory system by enhancing the circulation of air within the system.

C. Singing

Bird song has long captivated the imagination of man, yet our knowledge of the respiratory correlates of avian sound production is extremely limited. The acoustical characteristics of birds' songs have been examined and reviewed recently by Greenewalt (1968, 1969) and Stein (1968). While the anatomy of the syrinx, the primary sound-producing organ in birds, has been studied in detail (for references, see Gross, 1964; Chamberlain *et al.*, 1968), ornithologists do not agree on how various parts of the syrinx function during sound production.

Avian song differs fundamentally from human song. Many (perhaps most) bird vocalizations are produced solely at the syrinx by air flowing exteriorly (Chamberlain *et al.*, 1968; Berger and Hart, 1968; Calder, 1970) and are not modified by the trachea (Greenewalt, 1968). In man, on the other hand, song is produced with sustained exhalation at the upper end of the trachea (by the larynx) and is modulated by structures anterior to the larynx. Furthermore, while in man the volume of air inspired prior to the onset of singing limits the duration of song (Bouhuys *et al.*, 1966), this is not necessarily true in birds.

The extended duration and complicated acoustical properties of the songs of birds raise fascinating questions about avian respiratory physiology. The majority of workers have assumed implicitly that bird song is caused by a continuous flow of air to the exterior. Although this assumption may hold for certain types of bird song, it is untenable when applied to extended uninterrupted songs of small birds. For example, Greenewalt (1968) calculated that the Song Sparrow (*Melospiza melodia*) might expire 15 ml of air during its 2.5-second song, a figure incompatible with its 20–25 gm body weight (see Section III,A, Table I). Many small birds sing continuously for periods of 10 or more seconds, while others are capable of extended song during flight. Some birds have another unique vocal ability, which is the production of sound from two independent acoustical sources, presumably the two branches of the syrinx (Greenewalt, 1968).

Despite the limitations of the available data, at least three distinct patterns correlating sound production and respiration have been described. Some birds produce sounds during both expiration and inspiration, others during a continuous expiration. Still others vocalize during rapid, shallow respiratory movements, producing sound during expiration.

Shallenberger (1971) found that the call of the Wedge-tailed Shearwater (*Puffinus pacificus*) on its breeding grounds is composed of both an inspiratory and expiratory phase. Similarly, Kneutgen (1969) concluded that Shamas (*Copsychus malabaricus*) sing during both inspiration and expiration, and "that the rhythmic structure of the motifs is determined by respiration." The two other respiratory-sound production patterns involve song presumably caused by expiratory flow alone. One of the most familiar types of avian sound is produced by a continuous flow of air exteriorly through the syrinx, with the duration of the sound being limited by the expiratory volume. The crowing of a cock would seem to be a good example of this pattern (Gross, 1964), although it may be more complicated, since White (1968) provided radiographic evidence that the larynx could modify and conceivably produce sound in the crowing cock.

Evidence for the third pattern, in which sound is produced during rapid, shallow respiratory movements, is provided by the studies of Berger and Hart (1968) on the distress call of the Evening Grosbeak (*Coccothraustes vespertinus*), and by the studies of Calder (1970) on extended song in the Canary (*Serinus canaria*). The distress calls of the Evening Grosbeak were "produced in the expired airstream in pulsations of 55–60/second," with an average duration per call of 0.3 second. The calls were separated by short inspirations of 0.1 to 0.15 second which had increased V_T and air flow velocity relative to resting values. The records of respiration during distress calls studied by Berger and Hart show characteristics similar to the "mini-breaths" described by Calder. Calder (1970) succeeded in obtaining simultaneous impedence pneumographic recordings of respiratory movements and sound recordings during extended singing in the Canary. These demonstrated a 1:1 correspondence between song notes and "respiratory movement" (Fig. 21) even in trilled notes as rapid as 25/second. The trilled or warbled notes were accompanied by respiratory movements assumed to be "mini-breaths" which presumably provide expiratory flow through the syrinx for sound production. However, because of their presumed small tidal volumes, they would not be adequate for respiratory exchange. During an extended song, O_2 content would decrease in the respiratory system, while CO_2 content would increase. Thus the volume of air inspired prior to the onset of singing would not be limiting in this pattern of bird song (as it is in the

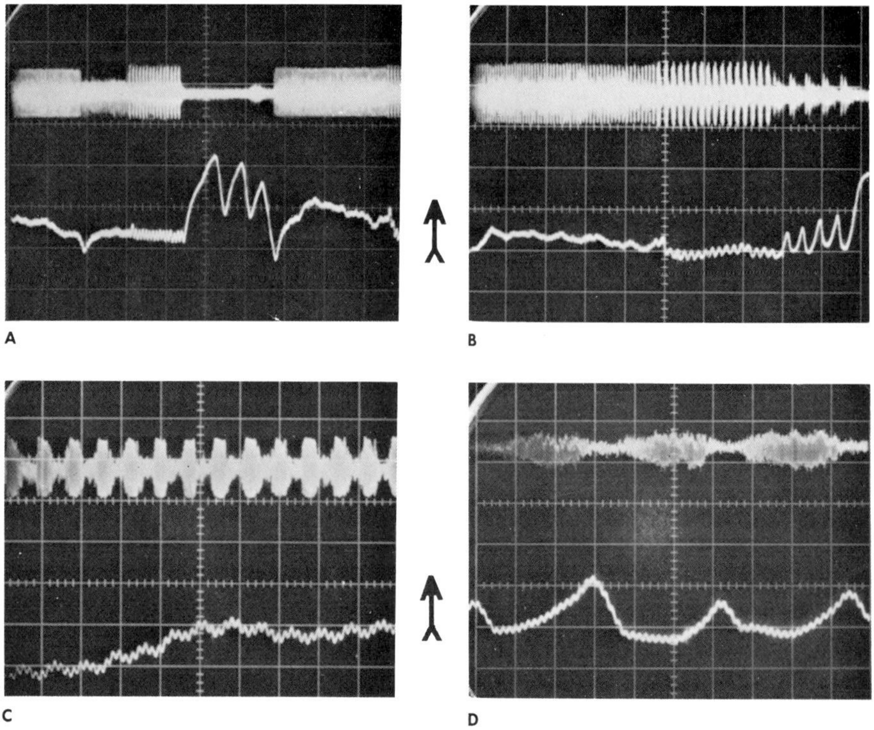

FIG. 21. Oscilloscope display of sound production (upper beam) and respiratory movements (lower beam) during waterslager Canary song. Arrows indicate expiratory direction. A: Fast (24 c/s), intermediate (17 c/s), and slow (11 c/s) trills, three resting breaths, fast trill resumed (scale: 1 sec/div). B: Fast, intermediate, and slow trills, terminal chirps, expiration (0.5 sec/div). C: Portion of fast trill (27 c/s; scale: 50 msec/div). D: Portion of terminal chirps (50 msec/div). The minor oscillation of the lower traces in C and D is an artifact. (From Calder, 1970.)

crowing cock or in man), but the composition of the air in the respiratory system might be. This conclusion is supported by Calder's evidence that the first respiratory movement at the end of a song is a vigorous expiration (Fig. 21B) which purges the system before bringing in fresh air.

Both Berger and Hart (1968) and Calder (1970) note that some types of sound production in birds may be related to resting respiratory frequencies. For example, the calling rate of the Whippoorwill (*Caprimulgus vociferus*) of 60/minute is the same as its predicted resting f (Calder, 1970). It seems unlikely, however, that a 1:1 relationship exists between breathing and sound production during the high frequency songs common in many birds. Greenewalt (1968) suggests that syringeal muscle tension will oscillate "within a given phrase, producing a rapid modulation in both frequency and amplitude about mean values." He proposes that, "The low mass of the syringeal muscles permits relatively high values for the modulating frequency, approaching 400 c/s."

Greenewalt (1968) indicates that birds cannot sing without application of air pressure external to the syrinx via the clavicular sac (which surrounds the syrinx). While this conclusion is supported by experimental evidence (see Gross, 1964), it should be noted that a sizeable pressure differential between the inside and outside of the syrinx is unlikely, due to the absence of anatomical valves (see Section III,B).

D. Hatching

Chicken and duck fetuses begin to breath air through their lungs about 24 and 72 hours before hatching, respectively (Windle *et al.*, 1938; Windle and Nelsen, 1938). Before taking air into its lungs, the chicken fetus periodically executes respiratory movements that appear as early as the 17th or 18th day (Windle *et al.*, 1938; Windle and Barcroft, 1938). Administration of CO_2 initiates rhythmic respiratory movements prematurely in the chicken fetus (Windle and Barcroft, 1938). Anoxemia and high CO_2 concentrations affect respiration in unhatched chickens and ducks, particularly during the stage when allantoic respiration is deteriorating and lung respiration becomes progressively more important.

Several recent studies have examined the exchange of respiratory gases by avian embryos (domestic fowl) in terms of gas permeabilities and diffusion rates. Eggs incubated in 60% O_2 had lower CO diffusing capacities when compared to those incubated in room air (Temple and Metcalfe, 1970). Wangensteen *et al.* (1970) concluded that the porosity of the egg shell, which limits gaseous exchange by the embryo through diffusive permeabilities to O_2 uptake and CO_2 and H_2O loss, is set when the egg is laid. Kutachai and Steen (1971) found that O_2 and CO_2 permeabilities of the egg shell and membranes of newly laid eggs are not sufficient to account for the gas exchange that takes place in later development, but that the drying of the shell membranes greatly increases their permeabilities. They found that the water content of shell membranes of fertilized eggs decreases markedly during incubation.

The gular flutter response is developed in newly hatched chicks of the Great Blue Heron, *Ardea herodias* (Bartholomew and Dawson, 1954), the Common Nighthawk, *Chordeiles minor* (Lasiewski and

Dawson, 1964), and the Masked Booby, *Sula dactylatra* (Bartholomew, 1966), but not in newly hatched Brown Pelicans, *Pelecanus occidentalis* (Bartholomew and Dawson, 1954). The hatchlings of all these species are shielded from intense insolation by the parents.

E. Respiration at High Altitude

Birds are known to fly at altitudes that cause severe respiratory stress in resting unacclimated humans (above 6000 m). At high altitudes, decreased density and oxygen content of air and low temperatures pose formidable respiratory and circulatory problems to birds performing flapping flight. Tucker (1968a) provides data relevant to these problems through studies on oxygen consumption and oxygen transport in resting House Sparrows exposed to ambient temperatures of 5°C and to hypobaric pressures simulating an altitude of 6100 m. He also provides comparative data for white mice (*Mus musculus*) and some observations on House Sparrows and Budgerigars flying in the hypobaric chamber.

When House Sparrows and mice were exposed to a simulated altitude of 6100 m for 1 hour, differences in their responses were marked, House Sparrows were normally active, had T_b's 2° lower than T_b's at sea level, and had higher rates of oxygen consumption, respiration, and heart beats. The mice were moribund, had T_b's 10° below normal, and had rates of oxygen consumption and respiration that were below those at sea level.

The effective ventilation of the sparrows' lungs at 6100 m was 77% greater than at sea level. They also removed a greater proportion of the oxygen from the ventilation air at high altitudes.

House Sparrows and Budgerigars could fly for short periods and gain altitude at the simulated altitude of 6100 m. A Budgerigar trained to fly wearing an oxygen mask in a wind tunnel could fly at pressures simulating an altitude of 6100 m, suggesting that these birds are aerodynamically but not physiologically capable of flight at that altitude.

Dunson (1965) found that montane American Robins, *Turdus migratorius,* had higher heart and lung weights than their lowland counterparts. Further studies of avian species that live at high altitudes would be profitable.

ACKNOWLEDGMENTS

I dedicate this chapter to my wife, Magi, for her encouragement, support, and understanding. I thank Joanna Miller, Jenifer Grohs, David Hudson, Robert Epting, and Peggy Caton for their assistance. Preparation of this manuscript was supported in part by NSF Grant GB 8445.

REFERENCES

Adolph, E. F. (1949). Quantitative relations in the physiological constitutions of mammals. *Science* **109,** 579–585.

Åkerman, B., Andersson, B., Fabricius, E., and Svensson, L. (1960). Observations on central regulation of body temperature and of food and water intake in the pigeon (*Columba livia*). *Acta Physiol. Scand.* **50,** 328–336.

Akester, A. R. (1960). The comparative anatomy of the respiratory pathways in the domestic fowl (*Gallus domesticus*), pigeon (*Columba livia*) and domestic duck (*Anas platyrhyncha*). *J. Anat.* **94,** 487–505.

Andersen, H. T. (1959). A note on the composition of alveolar air in the diving duck. *Acta Physiol. Scand.* **46,** 240–243.

Andersen, H. T. (1963). Factors determining the circulatory adjustments to diving. I. Water immersion. *Acta Physiol. Scand.* **58,** 173–185.

Andersen, H. T. (1966). Physiological adaptations in diving vertebrates. *Physiol. Rev.* **46,** 212–243.

Austin, G. T., and Bradley, W. G. (1969). Additional responses of the Poor-will to low temperatures. *Auk* **86,** 717–725.

Bartholomew, G. A. (1966). The role of behavior in the temperature regulation of the Masked Booby. *Condor* **68,** 523–535.

Bartholomew, G. A., and Dawson, W. R. (1954). Temperature regulation in young pelicans, herons, and gulls. *Ecology* **35,** 466–472.

Bartholomew, G. A., and Trost, C. H. (1970). Temperature regulation in the Speckled Mousebird, *Colius striatus*. *Condor* **72,** 141–146.

Bartholomew, G. A., Howell, T. R., and Cade, T. J. (1957). Torpidity in the White-throated Swift, Anna Hummingbird, and Poor-will. *Condor* **59,** 145–155.

Bartholomew, G. A., Hudson, J. W., and Howell, T. R. (1962). Body temperature, oxygen consumption, evaporative water loss, and heart rate in the Poor-will. *Condor* **64,** 117–125.

Bartholomew, G. A., Lasiewski, R. C., and Crawford, E. C. (1968). Patterns of panting and gular flutter in cormorants, pelicans, owls, and doves. *Condor* **70,** 31–34.

Bellairs, A. d'A., and Jenkin, C. R. (1960). The skeleton of birds. *In* "Biology and Comparative Physiology of Birds" (A. J. Marshall, ed.), Vol. 1, pp. 241–300. Academic Press, New York.

Berger, M., and Hart, J. S. (1968). Ein Beitrag zum Zusammenhang zwischen Stimme und Atmung bei Vögeln. *J. Ornithol.* **109,** 421–424.

Berger, M., Roy, O. Z., and Hart, J. S. (1970a). The co-ordination between respiration and wing beats in birds. *Z. Vergl. Physiol.* **66,** 190–200.

Berger, M., Hart, J. S., and Roy. O. Z. (1970b). Respiration, oxygen consumption and heart rate in some birds during rest and flight. *Z. Vergl. Physiol.* **66,** 201–214.

Bernstein, M. H. (1969). Cutaneous and respiratory evaporation in Painted Quail, *Excalfactoria chinensis*. *Amer. Zool.* **9,** 1099.

Bernstein, M. H. (1970). The development of temperature regulation in a small precocial bird (Painted Quail, *Excalfactoria chinensis*). Ph.D. Thesis, University of California, Los Angeles.

Bernstein, M. H. (1971a). Cutaneous and respiratory evaporation in Painted Quail *Excalfactoria chinensis* during ontogeny of thermoregulation. *Comp. Biochem. Physiol.* **38,** 611–617.

Bernstein, M. H. (1971b). Cutaneous water loss in small birds. *Condor* **73,** 468–469.

Biggs, P. M., and King, A. S. (1957). A new experimental approach to the problem of the air pathway within the avian lung. *J. Physiol. (London)* **138,** 282–299.

Blankart, R. (1960). Zur Charakterisierung der vagalen Atmungssteurung der Taube. *Helv. Physiol. Pharmacol. Acta* **18,** 35–42.

Bouhuys, A., Proctor, D. F., and Mead, J. (1966). Kinetic aspects of singing. *J. Appl. Physiol.* **21,** 483–496.

Bretz, W. L. (1969). Patterns of air flow in the avian lung. Ph.D. Thesis, Duke University, Durham, North Carolina.

Bretz, W. L., and Schmidt-Nielsen, K. (1971). Bird respiration: Flow patterns in the duck lung. *J. Exp. Biol.* **54,** 103–118.

Burton, R. R., and Smith, A. H. (1968). Blood and air volumes in the avian lung. *Poultry Sci.* **47,** 85–91.

Butler, P. J., and Jones, D. R. (1968). Onset of and recovery from diving bradycardia in ducks. *J. Physiol. (London)* **196,** 255–272.

Calder, W. A. (1968). Respiratory and heart rates of birds at rest. *Condor* **70,** 358–365.

Calder, W. A. (1970). Respiration during song in the Canary (*Serinus canaria*). *Comp. Biochem. Physiol.* **32,** 251–258.

Calder, W. A., and Schmidt-Nielsen, K. (1967). Temperature regulation and evaporation in the pigeon and the Roadrunner. *Amer. J. Physiol.* **213,** 883–889.

Calder, W. A., and Schmidt-Nielsen, K. (1968). Panting and blood carbon dioxide in birds. *Amer. J. Physiol.* **215,** 477–482.

Chamberlain, D. R., Gross, W. B., Cornwell, G. W., and Mosby, H. S. (1968). Syringeal anatomy in the Common Crow. *Auk* **85,** 244–252.

Chew, R. M. (1965). Water metabolism of mammals. *In* "Physiological Mammalogy" (W. V. Mayer and R. G. Van Gelder, eds.), Vol. 2, pp. 44–111. Academic Press, New York.

Chiodi, H., and Terman, J. W. (1965). Arterial blood gases of the domestic hen. *Amer. J. Physiol.* **208,** 798–800.

Clements, J. A., Nellenbogen, J., and Trahan, H. J. (1970). Pulmonary surfactant and evolution of the lungs. *Science* **169,** 603–604.

Cohn, J. E., and Shannon, R. (1968). Respiration in unanesthetized geese. *Resp. Physiol.* **5,** 259–268.

Cohn, J. E., Burke, L., and Markesbery, H. (1963). Respiration in unanesthetized geese. *Fed. Proc., Fed. Amer. Soc. Exp. Biol.* **22,** 397.

Comroe, J. H., Cournand, A., Ferguson, J. K. W., Filley, G. F., Fowler, W. S., Gray, J. S., Helmholtz, H. F., Otis, A. B., Pappenheimer, J. R., Rahn, H., and Riley, R. L. (1950). Standardization of definitions and symbols in respiratory physiology. *Fed. Proc., Fed. Amer. Soc. Exp. Biol.* **9,** 602–605.

Cook, R. D., and King, A. S. (1969). Nerves of the avian lung: Electron microscopy. *J. Anat.* **105,** 202–203.

Cook, R. D., and King, A. S. (1970). Observation on the ultrastructure of the smooth muscle and its innervation in the avian lung. *J. Anat.* **106,** 273–283.

Crawford, E. C. (1962). Mechanical aspects of panting in dogs. *J. Appl. Physiol.* **17,** 249–251.

Crawford, E. C., and Kampe, G. (1971). Resonant panting in pigeons. *Comp. Biochem. Physiol.* **40,** 549–552.

Crawford, E. C., and Lasiewski, R. C. (1968). Oxygen consumption and respiratory evaporation of the Emu and Rhea. *Condor* **70,** 333–339.

Crawford, E. C., and Schmidt-Nielsen, K. (1967). Temperature regulation and evaporative cooling in the Ostrich. *Amer. J. Physiol.* **212,** 347–353.

Dawson, W. R., and Fisher, C. D. (1969). Responses to temperature by the Spotted Nightjar (*Eurostopodus guttatus*). *Condor* **71,** 49–53.

Dawson, W. R., and Hudson, J. W. (1970). Birds. *In* "Comparative Physiology of Thermoregulation" (G. C. Whittow, ed.), Vol. 1, pp. 223–310. Academic Press, New York.

Dejours, P. (1966). "Respiration." Oxford Univ. Press, London and New York.

Dittmer, D. S., and Grebe, R. M. (eds.). (1958). "Handbook of Respiration." Saunders, Philadelphia, Pennsylvania.

Dotterweich, H. (1930). Die Bahnhofstauben und die Atmungsphysiologie der Atemluft in der Vogellunge. *Zool. Anz.* **90**, 259–262.

Dotterweich, H. (1933). Ein weiterer Beitrag zur Atmungsphysiologie der Vögel. *Z. Vergl. Physiol.* **18**, 803–809.

Dunson, W. A. (1965). Adaptation of heart and lung weight to high altitude in the Robin. *Condor* **67**, 215–219.

Eaton, J. A., Jr., Fedde, M. R., and Burger, R. E. (1971). Sensitivity to inflation of the respiratory system in the chicken. *Resp. Physiol.* **11**, 167–177.

Eliassen, E. (1960). Cardiovascular responses to submersion asphyxia in avian divers. *Arbok Univ. Bergen, Mat.-Natur. Ser.* **2**, 1–100.

Evans, H. E. (1969). Anatomy of the Budgerigar. *In* "Diseases of Cage and Aviary Birds" (M. L. Petrak, ed.), pp. 78–88. Lea & Febiger, Philadelphia, Pennsylvania.

Farner, D. S. (1970). Some glimpses of comparative avian physiology. *Fed. Proc., Fed. Amer. Soc. Exp. Biol.* **29**, 1649–1663.

Fedde, M. R. (1970). Peripheral control of avian respiration. *Fed. Proc., Fed. Amer. Soc. Exp. Biol.* **29**, 1664–1673.

Fedde, M. R. (1971). Personal communication.

Fedde, M. R., and Burger, R. E. (1963). Death and pulmonary alterations following bilateral, cervical vagotomy in the fowl. *Poultry Sci.* **42**, 1236–1246.

Fedde, M. R., and Peterson, D. F. (1970). Intrapulmonary-receptor response to changes in air-gas composition in *Gallus domesticus*. *J. Physiol. (London)* **209**, 609–625.

Fedde, M. R., Burger, R. E., and Kitchell, R. (1961). The influence of the vagus nerve on respiration. *Poultry Sci.* **40**, 1401.

Fedde, M. R., Burger, R. E., and Kitchell, R. L. (1963). Effect of anesthesia and age on respiration following bilateral, cervical vagotomy in the fowl. *Poultry Sci.* **42**, 1212–1223.

Feigl, E., and Folkow, B. (1963). Cardiovascular responses in "diving" and during brain stimulation in ducks. *Acta Physiol. Scand.* **57**, 99–110.

Fraenkel, G. (1934). Der Atmungsmechanismus der Vögel während des Fluges. *Biol. Zentralbl.* **54**, 96–101.

Frankel, H. M. (1965). Blood lactate and pyruvate and evidence for hypocapnic lacticacidosis in the chicken. *Proc. Soc. Exp. Biol. Med.* **119**, 261–263.

Frankel, H. M., and Frascella, D. (1968). Blood respiratory gases, lactate, and pyruvate during thermal stress in the chicken. *Proc. Soc. Exp. Biol. Med.* **127**, 997–999.

Frankel, H. M., Hollands, K. G., and Weiss, H. S. (1962). Respiratory and circulatory responses of hyperthermic chickens. *Arch. Int. Physiol. Biochim.* **70**, 555–563.

Fujiwara, T., Adams, F. H., Nozaki, M., and Dermer, G. B. (1970). Pulmonary surfactant phospholipids from Turkey lung: Comparison with rabbit lung. *Amer. J. Physiol.* **218**, 218–225.

George, J. C., and Shah, R. V. (1965). Evolution of air sacs in sauropsida. *J. Anim. Morphol. Physiol.* **12**, 255–263.

Gier, H. T. (1952). The air sacs of the loon. *Auk* **69**, 40–49.

Glass, N. R. (1969). Discussion of calculation of power function with special reference to respiratory metabolism in fish. *J. Fish. Res. Bd. Can.* **26**, 2643–2650.

Graham, J. D. P. (1939). The air stream in the lung of the fowl. *J. Physiol. (London)* **97,** 133–137.

Graham, J. P. D. (1940). Respiratory reflexes in the fowl. *J. Physiol. (London)* **97,** 525–532.

Greenewalt, C. H. (1968). "Bird Song: Acoustics and Physiology." Random House (Smithsonian Inst. Press), New York.

Greenewalt, C. H. (1969). How birds sing. *Sci. Amer.* **221,** 126–139.

Gross, W. B. (1964). Voice production by the chicken. *Poultry Sci.* **43,** 1005–1008.

Hainsworth, F. R., and Wolf, L. L. (1970). Regulation of oxygen consumption and body temperature during torpor in a hummingbird, *Eulampis jugularis. Science* **168,** 368–369.

Hamlet, M. P., and Fisher, H. I. (1967). Air sacs of respiratory origin in some procellariiform birds. *Condor* **69,** 586–595.

Haning, Q. C., and Thompson, A. M. (1964). Tissue CO_2 in vertebrates. *Comp. Biochem. Physiol.* **15,** 17–26.

Hart, J. S., and Roy, O. Z. (1966). Respiratory and cardiac responses to flight in pigeons. *Physiol. Zool.* **39,** 291–306.

Hazelhoff, E. H. (1951). Structure and function of the lung of birds. *Poultry Sci.* **30,** 3–10.

Henry, J. D., and Fedde, M. R. (1970). Pulmonary circulation time in the chicken. *Poultry Sci.* **49,** 1286–1290.

Herreid, C. F., and Kessel, B. (1967). Thermal conductance in birds and mammals. *Comp. Biochem. Physiol.* **21,** 405–414.

Hiestand, W. A., and Randall, W. C. (1941). Species differentiation in the respiration of birds following carbon dioxide administration and the location of inhibitory receptors in the upper respiratory tract. *J. Cell. Comp. Physiol.* **17,** 333–340.

Hinds, D. S., and Calder, W. A. (1971). Tracheal dead space in the respiration of birds. *Evolution* **25,** 429–440.

Jones, D. R. (1969). Avian afferent vagal activity related to respiratory and cardiac cycles. *Comp. Biochem. Physiol.* **28,** 961–965.

Kadono, H., and Okada, T. (1962). Electromyographic studies on the respiratory muscles of the domestic fowl. *Jap. J. Vet. Sci.* **24,** 215–222.

Kallir, E. (1930). Temperaturtopographie einiger Vögel. *Z. Vergl. Physiol.* **13,** 231–248.

Kanematsu, S., Kii, M., Sonoda, T., and Kato, Y. (1967). Effects of hypothalamic lesions on body temperature in the chicken. *Jap. J. Vet. Sci.* **29,** 95–104.

Kendeigh, S. C. (1944). Effect of air temperature on the rate of energy metabolism in the English Sparrow. *J. Exp. Biol.* **96,** 1–16.

Keskpaik, J., and Lyuleyeva, D. (1968). Temporary hypothermia in swallows. *Commun. Baltic Comm. Study Bird Migr., Tartu, 1968* No. 5, pp. 122–145. (In Russian.)

King, A. S. (1957). The aerated bones of *Gallus domesticus. Acta Anat.* **31,** 220–230.

King, A. S. (1966). Structural and functional aspects of the avian lungs and air sacs. *Int. Rev. Gen. Exp. Zool.* **2,** 171–267.

King, A. S., and Atherton, J. D. (1970). The identity of the air sacs of the Turkey (*Meleagris gallopavo*). *Acta Anat.* **77,** 78–91.

King, A. S., and Cowie, A. F. (1969). The functional anatomy of the bronchial muscle of the bird. *J. Anat.* **105,** 323–336.

King, A. S., and Molony, V. (1971). The anatomy of respiration. *In* "Physiology and Biochemistry of the Domestic Fowl" (D. J. Bell and B. M. Freeman, eds.), Chapter 5, pp. 93–169. Academic Press, New York.

King, A. S., Molony, V., McLelland, J., Bowsher, D. R., and Mortimer, M. F., (1968). Afferent respiratory pathways in the avian vagus. *Experientia* **24,** 1017–1018.

King, J. R., and Farner, D. S. (1964). Terrestrial animals in humid heat: birds. *In* "Handbook of Physiology" pp. 603–624. (D. B. Dill, E. F. Adolph, and C. G. Wilber, eds.), Sect. 4, Amer. Physiol. Soc., Washington, D.C.

Kneutgen, J. (1969). "Musikalische" Formen im Gesang der Schamadrossel (*Kittacincla macroura* Gm.) und ihre Funktionen. *J. Ornithol.* **110,** 246–285.

Kutachai, H., and Steen, J. B. (1971). Permeability of the shell and shell membranes of hens' eggs during development. *Resp. Physiol.* **11,** 265–278.

Lacy, R. A., Jr. (1965). Mechanical determinants of panting frequency in the domestic fowl. M.A. Thesis, University of California, Davis.

Lacy, R. A., Jr., and Burger, R. E. (1972). Personal communication.

Lasiewski, R. C. (1963). Oxygen consumption of torpid, resting, active and flying hummingbirds. *Physiol. Zool.* **36,** 122–140.

Lasiewski, R. C. (1964). Body temperatures, heart and breathing rate, and evaporative water loss in hummingbirds. *Physiol. Zool.* **37,** 212–223.

Lasiewski, R. C. (1969). Physiological responses to heat stress in the Poorwill. *Amer. J. Physiol.* **217,** 1504–1509.

Lasiewski, R. C., and Bartholomew, G. A. (1966). Evaporative cooling in the Poor-will and the Tawny Frogmouth. *Condor* **68,** 253–262.

Lasiewski, R. C., and Bartholomew, G. A. (1971). Unpublished data.

Lasiewski, R. C., and Calder, W. A. (1971). A preliminary allometric analysis of respiratory variables in resting birds. *Resp. Physiol.* **11,** 152–166.

Lasiewski, R. C., and Dawson, W. R. (1964). Physiological responses to temperature in the Common Nighthawk. *Condor* **66,** 477–490.

Lasiewski, R. C., and Dawson, W. R. (1967). A re-examination of the relation between standard metabolic rate and body weight in birds. *Condor* **69,** 13–23.

Lasiewski, R. C., and Dawson, W. R. (1969). Calculation and miscalculation of the equations relating avian standard metabolism to body weight. *Condor* **71,** 335–336.

Lasiewski, R. C., and Lasiewski, R. J. (1967). Physiological responses of the Blue-throated and Rivoli's Hummingbirds. *Auk* **84,** 34–48.

Lasiewski, R. C., and Seymour, R. S. (1972). Thermoregulatory responses to heat stress in four species of birds weighing approximately 40 grams. *Physiol. Zool.* (in press).

Lasiewski, R. C., and Snyder, G. K. (1969). Responses to high temperature in nestling Double-crested and Pelagic Cormorants. *Auk* **86,** 529–540.

Lasiewski, R. C., and Thompson, H. J. (1966). Field observations of torpidity in the Violet-green Swallow. *Condor* **68,** 102–103.

Lasiewski, R. C., Acosta, A. L., and Bernstein, M. H. (1966a). Evaporative water loss in birds. I. Characteristics of the open-flow method for determination and their relation to estimates of thermoregulatory ability. *Comp. Biochem. Physiol.* **19,** 445–457.

Lasiewski, R. C., Acosta, A. L., and Bernstein, M. H. (1966b). Evaporative water loss in birds. II. A modified method for determination by direct weighing. *Comp. Biochem. Physiol.* **19,** 459–470.

Lasiewski, R. C., Weathers, W. W., Bernstein, M. H. (1967). Physiological responses of the Giant Hummingbird, *Patagona gigas*. *Comp. Biochem. Physiol.* **23,** 797–813.

Lasiewski, R. C., Dawson, W. R., and Bartholomew, G. A. (1970). Temperature regulation in the Little Papuan Frogmouth, *Podargus ocellatus*. *Condor* **72,** 332–338.

Lasiewski, R. C., Bartholomew, G. A., and Ohmart, R. D. (1971a). Unpublished data.

Lasiewski, R. C., Bernstein, M. H., and Ohmart, R. D. (1971b). Cutaneous water loss in the Roadrunner and Poor-will. *Condor* **73,** 470–472.

LeFebvre, E. A. (1964). The use of D_2O^{18} for measuring energy metabolism in *Columba livia* at rest and in flight. *Auk* **81,** 403–416.

Leighton, A. T., Jr., Siegel, P. B., and Siegel, H. S. (1966). Body weight and surface area of chickens (*Gallus domesticus*). *Growth* **30**, 229–238.

Lepkovsky, S., Snapir, N., and Furuta, F. (1968). Temperature regulation and appetitive behavior in chickens with hypothalamic lesions. *Physiol. Behav.* **3**, 911–915.

Linsley, J. G., and Burger, R. E. (1964). Respiratory and cardiovascular responses in the hyperthermic domestic cock. *Poultry Sci.* **43**, 291–305.

Lustick, S. I. (1968). Energetics and thermoregulation in the Cowbird (*Molothrus ater obscurus*). Ph.D. Thesis, University of California, Los Angeles.

McAtee, W. L. (1947). Torpidity in birds. *Amer. Midl. Natur.* **38**, 191–206.

McLelland, J. (1965). The anatomy of the rings and muscles of the trachea of *Gallus domesticus*. *J. Anat.* **99**, 651–656.

Makowski, J. (1938). Beitrag zur Klärung des Atmungsmechanisms der Vögel. *Pfluegers Arch. Gesamte Physiol. Menschen Tiere* **240**, 407–418.

Odum, E. P., Rogers, D. T., and Hicks, D. L. (1964). Homeostasis of the nonfat components of migrating birds. *Science* **143**, 1037–1039.

Ohmart, R. D., Chapman, T. E., and McFarland, L. Z. (1970). Water turnover in Roadrunners under different environmental conditions. *Auk* **87**, 787–793.

Pattle, R. E., and Hopkinson, D. A. W. (1963). Lung lining in bird, reptile and amphibian. *Nature (London)* **200**, 894.

Payne, D. C. (1960). Observations on the functional anatomy of the lungs and air sacs of *Gallus domesticus*. Ph.D. Thesis, University of Bristol, England.

Pearson, O. P. (1950). The metabolism of hummingbirds. *Condor* **52**, 145–152.

Peiponen, V. A. (1966). The diurnal heterothermy of the nightjar (*Caprimulgus europaeus* L.). *Ann. Acad. Sci. Fenn., Ser. A4* **101**, 1–35.

Peterson, D. F. (1970). Peripheral control of avian respiration. Ph.D. Thesis, Kansas State University, Manhattan.

Peterson, D. F., and Fedde, M. R. (1968). Receptors sensitive to carbon dioxide in lungs of chicken. *Science* **162**, 1499–1501.

Pickwell, G. V. (1968). Energy metabolism in ducks during submergence asphyxia: Assessment by a direct method. *Comp. Biochem. Physiol.* **27**, 455–485.

Piiper, J., Pfeifer, K., and Scheid, P. (1969). Carbon monoxide diffusing capacity of the respiratory system in the domestic fowl. *Resp. Physiol.* **6**, 309–317.

Piiper, J., Drees, F., and Scheid, P. (1970). Gas exchange in the domestic fowl during spontaneous breathing and artificial ventilation. *Resp. Physiol.* **9**, 234–245.

Portmann, A. (1961). Sensory organs: Skin, taste and olfaction. *In* "Biology and Comparative Physiology of Birds" (A. J. Marshall, ed.), Vol. 2, pp. 37–48. Academic Press, New York.

Ramsey, J. J. (1970). Temperature changes in Chimney Swifts (*Chaetura pelagica*) at lowered environmental temperatures. *Condor* **72**, 225–229.

Rautenberg, W. (1969). Die Bedeutung der zentralnervösen Thermosensitivität für die Temperaturregulation der Taube. *Z. Vergl. Physiol.* **62**, 235–266.

Ray, P. J., and Fedde, M. R. (1969). Responses to alterations in respiratory P_{O_2} and P_{CO_2} in the chicken. *Resp. Physiol.* **6**, 135–143.

Richards, S. A. (1968). Vagal control of thermal panting in mammals and birds. *J. Physiol. (London)* **199**, 89–101.

Richards, S. A. (1969). Vagal function during respiration and the effects of vagotomy in the domestic fowl (*Gallus domesticus*). *Comp. Biochem. Physiol.* **29**, 955–964.

Richards, S. A. (1970a). Physiology of thermal panting in birds. *Ann. Biol. Anim., Biochim., Biophys.* **10** (*Sér. Suppl.* 2), 151–168.

Richards, S. A. (1970b). The role of hypothalamic temperature in the control of panting in the chicken exposed to heat. *J. Physiol. (London)* **211**, 341–358.

Richards, S. A. (1971). Brain stem control of polypnea in the chicken and pigeon. *Respir. Physiol.* **11,** 315–326.

Richet, C. (1899). De la résistance des canards à l'asphyxie. *J. Physiol. Pathol. Gen.* **1,** 641–650.

Rogers, F. R. (1928). Studies on the brain stem. XI. The effects of artificial stimulation and of traumatism of the avian thalamus. *Amer. J. Physiol.* **86,** 639–650.

Salt, G. W. (1964). Respiratory evaporation in birds. *Biol. Rev.* **39,** 113–136.

Salt, G. W., and Zeuthen, E. (1960). The respiratory system. *In* "Biology and Comparative Physiology of Birds" (A. J. Marshall, ed.), Vol. 1, pp. 363–409. Academic Press, New York.

Scharnke, H. (1934). Die Bedeutung der Luftsäcke für die Atmung der Vögel. *Ergeb. Biol.* **10,** 177–206.

Scharnke, H. (1935). Die Zusammensetzung des Gasgemisches in den Luftsäcken der Taube unter experimentellen Bedingungen. *Sitzungsber. Ges. Morphol. Physiol. Muench.* **44,** 78–81.

Scharnke, H. (1938). Experimentelle Beiträge zur Kenntnis der Vogelatmung. *Z. Vergl. Physiol.* **25,** 548–583.

Scheid, P., and Piiper, J. (1969). Volume, ventilation and compliance of the respiratory system in the domestic fowl. *Resp. Physiol.* **6,** 298–308.

Scheid, P., and Piiper, J. (1970). Analysis of gas exchange in the avian lung: theory and experiments in the domestic fowl. *Resp. Physiol.* **9,** 246–262.

Schmidt-Nielsen, K., Kanwisher, J., Lasiewski, R. C., Cohn, J. E., and Bretz, W. L. (1969). Temperature regulation and respiration in the Ostrich. *Condor* **71,** 341–352.

Schmidt-Nielsen, K., Hainsworth, F. R., and Murrish, D. E. (1970). Countercurrent heat exchange in the respiratory passages: Effect on water and heat balance. *Resp. Physiol.* **9,** 263–276.

Schorger, A. W. (1947). Deep diving of the Loon and Old-squaw and its mechanism. *Wilson Bull.* **59,** 151–159.

Seymour, R. S. (1972). Convective heat transfer in the respiratory systems of panting animals. *J. Theor. Biol.* **35,** 119–127.

Shallenberger, R. J. (1971). Personnal communication.

Shepard, R. H., Sladen, B. K., Peterson, N., and Enns, T. (1959). Path taken by gases through the respiratory system of the chicken. *J. Appl. Physiol.* **14,** 733–735.

Sinha, M. P. (1958). Vagal control of respiration as studied in the pigeon. *Helv. Physiol. Pharmacol. Acta* **16,** 58–72.

Sinha, M. P. (1959). Observations on the organization of the panting center in avian brain. *Int. Congr. Physiol. Sci., Symp. Spec. Lect., 21st, 1959* p. 254.

Smith, R. M. (1969). Cardiovascular, respiratory, temperature and evaporative water loss responses of pigeons to varying degrees of heat stress. Ph.D. Thesis, Indiana University, Bloomington.

Smith, R. M., and Suthers, R. (1969). Cutaneous water loss as a significant contribution to temperature regulation in heat stressed pigeons. *Physiologist* **12,** 358.

Soum, J. M. (1896). Recherches physiologiques sur l'appareil respiratoire des oiseaux. *Ann. Univ. Lyon* **28,** 1–126.

Spector, W. S. (1956). *In* "Handbook of Biological Data" (W. S. Spector, ed.), pp. 163–164. Saunders, Philadelphia, Pennsylvania.

Stager, K. E. (1965). An exposed nocturnal roost of migrant Vaux Swifts. *Condor* **67,** 81–82.

Stahl, W. R. (1967). Scaling of respiratory variables in mammals. *J. Appl. Physiol.* **22,** 453–460.

Stanislaus, M. (1937). Untersuchungen an der Kolibrilunge. *Z. Morphol. Oekol. Tiere* **33**, 261–289.

Stein, R. C. (1968). Modulation in bird sounds. *Auk.* **85**, 229–243.

Sturkie, P. D. (1966). "Avian Physiology." Cornell Univ. Press, Ithaca, New York.

Tcheng, K., Fu, S., and Chen, T. (1963). Supracardial encapsulated receptors of the aorta and the pulmonary artery in birds. *Sci. Sinica* **12**, 73–81.

Temple, G. F., and Metcalfe, J. (1970). The effects of increased incubator oxygen tension on capillary development in the chick chorioallantois. *Resp. Physiol.* **9**, 216–233.

Tenney, S. M., and Bartlett, D., Jr. (1967). Comparative quantitative morphology of the mammalian lung: Trachea. *Resp. Physiol.* **3**, 130–135.

Tomlinson, J. T. (1963). Breathing of birds in flight. *Condor* **65**, 514–516.

Tucker, V. A. (1968a). Respiratory physiology of House Sparrows in relation to high-altitude flight. *J. Exp. Biol.* **48**, 55–56.

Tucker, V. A. (1968b). Respiratory exchange and evaporative water loss in the flying Budgerigar. *J. Exp. Biol.* **48**, 67–87.

Tyler, W. S., and Pangborn, J. (1964). Laminated membrane surface and osmiophilic inclusions in avian lung epithelium. *J. Cell Biol.* **20**, 157–162.

Van Matre, N. S. (1957). Avian external respiratory mechanisms. Ph.D. Thesis, University of California, Davis.

von Saalfeld, E. (1936). Untersuchungen über das Hacheln bei Tauben. *Z. Vergl. Physiol.* **23**, 727–743.

Vos, H. J. (1935). Über den Weg der Atemluft in der Entenlunge. *Z. Vergl. Physiol.* **21**, 552–578.

Walter, W. G. (1934). Beiträge zur Frage über den Weg der Luft in den Atmungsorganen der Vögel. *Arch. Neer. Physiol.* **19**, 529–537.

Wangensteen, O. D., Wilson, D., and Rahn, H. (1970). Diffusion of gases across the shell of the hen's egg. *Resp. Physiol.* **11**, 16–30.

Wetherbee, D. K. (1951). Air sacs in the English Sparrow. *Auk* **68**, 242–244.

White, S. S. (1968). Movements of the larynx during crowing in the domestic cock. *J. Anat.* **103**, 390–392.

Wilson, W. O., Hillerman, J. P., and Edwards, W. H. (1952). The relation of high environmental temperature to feather and skin temperatures of laying pullets. *Poultry Sci.* **31**, 843–846.

Windle, W. F., and Barcroft, J. (1938). Some factors governing initiation of respiration in the chick. *Amer. J. Physiol.* **121**, 684–691.

Windle, W. F., and Nelson, D. (1938). Development of respiration in the duck. *Amer. J. Physiol.* **121**, 700–707.

Windle, W. F., Scharpenberg, L. G., and Steele, A. G. (1938). Influence of carbon dioxide and anoxemia upon respiration in the chick at hatching. *Amer. J. Physiol.* **121**, 692–699.

Zar, J. H. (1968). Calculation and miscalculation of the allometric equation as a model in biological data. *BioScience* **18**, 1118–1120.

Zar, J. H. (1970). On the fitting of equations relating avian standard metabolism to body weight. *Condor* **72**, 247.

Zeuthen, E. (1942). The ventilation of the respiratory tract in birds. *Kgl. Dan. Vidensk. Selsk., Biol. Medd.* **17**, 1–50.

Zimmer, K. (1935). Beiträge zur Mechanik der Atmung bei den Vögeln in Stand und Flug. *Zoologica (New York)* **88**, 1–68.

Chapter 6

DIGESTION AND THE DIGESTIVE SYSTEM

Vinzenz Ziswiler and Donald S. Farner

I. Introduction 343
II. The Buccal Cavity, Buccal Glands, and Pharynx 344
III. The Esophagus and Crop 352
IV. The Gastric Apparatus 360
A. The Glandular Stomach 362
B. The Muscular Stomach 372
C. Gastric Digestion 378
D. The Motor Activities of the Gastric Apparatus 382
E. The Pyloric Stomach 384
V. The Intestine 384
A. The Small Intestine 384
B. The Large Intestine 388
C. The Cloaca 389
D. The Intestinal Ceca 390
E. Intestinal Digestion 392
F. Motor Functions of the Intestine 396
G. Microbial Digestion 398
VI. The Liver 399
VII. The Pancreas 402
References 405

I. Introduction

The avian digestive system has evolved to a level of performance consistent with the energy requirements of homoiothermy and flight; many of its distinctive features represent adaptations for flight.

The characteristics of the general structural plan of the digestive system are: The presence of a horny bill and the lack of true teeth in the jaw region, the division of the stomach into at least two parts, and, in many forms, the development of a storage diverticulum in the esophagus. Whereas the embryonic origin and differentiation of the digestive tract do not differ qualitatively from that of other vertebrates, there are important temporal differences between the precocial and altricial birds in the development of the digestive system into a completely functional organ system. In the altricial species, the embryonic digestive tract has such a strongly positive allometric growth with respect to general body growth that at hatching it is relatively large and extraordinarily functional; this gives the young bird the capacity to receive and to digest rapidly the large amounts of food supplied by the adults thus permitting a rapid postembryonic development during a brief growth period (Joos, 1952; Portmann, 1935, 1938, 1942; Sutter, 1943).

Corresponding to the diverse adaptive radiation in types of nutrition, the individual parts of the digestive tract vary extensively in form and function (Bujard, 1909; Eber, 1956; Groebbels, 1932b; Magnan, 1910a,b, 1911a–e, 1912a–d, 1913; Nekrasov, 1958; Stresemann, 1934; Ziswiler, 1964, 1965, 1967a,b). Since the digestive system clearly reflects evolutionary trends, it is useful in certain cases for systematic purposes (Naik, 1962; Ziswiler, 1964, 1965, 1967a,b,c, 1968, 1969).

There is an enormous literature, mostly published between 1870 and 1930, concerning the morphology and histology of the digestive system. On the basis of this literature, largely from investigations on the domestic fowl and the domestic pigeon, rests predominantly the available information in text and reference books.

II. The Buccal Cavity, Buccal Glands, and Pharynx

The functions of the buccal cavity and pharynx include the grasping, intake, testing, often some mechanical processing, moistening, and further dispatch of the food into the esophagus. A basic characteristic of birds is the horny beak, which develops as a rhamphotheca, enclosing the upper and lower jaws.

The great spectrum of dietary adaptations of bills, a well-worn subject in treatises of ornithology and evolution, lies beyond the limits of this chapter; the best general references are the treatises of Groebbels (1932b), Mayaud (1950), Mountfort (1964), and Newton (1967). (See also Chapter 4, Volume I.)

The upper "mandible" forms the anterior part of the palate, the so-called horny palate (Fig. 1A,B). This horny palate is especially well developed among the groups of graminivorous birds (Fig. 2). Among

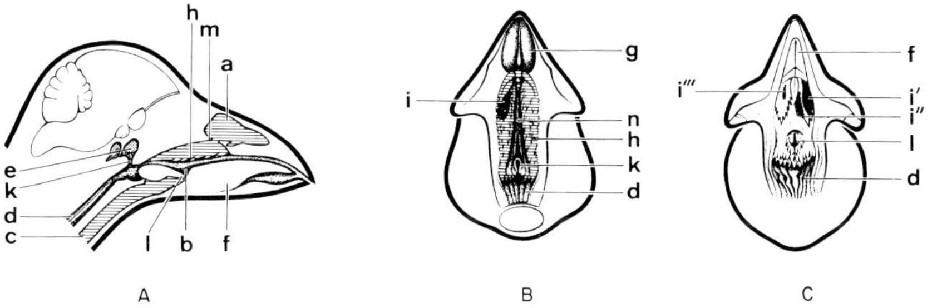

FIG. 1. Pharyngeal and esophageal region of a passerine bird. (A) Sagittal section; (B) viewed from below with lower part cut away; (C) viewed from above with upper part cut away. a, nasal cavity; b, mouth cavity; c, trachea; d, esophagus; e, antrum tubarum; f, tongue; g, horny palate; h, palatal folds; i, Glandula maxillaris; i′, Glandula mandibularis externa; i″, Glandula mandibularis medialis; i‴, Glandula mandibularis interna; k. infundibular slit; l, larynx; m, choanae; n, orbital slit.

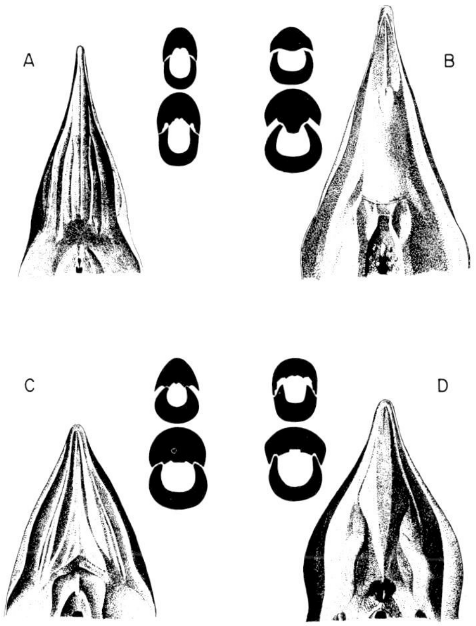

FIG. 2. View of the horny palate of graminivorous passerine birds with accompanying cross sections through the upper and lower bills. (A) *Carduelis carduelis* (Fringillidae), (B) *Emberiza hortulana* (Emberizinae), (C) *Sporophila albogularis* (Emberizinae), (D) *Chloebia gouldiae* (Estrildidae).

the graminivorous passerine species it serves especially as a firm base or holding device in the crushing or cutting of seeds and displays a species characteristic system of longitudinal furrows and longitudinal and cross tuberosities (Ziswiler, 1965). The caudal part of the roof of the mouth is covered by the mucous membrane of the orbital field. It contains a longitudinal groove of the precise configuration of the tongue. In this groove lies the orbital slit, flanked by the two palatine folds, which usually have a series of posteriorly directed papillae (Fig. 1B). The orbital slit lies ventral to the choanae (Fig. 1A) through which the respiratory air stream passes. Opposite this opening, in the region of the floor of the pharynx, lies the slitlike opening to the larynx and trachea, the glottis (Fig. 1A,C).

The mouth region continues posteriorly as the pharynx. In its upper part, immediately posterior to the orbital slit, lies the infundibular slit which connects the pharynx with a sac-shaped chamber, the antrum tubarum, into which the two Eustachian tubes open. The pharynx continues posteriorly into the esophagus.

The membranous floor of the mouth cavity is highly extensible in some groups and serves as a food-storage organ in some species such as in the pelicans and in some passerine species which have, in addition, paired gular sacs, as in *Leucosticte* (A. H. Miller, 1941), *Rhodopechys sanguinea* (Niethammer, 1966a; Kovshar and Nekrasov, 1967), *Pyrrhula pyrrhula* (Nicolai, 1956), and *Pinicola* (N. R. French, 1954).

The mucosa of the buccal and pharyngeal cavities is cutaneoid, invariably with a stratified epithelium (Calhoun, 1954; Greschik, 1917a,b; Schumacher, 1927a,b; Stresemann, 1934; Warner *et al.*, 1967; Zietschmann *et al.*, 1943). In some areas, such as the roof of the buccal cavity and the pharynx, and the surface of the tongue, there is a well-developed stratum corneum. The lamina propria, which is often indistinguishable from the submucosa, contains lymphoid tissue, either diffuse or nodular.

The adaptive variations of the tongues of birds are as manifold as those of the bill (Gardner, 1925; Groebbels, 1932b; Lucas, 1897; Nekrasov, 1961). Among the functional adaptations are long, sticky tongues, as in the woodpeckers (Leiber, 1907; Lucas, 1895; G. Steinbacher, 1934; J. Steinbacher, 1941, 1955, 1957); licking tongues, as in the trichoglossine parrots (G. Steinbacher, 1951), certain honeyeaters (Scharnke, 1931, 1932), and some dicaeids (Rand, 1961); tubular and semitubular sucking tongues, as in the hummingbirds (G. Steinbacher, 1935), sunbirds, and some honeyeaters (Dorst, 1952; Moller, 1931; Scharnke, 1931, 1932); tongues with posteriorly directed horny hooks for holding food tightly, as in the penguins (Bartram, 1901); tongues with horny marginal processes providing a filtering function, as in some of the ducks and geese; tongues adapted for testing special foods like the club-shaped tongues of the parrots; and tongues with a spoonlike tip, as in many graminivorous song birds (Ziswiler, 1965). In some species the tongue is very short and reduced, as in *Struthio*, the Bucerotidae, the Upupidae, and the Pelecaniformes.

Embryologically the tongue develops from two *Anlagen*, the anterior tongue body which develops from the ventral part of the mandibular arch, and the tongue base which arises from parts of the hyoid and first gill arches.

In most birds, the parrots being an exception, there is no intrinsic tongue musculature. The muscles that enter the posterior part of the tongue arise entirely from the hyoid region. The movement of the bird tongue is due very largely to the great mobility of the hyoid apparatus which attains its greatest degree of specialization in the woodpeckers (Leiber, 1907). According to Kallius (1906), birds lack a sensory branch of the trigeminus which ordinarily supplies the tongue. Its function is replaced partially by the strongly developed glossopharyngeal nerve.

The surface of the tongue is covered with an especially strongly cornified epithelium which has its maximum development on the surface of the tongue directed towards the roof of the mouth. This tongue epithelium, according to the kind of specialization involved, may have horny papillae. The tongue papillae correspond in their structure with the other papillae of the mucosa of the mouth cavity and are not homologous to the fungiform papillae of mammals (Zietschmann, 1911). The lamina propria of the mucosa is strongly developed and well vascularized. The body of the tongue itself, in addition to the muscle components already mentioned, contains fat deposits, glands, and the entoglossal bone which articulates posteriorly with the hyoid bone. The entoglossal bone terminates at the tip of the tongue in a hyaline cartilage process. In some groups, such as the ducks and geese, elastic lamellae extend from the entoglossal bone into the submucosa. These are the bases for the fanlike arrangement of cavities in the surface of the tongue.

The distribution of taste buds varies extensively. Generally they are most abundant in the soft part of the palate and around the glottis; they also may occur beneath the tongue (especially in species with round tongues), on the posterior surface of the tongue, and scattered in areas about the salivary glands (Bath, 1906; Botezat, 1904, 1910; Greschik, 1917a,b; Moore and Elliott, 1944; Stresemann, 1934;

Warner *et al.*, 1967; Ziswiler, 1970). Pressoreceptors also occur widely in the buccal cavity and pharynx (Jobert, 1872); Grandry corpuscles have been found on the palate and tongue of ducks and owls; Merkel and Herbst bodies occur widely in the buccal and pharyngeal cavities of birds (A. E. Anderson and Nafstad, 1968; Botezat, 1906; Merkel, 1878, 1880; Szymonowicz, 1897; Stresemann, 1934; Ziswiler and Trnka, 1972). Herbst corpuscles have been described from the tongue (especially in woodpeckers and parrots), and from the palate and beak (especially in ducks and *Scolopax*) (A. E. Anderson and Nafstad, 1968; Botezat, 1906; Clara, 1926; Schildmacher, 1931). Merkel and Herbst bodies play an especially important role in graminivorous birds that remove the hulls of seeds before swallowing. In species that cut seeds open, it is, above all, important to be able to feel the kernel before the seed is completely opened. In this connection, in many species of fringillids, dense receptor fields with Herbst bodies can be detected in the middle of the tongue as well as in the angles of the mouth (Ziswiler, 1970). A. E. Anderson and Nafstad (1968) have demonstrated free nerve endings in the hard palate region of the domestic fowl.

The comparative morphology and microanatomy of the salivary glands have been a subject of interest to investigators for more than a century (Antony, 1920; Batelli and Giacomini, 1883, 1889; Breteau, 1962; Bock, 1961; Cholodowsky, 1892; Foelix, 1970; Giacomini, 1890; Greschik, 1913; Hoelting, 1912; Meckel, 1829; Reichel, 1883; Tichomirov, 1925; Tiedemann, 1810–1814). The variation approaches, in extent, that of the tongue and beak (Groebbels, 1932b). Because of this, and because of differences of opinion concerning homologies, the nomenclature of avian salivary glands is both inconsistent and confusing. The following scheme is based on that of Antony (1920) and Foelix (1970).

1. *Large glands* (The description is based on investigations of passerine species.)
   (a) *Glands of the angle of the jaw (Glandula angularis oris).* There is a compact gland on each side of the choanal slit directly beneath the mucosa of the roof of the mouth at the level of the palate. The opening is through a duct at the margin of the mouth cavity with the horny beak.
   (b) *Mandibular glands*
      Glandula mandibularis externa (Fig. 1C): These are compound glands with two, rarely three, ducts. They lie very close to the dentale at the margin of the floor of the mouth and extend from the angle of the lower jaw to the level of the glottis. Each duct opens entirely anteriorly to the angle of the lower jaw.

Glandula mandibularis medialis (Fig. 1C): This is a group of about ten serially arranged glands lying between the base of the tongue and the glandula mandibularis externa, and extending caudally to the rostral third of the floor of the mouth, and to the level of the glottis as small individual glands. This complex of glands is vertical to the floor of the mouth; the individual ducts course diagonally from caudal to rostral and enter in a linear series of openings through the mucosa of the floor of the mouth.

Glandula mandibularis interna (Fig. 1C): These smaller glands are lacking in many species. They consist of 1–3 individual glands located on both sides of the base of the tongue. The openings of the ducts lie near the midline of the lower jaw in the floor of the mouth cavity.

2. *Small Glands*
   (c) *Lingual glands (glandulae linguales superiores et inferiores)*
   (d) *Sublingual glands (glandulae mandibulares posteriores)*
   (e) *Palatine glands (glandulae palatinae posteriores internae et externae)*
   (f) *Cricoarytenoid glands (glandulae cricoarytenoideae)*
   (g) *Sphenopterygoid glands (glandulae pterygoideae)*

The salivary glands are usually tubular. Individual elements may be simple, branched, or compound. Clearly differentiated duct systems have been demonstrated in the mandibular glands of *Picus viridis* and *Coccothraustes* (Giacomini, 1890) and in numerous graminivorous passerine species (Foelix, 1970). In a few cases the salivary glands are compound tubuloacinar or compound acinar (Batelli and Giacomini, 1883).

The adaptation of salivary glands to type of food has been treated extensively by Antony (1920). In general, species that take relatively slippery aquatic food have poorly developed salivary glands. For example, she found no salivary glands in *Anhinga anhinga* and only inferior lingual glands in *Ardea cinerea*. On the other hand, ducks and geese, whose food has little natural lubrication, have a full complement of salivary glands. The hawks and owls studied by Antony had no cricoarytenoid glands. In wookpeckers there is a large distinctive gland (glandula picorum of Antony) whose ducts open beneath the tongue; it is differentiated into two sections, the posterior part producing the extremely sticky fluid that occurs characteristically on the tongue of woodpeckers (Antony, 1920; Greschik, 1913; Meckel, 1829; Tiedemann, 1810–1814). Graminivorous passerine species have highly differentiated salivary glands (Antony, 1920; Foelix, 1970). Among the Fringillidae, Estrildidae, and Ploceidae, some food specialists have extraordinarily developed glandular systems that are extensively sacular with a consequent enlargement of the area of secretory epithelium as well as extensive branching of the duct system.

Despite the usual absence of serous elements in avian salivary glands, the occurrence of amylolytic activity has been reported in salivary extracts or in the saliva of turkeys (Fedorovskii, 1951), geese (Nikulina, 1919), domestic fowl (Leasure and Link, 1940; Ponirovskii, 1915; Shaw, 1913), and pigeons (Ponirovskii, 1915).

The primary function of avian salivary glands is mucogenesis and the primary function of the saliva is that of lubrication. In conjunction with this it should be noted that most cytologic and histologic investigations have revealed only mucous cells (Calhoun, 1954; Chodnik, 1948; Greschik, 1913; Heidrich, 1908; McCallion and Aitken, 1953; Pilliet, 1893) (Fig. 6a). The only known exceptions to this are the seromucous and serous cells described for some fringillid species by Antony (1920) and Foelix (1970), and by Warner *et al.* (1967) for the Japanese Quail. The serous elements have been investigated carefully by Foelix (1970) and found to occur always in the duct system and never in the terminal parts of the glands.

In contrast to the experimental results cited above, other investigators have denied the existence of amylase in the salivary glands or saliva of pigeons (Alekseev, 1913) and chickens (Schwarz and Teller, 1924). Jung and Pierre (1933) concluded that salivary amylase in the domestic fowl, if it exists at all, is of negligible importance. Furthermore, in the serous epithelium of the duct system of certain salivary glands, investigated by Foelix (1970), no clear indication of amylase secretion could be detected. If an amylase is present in the saliva, it could be of functional importance only in the crop. The question of the digestion of starch in the crop is discussed below.

Among the secretory cells in the duct systems of the salivary glands in *Fringilla*, *Pyrrhula*, and *Euplectes*, Foelix (1970), in agreement with Tucker (1958), was able to demonstrate cells with holocrine secretion, in addition to cells with apocrine secretion.

Secretory activity of salivary glands increases during feeding (Chodnik, 1948; Nikulina, 1919) and decreases, with accumulation of granules in the cells, during starvation (Antony, 1920; Bernstein, 1859).

Among the swifts, the sublingual salivary glands may be specialized to produce an adhesive material that is used in the construction of the nest (Antony, 1920; Batelli, 1890a; Bernstein, 1859; Johnston, 1958; Lack, 1956; Marshall and Folley, 1956; Medway, 1962, 1969). Although Lack (1956) regarded this adaptation as a common characteristic of the Apodidae, at least the Black Swift (*Cypseloides niger*) appears to be an exception (Johnston, 1961). These cement-producing glands enlarge seasonally in both sexes. In *Chaetura pelagica* the enlargement is approximately twelvefold (Johnston, 1958); in *Collocalia maxima* it appears to be of the order of fiftyfold (Medway, 1962). The adhesive material is a glycoprotein resembling mucin (Medway, 1969). Although the cycle has been suggested to be hormonally controlled (Barnawall, 1968; Johnston, 1958; Medway, 1962), acceptable experimental evidence is still lacking.

The jays of the genus *Perisoreus* have extremely specialized mandibular glands (Bock, 1961). These enlarged glands produce a sticky secretion similar to that of the woodpeckers. With the help of the tongue and this secretion very compact food balls can be formed; these can be attached to twigs or deposited in other places as a device for storing food (Dow, 1965).

The process of deglutition in birds is not completely understood. In general, food is propelled to the rear of the pharynx by raising the head, or by a quick forward thrust of the head, or by both together. There is a simultaneous reflex closure of the choanal slit (including the infundibular slit) and a forward movement of the larynx and trachea so that the glottis is closed off by coming in contact with the base of the tongue (Mangold, 1950). Presumably, the food is then caught up by peristaltic activity and moved thereby into and through the esophagus.

The intake of water is somewhat different. In most birds the fluid is allowed to flow passively into the mouth cavity; the mouth is then closed and the head raised, causing the fluid to flow by gravity to the esophagus. Presumably this involves reflex closure of the choanal slit and the glottis as described above; others, including pigeons, hummingbirds (G. Steinbacher, 1935), and some estrildids (Immelmann, 1964) can drink actively without raising the head.

Little is known concerning the mechanisms of manipulation of food in the bill and mouth-cavity region. In most species food items are swallowed entire. However, mechanical reduction of food by the bill occurs in those graminivorous that crush large seeds on the cross-furrowed, horny roof of the mouth, and in some Phytotomidae which, with the toothed edges of their bills, not only can cut off parts of plants, but can also effect chewing movements (Küchler, 1936). A simple form of mechanical reduction of food is the crushing of berries by many frugivores, as in *Pyrrhula*. A special type of food handling has developed in some families of graminivorous oscinine species, and in some parrots, in the removal of the husks from the seeds before swallowing them.

All members of the Emberizinae, Cardinalinae, Ploceidae, and most of the Estrildidae open the seed hull by crushing. In doing this the seed is pressed by one or both margins of the bill against a definite abutment in the horny roof of the mouth and the kernel is thereby pressed out of the shell (Fig. 3A). This method works best in opening of elongated seeds. The members of the family Fringillidae (including the Carduelinae), together with members of the genus *Erythrura* of the Estrildidae, cut up seeds (Fig. 3B). To effect this

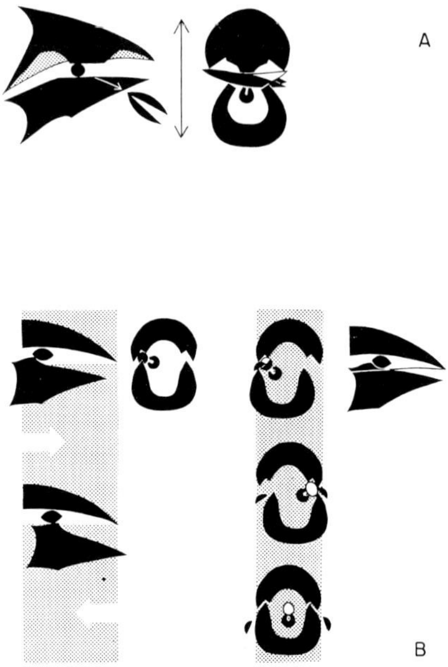

FIG. 3. Schematic demonstration of seed opening by a passerine bird. (A) Pressing out a kernel (*Emberiza*), (B) shearing of a kernel (*Carduelis*).

the seed is lodged in special furrows of the horny palate and is cut by rapid forward and backward movements of the sharp margin of the mandible. This process represents an adaptation for the opening of compact dicotyledenous seeds (Ziswiler, 1964, 1965, 1967a,b, 1969).

III. The Esophagus and Crop

The basic general function of the esophagus is that of passing material from the pharynx to the stomach, and the reverse in species that regurgitate pellets or food for young. Associated with this is the production of a lubricating fluid by the esophageal glands. In addition, there may be an important storage function, effected simply by a temporary expansion of the esophagus or by a more specialized and permanent diverticulum or expansion of the esophagus, the *crop*. Other functions include the production of milk by the crops of doves and pigeons, a role in resonance and display by esophageal diverticula in a few species, and a possible role in chemical digestion in galliform, columbiform, and other species.

Compared generally with vertebrates, the esophagus of birds is relatively long and of relatively large caliber, the latter varying, however, with the mode of nutrition. Usually the esophagus lies to the right of the midline of the neck and, unstretched, is shorter than the cervical section of the vertebral column (Böker, 1931; Niethammer, 1933). The section of the esophagus between the crop and head is designated as the pars cervicalis and that between the crop and the stomach as the pars thoracica.

The comparative morphology and microanatomy of the esophagus and crop have been the subject of numerous investigations, including those of Allenspach (1964, 1966), Allenspach and Hamilton (1962), Bartels (1895), Ivey and Edgar (1952), Kaden (1936), Malewitz and Calhoun (1958), Masahito and Fuji (1959), Niethammer (1933, 1937, 1961, 1966a), Postma (1887), Schreiner (1900), and Ziswiler (1967a,b). In general the esophagus has a wider diameter and a greater degree of folding in species whose food consists of bulky items and also in species in which the esophagus as a whole has a storage function. Thus the diameter is especially great among the grebes, loons, auks, puffins, petrels, gulls, cormorants, storks, herons, coots, gallinules, hawks, owls, and the piscivorous kingfishers (Gadow, 1891; Meckel, 1829; Swenander, 1899, 1902). It tends to be small in birds whose food consists of small items (many insectivores and graminivorous) or which reduce the food to small pieces (many of the parrots). Extensive folding occurs in the esophagus of penguins (Bartram, 1901), auks, petrels, gulls, gannets, pelicans, ducks, geese, grouse, hawks, owls, goatsuckers, crows, and waxwings (Swenander, 1899, 1902).

In its histologic structure, the esophagus of birds, which has the same general organization as the other vertebrates, shows the following characteristics.

1. A very slightly developed external longitudinal muscle layer (stratum longitudinale of the tunica muscularis).
2. A powerful circular muscle layer (stratum circulare of the tunica muscularis).
3. A very weakly developed tela submucosa.
4. A clearly established, more or less strongly developed inter-

longitudinal muscle layer (lamina muscularis mucosae) (Bartels, 1895; Kaden, 1936; A. Oppel, 1895; Schreiner, 1900; Warner *et al.*, 1967; Ziswiler, 1967a). The tunica muscularis, like the muscularis mucosae, contains connective tissue fibers, blood vessels, and numerous lymph follicles; the latter are specially numerous in the vicinity of the glands (Calhoun, 1954; Schreiner, 1900; Schumacher, 1927c; Warner *et al.*, 1967).

5. A very thick epithelial layer (lamina epithelialis mucosae).

The uppermost cells of this layer can be cornified into a stratum corneum. This cornification is especially well developed in pigeons and ducks (Kaden, 1936; Zietschmann, 1911). The layer next beneath consists of flattened pavement epithelium which lies on a mostly cuboidal stratum germinativum. The thickness of the epithelium varies enormously from as little as 52 μ in *Cacatua sulphurea* to as much as 570 μ in *Picus viridis* (Bartels, 1895). The thickness of the individual connective tissue parts is dependent on location. In graminivorous passerine species the lamina epithelialis is, as a rule, thickest in the crop region; the extent of cornification is greatest in the pars cervicalis. The esophageal glands, in contrast to the situation in mammals, lie in the tunica mucosa. The cellular lining of the terminal portions of the glands, and often also those of the ducts, is of a mucous type (Fig. 6b). The form of the glands is tubular (Fig. 6b) to alveolar (Fig. 6c). Their secretion is probably merocrine (Niethammer, 1933; Ziswiler, 1967a) and not holocrine as Kaden (1936) asserted. The terminal portions are simple or compound and can lie entirely in the lamina propria. If they occur in the lamina epithelialis, as, for example, in swallows, they are then surrounded by a fine net of connective tissue that is derived from the lamina propria. In position and structure of the esophageal glands there are very great differences among different species and groups (Kaden, 1936); no clear correlation can be made either between structure and position, on one hand, or between structure and type of nutrition on the other. Also the structure and density of glands may differ in the various sections of the esophagus. Ziswiler (1967a), in three families of graminivorous passerine species, distinguished the following types of esophageal glands.

1. Long tubular glands with a uniformly polygonal secretory epithelium that covers the fundus as well as the ducts of the gland: *Emberiza*.

2. Flask-shaped tubulo-alveolar glands in the pars cervicalis and the crop region; tubular glands in the pars thoracica. The fundus of

the gland is lined with columnar secretory epithelium, the lumen with polygonal secretory epithelium: Cardinalinae, *Dinemellia*.

3. Alveolar glands in the pars cervicalis and the crop region; tubular glands in the pars thoracica. The fundus of the gland is lined with columnar secretory epithelium, the secretory duct with polygonal secretory epithelium: Ploceidae.

3a. Like 3, but with compound glands in the pars cervicalis and the crop region: *Passer*.

4. Alveolar glands throughout the entire esophagus. Columnar secretory epithelium lines the fundus of the gland, nonsecretory pavement epithelium the ducts: Fringilladae.

4a. Like 4, but with compound glands: *Pyrrhula*.

The esophagus of many, perhaps most, species of birds has a storage function. This may be effected without a special differentiation of any part of the esophagus, as in the grebes, penguins, petrels, gulls, ducks and geese, owls, woodpeckers, Pelicaniformes, and many passerine species (Groebbels, 1932b; Swenander, 1899, 1902). A clearly differentiated expansible portion of the esophagus in which food is stored is a crop.

The outer form of the crop can be spindle-shaped (Fig. 4a) as in *Casuarius, Uria, Fratercula, Phalacrocorax, Anas, Ciconia, Philomachus, Tringa, Numenius, Scolopax, Haematopus, Otis, Anhima*, and also in Trochilidae, Fringillidae, Emberizidae, Paradiseidae, and Estrildidae.

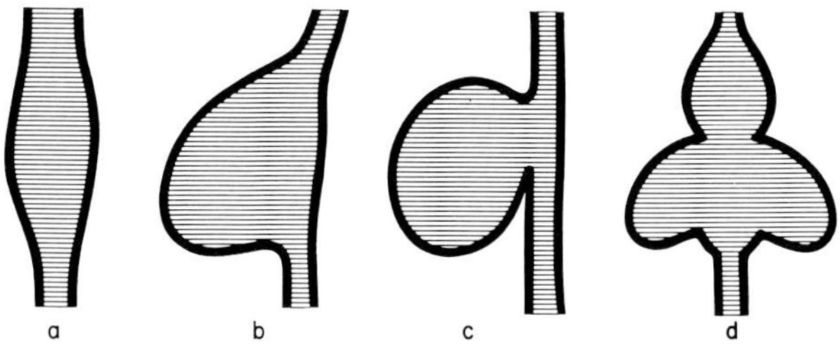

FIG. 4. Various forms of crops. (a) spindle-shaped crop (*Phalacrocorax*), (b) unilateral out-pocketing (*Gyps*), (c) unilateral sac (*Pavo*), (d) double sac (*Columba*).

The maximally developed storage crops are saclike diverticula (Fig. 4b) which in the filled condition protrude outwardly and can store a large quantity of food. In the Ploceidae, Falconiformes, and Thinocoridae (Hanke and Niethammer, 1955), for example, the sac is unilobular; among the galliform and columbiform species it is bilobed.

Of interest is the relative position of the full crop. In hawks and parrots it extends ventrally whereas in many passerine species it protrudes laterally or dorsally. In galliform species and in the Thinocori (Hanke and Niethammer, 1955) the full crop rests on the furcula. In two species of hummingbirds, the crop is a sinistrally lateral expansion of the esophagus in the S-shaped bend of the vertebral column and, when filled, lies dorsal to it (Niethammer, 1933). Also in the nestlings of the Carduelinae and Estrildidae the full crop may extend dorsally over the vertebral column (Eber, 1956; Farner, 1960; Ziswiler, 1967a,b).

As an additional function the crop of some species serves as a storage chamber for food for the young; in feeding the young the food is regurgitated by a special reverse peristalsis. Such a storage function occurs in the esophagi or crops, for example, of hawks, storks, penguins, Pelicaniformes, pigeons, parrots, some Estrildidae, and some Fringillidae (Fisher and Dater, 1967). Since the process of food storage for young can differ widely among groups, the structure of this type of crop is very diverse. Its structure and function in the Fringillidae and Estrildidae have been investigated in detail by Ziswiler (1967a). Thus in the Estrildidae, in which the process of feeding of the stored material involves a prolonged pumping action with the bills of the old and young bird side by side, the presence of an antagonistically working crop musculature can be detected. In the Estrildidae not only is the circular muscle layer of the muscularis propria, necessary for the contraction of the crop, well developed, but also there is a powerful interlongitudinal muscle layer of the muscularis mucosae as the antagonist. In the Fringillidae and Estrildidae there are no suggestions whatsoever of differences in histological structure or in secretory activity between those that store food for young and those that do not.

A striking modification of the esophagus is seen in the Opisthocomidae (Pernkopf, 1937). Here the crop assumes the function of the stomach. The principal enlargement of the esophagus, which is abundantly supplied with glands and with a powerful musculature, is followed by further croplike diverticula, also with gastric functions; the true gastric apparatus is only weakly developed and obviously of secondary importance. In *Opisthocomus* the esophageal region is about 50 times as heavy as the stomach region. A similar relationship occurs in the Owl Parrot *Strigops habroptilus*.

Perhaps the most remarkable crops are those of the doves and pigeons (Fig. 4d). In both sexes this highly developed crop produces a nutritive fluid, the "crop milk." This function is discussed in greater detail below.

As a storage organ the crop functions in such a manner as to supply food to the gastric apparatus as constantly as possible. Thus when a fasted chicken is fed, the first food taken is swept by peristalsis through the esophagus directly into the gizzard (Ashcraft, 1930; Henry *et al.*, 1933; Ihnen, 1927, 1928; Vonk and Postma, 1949) while the opening to the crop is closed (Ihnen, 1927, 1928). The rate of progression of the peristaltic wave is of the order of several centimeters per second (Vonk and Postma, 1949). After the first food is swept directly into the stomach, the crop sphincter relaxes (Ashcraft, 1930; Ihnen, 1927, 1928) so that further food taken in is then stored largely therein (Ashcraft, 1930; Ihnen, 1927, 1928; Vonk and Postma, 1949). Thereafter food is moved to the stomach at irregular intervals, apparently controlled reflexively by the degree of fullness of the remainder of the digestive tract (Ihnen, 1927, 1928). The time required for a bolus to pass from the crop to the gizzard is on the order of 5–30 seconds (Henry *et al.*, 1933). Bearing in mind this reflexive control of emptying, which is dependent on the state of other organs of the digestive tract, the variation in time (from a few minutes to a day) that food spends in the crop (Heuser, 1945; Ihnen, 1928; Schwarz and Teller, 1924) is understandable. In *Accipiter gentilis*, which takes food in large pieces, the crop fills first and is initially immobile. Peristaltic movement of the crop develops within a few minutes and begins to push food into the stomach; a portion of the food may remain in the crop as long as 24 hours after feeding (Dedič, 1930).

It has long been known that vagal fibers are involved in the normal movements of the esophagus and crop (Bernard, 1858; Chauveau, 1862; Doyon, 1894b; Everett, 1966; Hanzlik and Butt, 1928; Ihnen, 1928; Nolf, 1925c; Rossi, 1904a,b; Zander, 1879); also involved are sympathetic efferent fibers and the myenteric nervous systems of these organs (Hanzlik and Butt, 1928; Ihnen, 1928). Apparently peristalsis requires efferent vagal impulses and an intact functional myenteric nervous system (Hanzlik and Butt, 1928; Ihnen, 1928; Nolf, 1925d). The crop sphincter in pigeons and chickens is apparently also controlled by the vagus (Ihnen, 1928).

The crops of pigeons and chickens have long been known to exhibit more or less rhythmic contractions (Brown-Séquard, 1850; Doyon, 1894a; Kratinoff, 1929; Lieberfarb, 1927; Patterson, 1927, 1933; Rogers, 1916; Rossi, 1904a,b; Winokurow, 1925); these are stronger and more regular when the crop is empty (Patterson, 1927, 1933; Rogers, 1916) and are subject to cerebral inhibition (Patterson, 1927, 1933). The period of contraction is of the order of 10–20 seconds for the empty crop of the pigeon (Patterson, 1927, 1933) and 50–60 seconds in the chicken (Lieberfarb, 1927); feeding causes a cessation of crop movements for a period of a half hour or more, after which they are irregular (Lieberfarb, 1927; Patterson, 1927, 1933). These rhythmic crop movements also apparently require an intact vagus and a functional myenteric nervous system (Hanzlik and Butt, 1928; Ihnen, 1927, 1928; Nolf, 1925c; Rossi, 1904a,b).

The results of more recent investigations (Bowman and Everett, 1964; Hassan, 1968, 1969; Ohashi and Ohga, 1967) can best be interpreted as demonstrating that the vagi contain both pre- and postsynaptic cholinergic excitatory fibers to the esophagus although the ratio of sympathetic to parasympathetic fibers is not clear (Bennett, 1970). The esophagus and crop receive few adrenergic fibers; those present are associated with the myenteric plexus rather than with the musculature. (For further details, see discussion of the autonomic nervous system in Volume IV.)

The role of the crop as a digestive organ in the domestic fowl has long been controversial. It is a chamber in which storage, softening, and swelling of food occurs (Ihnen, 1927, 1928; Mangold, 1934a, 1950; Schwarz and Teller, 1924). However, its role as an organ of chemical digestion has been extensively debated. The mucosa has been reported to contain amylase (Alekseev, 1913; Bernardi, 1926; Farner, 1941, 1942a; Fedorovskii, 1951; Hewitt and Schelkopf, 1955; Plimmer and Rosedale, 1922), and amylolytic activity has been reported within the lumen (Shaw, 1913). Although it is possible that this amylase may come from the salivary glands (Leasure and Link, 1940; Ponirovskii, 1915; Shaw, 1913), it appears more probable that it comes from the food or is microbial in origin (Farner, 1941; Ivorec-Szylit *et al.*, 1965; Mangold, 1934a; Schwarz and Teller, 1924). It should be noted that the hydrogen-ion concentration within the crop, ca. pH 4.5–6.7 (Buckner *et al.*, 1944; Farner, 1942b; Heller and Penquite, 1936; Herpol and van Grembergen, 1961; Herpol, 1966b; Hewitt and Schelkopf, 1955; Kerr and Common, 1935; Mayhew, 1935; Vonk *et al.*, 1946), is favorable for activity of plant and microbial amylases. The most conclusive experiments are those of Ivorec-Szylit *et al.* (1965) which lead to the conclusion that starch digestion therein is largely microbial and that saliva and crop secretion have little or no role; lactic acid is produced in significant quantities (Ivorec-Szylit *et al.*, 1965; Mangold, 1934a). Bernardi (1926) has suggested that some hydrolysis of sucrose may also occur. The acid proteinase reported from the crop is almost certainly of gastric origin. Certainly the role of the crop as an organ of chemical digestions in this species must be relatively inappreciable (see Mangold, 1927a, 1934a). We suspect that the same is true for the domestic turkey, although Fedorovsky (1951) has suggested that salivary amylase has a role in digestion of starch in the crop.

In the domestic pigeon it may be more probable that the crop has some role as an organ of chemical digestion. The mucosa has been reported to contain an amylase (Alekseev, 1913; Dulzetto, 1927a,b, 1928a, 1930; Iljin, 1913) and an invertase (Dulzetto, 1927a, 1928a,b); these enzymes also occur in the crop fluid (Dulzetto, 1929, 1930, 1938). The glands occur throughout the crop and, unlike the crop glands of other birds, contain both serous and mucous elements (Dulzetto, 1928a, 1932b, 1938). The proteinases reported from the crop (Teichmann, 1889) are either intracellular or are regurgitated from the proventriculus (Dulzetto, 1928b, 1938; see also Herpol, 1967b). Contrary to the opinion of Mangold (1925), Dulzetto regarded the role of the crop in chemical digestion to be of sufficient importance to designate the crop glands as parasalivary glands. The hydrogen-ion concentration (ca. pH 6.0) in the crop would be reasonably favorable for such activity (Herpol, 1967a). It must be borne in mind, however, that any chemical digestion in the crop would precede the important physical digestive processes of the gizzard; this almost certainly restricts chemical digestion in the crop to a relatively minor role.

A spectacular adaptation of the avian crop is the production of milk in doves and pigeons, a phenomenon described in considerable detail as early as 1786 by Hunter, and well known in the early physiologic literature (Bernard, 1859; Charbonnel-Sallé and Phisalix, 1886; Hasse, 1865; Milne-Edwards, 1860; Arcangeli, 1904). As observed in the domestic pigeon (Beams and Meyer, 1931; Litwer, 1926; Niethammer, 1931, 1933; Teichmann, 1889; Weber, 1962), the milk is produced by a desquamation of fat-laden cells from the proliferated squamous epithelium of the crop. In the domestic pigeon the proliferation of the crop mucosa begins about the eighth day of incubation in both sexes, with milk "secretion" beginning at the fourteenth to eighteenth day and lasting until about 16 days after hatching (Beams and Meyer, 1931; Champy and Colle, 1919; Litwer, 1926; Niethammer, 1931, 1933). The desquamating superficial "nutritive" layer is distinct from the deeper proliferating layer of the epithelium (Beams and Mayer, 1931; Litwer, 1926; Weber, 1962). Crop milk is rich in desquamated cells and their fragments and usually contains also small food fragments (Beams and Meyer, 1931; Bernard, 1859; Dulzetto, 1932a; Niethammer, 1931; Teichmann, 1889). It is rich in lipids and protein, but contains no sugar (Bernard, 1859; Carr and James, 1931; Davies, 1939; Dulzetto and di Volsi, 1934; Dulzetto, 1938; Vandeputte-Poma, 1968). (See Chapter 7 on nutrition for further details on nutritive properties of pigeon milk.) The proliferation of the crop epithelium and the formation of milk is stimulated by prolactin and is sufficiently sensitive and uniform therein to serve as a basis for an assay of prolactin (see Riddle, 1963; also Nicoll, 1969, for review).

Greater Flamingos (*Phoenicopterus ruber*) produce, by merocrine secretion, an esophageal nutritive juice that is fed to the young (Lang, 1963; Lang *et al.*, 1962; Studer-Thiersch, 1967; Wackernagel, 1964). Male Emperor Penguins (*Aptenodytes forsteri*) which incubate the single egg and care for the chick initially, feed the chick a nutritive fluid also produced in the esophagus (Prévost, 1961; Prévost and Vilter, 1963). The nutritional properties of these fluids are also discussed in Chapter 7.

In the males of a number of species, inflatable esophageal diverticula or enlargements have a conspicuous role in courtship as resonating chambers and/or as integral parts of display devices. Such diverticula are known to occur, for example, in several species of grouse (Clarke *et al.*, 1942; Gross, 1926, 1930; Honess and Allred, 1942; Lehmann, 1941), in several species of pigeons (Stresemann, 1934), in *Cephalopterus ornatus* (Sick, 1954), in *Struthio camelus* (Stresemann, 1934), in the female of *Rostratula benghalensis* (Niethammer, 1966a), in the female of *Turnix sylvatica* (Niethammer, 1961), and in *Eupodotis australis* (Garrod, 1874a,b). With respect to the last, it should be noted that in another species of bustard, *Otis tarda*, there is a gular pouch that extends into the neck and serves the same function (Garrod, 1874a,b; Murie, 1868, 1869; Niethammer, 1937). In the Sage Grouse, *Centrocercus urophasianus*, according to Clarke *et al.* (1942),

> the enlargement in the esophagus occurs as a result of the transmission of air from the lungs and air sacs through the glottis into the pharynx and then into the esophagus. This is accomplished with the aid of two fleshy pads, one on either side of the glottis, which are connected anteriorly and definitely separated posteriorly, thus forming a groove which extends into the esophagus. Upon elevation of the base of the tongue, these pads are brought into direct contact with similar but less well defined pads in the roof of the mouth. As a result, the pharynx is occluded anteriorly into the mouth and opened posteriorly into the esophagus. Accompanying the movement of air into the esophagus, there is apparently a contraction of the involuntary circular muscles anterior to the crop, thus trapping the air in the esophagus. A swallowing type action of pharyngeal and tongue muscles directs the expired air into the esophagus.

IV. The Gastric Apparatus

Interest in the structure and function of the gastric apparatus arose in the very early days of scientific biological investigation and played a prominent role in the development of a general understanding of the nature of digestive processes in animals (Home, 1810, 1812, 1814–1828; de Réaumur, 1756a,b,c; Spallanzani, 1783). Usually the gastric apparatus consists of two relatively distinct chambers, an anterior glandular stomach or proventriculus and a posterior muscular stomach, known also as the gizzard (Fig. 5). The former functions principally in secretion of the gastric juice and as a passageway from the esoph-

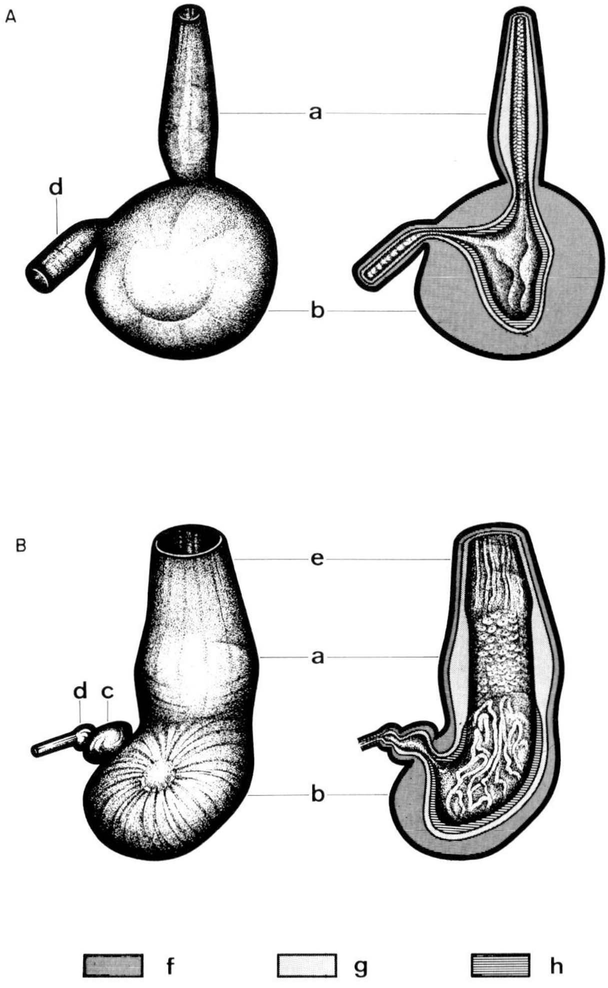

FIG. 5A,B. The parts of the foregut: left, external view; right, sagittal section. (A) A graminivorous passerine species (*Carduelis*), (B) a heron (*Ardea*). a, Glandular stomach or proventriculus; b, muscular stomach or gizzard; c, pyloric chamber; d, pylorus; e, esophagus; f, muscular layer; g, glandular layer, h, koilin layer.

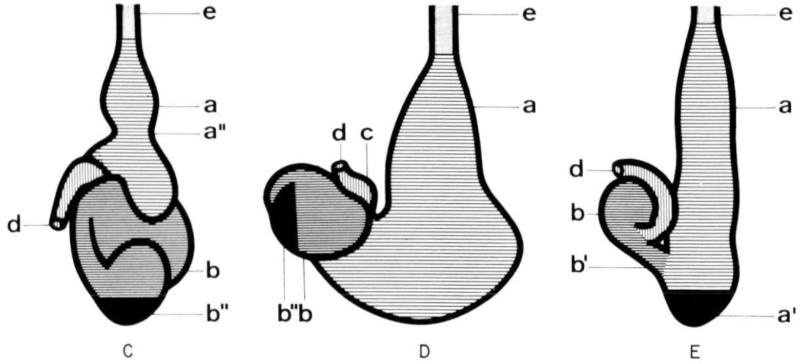

FIG. 5C–E. Schematic representation of the parts of the stomach in (C) *Pavo cristatus*, (D) *Struthio camelus*, (E) *Procellaria parkinsoni* (modified from Pernkopf, 1937). a, Glandular stomach; a′, isthmus; a″, diverticulum of the stomach; b, muscular stomach; b′, intermediate zone; b″, diverticulum; c, pyloric stomach; d, pylorus.

agus to the muscular stomach; it may also function as a storage organ and as an organ of acid proteolysis. The muscular stomach functions as an organ of mechanical digestion in many species; to this end it is often the functional analog of the teeth. It also serves as a chamber for acid proteolysis and, in many species, also as a storage organ. In some groups, such as herons, storks, cormorants, and snakebirds, there is an additional chamber, the pyloric stomach, between the muscular stomach and intestine.

The functional and morphological differentiation of the stomach has the greatest range of variation of all internal organs of birds.

With respect to the homologies of the individual parts of the stomach with corresponding parts of the stomach of other vertebrate animals, there is voluminous literature, in part with very contradictory opinions. Solely on the basis of comparisons of muscle structure Pernkopf (1930) presented a convincing homology of the muscular stomach with the part of the reptilian stomach lying just forward from the pyloric section.

A. The Glandular Stomach

The glandular stomach of birds is spindle- or cone-shaped (Fig. 5). The transition from the esophagus is usually gradual whereas the transition to the gizzard involves a distinct intermediate section or isthmus. The size of the glandular stomach is significantly correlated with the volume of food intake. Thus the lumen is especially large in piscivorous or frugivorous species; also, however, the graminivorous oscinine species that eat large seeds have glandular stomachs with large lumina (Ziswiler, 1967a). Cymborowski (1968) has demonstrated in *Sterna hirundo* that the histologic form of the proventriculus is dependent on the type of food.

With respect to the histology of the glandular stomach, reference may be made to the extensive accounts of Batt (1924), Cazin (1887), Clara (1934), G. H. French (1898), Groebbels (1932b), Luppa (1962), Plenk (1932), Postma (1887), Schreiner (1900), Wilczewsky (1870), Swenander (1899, 1902), Toner (1963), and Ziswiler (1967a,b).

The tunica mucosa of the glandular stomach is characterized by an abundance of glands of two principal types, the centrally located tubular glands and the peripheral compound glands.

The tubular glands are mostly simple, infrequently compound. The courses may be straight or tortuous; their epithelial cells are high cuboidal to cylindrical, often in multiple layers, with round or highly elliptical nuclei. Their secretion is mucous and merocrine. There is a high density of goblet cells. Clara (1934) was able to demonstrate oxyphilic granules in the epithelial cells of thrushes. The epithelial cells of the tubular glands penetrate to a greater or lesser extent into the duct system of the compound glands as they follow the folds of the inner lumen of the glandular stomach.

The dominant elements of the glandular stomach are the compound glands, which have been known for more than a century (Berlin, 1853). The form of these compound glands is highly variable; according to the group involved, they may be cylindrical, spherical, or ellipsoidal. The compound glands penetrate into the muscularis mucosae, separating it into inner and outer layers (A. Oppel, 1895; Plenk, 1932; Schreiner, 1900; Swenander, 1899, 1902).

The ducts from the compound glands open in papillae in the lumen of the proventriculus. These papillae are regularly arranged in group-typical patterns. Just as diverse can be the arrangement of the glandular bodies of the compound glands in the proventriculus. They may be more or less uniformly distributed throughout the Gruiformes and Podicipediformes, as well as in the domestic pigeon, *Apus apus, Caprimulgus europaeus, Picoides tridactylus, Ardea cinerea,* and many passeriform species. They may be in more or less discrete areas or juga as in several species of ducks, *Nyctea scandiaca, Falco columbarius, Grus grus* (Swenander, 1902), *Eudyptes crestatus* (Bartram, 1901), *Struthio camelus, Rhea americana, Anhinga melanogaster* (Cazin, 1886; Forbes, 1882), *Macrocephalon maleo* (Garrod, 1878b), *Buteo buteo, Buteo lagopus,* and *Haliaeetus albicilla* (Giebel, 1957). In some species they occur in a relatively restricted band as in *Jynx torquilla* (Swenander, 1902) and *Dryocopus martius* (Swenander, 1902). In *Anhinga anhinga*, the compound glands occur in a separate diverticulum (Garrod, 1876).

In the passerine species the compound glands are arranged in one, one and a half, or two layers (Fig. 6d). According to the arrangement and form of the glands, the lumen of the proventriculus, which in cross section is bounded by the above-mentioned band of tubular glands, appears more or less folded (Ziswiler, 1967a,b). In the flower-seeking parrots the compound glands are arranged in longitudinal fields between which there are extensive gland-free zones. Through this arrangement the glandular stomach is extensible (G. Steinbacher, 1934).

In a compound gland it is always possible to differentiate clearly between a terminal portion and the duct system. The duct system is partly or entirely lined with columnar or cuboidal epithelial cells arranged in one or more layers. The cells of the duct system are not appreciably different from the secretory epithelium of the tubular glands. The cells are mucous and the presence of goblet cells, especially in the central lumenward-oriented sections of the duct system, suggests an intensive secretory activity (Ziswiler, 1967a).

The terminal system consists of polygonal, granular cells with spherical nuclei. They are comparable morphologically with the parietal cells of mammals. Clara (1934), Groebbels (1924, 1932b), and Toner (1963), on the basis of tinctorial differences, were able to identify two types of cells which they considered to be identical with the chief and parietal cells of mammals. In the graminivorous passerine birds, however, it is not possible to detect these two types (Ziswiler, 1967a). By electron microscopy Selander (1963) has described the acidophilic cells of the proventriculus of the domestic fowl. The prominent ultrastructures are numerous, large, and dense mitochondria, a striking ergastoplasma, many vesicles and tubules with smooth surfaces in the apical cytoplasm, and zymogen granules that are often associated with the well-developed Golgi apparatus.

In ultrastructure these cells appear equivalent to the parietal and chief cells of mammals and to the acidophilic cells of the amphibians.

The polygonal cells are arranged regularly in two or more tiers or irregularly along connective tissue strands with fibrocytes and blood vessels from the lamina propria. In the multilayer arrangement intercellular secretion tubules are clearly recognizable.

The terminal system, like the duct system, can be variously organized.

The duct system in the simplest case consists of a principal chamber into which the individual lobules of the terminal system open. From

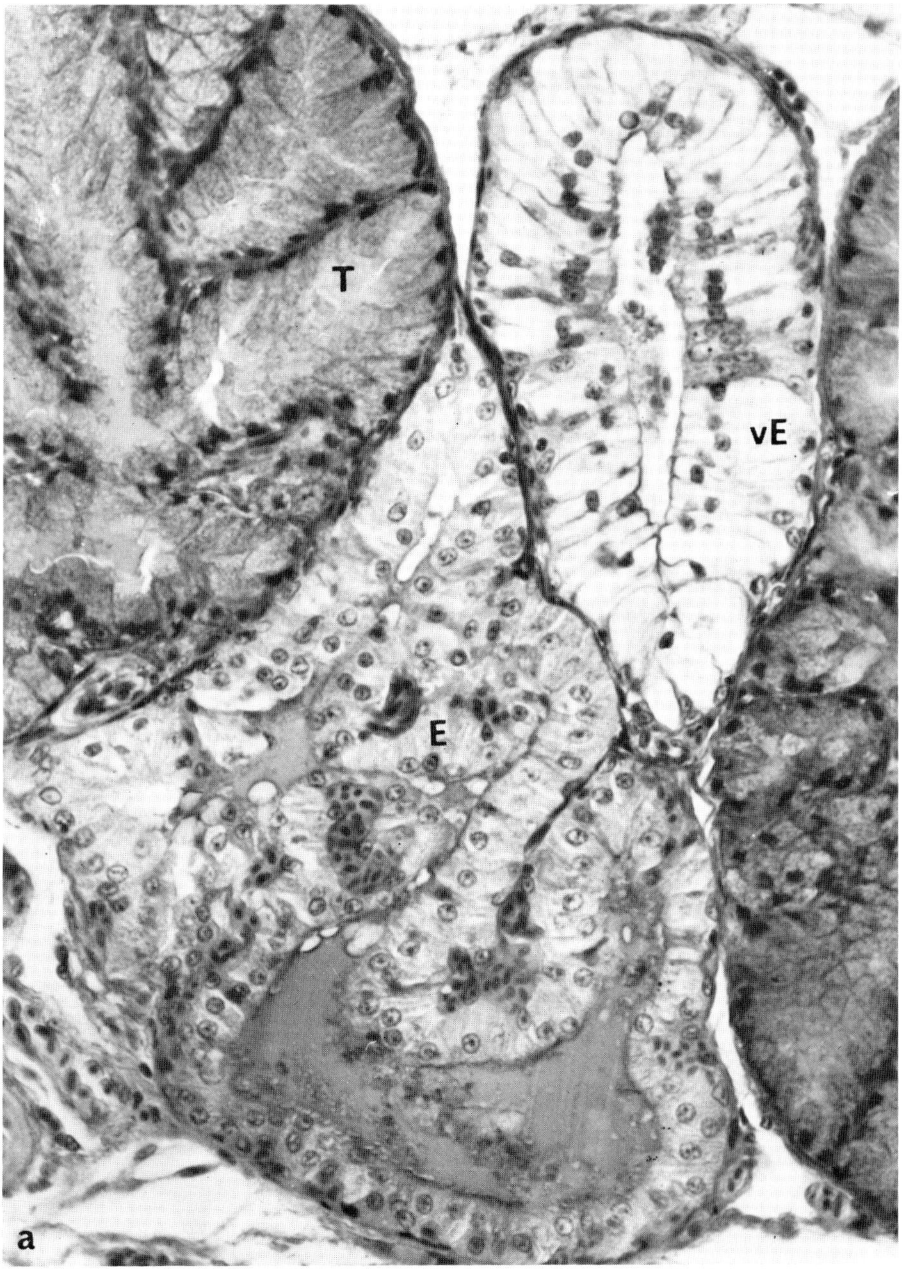

FIG. 6a. Histology of the digestive system. Salivary gland. Glandula mandibularis externa of a passerine species (*Euplectes gierowii*). A lobule with mucigenic tubule (T), acidophils, epithelium of a glandular duct (E), vacuolated epithelium of glandular duct (vE). Alcian-gold-Goldner stain, ×353 (photograph by R. Foelix).

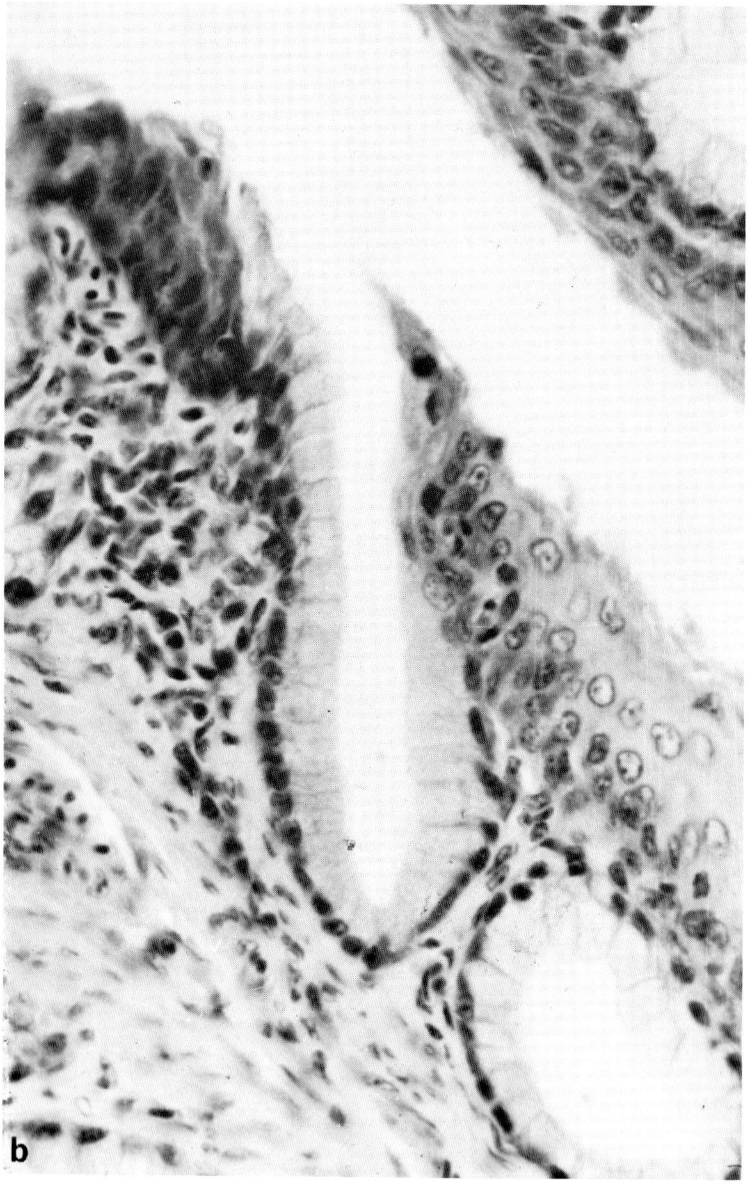

FIG. 6b. Histology of the digestive system. Tubular esophageal glands of a passerine species (*Ploceus manyar*). Hematoxylin–eosin, ×800.

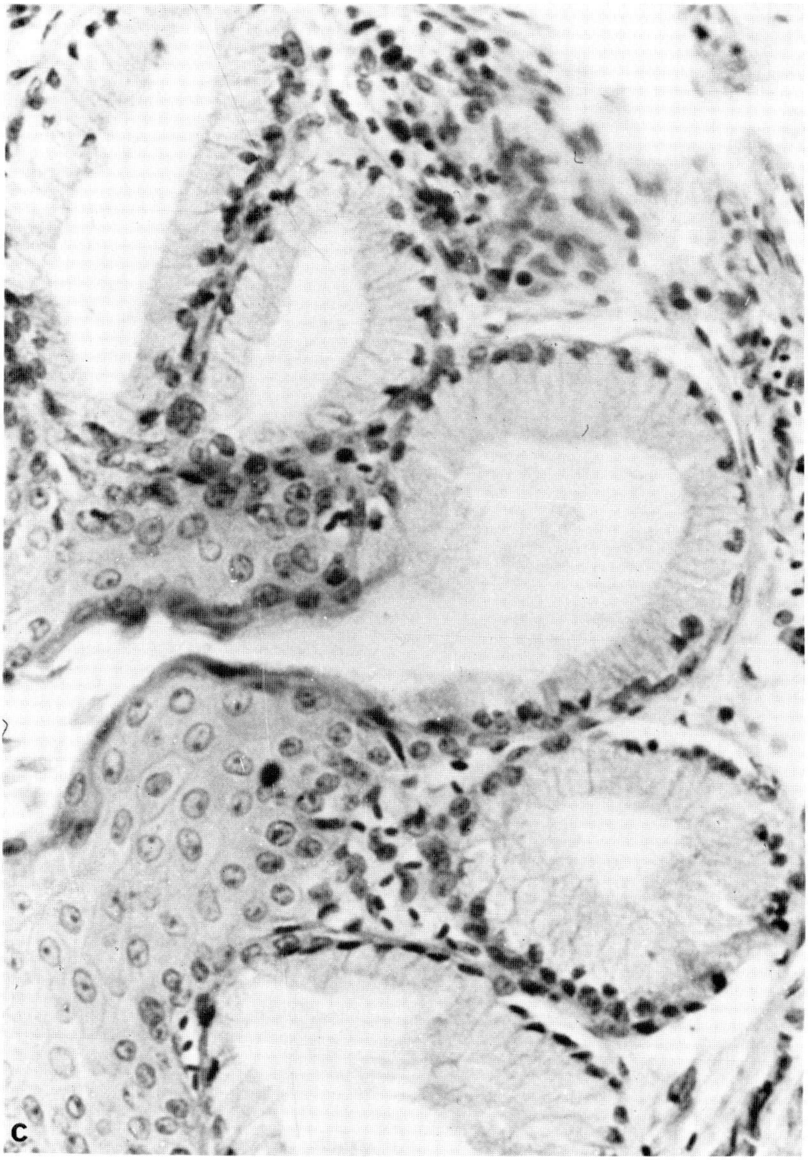

FIG. 6c. Histology of the digestive system. Alveolar esophageal glands of a passerine species (*Carduelis chloris*). Hematoxylin–eosin, ×800.

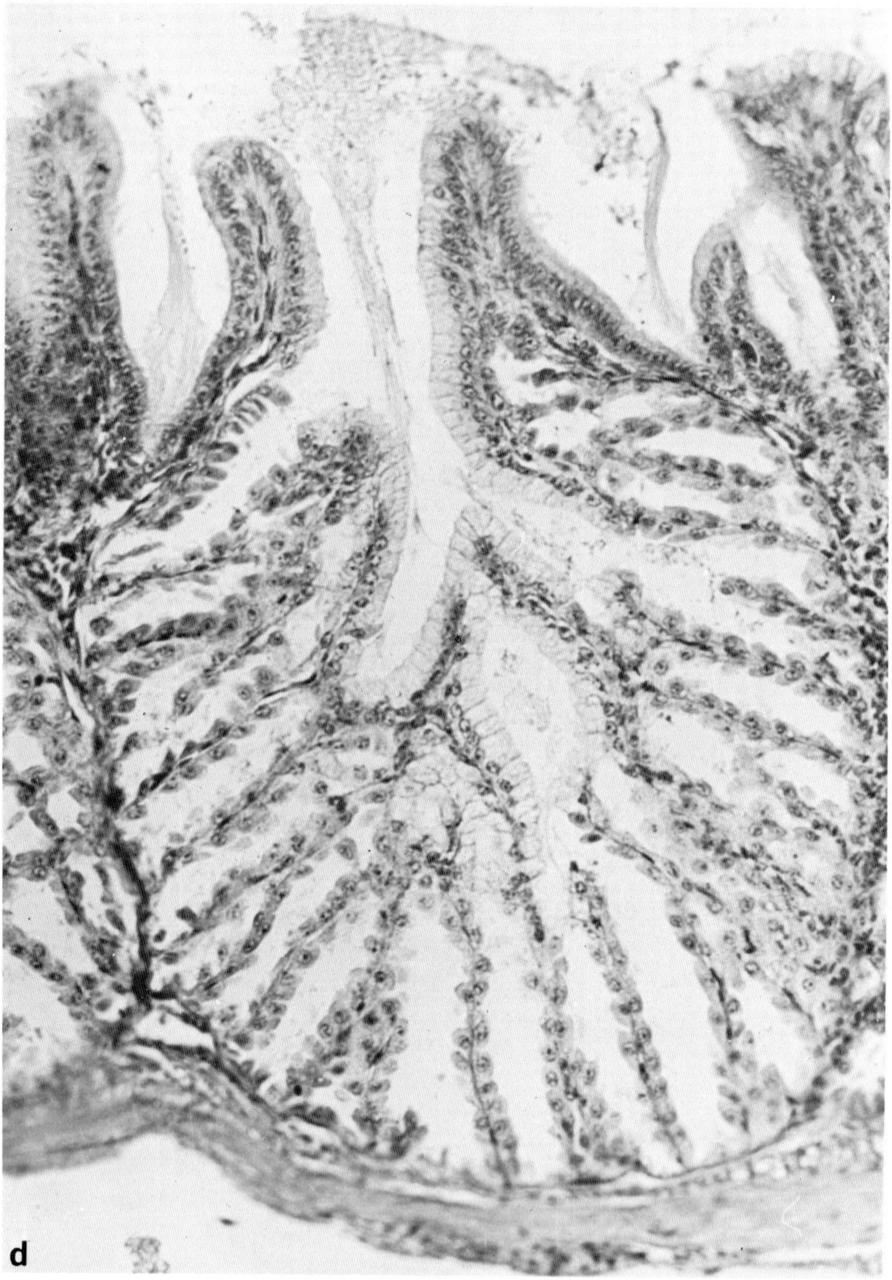

FIG. 6d. Histology of the digestive system. Compound glands from glandular stomach of a passerine species (*Lonchura cucullata*). End-piece system with polygonal cells and duct system with cylindrical cells. Hematoxylin–eosin, ×200.

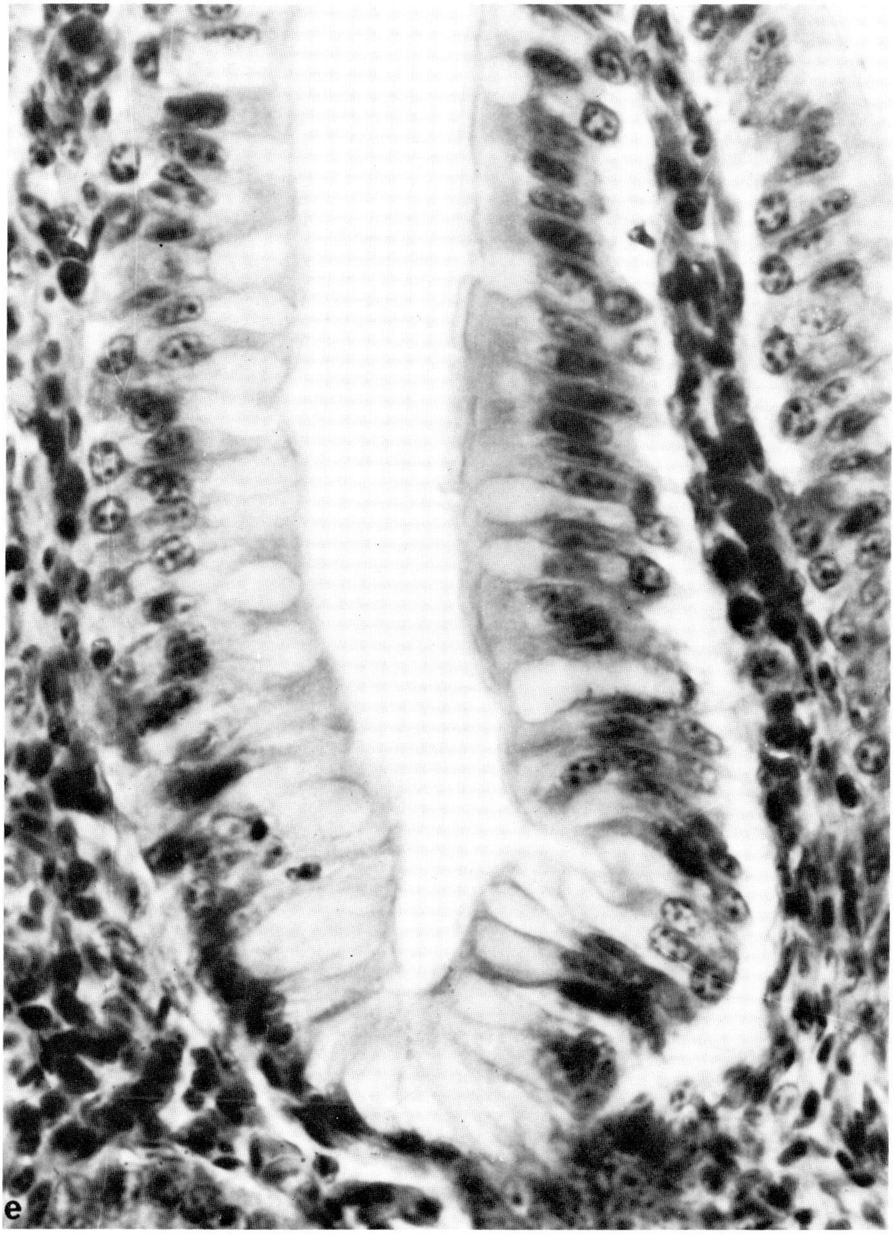

FIG. 6e. Intestinal glands. Small intestine of a passerine species (*Ploceus manyar*). Epithelial cell from end-piece of a crypt of Lieberkühn. Hematoxylin–eosin, ×1500.

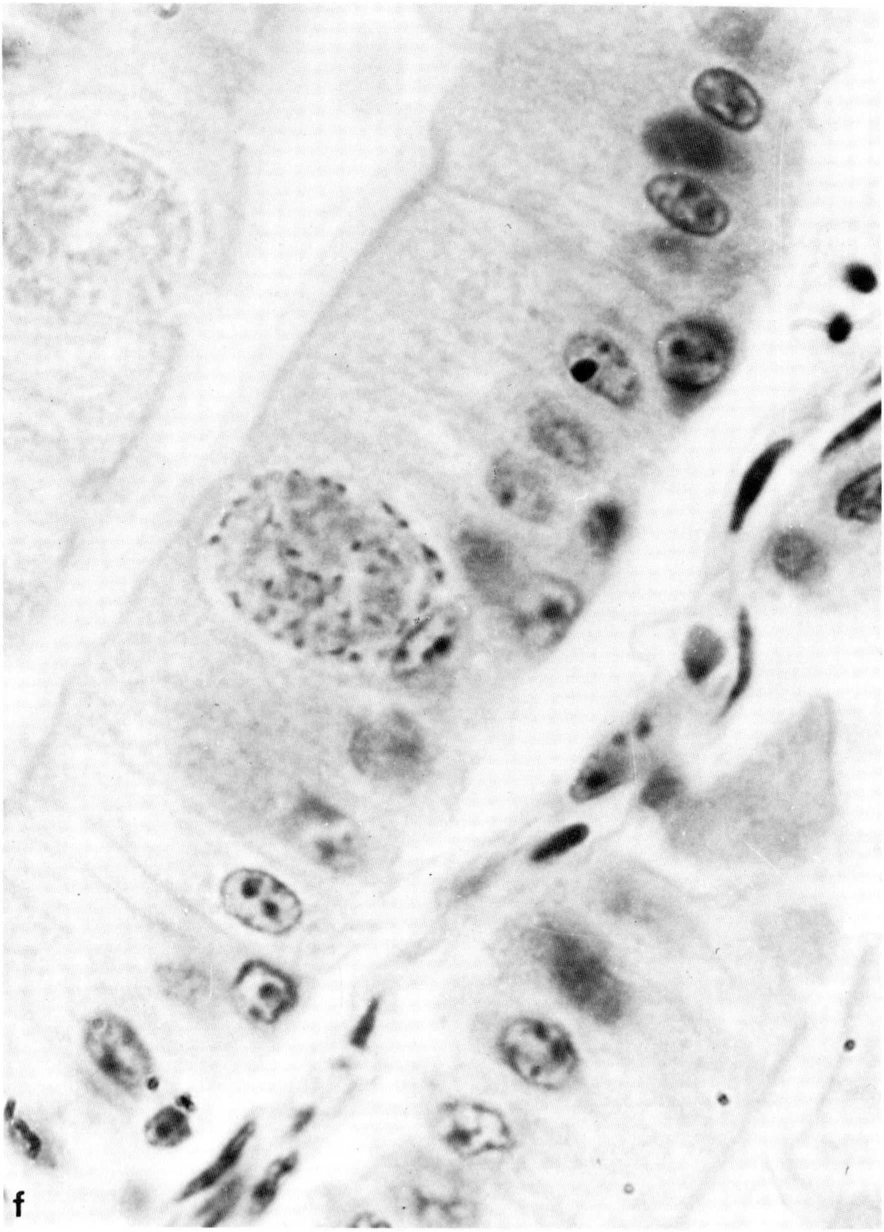

FIG. 6f. Small intestine of a passerine species (*Ploceus manyar*). Paneth cell in folded epithelium. Hematoxylin–eosin, ×2000.

this chamber leads then a duct into the lumen of the glandular stomach. This system, with minor variations, occurs in the auks, gulls, parrots, cuckoos, woodpeckers, storm-petrels, ducks, and geese (Groebbels, 1932b).

Beyond this there is a somewhat more complicated duct system in which the terminal parts of the glands open first into a collecting chamber or collecting ducts, which, in turn, lead into the principal chamber. Such a system is known to occur principally in passerine species (Bergmann, 1862; Schreiner, 1900; Ziswiler, 1967a).

The individual terminal parts of the gland are formed from polygonal cells and can be either tubular or alveolar. Whereas in ducks they are unilobar (Cazin, 1887; Plenk, 1932; Schreiner, 1900; Swenander, 1902), they are multilobal in the domestic fowl. Among oscinine birds these glands are tubular with first- and second-order lobules in the Estrildidae, tubular with first- to third-order lobules in the Ploceidae, tubular with first- to fourth-order lobules in the Fringillidae, and tubulo-alveolar with first- to third order lobules in the Emberizidae.

A singularly interesting arrangement occurs in *Apus* (Bergmann, 1862; Postma, 1887; Swenander, 1902) in which the secretory tubes of each gland are arranged in bundles, almost perpendicularly to the lumen of the proventriculus. These communicate by individual ducts to a common opening into the lumen of the glandular stomach.

The compound glands are surrounded by muscle fibers that are associated with those of the muscularis mucosae. The inner muscle layer between the tubular glands and the compound glands varies from thin to well developed among the various systematic groups.

The submucosa of the glandular stomach is generally poorly developed. The muscularis consists of a relatively thick inner, circular layer and a much more poorly developed outer, longitudinal layer (Clara, 1934; A. Oppel, 1895; Schreiner, 1900; Swenander, 1902). This layer of longitudinal muscle surrounds the glandular stomach either as a thin compact layer or in the form of isolated longitudinal strands of muscle.

Between the glandular stomach and the muscular stomach there is an intermediate zone, which is characterized by the absence of the compound glands of the glandular stomach and the koilin layer of the muscular stomach. However, the tubular glands of the glandular stomach are present. The epithelium attains a greater thickness than in the glandular stomach (Cazin, 1887). In the lamina propria there are increased numbers of lymph follicles. The development of the individual muscle layers is similar to that of the glandular stomach.

The length of the intermediate zone is highly variable. In the domestic fowl (Nussbaum, 1882) it is relatively short whereas in ducks and pigeons (Hasse, 1866) it is relatively long. A special adaptation of this zone occurs in the flower-seeking and frugivorous parrots in which it is developed into a highly distensible storage organ with very few glands; its lumen can be several times as great as those of the glandular and muscular stomachs (G. Steinbacher, 1934).

B. The Muscular Stomach

The muscular stomach or gizzard has long been an organ of interest to zoologists and physiologists; the literature contains hundreds of research papers concerned entirely or in part with this organ. The strong force generated by its musculature in domestic galliform birds was studied by Borelli (1743), Spallanzani (1783), and de Réaumur (1756a,b,c,) in the eighteenth century. As noted above, the latter two investigators first distinguished clearly between the chemical and mechanical roles of the gizzard in digestion. There was an extensive early interest in the question of metamorphosis of the gizzard with changes in type of food. Charles Darwin (1896) discussed the question at some length, citing the early writings of Hunter (1786). Although the studies of Holmgrén (1867, 1872) and Brandes (1896) had demonstrated well the impossibility of an actual metamorphosis from one gizzard type to another, nevertheless, interest in modification of the gizzard by altered diet, and the possible inheritance of such modifications, persisted for several decades thereafter (Houssay, 1902, 1907; Roux, 1906; Schepelmann, 1906).

The muscular stomach of birds is frequently considered only as a compensatory organ for the lack of a chewing apparatus. Actually it has at least five major functions.

1. It may serve as a storage organ until such time as the gastric juice penetrates sufficiently to effect the preliminary acid proteolytic digestion or until it passes into the small intestine. This storage function has importance primarily among carnivores, and is shared with the proventriculus and esophagus.
2. It is the site of preliminary acid proteolytic digestion, a function that may be shared with the proventriculus in carnivores, but not in gramnivores. It may be the site of some chitinolytic activity (Jeuniaux, 1962, 1963).
3. The muscular stomach also has a role in the mechanical or physical phase of digestion. In this respect the grinding function of the organ in graminivores, herbivores, and some insectivores is the most

striking. As an accessory device in this function, the muscular stomach of many species contains small stones (L. Miller, 1962; Lienhart, 1953).

4. The muscular stomach has a function of propelling food into the intestine.

5. Finally, in some species, the muscular stomach has the role of a filter for undigestible materials such as fish skeletons, bones, feathers, hair, and chitinous parts; these can be formed into pellets which are from time to time regurgitated, as, for example, in owls, hawks, kingfishers, and shrikes (Chitty, 1938; R. J. Grimm and Whitehouse, 1963; Howard, 1958).

In comparing forms of the muscular stomach in various groups of birds (Fig. 5), it is possible, to some degree, to recognize certain generalizations (Cattaneo, 1885; Cornselius, 1925; Gadow, 1879a; Milne-Edwards, 1860; Pernkopf, 1930; Portmann, 1950; Postma, 1887; Swenander, 1902): Species whose food is soft, requiring only the action of digestive juices for adequate breakdown, have a distensible muscular stomach, which resembles an elongated sac, with a weak musculature; such is the case, for example, in ducks, cuckoos, hawks, Pelecaniformes, and loons. The tanager genus *Euphonia*, which has become specialized for feeding on loranthacid fruit, has evolved a similar expansible, storage muscular stomach, with a weak musculature (G. Steinbacher, 1935). In other frugivores and nectivores of the Dicaeidae, Thraupinae, and Psittacidae, as well as in the Procellariiformes (Cadow, 1933; Desselberger, 1931; Garrod, 1874a, 1878a; G. Steinbacher, 1934; Wood, 1924), the muscular stomach is very much reduced and the storage function is taken over by the glandular stomach.

An interesting situation exists among the frugivorous flowerpeckers (Desselberger, 1931); the more primitive species (with respect to gizzard), such as *Melanocharis versteri*, have a somewhat reduced but otherwise normal passerine gizzard, whereas the more specialized species *Dicaeum aureolimbatum*, *Dicaeum celebicum*, and *Dicaeum nehrkorni*, which feed on *Loranthus* berries and small arthropods, have a gizzard which is really a diverticulum at the junction of the stomach and intestine. According to Desselberger (1931), a separation of food is effected in this scheme; berries go directly into the intestine from the proventriculus, whereas arthropods go into the gizzard diverticulum for mechanical and peptic digestion.

The "classical form" of the muscular stomach, as a lens-shaped organ with powerful muscular layers, occurs in the graminivorous species (Fig. 5A). Among these species are generally those groups that feed on unshelled seeds; included are galliform birds, pigeons, certain weaver-finches, and thrushes that swallow seeds before removing the shell; these have more powerfully developed muscular stomachs than the members of the families Fringillidae and Estrildidae that remove the shells before swallowing. Among the graminivorous birds the most poorly developed gizzard musculature occurs in the parrots, which not only remove the shells but also fragment the seeds mechanically with the bill.

The stomach lies in the left median part of the abdominal cavity which it fills to a great extent. In the graminivorous birds the gizzard has a form similar to a biconvex lens. The opening from the glandular stomach lies dorsally; the opening to the intestine is somewhat caudal therefrom. In the middle, on both sides of the gizzard, are conspicuous tendinous aponeuroses. The epithelial layer of the gizzard consists of cylindrical to polygonal cells. Within the lamina mucosa tubular glands with merocrine secretion are more or less numerous. They are, together with the other epithelial cells, responsible for the formation of the koilin ("keratinoid") layer.

The koilin lining is perhaps the most striking general feature of the muscular stomach (Fig. 5). This peculiar layer has been the subject of extensive investigation and speculation. Although its origin has often been regarded incorrectly as analogous to the formation of epidermal derivatives, the role of secretory activity by tubular glands of the mucosa in its formation was suggested more than a hundred years ago (Berlin, 1853; Curschmann, 1866; Cuvier, 1805; Flower, 1860; J. D. Grimm, 1866; Hasse, 1866; Leydig, 1854; Neergaard, 1806; Norris, 1961; Toner, 1964). Actually this layer appears to be formed both by secretory activity of the tubular glands, giving the clear perpendicular striations, and by entrapped sloughed epithelial cells and cellular debris, giving the alternate darker laminated perpendicular striae (Calhoun, 1954; Capobianco, 1902; Cazin, 1886, 1887; Clara, 1934; Cornselius, 1925; Joos, 1952; Malewitz and Calhoun, 1958; Olivo, 1947; Plenk, 1932; Postma, 1887; Schreiner, 1900; Swenander, 1902; Vigorita, 1906); however, Greschick (1914a) has insisted that the layer is entirely of secretory origin and that the alternating perpendicular striae are obtained simply by differentiation of a single substance. The koilin lining in the gizzards of birds that feed on soft foods is usually thin and lacks the distinct organization into striae which is so conspicuous in herbivores and graminivores. The green, yellow, or brown color in the koilin sheath is, as suggested originally by Spallanzani (1785), due to regurgitated bile pigments (Groebbels, 1929; Kostenko, 1957; Mangold, 1906; Norris, 1961).

The koilin layer, as the inner lining of the muscular stomach, is especially well developed in graminivorous species (Galliformes, pigeons, parrots, and graminivorous passerine birds) or in those species that eat shelled mollusks (many gulls and waders) or insects with hard exoskeletons; in these the luminar surface of the koilin layer is distinctly structured. In its general course it reflects more or less the underlying structure of the lamina epithelialis. It can have a more or less pleated pattern or it can be folded into coarse, irregular longitudinal folds, or it can form a system of grooves and ridges. In some species the koilin layer above the adjoining muscle is thickened into distinctive grinding plates; this occurs, for example, in the Anseriformes, the Rallidae, the Columbiformes, and *Struthio*. In the frugivorous doves of the genus *Ducula* the grinding plates even have toothlike processes (Garrod, 1874b). In the frugivorous doves of the genus *Ptilinopus* (Garrod, 1874b), as well as in the Fringillidae (Ziswiler, 1967b), the koilin layer has two pairs of grinding plates. In the Cardinalinae and Estrildidae there is only a single pair of grinding plates which, however, are ribbed (Ziswiler, 1967a,b).

In most species the koilin layer is from time to time sloughed off and is removed either through the mouth, for example in the Bucerotidae, Cuculidae, Falconidae, Sturnidae, Corvidae, Anseriformes, Laridae, Anhingidae, Phalacrocoracidae (McAtee, 1906), or by passing on through the remainder of the digestive tract.

The mucosa of the muscular stomach consists largely of simple tubular glands with funnel-shaped openings. While Chodnik (1947) described only a single cell-type in the epithelium of these glands, it is now known that there are others. Thus Dawson and Moyer (1948) described argyrophilic cells in the interior of the gland. Our knowledge of the epithelial cells of the gizzard has been extended substantially by the electron microscope investigations of Aitken (1958), Eglitis and Knouff (1962), and Toner (1966). Toner characterizes the various epithelial cells in the domestic fowl as follows:

1. *Chief Cells*

In the lower parts of the glands the chief cells are cuboidal, with nuclei that are more or less rounded in outline, while higher in the gland the cells become flatter and elongated, and with oval nuclei. The chief cell is characterized by the extreme density of its cytoplasm and its fine structural features. The chief cells have granules that stain specifically with phosphotungstic acid, and have the characteristic of protein-secreting cells. These cells also have unusual microvilli.

2. *Surface-Epithelium Cells*

The cells lining the openings of the crypts and the mucosal surface are taller than the chief cells. The apex of the surface cell bulges into the lumen of the gland; there are microvilli with no extracellular covering such as that seen in the chief cell. The membrane at the base of the cell is irregular in contour. The surface cell may be an unusual form of mucoid cell. The surface cells show a sequence of degenerative changes, terminating in the sloughing of degenerate cells at the mucosal surface.

3. *Basal Cells*

The basal cell has distinctive microvilli, mitochondria, and nucleus, and lacks the elaborate granular endoplasmatic reticulum of the chief cell. The basal cells are situated in the deep parts of the gland. It is possible that the basal cells supply the carbohydrate component of the secretion in the gizzard gland. It is also possible that they are stem cells from which chief cells are differentiated.

4. *Intermediate Cells*

In the deep parts of the glands, there are a few cells that combine some of the structural features of both basal and chief cells. There is the possibility that they represent a stage of differentiation of basal cells into chief cells.

The lamina propria of the muscular stomach is lymphoreticular and contains eosinophilic granulocytes with spherical nuclei (Greschik, 1914a; Schreiner, 1900); in the thrushes it contains collagenous and argyrophilic fibers (Clara, 1934). The muscularis mucosae can be absent (Calhoun, 1954; Plenk, 1932; Schreiner, 1900), or it may be weakly developed at the cranial end of the gizzard. Greschik (1914b) described a well-developed muscularis mucosae in *Corvus frugilegus;* Swenander (1902) found it to be at least as strongly developed as in the glandular stomach in some species. According to Cornselius (1925) muscle fibers occur between the tubular glands if there is no thick koilin layer. Attached to the layer of the tubular glands, the lamina propria forms a stratum compactum consisting of elastic and collagenous fibers (Plenk, 1932). This stratum compactum is especially well developed in species with a well-developed koilin layer. The submucosa consists of areolar connective tissue. In most species it is poorly developed or lacking. The submucosa in the domestic fowl is well developed (Calhoun, 1954).

The muscularis of the gizzard consists of an inner circular muscle layer and an outer longitudinal muscle layer (Bauer, 1901; Calhoun, 1954; Clara, 1934; Cornselius, 1925; Greschik, 1914b, 1917a,b; Pernkopf, 1930; Plenk, 1932; Swenander, 1902); the latter is poorly developed and has been reported as absent in some species (Swenander,

1902). The circular layer varies enormously in its development and organization from the relatively unspecialized, uniform distribution in carnivores to differentiation into distinct semiautonomous masses as in many graminivores and herbivores. This development attains a maximum in the galliform birds (Pernkopf, 1930), as well as in the Fringillidae, Ploceidae, and Estrildidae among the passerine species (Ziswiler, 1967a) in which relatively thin upper and lower musculi intermedii and two oppositely placed musculi laterales provide an antagonistic system for the grinding motions of the gizzard. The fibers of the musculi laterales are inserted in the connective tissue plate on the side of the muscle, the facies tendineae.

The fine- and ultrastructure of the gizzard and its innervation have been investigated intensively by Bennett (1969a,b), Bennett and Cobb (1969a,b,c), Choi (1962), and others. The smooth muscle is organized into extensive interlocking bundles, with frequent anastomoses between adjacent bundles. Nexus are common between apposing smooth muscle cell membranes. The innervation is comparable to that of other gastrointestinal muscle. Auerbach's plexus in the domestic fowl and domestic pigeon is a ganglionated nerve trunk. Since the outer longitudinal muscle coat of the gizzard is lost during development, the main body of Auerbach's plexus lies immediately below the serosa. Frequently there are extensions from the plexus into the underlying musculature. The ganglion cells and the associated nerve fibers that derive from these extensions are scattered throughout the lamina muscularis. Large numbers of myelinated axons are present in Auerbach's plexus. There are synapses between unmyelinated fibers and the ganglion cells, either on the cell soma or on processes from it. All synapses are characterized by a wide (250–300 Å) cleft; presynaptic specializations are resolvable into regular densities that project into the cytoplasm. A plexus comparable to Meissner's plexus of mammals is absent. (The gastric innervation is discussed in detail in the chapter on the autonomic nervous system in Volume IV.)

There have been many investigations concerning the effects of the nature of the diet on the musculature of the gizzard. Many of the older studies (Brandes, 1896; Holmgrén, 1867, 1872, Houssay, 1902, 1907; Magnan, 1912a,c, 1913; Roux, 1906; Schepelmann, 1906) were doubtless motivated by interest in the Lamarckian theories of evolution. Many of these studies, and also more recent investigations (Broussy, 1936; Lenkeit, 1934; Mangold, 1934c), show quite clearly that the mass of musculature in many species is fairly readily modifiable as a direct function of the hardness of the food.

The gizzards of most graminivores, herbivores, and some omnivores contain small stones or grit. The presence and function of this material has been the subject of an extraordinary number of publications. Much of the information has been reviewed by Jaeckel (1925), Mangold (1927b,c), Mangold and Rüdiger (1934), and Salques (1934). Although the occurrence of gizzard stones is widespread in birds, experimental observations are restricted largely to the domestic fowl. If grit is unavailable, it may be retained in the gizzard for several months (Buckner and Martin, 1922; Lienhart, 1953; Zaitschek, 1904); the same appears to be true in *Lagopus lagopus* (Kolderup, 1924). Furthermore, it has been shown well in the domestic fowl that grit is not essential for normal digestion (Buckner *et al.*, 1926), although without it the efficiency of digestion is somewhat decreased (Fritz, 1937; Lienhart, 1953).

It should be noted that gizzardectomy in the domestic fowl (Burrows, 1936; Fritz *et al.*, 1935) reduces significantly the digestibility of coarse foods but that of fine foods only slightly.

The period of retention of food in the gizzard is highly variable from species to species and also depends on the functional state of the digestive tract. In domestic ducks and geese, the *mean time of retention* is about 2 hours (Groebbels, 1930, 1932a); the situation is similar in the domestic fowl (Steinmetzer, 1924) although a portion of each food mass entering the gizzard leaves in much less time since some food may pass through the entire tract in 2.5 hours (Habeck, 1930; Henry *et al.*, 1933). Consistency is obviously a factor in determining the period of retention (Vonk and Postma, 1949).

C. Gastric Digestion

Gastric digestion in birds differs generally from the common vertebrate pattern in that gastric juice is produced in one organ, the proventriculus, whereas gastric proteolysis usually occurs largely, or entirely, in a second organ, the gizzard.

The acid and proteolytic properties of the gastric juice of birds were recognized in the eighteenth century by de Réaumur (1756a,b,c), who performed experiments *in vitro* with gastric juice from vultures, hawks, and owls. His experiments were confirmed and extended by those of Spallanzani (1783) with ducks, geese, chickens, and pigeons, and by those of Tiedemann and Gmellin (1831). The occurrence of hydrochloric acid in the gastric juice is apparently general in birds, although little is known about its actual concentration. The available data suggest, however, that pure gastric juice must have a hydrogen-ion concentration of the order of pH 0.2–1.2 (van Dobben, 1952; Farner, 1943a; Fedorovskii, 1951; Groebbels, 1930; Herpol and van Grembergen, 1967; Popov and Kudriavtsev, 1932).

Although it is clear that the gizzard provides an acid environment for gastric proteolysis, relatively few useful quantitative data on hydrogen-ion concentration therein are available; most useful are the very extensive measurements of Herpol (1965, 1966b, 1967a,b) and Herpol and van Grembergen (1961, 1967b). Many of the published data are from *post mortem* measurements which generally give pH readings that are higher than *in vivo* measurements or than from measurements of gizzard contents removed from the live bird (Cheney, 1938; Farner, 1943a; Herpol, 1966c, 1967a; Herpol and van Grembergen, 1961, 1967d). Gastric pH is also a function of the composition of the diet (Farner, 1943b); the site within the gizzard, especially in species that swallow large food items (Mennega, 1938); and time after food intake (van Dobben, 1952; Mennega, 1938); it may also have a marked daily cycle (Farner, 1943a). These variables must be borne in mind in the necessarily cautious interpretation of the measurements compiled in Table I.

Acid gastric proteolytic activity (optimum pH ca. 1) has been demonstrated in gastric juice or in extracts of the proventriculus in more than 20 species in seven orders (Bernardi, 1926; Bernardi and Rossi, 1923; Braitmaier, 1904; Danysz and Koskowski, 1922; Fedorovskii, 1951; Friedman, 1939; Galvyaalo and Goryukhina, 1937; Herpol, 1964, 1965; Herpol and van Grembergen, 1967a; Hewitt and Schelkopf, 1955; Karpov, 1919; Klug, 1891, 1892; Langendorff, 1879b; H. Meyer, 1929; Paira-Mall, 1900; Popov and Kudriavtsef, 1931; Plimmer and Rosedale, 1922; Reed and Reed, 1928; Shaw, 1913). The properties of the proteolytic activity indicate that the enzyme is a pepsin that can be assumed to occur generally in birds. The few comparative studies that have been effected (Fruton and Bergmann, 1940; Herriott, 1938, 1941) suggest that its action is similar to that of the mammalian pepsin. The inactive form, pepsinogen, has been prepared from the proventriculus of the domestic fowl (Herriott, 1938). Although it has been generally assumed (e.g., Farner, 1960) that pepsin is produced only by the proventriculus, it now appears that this generalization must be reexamined since Herpol (1964) found that extracts of the gizzards of *Athene noctua, Falco tinnunculus,* and *Buteo buteo* have relatively high proteolytic activity (optimum at ca. pH 1). The activity appears too great to be attributable to contamination by proventricular pepsin; furthermore, with identical methods, no such activity was found in the gizzard of the domestic fowl and very little in the gizzard of the domestic pigeon.

TABLE I
HYDROGEN-ION CONCENTRATION IN THE MUSCULAR STOMACH OF SELECTED SPECIES

Species	No. of Individuals	No. of Measurements	Range of pH[a]	Method[b]	Reference
Falco tinnunculus	5	500	1.2–1.8	I	Herpol (1967b)
Buteo buteo	3	213	0.7–1.5	I	Herpol (1967b)
Tyto alba	1	100	1.3	I	Herpol (1967b)
Athene noctua	2	23	1.0–1.5	I	Herpol (1967b)
Phalacrocorax carbo	2	25	1.6–1.9	I	Herpol (1967b)
sinensis	1		0.9–4.6	C	van Dobben (1952)
Haematopus ostralegus	3	59	0.7–1.6	I	Herpol (1967b)
Larus ridibundus	8	503	1.1–1.8	I	Herpol (1967b)
Gallinula chloropus	4	221	1.4	I	Herpol (1967b)
Domestic mallard	3	300	1.2–1.3	I	Herpol (1967b)
	10	10	2.3–2.4	D	Farner (1942b)
Domestic pigeon	100	100	1.9–2.0[c]	D	Herpol (1967a)
	15	15	1.9–2.1[d]	D	Farner (1942b)
	100	100	0.9–1.1[c]	I	Herpol (1967a)
Domestic fowl	20	20	2.5–2.8[d]	D	Farner (1942b)
	81	81	1.5–2.5[c]	D	Herpol and van Grembergen (1961)
	450	450	2.5[c]	D	Herpol and van Grembergen (1967b)
	22	130	1.9–2.4	C[e]	Farner (1943a)
	450	450	1.4[c]	I	Herpol and van Grembergen (1967a)
Phasianus colchicus	11	11	1.9–2.3[d]	D	Farner (1942b)
Corvus corone	5	225	1.0–1.6	I	Herpol (1967b)
Sturnus vulgaris	15	500	1.6–2.3	I	Herpol (1967b)

[a] Approximate range from data given, taking into account standard errors of measurements when given.
[b] D, contents, *post mortem;* C, contents removed from live bird; I, *in vivo* gastric electrode.
[c] After fasting for 24 hours.
[d] Food *ad libitum;* at least 5 hours after feeding.
[e] At intervals throughout the day and night; means from time of highest (3:00 A.M.) and lowest (12:00 noon) are given.

De Rycke (1962) has reported significant amounts of dipeptidase activity (optimum pH, 8.5) as well as smaller amounts of aminopeptidase and carboxypeptidase activities in the proventricular mucosa of the domestic fowl; if these have a digestive function it presumably occurs in the small intestine.

The gastric lipases reported for several species (Haurowitz and Petrou, 1925; Willstätter *et al.*, 1924) could be the result of contamination from regurgitated intestinal contents; however, regardless of their source, their hydrogen-ion requirement precludes significant activity in the stomach. The same observations hold for the reports of gastric invertase and amylase (Bernard, 1859; Bernardi and Schwartz, 1932a,b, 1933a,b; Bernardi and Zanini, 1930). The report that the "crop fluid" of carnivorous birds has keratolytic activity (Stanković *et al.*, 1929) has not been sustained by careful experiments by Mangold (1931b).

Jeauniaux (1962) has presented evidence from three species of birds (*Passer domesticus*, *Turdus merula*, and *Leiothrix lutea*) for the occurrence of a gastric chitinolytic system consisting of two enzymes, a chitinase and a chitobiase (N-acetylglucosaminidase) whose combined activities hydrolyze chitin to acetylglucosamine. Smaller amounts of activity were found in the domestic fowl but none in *Columba palumbus*. The role of this enzyme system in the nutritional economy has not been assessed. An assessment must bear in mind that chitinolysis may well free chitin-bound protein for proteolysis.

Although from many investigations it is clear that the secretory functions of the proventriculus are adjustable to the amount and kind of food present in the anterior portion of the digestive system, the nature of the mechanisms involved is by no means clear. As early as 1859, Brücke demonstrated that secretory activity continued in the excised proventriculus but that this activity required intact cells. Braitmaier in 1904 demonstrated nicely that the "proferment" of pepsin (pepsinogen) accumulated as granules in the cells of the compound glands of pigeons during periods of hunger and that the cells became relatively free of these pepsinogen granules within 6 hours after feeding. These observations were confirmed by Michalovsky (1909) in similar experiments with the domestic fowl and with observations on *Accipiter gentilis*, *Corvus corone cornix*, *Corvus corax*, domestic pigeons, domestic ducks, and other species. In turkeys there is an increase in the rate of secretion of gastric juice and in its proteolytic activity after feeding (Fedorovskii, 1951); in ducks (Kostenko, 1957) there is a constant production of highly acid gastric juice in hunger with an increase after feeding.

Both neural and hormonal mechanisms are involved in the regulation of secretory activity of the proventriculus. Neural mechanisms apparently involve efferent fibers in the vagus whose impulses cause the secretion of increased volume of gastric juice and increased pepsin and hydrochloric acid (Friedman, 1939; Kostenko, 1957). Possibly more direct evidence of hormonal regulation is the demonstration that mammalian gastrin (Keeton *et al.*, 1920) and extracts of duodenum and proventriculus (Collip, 1922) stimulate secretion.

The rate of gastric secretion is also affected by the quality of the food. In ducks and geese Groebbels (1930, 1932a) found that gastric juice is produced at a greater rate with feeding of potatoes and meat than with feeding of bread and grain. The proteolytic activity and the rate of production of gastric juice in turkeys appear to be related directly to the protein content of the food (Fedorovskii, 1951). Collip (1922) has reported that sham feeding causes increased secretory activity in the domestic fowl, whereas distention of the crop was not effective. However, Farner (1941, 1942a) found that sham feeding with an esophageal fistula above the crop gave equivocal results, whereas sham feeding with esophageal fistula below the crop and food in the crop was markedly stimulatory. The sight of food, even following prolonged periods of hunger, failed to cause increased secretory activity. It would thus appear that a true *cephalic phase* of gastric secretion is lacking in the domestic fowl. On the other hand, Walter (1939) has reported psychic stimulation of gastric juice in ducks and conditioning thereof to auditory stimuli.

D. The Motor Activities of the Gastric Apparatus

The activity of the muscular stomach of birds is characterized by rhythmic contractions of variable frequency and amplitude; it was observed in the domestic fowl by de Réaumur (1756a) and Spallanzani (1785). This phenomenon has been studied most extensively in the domestic fowl, whose gizzard has a frequency of one to four contractions per minute (Doyon, 1894a; Mangold, 1906, 1911a; Rossi, 1905; Vonk and Postma, 1949). More recently Mangold (1950) has described the contraction of the musculi laterales as the first phase of the cycle with narrowing of the lumen; the second phase is then the contraction of the musculi intermedii; he describes the contractions of the musculi laterales as asymmetric, so that they produce rubbing, grinding, and pressing effects. The pressure generated in a contraction, as measured with ballons inserted in the lumen of the gizzard, is of the order of 100–200 mm Hg, apparently being greater in the presence of hard food. Kato (1914) has found the mean pressure generated by the muscular stomachs of domestic geese and ducks to be 257 and 178 mm Hg, respectively, with frequencies somewhat higher than those observed in the domestic fowl. Schorger (1960) has shown that the gizzard of the domestic turkey can crush hickory nuts that require ca. 75 kg mechanical pressure for crushing.

The periodic contractions of the gizzards of *Corvus corone cornix* and *Corvus monedula* resemble those of the domestic fowl although they have slightly greater frequencies and apparently do not vary significantly with respect to presence or absence of food (Mangold, 1911b). In *Buteo buteo* the frequency of contraction is similar to that of the domestic fowl, apparently being slightly lower shortly after feeding. The maximum pressure generated is of the order of 20 mm Hg; 48 hours after feeding it is reduced to a few mm Hg, indicating very feeble movements. Extracts of meat have a marked stimulatory effect on gizzard movements (Mangold, 1911b).

The gastric apparatus is innervated extrinsically by preganglionic and postganglionic cholinergic excitatory fibers in the branches of the vagi and in the perivascular trunks. Nonadrenergic inhibitory fibers, perhaps of vagal origin, occur in both the vagal and perivascular trunks; however, the sympathetic adrenergic fibers occur principally in the perivascular trunks (Bennett, 1970b). Auerbach's plexus, at least in the galliform and columbiform species, is essentially superficial, lying just below the serosa because of the absence of a longitudinal muscle layer; it has the structure of a condensed, ganglionated trunk, plexiform in certain regions, receiving fibers from both the vagal and perivascular trunks (Nolf, 1938a; Bennett and Cobb, 1969c). Freed of all extrinsic innervation, the gizzard of the domestic fowl retains a regular automatism that is dependent on the intrinsic myenteric nervous system (Nolf, 1938a,b,c, 1939a,b). The proventriculus also receives excitatory and inhibitory fibers from both the vagi and perivascular (coeliac) trunks (Nolf, 1925c); it is partially under the influence of the myenteric system of the gizzard; thus chronotropic effects in the proventriculus can be obtained by stimulation of vagal fibers to the gizzard, whereas inotropic effects can be obtained by stimulation of its own vagal fibers (Nolf, 1936). Actually Nolf (1938a) found the proventriculus to be completely inert when separated from the gizzard unless the proventricular vagal fibers are stimulated.

As a general but incomplete picture, the following emerges from the extensive morphological, ultrastructural, physiological, and pharmacologic investigations that have now been published concerning the innervation of the gastric apparatus in relationship to its motor functions (Bennett, 1969a,b,c, 1970a,b; Bennett and Cobb, 1969a,b; Bogdanov and Kibiakov, 1955; Cobb and Bennett, 1969; Doyon,

1894b, 1925; Henry *et al.*, 1933; Iwanow, 1930; Mangold, 1906, 1911a,b; Nolf, 1925a,b,c,d, 1927a, 1935, 1936, 1938a,b,c,d, 1939a,b): (1) There is an extensive, intrinsic myenteric nervous system, consisting of Auerbach's plexus and the fibers extending therefrom, which is responsible for a basic, rhythmic motility pattern. (2) This basic rhythm is modified, both inotropically and chronotropically, by discharges from the vagal branches and sympathetic (perivascular) trunks, both of which contain both excitatory and inhibitory fibers. Details of the myenteric nervous system and its relationship to the extrinsic innervation will be discussed in the chapter on the Autonomic Nervous System in Volume IV.

E. The Pyloric Stomach

From the gizzard chyme passes through the pyloric opening into the intestine. The pyloric segment is a histologically clearly identifiable zone in most species (Zietschmann, 1908). In some groups the pylorus is developed into a third gastric chamber (Fig. 5C) which was designated by Stannius (1846) as the portio pylorica. Such pyloric chambers have been demonstrated in the penguins, grebes, pelicans, many storks (Leuckart, 1841), as well as in some ducks, geese, and rails (Gadow, 1879a,b). Swenander (1902) correlated the presence of distinct pyloric sections with food that contains a large amount of water; Gadow (1879a,b) associated it with a fish diet. Contrary to these conclusions is the fact that other groups, such as most geese, hawks, and cuckoos, have pyloric chambers.

In the Ardeidae the pyloric chamber is separated from the gizzard by two strong folds (Swenander, 1899). In grebes the pyloric stomach regularly contains feathers which possibly function as a filter (Wetmore, 1920). Among the snakebirds the pyloric mucosa has long, hairlike processes that appear also to function as a filter (Garrod, 1876, 1878c).

V. The Intestine

A. The Small Intestine

The small intestine is the principal site of chemical digestion. Of equal importance is its role in the transfer of nutrient materials from the lumen into the blood and lymph of the mucosa. The small intestine has also a role in the propulsion of the contents of its lumen. Furthermore, some of its cells, presumably in the mucosa, have an endocrine function.

It is customary to designate the first section of the small intestine as the duodenum although it is not a histologically clearly definable section as in mammals. Its definition in birds, therefore, depends on its topographic position, i.e., as the first loop of the small intestine after its origin from the pylorus. The shanks of this loop lie side by side and are oriented towards the pelvis. The pancreas is enclosed within the duodenal loop. Following the duodenum is that part of the small intestine designated as the ileum.

The gross morphology of the intestinal loops and their mesenteries (Fig. 7) involves substantial variation (Gadow, 1879a,b, 1889; Mitchell,

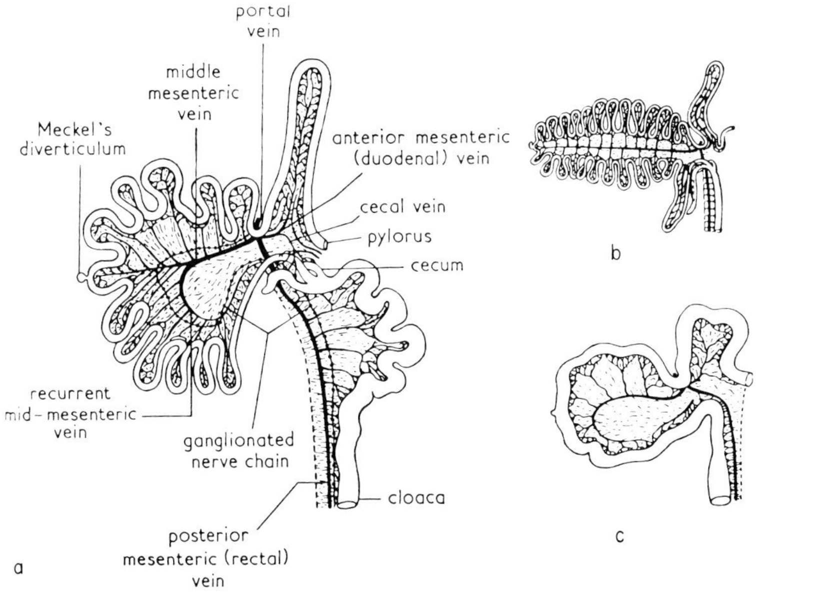

FIG. 7. Selected examples of gross morphology of the intestine. (a) *Anhima cornuta* as an example of the archicentric type of Mitchell (1901), (b) *Phoenicopterus ruber* in which Meckel's tract is elongated and the duodenum is a simple loop, (c) *Apus apus* with a short Meckel's tract (modified from Mitchell, 1901).

1901) which can be explained partially on the basis of nutritional adaptations (Magnan, 1910b, 1912a; Eber, 1956) and partially on phylogenetic relationships (Gadow, 1889; Herpol, 1966a; Mitchell, 1901; Eber, 1956; Ziswiler, 1967a,b). Usually the duodenal loop extends posteriorly to about the level of the large intestine, with the ascending loop returning approximately to the level of the muscular stomach. The intestine shows many nutritional adaptations; it tends to be relatively long in herbivores and graminivores and relatively short in carnivores, insectivores, and frugivores, although it is difficult to separate these relations from those with body size and systematic position (Herpol, 1967b). The ileum often bears the remnant of the embryonic yolk sac as Meckel's diverticulum. In adults this diverticulum lies approximately in the middle of the duodenum. It is highly variable in size. In passeriform species, woodpeckers, parrots, and doves and pigeons (Mitchell, 1901) it is usually an invisible rudiment that contains numerous Peyer's patches and lymph follicles, similar to the cecum. In *Phalacrocorax* Meckel's diverticulum postembryonically becomes a conspicuous lymphoid organ.

The mucosa of the small intestine has a single-layer epithelium of elongate cuboidal cells with round or oval nuclei. It consists of cells with brush orders, goblet cells, and cells with basophilic granules. Strongly granulated cells, which are probably identical with the Paneth cells of mammals, have been found not only in the epithelial crypts of various passerine species (Chodnik, 1947; Clara, 1927, 1928, 1934; Greschik, 1922; Patzelt, 1936; Rosenberg, 1941), but also in the epithelium of the folds, as in the Ploceidae (Ziswiler, 1967a).

To provide increased surface area the mucosa of the intestine is modified into a system of folds, lamellae or villi, which give to the interior surface of the intestine a definite relief (Bujard, 1909). The simplest form is the zigzag fold relief that occurs in thrushes (Clara, 1934; S. Müller, 1922), some Emberizinae and Ploceidae (Ziswiler, 1967a), as well as in many nonpasserine species. From this basic pattern, which occurs in taxonomically diverse groups, a number of differentiated structures can be derived—the systems of zigzag lamellae in the Estrildidae (Ziswiler, 1967a); the displaced lamellar system of some Charadrii (Müller, 1922), Phasianinae (Biedermann, 1911), and the Viduinae (Ziswiler, 1967a); the leaflife villi as in the Tetraoninae and of the genera *Passerina* (Ziswiler, 1967a) and *Phoenicopterus* (Müller, 1922), hawks (Müller, 1922), and the Emberizinae (Ziswiler, 1967a). There is no clear relationship between the form of relief of the intestinal surface and the type of diet; on the contrary, the nature of this relief in some cases is a useful criterion for taxonomic relationships.

The distinctiveness of the relief due to folding or to villi is always most prominent in the duodenum whereas, as a rule, the pattern becomes simplified into a longitudinal zigzag fold pattern toward the large intestine.

In addition to folds and villi the *lamina epithelialis* forms crypts of Lieberkühn which also can vary extensively among different taxonomic groups. These crypts extend to the *lamina propria* forming a distinct crypt layer.

Alone among the gramnivorous oscinine species, four different types of crypts can be differentiated (Ziswiler, 1967a).

1. Crypts with short tubular-shaped terminal pieces and a straight duct, lying closely together: Ploceidae.
2. Crypts with the terminal parts with wide, vesicle-shaped lumina and short ducts that are mostly straight and lying close together: Fringillidae.
3. Crypts with vesicle-shaped terminal parts. Crypt ducts with one or two turns, very loosely arranged in the crypt layer: Estrildidae.
4. Crypts with end pieces with wide lumina and long ducts with three or four turns, lying close together: Cardinalinae, Emberizinae.

The crypt epithelium is formed differently according to systematic group. In the Estrildidae and Ploceidae, for example, the epithelium in the crypts differs very little from that of the folds and lamellae (Fig. 6e). It contains, like the epithelium of the folds in lamellae, goblet cells and striated-border cells; only in the terminal parts is the striated border lower than on the epithelium of the folds. In its stainability with eosin this epithelium is not different from that of the villi.

In other groups, as for example in the Fringillidae, the striated-border cells and goblet cells extend into the ducts of the crypts but not into the terminal parts whose epithelium has no striated border and consists almost exclusively of cells with basic granules. This epithelium stains differently with eosin than that of the villi.

Finally, there is a type of crypt in which neither the ducts nor the terminal parts have goblet cells or striated-border cells, as in the Cardinalinae and in the Emberizinae.

Although it has often been stated that Brünner's glands are lacking in birds, there is, in at least many species, a narrow zone between the muscular stomach and duodenum that contains tubular glands (Clara, 1934; Patzelt, 1936) that may be considered similar to, if not homologous with, Brünner's glands of mammals. The lamina propria forms a villus stroma and contains basal connective tissue that, next to the muscularis mucosae, is thickened into a stratum compactum. The lamina propria also contains lacteals. The lymphocytes, which are present in large numbers in the lamina propria, can be distributed diffusely or are in isolated lymph follicles (folliculi lymphatici solitarii) or in Peyer's patches (Clara, 1934; Patzelt, 1936; Rosenberg, 1941; Ziswiler, 1967a).

The muscularis mucosae can be a compact functionally effective layer. The submucosa is reduced to a few connective tissue fibers (Calhoun, 1954; Clara, 1934; Patzelt, 1936; Rosenberg, 1941). The tunica muscularis consists of a well-developed inner circular muscle layer and a weakly developed outer longitudinal muscle layer. The circular muscle layer is further divided into a loose outer zone and a dense inner zone.

If one follows the nature of the individual histological structures along the length of the intestinal tract, there are no major differences between the duodenum and ileum. Structural differences are only of a quantitative nature and change continuously through the length of the small intestine. In addition to the already mentioned simplification of the intestinal relief caudally from the zigzag fold relief of the rostral part, the ducts of the crypts become shorter. The special crypt epithelium is lacking toward the end of the small intestine and is replaced by a striated-border epithelium. In certain species, for example in the Fringillidae (Ziswiler, 1967a), the striated-border epithelium in the rostral part of the small intestine extends into the terminal parts of the crypts. In graminivorous passerine species the diameter of the small intestine becomes about 50% smaller from rostral to caudal; the height of the folds in the region of the ceca is about one-third of that of the duodenum; the same is true for the thickness of the crypt layer. The thickness of the muscle layers, on the other hand, remains unchanged throughout the length of the small intestine, or may even increase caudally.

B. The Large Intestine

The large intestine of birds is a terminal enlargement of the small intestine that extends from the level of the openings of the ceca to the cloaca. The course of the large intestine is relatively straight; it is short in comparison with the small intestine. In passerine birds its length is 3–10% of the small intestine; the extremely specialized seed-eaters have a relatively short large intestine whereas it is longer in the more omnivorous forms. The outer diameter of the large intestine in passerine birds is about 25% greater than that of the ileum; in no case does it attain the diameter of the duodenum, which is 50–100% greater. A very long large intestine has been described in *Rhea americana* and *Anhima cornuta* (Mitchell, 1901).

Actually, little is known about its functions other than those of movement and temporary storage of intestinal contents. Presumably it has very little function in digestion. The lower moisture content of the material in its lumen suggests a role in resorption of water.

Histologically the large intestine does not differ, in principle, from the small intestine. The inner relief is always a fold relief with the folds gradually becoming more extended and lower towards the cloaca. Also, the thickness of the crypt layer and the diameter of the individual crypts decrease continuously. The thickness of the individual muscle layers remains approximately the same as in the ileum (Greschik, 1912; S. Müller, 1922; Ziswiler, 1967a).

C. The Cloaca

The terminal part of the large intestine is known as the cloaca (Fig. 8). It serves as a storage chamber for both feces and urine and is also

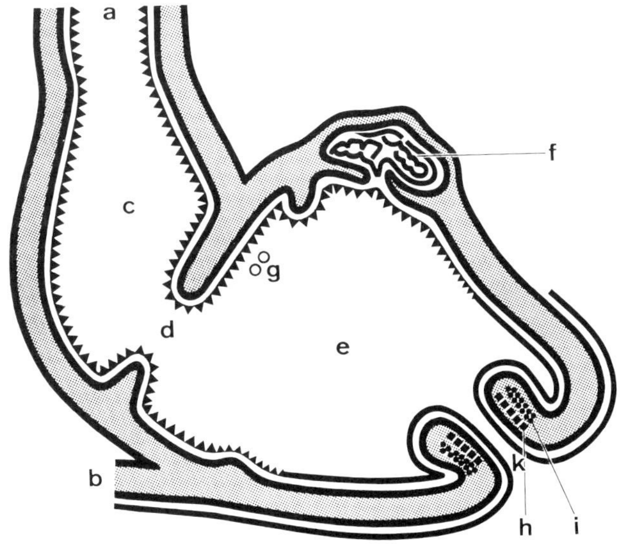

FIG. 8. Schematic of longitudinal section through the cloaca of *Columba livia* (modified after Clara, 1926). a, Large intestine; b, epidermis; c, coprodaeum; d, urodaeum; e, proctodaeum; f, Bursa fabricii; g, papillae of the ureter and vas deferens; h, mucosal glands; i, sphincter muscles; k, cloacal opening.

the pathway for the products of the sex glands. The cloaca is divided into three sections (Gadow, 1879a,b) designated as the coprodaeum, urodaeum, and proctodaeum. Concerning the embryological development of these sections there is still no detailed concept (Pomayer, 1902; Hafferl, 1926; Boyden, 1922). The coprodaeum is continuous with the large intestine, although often of greater diameter. Next posterior is the urodaeum, into which open the oviduct, or vasa deferentia, and the ureters. It in turn continues into the proctodaeum which opens externally through the anus. Histologically, the cloaca is basically similar to the large intestine; however, there are more goblet cells in the mucosa, the surface relief may be different, and caudally the cuboidal cells of the mucosa change to stratified squamous epithelium (Clara, 1928, 1934; Patzelt, 1936). The musculature of the caudal part of the proctodaeum, including the anal sphincter, has striated fibers (Clara, 1926; Greschick, 1912, 1914b).

There is typically in young birds at the junction of the large intestine and the cloaca a dorsal diverticulum, the bursa of Fabricius. Subsequently this diverticulum loses its lumen and becomes lymphoid. It has no known or apparent function in digestion and will therefore not be considered further in this treatise despite the substantial amount of research to which it has been subjected (e.g., Al-Hussaini and Amer, 1959; Fennell and Pearse, 1961; Glick, 1956, 1957; Jolly, 1911, 1914; Klima, 1956, 1958; Linduska, 1943; Märk, 1944; Retterer, 1885); its function as an organ in the antibody-producing system of the domestic fowl has been reviewed by Glick (1969).

D. The Intestinal Ceca

Birds generally have a pair of ceca at the junction of the small and large intestine, although in many groups they are not developed sufficiently to become functional parts of the digestive system. The parrots apparently constitute an exception in that there appears to be no trace of these structures in the adult birds (Mitchell, 1901) and, according to Abraham (1901), at least in *Melopsittacus undulatus*, they do not develop in the embryo.

In the adult stage they are lacking in all of the Alcedinidae (Naik and Dominic, 1962). They are present as rudiments or lacking in the Trochilidae and Apodidae (Shufeldt, 1890). Ordinarily small ceca occur in the falcons, the woodpeckers, the Upupidae, and the insectivorous passerine species (Maumus, 1902; Naik, 1962; Naik and Dominic, 1962). The ceca of doves and pigeons (Markus, 1964) and the graminivorous passerine species are variably small (Ziswiler, 1967a). On the contrary, herbivorous and omnivorous species, such as cranes, galliform birds, ducks and geese, and mollusk-feeding Charadriiformes usually have well-developed ceca (Naik and Dominic, 1962). The ceca of fish-feeders, such as ibises and procellariiform species (Naik and Dominic, 1962), are rudimentary. Among the herons, at least some species have only a single functional cecum, the second existing only as a rudiment in the wall of the intestine (Corti, 1923). Well-developed ceca occur also in the Cuculidae, the Meropidae, and the Coraciidae (Naik and Dominic, 1962). The most extensive development is to be noted in *Struthio camelus* (Fig. 9a) and in the Tetraoninae (Schumacher, 1923). Functional ceca are oriented anteriorly and are connected, at least in part, to the intestine by means of

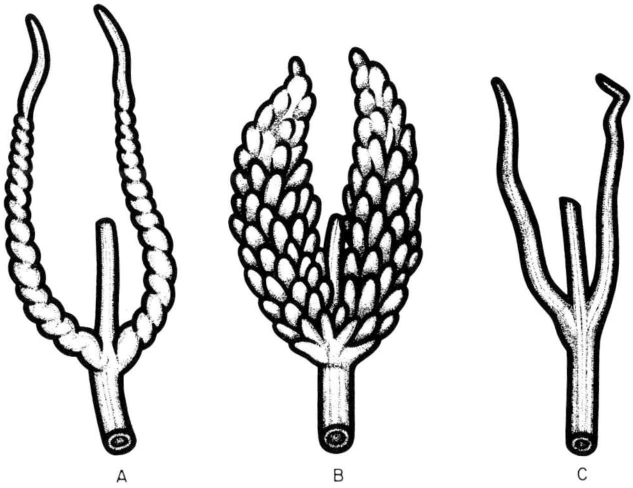

FIG. 9. Selected examples of intestinal ceca. (A) *Struthio camelus*. (B) *Eudromia elegans*. (C) domestic fowl (modified from Portmann, 1950).

mesenteries (Gadow, 1891; Maumus, 1902; Mitchell, 1901). Histologically, they are basically similar to the intestine (Schumacher, 1922, 1925; Naik and Dominic, 1969b; Ziswiler, 1967a), although lymphoid tissue is often much more abundant, especially among the galliform birds, tinamous, ducks and geese, waders, and others. Mitchell (1901) and Naik and Dominic (1963) have proposed the following categories of ceca.

1. The *intestinal type,* in which these organs resemble the intestine in their histology. This type is a well-developed, thin-walled cecum containing intestinal material and having apparently some role in the digestive processes. This type occurs in the ducks and geese, some ratites, grebes, loons, the Pelecanidae, goatsuckers, trogons, rollers, bee-eaters, and others. An enlarged intestinal type appears sometimes to be an adaptation to increased cellulose content in the diet. It is characterized by increased amounts of lymphoid tissue. This type occurs in the galliform species, the tinamous, screamers, rails, sandgrouse, many waders, and others.

2. The *glandular type,* in which these organs are long and show profuse secretory activity (Naik and Dominic, 1963). This type is typical

of the owls. Conspicuous goblet cells occur in the epithelium (Naik and Dominic, 1969a).

3. The *lymphoepithelial type,* in which these organs are much reduced and infiltrated by lymph cells. They are characteristic of some doves and pigeons, and some waders. Also most of the passerine species have this type although among the oscinine species there are some in which the lumina of the ceca are still clearly connected with lumen of the large intestine and whose epithelial linings have marginal and goblet cells and isolated crypt terminal pieces as in the Emberizinae, a few Cardinalinae, and Ploceidae (Ziswiler, 1967a).

4. The *nonfunctional and vestigial* (or completely absent) *type* is characteristic of some penguins, petrels, hawks, parrots, turacos, most doves and pigeons, woodpeckers, swifts, hummingbirds, and kingfishers.

E. Intestinal Digestion

As a component of the digestive system, the small intestine functions both as a secretory organ and as a digestive chamber in which the vast bulk of the hydrolytic processes occurs. Actually, little is known about the succus entericus of birds. In the domestic fowl it is slightly acid to slightly alkaline (Farner, 1941; Kokas *et al.,* 1967; Polyakov, 1958) and iso-osmotic (Kokas *et al.,* 1967). It has a substantially higher buffering capacity than either pancreatic juice or bile (Farner, 1941). The secretion of succus entericus is increased by duodenal distension, vagal stimulation, secretin (mammalian), extracts of duodenal mucosa from the domestic fowl (Kokas *et al.,* 1967), and by bile (Lavrenteva, 1963).

There is relatively little definite information about the enzymes of the succus entericus, primarily because of the difficulty in obtaining pure samples. Studies involving extracts of intestinal mucosa are subject to possibly incorrect conclusions because of contamination with enzymes of pancreatic, hepatic, or bacterial origin, and also by tissue enzymes that are not secreted into the lumen of the intestine.

Mucosal extracts from the intestine of the domestic fowl have been reported to hydrolyze starch (Bernardi and Schwartz, 1933b; Farner, 1941; Fischer and Niebel, 1896; Hewitt and Schelkopf, 1955; Plimmer and Rosedale, 1922; Shaw, 1913), maltose (Farner, 1941; Karrer and Parsons, 1968; Laws and Moore, 1963a), sucrose (Bernardi and Schwartz, 1933b; K. M. Brown and Moog, 1967; Plimmer and Rosedale, 1922), fat (Hewitt and Schelkopf, 1955), protein (Antonino, 1923; Bernardi and Schwartz, 1933b; Herpol and van Grembergen, 1967b; Hewitt and Schelkopf, 1955; Laws and Moore, 1963b; Plimmer and Rosedale, 1922), and dipeptides (Berger and Johnson, 1940; De Rycke, 1962). Alkaline proteinase activity has been demonstrated in the intestinal mucosa of *Falco tinnunculus*, *Larus ridibundus*, *Sturnus vulgaris*, *Gallinula chloropus*, and the domestic pigeon (Herpol, 1967b) and in the succus entericus of the domestic fowl (Kokas *et al.*, 1967). That the proteolytic enzyme(s) of the succus entericus may have a significant function is suggested by the observation that depancreatized chickens utilize about 25% of dietary protein (Ariyoshi *et al.*, 1964). Aminopeptidase and smaller amounts of carboxypeptidase have been reported from duodenal mucosa (De Rycke, 1962) but their significance in digestion has not been ascertained.

Intestinal maltase activity has been reported in *Coturnix chinensis* and *Grus grus* (Zoppi and Schmerling, 1969), in *Aptenodytes patagonica*, *Eudyptes crestatus*, *Catharacta skua*, and *Larus dominicanus* (Kerry, 1969), and in the domestic fowl (succus entericus) (Polyakov, 1958); intestinal sucrase in *Coturnix chinensis* and *Grus grus* (Zoppi and Schmerling, 1969), in *Catharacta skua* and *Larus dominicanus* (Kerry, 1969), and in the domestic fowl (succus entericus) (Kokas *et al.*, 1967; Polyakov, 1958); and oligo-1,6-α-glucosidase in *Coturnix chinensis* and *Grus grus* (Zoppi and Schmerling, 1969). The evidence concerning an intestinal lactase is equivocal (Hamilton and Mitchell, 1924; Kerry, 1969; Plimmer, 1906; Plimmer and Rosedale, 1922; Portier, 1898; Portier and Bierry, 1901; Semenza, 1968; Weinland, 1899) both in the domestic fowl and domestic duck, although lactose can be utilized in appreciable quantities by the domestic fowl (Hamilton and Card, 1924). Proteolytic, amylolytic, and invertase activities have been reported also from the intestinal mucosa of *Vanellus vanellus* (Bernardi and Schwartz, 1933b).

Our own investigations of the domestic fowl (Farner, 1941) suggest that intestinal amylase, if such actually exists, must have a relatively small role in the digestion of starch.

It seems very probable that maltase in the succus entericus may have a trivial role in digestion; it is far more likely that most of the maltose is "absorbed" as such (Laws and Moore, 1963a) and hydrolyzed in the brush-border complex or within the cell. This type of mechanism has been demonstrated for mammals. The same may be true for sucrase despite its demonstrated occurrence in the succus entericus. This concept is not inconsistent with the monosaccharide transport system discussed below.

Laws and Moore (1963b) have expressed misgivings concerning the occurrence of an intestinal lipase in the sense of Desnuelle (1961) although their investigations indicate quite clearly the presence of an intestinal esterase that catalyzes the hydrolysis of more water-soluble esters. Kokas *et al.* (1967) consistently failed to find lipase in succus entericus. However, it must be noted that depancreatized chickens are able to use about 25% of the dietary fat intake (Ariyoshi *et al.*, 1964).

In considering the chemical phase of digestion in the small intestine, an important function must be assigned to the carbohydrases, proteinases, and lipase of the pancreas. Also it must be borne in mind that the small intestine harbors a microbial flora that may affect the availability of some nutrients for absorption (Coates and Jayne-Williams, 1966). Although it is not possible at this time to present a detailed scheme of chemical digestion in the small intestine, suffice it to say that the many studies on the digestibility of food compounds, primarily in domestic birds, make it amply clear that the avian digestive system effects the chemical digestion of starch, disaccharide sugars, most proteins, and fats at least as effectively as the digestive system of mammals (E. W. Brown, 1904; Diakow, 1932; Engler, 1933; Günthenberg, 1930; Halnan, 1928; Heller *et al.*, 1930; MacDonald and Bose, 1944; Olsson *et al.*, 1950; Tscherniak, 1936). The chemical phase of digestion, except for cellulose digestion, discussed below, certainly occurs largely in the small intestine (Herpol and van Grembergen, 1967a; Heupke and Franzen, 1957; Mangold, 1925, 1931a, 1950; W. K. Meyer, 1927).

The pancreatic and intestinal enzymes, whose digestive functions occur in the small intestine, have their optimum rates of activity at hydrogen-ion concentrations in the range of pH 6–8 (Farner, 1941; Herpol and van Grembergen, 1967a,b; Laws and Moore, 1963a,b). It is of interest, therefore, to consider the intestinal environment in this respect. It should be emphasized that the published data have been obtained by *post mortem* measurement of the pH of intestinal contents. In view of the fact that the contents of the gizzard become less acid after death (Farner, 1943a), it is very probable that the contents of the small intestine, and especially the duodenum, may become more acid. Most of the available data on intestinal pH concern domestic species (Table II).

Although is is clear that the small intestine is the major organ for absorption of nutrients (e.g., Tilgner-Peter and Kalms, 1957), the processes involved therein have received relatively little attention in birds. However, it has been demonstrated (Bogner and Haines, 1961, 1964; Bogner, 1966) that D-glucose, D-galactose, D-xylose, and possibly D-fructose are absorbed at rates that are significantly higher than those for other monosaccharides and disaccharides. The active-

TABLE II
APPROXIMATE RANGE OF HYDROGEN-ION CONCENTRATION IN THE SMALL INTESTINE

Species	Number of Individuals	pH	Reference
Domestic fowl	30	6.0–6.8	Hewitt and Schelkopf (1955)
	20	5.8–6.5[a]	Farner (1942b)
	450	6.4–7.2[b]	Herpol and van Grembergen (1967b)
Domestic turkey	4	5.8–6.9[a]	Farner (1942b)
Phasianus colchicus	11	5.6–6.8[a]	Farner (1942b)
Domestic pigeon	100	6.4–6.9[b]	Herpol (1967b)
Domestic mallard	10	6.0–6.9[a]	Farner (1942b)

[a] Food *ad libitum;* at least 5 hours after beginning of feeding.
[b] After fasting for 24 hours.

transport mechanism develops before hatching (Holdsworth and Wilson, 1966). Glucose absorption is sharply reduced by phlorizin; absorption of sorbose, which is passive, is unaffected thereby. The middle third of the chick intestine appears to have the greatest capacity for active absorption (Bogner, 1966). Investigations of the transport of D-xylose (Alvarado, 1967b; Wagh and Waibel, 1967) and phenylglucosides (Alvarado and Monreal, 1967) lead to the conclusion that sugars are absorbed in the small intestine of the domestic fowl by a sodium-dependent, mobile-carrier system similar or identical to that of mammals, a system that may be widespread among the vertebrates (see, also, Holdsworth and Wilson, 1966; Holdsworth and Hastings, 1967; Sato *et al.*, 1960a,b). This is not inconsistent with the concept of hydrolysis of disaccharides in the brush-border complex of the cells of the intestinal epithelium or within these cells themselves. In comparison with mammals of equal size, the overall rate of intestinal absorption of glucose in birds is very similar (Tilgner-Peter and Kalms, 1957).

The rates of intestinal absorption of amino acids in the domestic fowl are similar to those of the white rat (Kratzer, 1944) but the relative rates are not functions of molecular weight (Tasaki and Takahashi, 1966). As in mammals, it is clear that active transport is involved (e.g., Paine *et al.*, 1959; Lerner *et al.*, 1968; Tasaki and Takahashi, 1966). A glycine transport mechanism exists as early as the fifteenth day after hatching. The number of transport mechanisms is unknown. It seems probable that there is a separate transport system for glycine (Lerner and Taylor, 1967). Methionine, cysteine, cystine, leucine, and phenylalanine appear to share, at least in part, a common transport system that does not involve lysine, histidine, and aspartic acid (Lerner *et al.*, 1968; Lerner and Taylor, 1967; Paine *et al.*, 1959; Tasaki and Takahashi, 1966). The overall picture may be similar to that of mammals (see, for example, Wiseman, 1968).

Although it is clear that a substantial fraction of fat—up to 75% in the domestic fowl (e.g., Cole and Boyd, 1967)—is absorbed, the processes involved have not yet been adequately characterized.

Other than its obvious role in movement and temporary storage of intestinal contents, little is known about the function of the large intestine. The lower moisture content of its contents suggests a function in resorption of water. According to Skadhauge (1967, 1968) it, like the cloaca (Nechay and Lutherer, 1968; Schmidt-Nielsen *et al.*, 1963; Skadhauge, 1967, 1968), has a role in recovery of water and salts.

The digestive functions of the intestinal ceca are still largely a matter of speculation. Most of the attention has been directed toward the galliform species. In these it appears obvious that only a fraction of the total intestinal contents actually enters the ceca and that the time of retention is relatively long. In the domestic fowl and in the grouse the ceca are emptied no more than once every 24 hours. There are several possible functions:

1. Absorption of water (Radeff, 1928).
2. Absorption of nonprotein nitrogen (Mangold, 1929; Röseler, 1928).
3. Digestion of carbohydrate and protein (Maumus, 1902; Maumus and Launoy, 1901).
4. Microbial decomposition of cellulose (see part G, below).
5. Microbial synthesis and absorption of vitamins.

All of these suggestions require further critical investigation. Recent investigations, including studies with gnotobiotic chicks, seem to cast some doubt on an important role in synthesis of vitamins (Coates *et al.*, 1968; Couch *et al.*, 1950; Jackson *et al.*, 1955; Sunde *et al.*, 1950). It must be noted also that chickens and turkeys survive well after cecectomy or occlusion of the ceca (Beattie and Shimpton, 1958; Durant, 1929; Röseler, 1929; Schlotthauer *et al.*, 1933, 1935; Sunde *et al.*, 1950).

F. Motor Functions of the Intestine

The rate of passage of material through the intestine, of course, varies substantially in different species so that no very useful generalizations can be made (e.g., Löffler and Leibetseder, 1966) although the period of retention is generally shortest in the carnivores and frugivores and longest in the herbivores. Thus Groebbels (1932b) found that berries could pass through the digestive system of *Sylvia atricapilla* in as little as 12 minutes although the time for an entire meal of berries was over 2 hours. This is consistent with the observations of Stevenson (1933) for other passerine species. Oats can pass through the entire alimentary canal of a goose in as little as 4 hours, whereas the time of passage for an entire meal is about 17 hours (Groebbels, 1932b). X-ray investigations of the domestic fowl (Steinmetzer, 1924) suggest that food may reach the small intestine in as little as 10 minutes after feeding and that the undigested residue may appear in the feces in as little as 3.5 hours after feeding (see, also, Browne, 1922; Habeck, 1930; Kaupp and Ivey, 1923), although it is obvious that passage of the entire meal may require more than a day. The times involved in domestic turkeys are similar to those of the domestic fowl (Hillerman *et al.*, 1953).

The movements of the avian intestine have been studied only to a limited extent; obviously both peristaltic and segmenting movements are involved (Vonk and Postma, 1949). Peristaltic waves travel at the rate of several centimeters per second. Intestinal motility is a complex function of the intrinsic automaticity of the myenteric nervous system, extrinsic neural controls, hormonal control, and local chemical and mechanical influences. The myenteric nervous system, including the myenteric (Auerbach) and submucosal (Meissner) plexus, is well developed (Ábrahám, 1936; Kolossow, 1959; Kolossow *et al.*, 1932; Nolf, 1927a,b, 1928; Patzelt, 1936).

The neuronal aspects of the motor functions of the intestine will be considered in some detail in Volume IV. The following brief résumé is based largely on Burn (1968), Burnstock (1969), Enemar *et al.* (1965), Everett and Mann (1968), Nolf (1927a,b, 1928, 1934a,b,c, 1938a,b,c), and, especially, on the review of Bennett (1970b). The extrinsic innervation consists of pre- and postganglionic cholinergic fibers which may be vagal or sympathetic in origin. These reach the intestine via the celiac plexus and, in part, then via Remak's nerve, and via the mesenteric plexus. A predominantly inhibitory adrenergic supply terminates in the enteric plexus rather than in the musculature. These adrenergic fibers may arise from the paravertebral ganglia, the celiac and mesenteric plexus, or from ganglionic cells in Remak's nerve. Nonadrenergic inhibition of the intestine occurs in response to stimulation of the perivascular trunks; the neurons involved are probably vagal in origin. The vagus has a generally excitatory effect on the entire small intestine, an effect that is primarily a reinforcement of the fundamentally automatic movements of the musculature. Less is known about the control of the movements of the cloaca. It is, at least, innervated by cholinergic excitatory and adrenergic inhibitory fibers although the precise origins of these fibers are not definitely known.

Yasukawa (1959) has described in some detail the movements of the ceca of the domestic fowl. Antiperistaltic movements begin in the large intestine and pass alternately into the right and left ceca. Peristaltic movements begin at the apices and may cease at the proximal ends of the ceca or may proceed on into the large intestine. The latter, mass peristaltic waves occur simultaneously in both ceca and are propagated in the large intestine as a single wave; this type of mass peristalsis is involved in the formation of the cecal droppings.

G. Microbial Digestion

It is well known that the digestive glands of vertebrate animals do not produce carbohydrases capable of hydrolyzing the β-linkages of cellulose and other related structural carbohydrates of plants. *A priori*, it therefore would appear likely that herbivorous birds may rely on microbial digestion of cellulose and utilization of the resulting metabolites as is the case in many herbivorous mammals. Actually there is some evidence to suggest that this may indeed be the case. Unfortunately most of the investigations have been conducted with the domestic fowl, which may be a poor experimental species. It has been shown that some cellulose actually disappears in the course of its passage through the digestive tract, indicating the occurrence of microbial decomposition of cellulose in the domestic fowl (Brüggemann, 1931; Chelbnikow, 1936; Engler, 1933; Halnan, 1928; Heller *et al.*, 1930; Katayama, 1924; Olsson *et al.*, 1950; Stotz and Brüggemann, 1933), domestic ducks (Brüggemann, 1931; Troitzkaja *et al.*, 1936), and domestic geese (Brüggemann, 1931). On the other hand, Weiser and Zaitschek (1902) were unable to demonstrate cellulose utilization in ducks, geese, or chickens, and Olsson *et al.* (1950) suggest that the amounts decomposed in chickens may be trivial. It must be pointed out that there are technical difficulties in studying the rate of cellulose utilization and further that there are obvious differences in utilization depending on the source. Nevertheless, the extensive studies of Mangold (1928, 1929, 1931a, 1934b, 1943), Radeff (1928), Henning (1929), and Röseler (1929), including experiments with cecectomized birds, led them to the firm conclusion that cellulose is decomposed in the digestive tract and that the principal site of decomposition is the cecum. More recently Beattie and Shrimpton (1958) have shown that the microbial metabolism in the cecum of the domestic fowl is primarily fermentative; like Sunde *et al.* (1950) they found, however, that cecectomy was not detrimental to growth.

Further evidence of microbial decomposition of cellulose comes from the investigations of Suomalainen and Arhimo (1945) who found cellulose-decomposing activity in cultures from the ceca of *Tetrao urogallus*, *Lyrurus tetrix*, *Tetrastes bonasia*, and *Lagopus lagopus*. The cultures from young birds were less active than those from adults. Fermentation in the ceca of *Lagopus lagopus*, on their winter diet of willow buds and twigs, has been demonstrated by McBee and West (1969); the principal fermentation products were acetic, propionic, butyric, and lactic acids and ethanol. However, the extent to which the products of cecal fermentation contribute to the energy requirement of the birds is not yet clear.

It must be emphasized strongly that neither the demonstration of cellulolytic bacteria nor the disappearance of cellulose in the digestive tract proves that the bird is actually deriving nutrient material from the decomposed cellulose. Only the demonstration of actual absorption of usable decomposition products into the blood can be accepted as conclusive evidence.

It must be added that the comparative morphology of galliform ceca argues in support of a role in microbial digestion of cellulose, since the browsing herbivorous species have much more extensively developed ceca (Leopold, 1953; Schumacher, 1925) than the graminivorous species.

Evidence of a fascinating type of microbial digestion has emerged from the studies of Friedmann and Kern (1956) on the honeyguides. These birds eat substantial quantities of beeswax. *Indicator minor* can maintain itself on a diet of beeswax for a considerable period of time and certainly, therefore, must derive energy from it. The evidence is highly suggestive that breakdown of the beeswax to usable compounds is effected by the microflora of the intestine (Friedmann and Kern, 1956).

VI. The Liver

The avian liver is a large, conspicuous bilobed organ, the right lobe usually being the larger (Fig. 10). Its role in digestion is associated primarily with the production of bile. Among its numerous nondigestive functions are the storage of lipids and glycogen, intermediary metabolism and synthesis of proteins and glycogen, and the formation of uric acid. It is a hematopoietic organ during the embryonic and immediate postembryonic periods (Sandreuter, 1951). Attention here is directed solely to its role as a part of the digestive system. According to Magnan (1910a), the liver is relatively smallest in carnivores and graminivores and largest in piscivores and insectivores;

he (Magnan, 1912b) reports that ducks fed on a fish diet had much larger livers than ducks fed on a grain or meat diet.

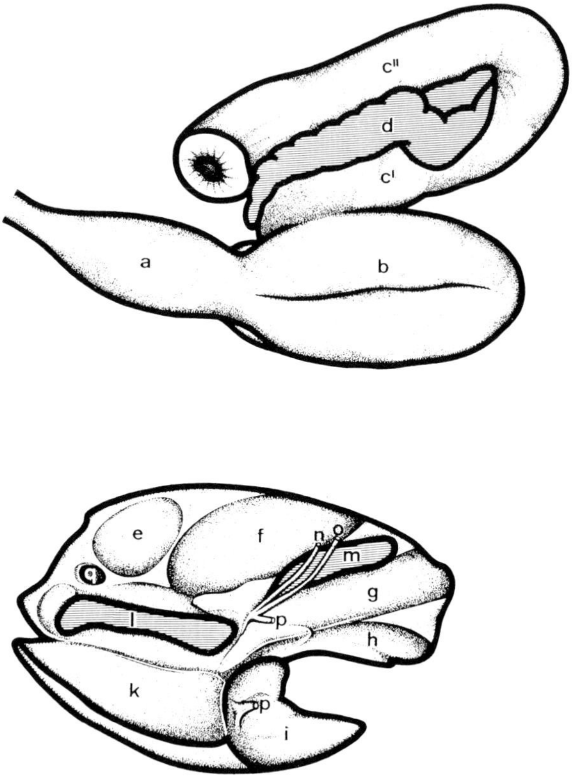

FIG. 10. Topography of the liver and adjacent organs. Above, digestive tract with liver removed; below, liver. a, Glandular stomach; b, gizzard; c', descending duodenum; c'', ascending duodenum; d, pancreas; e, impression of the right testis; f, Impressio duodenalis distalis; g, Impressio duodenalis proximalis; h–i, impression of the gizzard; k, impression of the proventriculus; l, spleen; m, gall bladder; n, Ductus cysticoentericus; o, Ductus hepatoentericus; p, Vena portae; q, point of entry of the Vena cava caudalis.

In situ the liver has a convex and more or less uneven visceral surface that faces the abdominal wall. A deep caudal and a shallower cranial cleft divide the liver into its right and left lobes. The craniodorsal parts of the liver extend to the lungs. The other adjacent organs make typical impressions in the liver; thus that of the heart lies in the cranial part of the parietal surface. On the visceral surface of the left lobe are the impressions of the glandular and muscular stomach and on the right lobe is the impression of the cranial part of the duodenal loop with the pancreas. The spleen is embedded in the medial part of the visceral surface. The impression of the right testis can be distinguished on the surface of the liver of males (Martin, 1970). In basic organization the avian liver is a muralium (Eberth, 1867; Elias, 1955; Elias and Bengelsdorf, 1952). The walls or plates that separate the lacunae are either one or two cells thick. One-cell-thick plates are apparently general among passerine species (Hickey and Elias, 1954) and also in *Tympanuchus cupido* and *Podiceps grisegena;* two-cell-thick plates occur in the domestic fowl, *Aix sponsa,* and *Fulica americana.*

The vena cava passes through the cranial section of the right lobe; the hepatic veins and the vena cava enter and leave, respectively, at the same site. The two hepatic portal veins and the two hepatic arteries enter through a groove, the fossa transversa, in the middle of the visceral surface. Through this same groove the two bile ducts leave the liver (Martin, 1970).

A hepatic duct leads from each lobe of the liver into the duodenum. The left hepatic duct communicates directly into the duodenum whereas the right may have a branch into the gall bladder, or may be enlarged locally as a gall bladder; hence, in its terminal part, it serves as a cystic duct. The gall bladder is lacking in many species, a condition without apparent phylogenetic significance (Gorham and Ivy, 1938). For example, the gall bladder is absent in *Struthio camelus,* the Rheidae, *Falco peregrinus,* Opisthocomidae, *Turnix tanki,* many doves and pigeons, many parrots, at least two species of hummingbirds, some woodpeckers, and some passerine species such as *Bombycilla garrulus.* The gall bladder is both a storage and concentrating organ. The latter function is clearly demonstrated by the studies of Schmidt and Ivy (1937) with the domestic fowl and mallard. Presumably, this function gives the advantage of delivering into the intestine a very concentrated bile at the beginning of a digestive period. Concentration is effected primarily with respect to bile pigments and bile salts (Farner, 1941; Schmidt and Ivy, 1937), whereas the concentration of buffering compounds is lower in gall-bladder bile than in hepatic-duct bile.

The principal functions of bile in digestion are the neutralization of acid in the chyme and the emulsification of fats. The latter is effected by the bile salts, of which at least six have been identified from eight different species (I. G. Anderson *et al.,* 1957; Hosizima *et al.,* 1930; Takahashi, 1938; Windaus and von Schoor, 1926a,b; Yamasaki, 1933, 1951; Yonemura, 1926, 1928). Bile stimulates the secretion of succus entericus (Lavrenteva, 1963). The bile of the domestic fowl also has appreciable amylolytic activity (Farner, 1943c). The rate of bile secretion increases during feeding (Lavrenteva, 1963).

VII. The Pancreas

The pancreas is both an endocrine and exocrine gland. Typically, it lies in, or associated with, the duodenal loop. As an endocrine gland it produces two hormones, glucagon and insulin, presumably in the A and B cells, respectively. This discussion will be directed primarily toward the exocrine function of the pancreas in the production of pancreatic juice.

The avian pancreas arises embryologically from the buds from the gut and consists of at least three morphologically identifiable lobes (Clara, 1924; Frei, 1966; Hill, 1926; Widmer, 1966), which can be designated as dorsal, ventral, and intermediate. The intermediate lobe, especially, can be subdivided into further lobes.

The organization of the individual lobes, their relationship to each other, and the arrangement of the duct system are highly variable among the various groups of birds; however, within individual groups, for example the families of passerine birds, the relationships are relatively constant (Frei, 1966; Widmer, 1966). In the Fringillidae there is always a connection between the dorsal and intermediate lobes and between the ventral and intermediate lobes whereas the two principal lobes are directly connected. There are three pancreatic ducts. The ducts from the dorsal and ventral lobes always open at the same level near the bend of the ascending duodenum. The third duct leaves the dorsal lobe more cranially and opens together with the bile ducts into the small intestine. The intermediate lobe has no direct connection to the small intestine but empties its secretion into the duct system of the dorsal and ventral lobes (Frei, 1966). In many of the Cardinalinae, on the contrary, the dorsal and ventral lobes are fused together and the intermediary lobe is organized into two independent sections that always follow the ascending and descending parts of the duodenum as delicate bands. From the dorsal and ventral lobes a pancreatic duct always opens near the bend of the duodenum. A third duct leads from the part of the intermediate lobe on the ascending part of the duodenum to the site of the opening of the bile ducts and from there into the small intestine (Widmer, 1966).

The relatively weakly staining Islets of Langerhans occur with variable frequency in the pancreas. In contrast to the relationships in mammals, the individual islets are not separated by connective tissue capsules from the other tissue of the pancreas (H. Oppel, 1900; Clara, 1924) so that in certain cases it even becomes difficult to identify individual cells as exocrine or endocrine. Moreover, exocrine parenchyma, connected with the duct system, often lies within the islets (I. Müller *et al.*, 1956; Hellman and Hellerström, 1960). The detailed histology and function of the islets will be presented in the chapter on peripheral endocrine glands.

The exocrine part of the pancreas is tubulo-acinar; the secretion is serous. Between the tubules the smallest blood vessels—capillaries, venules, arterioles—occur in large numbers. The tissue also contains numerous ganglia of the intramural type. The connective tissue between the tubules is sparse. The tubules can be very long and branched and can be interconnected through anastomoses. Zymogen granules are numerous in the vicinity of the lumen. The cell nuclei are spherical, with very conspicuous nuclear membranes, and contain very large nucleoli. In addition to the principal nucleolus there may be a few to many accessory nucleoli.

The secretion reaches the lumen through intercellular secretory ductules. Into these open intracalated ducts, whose lumina are formed from centro-acinar cuboidal to flat epithelium. Intercalated ducts can intercommunicate and open finally directly into the secretory duct which is enclosed with a well-developed connective tissue layer.

The secretory ducts are lined with columnar epithelium. This epithelium, near the opening of the main duct into the intestine, begins to become intensively folded (Frei, 1966).

The pancreatic juice has two principal roles in digestion. Because of its relatively high concentration of buffering compounds (Farner, 1941), it has an important role in the neutralization of chyme. Its second role, as a consequence of its digestive enzymes, is in the chemical phase of digestion in the small intestine. Amylolytic activity has been demonstrated in the pancreas of the domestic fowl (Farner, 1941, 1942a; Hewitt and Schelkopf, 1955; Hirata, 1910; Paira-Mall, 1900; Plimmer and Rosedale, 1922; Polyakov, 1958; Salman *et al.*, 1967; Shaw, 1913; Younathan and Frieden, 1956), domestic pigeon (Braitmaier, 1904; Dandrifosse, 1970; L. E. Hokin, 1951a,b; Langendorff, 1879a; Rothlin, 1922; Webster, 1969), domestic goose (Klug, 1891, 1892), and *Passer domesticus* (Langendorff, 1879b). As early as 1856, Bernard had found that depancreatized birds failed to digest starch, an observation repeated by Langendorff (1879a). In the domestic fowl there can be little doubt that the pancreas is the major source of amylase (Farner, 1941). The pancreas was known by Bernard

(1849) as early as 1849 as a source of lipase; it has been demonstrated specifically to produce lipase in the domestic fowl (Hewitt and Schelkopf, 1955; Laws and Moore, 1963b; Plimmer and Rosedale, 1922; Polyakov, 1958; Shaw, 1913) and in the domestic pigeon (Braitmaier, 1904; Langendorff, 1879a; Schleucher and Hokin, 1954).

The production of proteolytic enzymes by the pancreas has been reported for at least 20 species in six different orders (Braitmaier, 1904; Herpol, 1964, 1965; Herpol and van Grembergen, 1967b; Hewitt and Schelkopf, 1955; Klug, 1891, 1892, Langendorff, 1879a; Paira-Mall, 1900; Plimmer and Rosedale, 1922; Polyakov, 1958). Beyond the fact that the proteinases involved have their optimum activity under slightly alkaline conditions, and that they can use whole protein as substrate, little is known about them. Chymotrypsin A and B have been demonstrated in the domestic fowl (Pfleiderer *et al.*, 1970; Ryan, 1965; Ryan *et al.*, 1965; Salman *et al.*, 1967) as has been trypsinogen (Salman *et al.*, 1967). A trypsin from the domestic turkey has been prepared and partially characterized (Ryan, 1965; Ryan *et al.*, 1965). It has been demonstrated (Ivanov and Gotev, 1962; Dal Borgo *et al.*, 1968; Salman *et al.*, 1967) that the presence of inhibitors of trypsin and chymotrypsin in the diet, with accompanying incomplete digestion, results in hyperplasia of the pancreas, an increase in volume rate of secretion, and a corresponding increase in total production of trypsinogen and chymotrypsinogen. It appears that the pancreas can adjust its secretion of trypsinogen, amylase, and lipase in accordance with the composition of the diet (Ivanov and Gotev, 1962). Trypsin and intestinal enterokinase, which activates inactive trypsinogen, have been reported from the "sea gull" by Waldschmidt-Leitz and Shinoda (1928). Dipeptidase, aminopeptidase, and carboxypeptidase activities have been demonstrated in pancreatic extracts of domestic fowl (De Rycke, 1962). Thus the fragmentary information available suggests that the digestion of proteins in birds does not deviate significantly from that of higher invertebrates in general. However, much remains to be learned.

The rate of secretion of enzymes is apparently controlled by both humoral and neural mechanisms. In experiments with pigeon-pancreas slices, it has been shown that the processes of synthesis and secretion are not interdependent, i.e., synthesis is not affected by secretion (L. E. Hokin and Hokin, 1961; Fernandez and Junquiera, 1955; Sahba *et al.*, 1970; Schleucher and Hokin, 1954). Experiments concerning the role of pancreozymin have not yet given a completely satisfactory picture. It is clear that the rate of secretion of amylase, and presumably other enzymes, is increased markedly by pancreozymin. There is evidence that it can increase the rate of synthesis experimentally, especially with large doses given to previously fasted animals, but an assessment of its real role, if any, in the control of synthesis cannot be made at this time (L. E. Hokin and M. R. Hokin, 1956; Sahba *et al.*, 1970; Webster and Tyor, 1966). The evidence of neural (cholinergic) control rests primarily on the use of acetylcholine and cholinergic drugs on pigeon-pancreas slices. Collectively the evidence indicates that both synthesis and secretion are under cholinergic control (L. E. Hokin, 1951a; M. R. Hokin and L. E. Hokin, 1953; Webster, 1968, 1969). Cholinergic effects are blocked by atropine whereas those of pancreozymin are not. Although secretin is doubtless involved in the control of the volume rate of secretion of pancreatic juice, good evidence appears still to be lacking (Heatley *et al.*, 1965).

REFERENCES

Ábrahám, A. (1936). Beiträge zur Kenntnis der Innervation des Vogeldarmes. *Z. Zellforsch. Mikrosk. Anat.* **23,** 737–745.

Abraham, K. (1901). Beiträge zur Entwicklungsgeschichte des Wellensittichs (*Melopsittacus undulatus*). *Anat. Hefte Abt. 2* **17,** 589–669.

Aitken, R. N. C. (1958). A histochemical study of the stomach and intestine in the chicken. *J. Anat.* **92,** 453–466.

Alekseev, S. N. (1913). O rasprostranenii diastaticheskovo fermenta u nokotorych zernoyadnykh ptits. *Izv. St. Petersb. Biol. Lab.* **13.** (Original not seen. See Karpov, 1919.)

Al-Hussaini, A. H., and Amer, F. I. (1959). Studies on the developmental anatomy of the bursa Fabricii of the fowl. *Bull. Zool. Soc. Egypt* **14,** 60–96.

Allenspach, A. L. (1964). Experimental analysis of closure and reopening of the esophagus in the developing chick. *J. Morphol.* **114,** 287–302.

Allenspach, A. L. (1966). The reopening process of the esophagus in the normal chick and the crooked neck dwarf mutant. *J. Embryol. Exp. Morphol.* **15,** 67–76.

Allenspach, A. L., and Hamilton, H. L. (1962). Histochemistry of the esophagus in the developing chick. *J. Morphol.* **111,** 321–343.

Alvarado, F. (1967a). D-Xylose transport in the chicken small intestine. *Comp. Biochem. Physiol.* **20,** 161–170.

Alvarado, F. (1967b). *d*-Xylose transport in the chicken small intestine. *Comp. Biochem. Physiol.* **20,** 461–470.

Alvarado, F., and Monreal, J. (1967). Na^+-dependent active transport of phenylglucosides in the chicken small intestine. *Comp. Biochem. Physiol.* **20,** 471–488.

Anderson, A. E., and Nafstad, P. H. J. (1968). An electron microscope investigation of the sensory organs in the hard palate region of the hen (*Gallus domesticus*). *Z. Zellforsch. Mikrosk. Anat.* **91,** 391–401.

Anderson, I. G., Haslewood, G. A. D., and Wooton, I. D. P. (1957). Comparative studies of bile salts. X. Bile salts of the King Penguin, *Aptenodytes patagonica*. *Biochem. J.* **67,** 323–328.

Antonino, C. (1923). Die caseinolytische Wirkung im Darmsaft und ihre allgemeine Verbreitung in den Geweben des tierischen Organismus. *Biochem. Z.* **136,** 71–77.

Antony, M. (1920). Über die Speicheldrüsen der Vögel. *Zool. Jahrb., Abt. Anat. Ontog. Tiere* **41,** 547–660.

Arcangeli, A. (1904). Ricerche istologiche sopra il gozzo del colombo all'epoca del cosidetto "allattamento." *Monit. Zool. Ital.* **15,** 218–232.

Ariyoshi, S., Koike, T., Furuta, F., Ozone, K., Matsumura, Y., Dimick, M. K., Hunter, W. L., Wang, W., and Lepkovsky, S. (1964). Digestion of protein, fat, and starch in the depancreatized chicken. *Poultry Sci.* **43,** 232–238.

Ashcraft, D. W. (1930). The correlative activities of the alimentary canal of the fowl. *Amer. J. Physiol.* **93,** 105–110.

Barnawall, E. B. (1968). Hormonal influence on salivary glands of Chimney Swift (*Chaetura pelagica*) in organ culture. *J. Exp. Biol.* **169,** 161–172.

Bartels, P. (1895). Beitrag zur Histologie des Ösophagus der Vögel. *Z. Wiss. Zool.* **59,** 655–689.

Bartram, E. (1901). Anatomische, histologische und embryologische Untersuchungen über den Verdauungstraktus von *Eudyptes chrysocome*. *Z. Naturwiss.* **74,** 173–236.

Batelli, A. (1890a). Delle glandule salivari del *Cypselus apus*. *Atti. Accad. Med.-Chir. Perugia* **2,** 27–35.

Batelli, A. (1890b). Glandule salivari del Trampolieri. *Atti. Accad. Med.-Chir. Perugia* **2,** 94–102.

Batelli, A., and Giacomini, E. (1883). Contributo delle glandulae salivari degli uccelli. *Mem. Soc. Toscana Sci. Natur.* **6,** 385–444.

Batelli, A., and Giacomini, E. (1889). Struttura istologica delle glandule salivari degli uccelli. *Atti. Accad. Med.-Chir. Perugia* **1,** 57–64 and 87–99.

Bath, W. (1906). Die Geschmacksorgane der Vögel. (Teil I der Arbeit: Die Geschmacksorgane der Vögel und Krokodile.) Inaugural-Dissertation, Friedrich Wilhelms University, Berlin.

Batt, H. E. (1924). A study of normal histology of the fowl. *Rep. Ontario Vet. Colloq.* p. 122.

Bauer, M. (1901). Beiträge zur Histologie des Muskelmagens der Vögel. *Arch. Mikrosk. Anat. Entwicklungsmech.* **57,** 653–676.

Beams, H. W., and Meyer, R. K. (1931). The formation of pigeon milk. *Physiol. Zool.* **4,** 486–500.

Beattie, J., and Shrimpton, D. H. (1958). Surgical and chemical techniques for *in vivo* studies of the metabolism of the intestinal microflora of domestic fowls. *Quart. J. Exp. Physiol.* **43,** 399–407.

Bennett, T. (1969a). Nerve-mediated excitation and inhibition of the smooth muscle cells of the avian gizzard. *J. Physiol. (London)* **204,** 669–686.

Bennett, T. (1969b). The effects of hyoscine and anticholinesterases on cholinergic transmission to the smooth muscle cells of the avian gizzard. *Brit. J. Pharmacol.* **37,** 585–594.

Bennett, T. (1969c). Studies on the avian gizzard: Histochemical analysis of the extrinsic and intrinsic innervation. *Z. Zellforsch. Mikrosk. Anat.* **98,** 188–201.

Bennett, T. (1970a). Interaction of nerve-mediated excitation and inhibition of single smooth muscle cells of the avian gizzard. *Comp. Biochem. Physiol.* **32,** 669–680.

Bennett, T. (1970b). Autonomic neuro-effector systems with particular reference to birds. Doctoral Thesis, University of Melbourne.

Bennett, T., and Cobb, J. L. S. (1969a). Studies on the avian gizzard: The development of the gizzard and its innervation. *Z. Zellforsch. Mikrosk. Anat.* **98,** 599–621.

Bennett, T., and Cobb, J. L. S. (1969b). Studies on the avian gizzard: Auerbach's plexus. *Z. Zellforsch. Mikrosk. Anat.* **99,** 109–120.
Bennett, T., and Cobb, J. L. S. (1969c). Studies on the avian gizzard. *Z. Zellforsch. Mikrosk. Anat.* **91,** 173–185.
Berger, J., and Johnson, M. J. (1940). The occurrence of leucylpeptidase. *J. Biol. Chem.* **133,** 157–172.
Bergmann, C. G. L. C. (1862). Einiges über den Drüsenmagen der Vögel. *Arch. Anat. Physiol.* **1862,** 581–587.
Berlin, W. (1853). Bijdrage tod de spijsvertering der Vogels. *Lancet* **2,** 57–68.
Bernard, C. (1849). Recherches sur les usages du suc pancréatique. *C. R. Acad. Sci.* **28,** 283–285.
Bernard, C. (1858). "Leçons sur la physiologie et la pathologie du système nerveux," 2 vols. Baillière, Paris.
Bernard, C. (1859). "Les propriétés physiologiques et les altérations pathologiques des liquides de l'organisme," Dixième Leçon, pp. 220–238. Baillière, Paris.
Bernardi, A. (1926). I fermenti nell'intestino anteriore del polli. *Biochim. Ter. Sper.* **13,** 287–295.
Bernardi, A., and Rossi, G. (1923). Di un fermento invertente riscontrato nel ventriglio dei polli. *Biochim. Ter. Sper.* **10,** 290–298.
Bernardi, A., and Schwartz, M. A. (1932a). Beitrag zur Kenntnis des Ursprungs der Amylase im Kaumagen der Hühner. *Biochem. Z.* **253,** 382–386.
Bernardi, A., and Schwartz, M. A. (1932b). Das Vorkommen einer Invertase im Kaumagen der Hühner. *Biochem. Z.* **256,** 406–410.
Bernardi, A., and Schwartz, M. A. (1933a). Über einige Fermente der *Hydrochelidon nigra* L. *Biochem. Z.* **260,** 369–375.
Bernardi, A., and Schwartz, M. A. (1933b). Über das Vorkommen von Amylase, Invertase, und Proteinase in Huhn, *Hydrochelidon nigra* L., und in *Vanellus vanellus* L. *Biochem. Z.* **262,** 175–180.
Bernardi, A., and Zanini, M. (1930). Contributo allo studio die fermenti contenuti nel ventriglio di pollo e di rondine di mare (*Hydrochelidon nigra* L.). *Biochim. Ter. Sper.* **17,** 6–15.
Bernstein, H. A. (1859). Über die Nester der Salanganen. *J. Ornithol.* **7,** 111–119.
Biedermann, W. (1911). Die Aufnahme, Verarbeitung und Assimilation der Nahrung. *Wintersteins Handbuch der vergleichenden Physiologie,* Vol. 2, Jena.
Bock, W. J. (1961). Salivary glands in gray jays (*Perisoreus*). *Auk* **78,** 355–365.
Bogdanov, R. Z., and Kibiakov, A. V. (1955). Kmekhanizmu innervatsii zheludka ptits. *Fiziol. Zh. SSSR im I. M. Sechenova* **41,** 239–242.
Bogner, P. H. (1966). Development of sugar transport in the chick intestine. *Biol. Neonatorum* **9,** 1–7.
Bogner, P. H., and Haines, I. A. (1961). Development of intestinal selective absorption of glucose in newly-hatched chicks. *Proc. Soc. Exp. Biol. Med.* **107,** 265–267.
Bogner, P. H., and Haines, I. A. (1964). Functional development of active sugar transport in the chick intestine. *Amer. J. Physiol.* **207,** 37–41.
Böker, H. (1931). Abnorme Linkslage der Halseingweide bei Vögeln und ihre Entstehung. *Gegenbaurs Jahrb.* **66,** 220–230.
Borelli, G. A. (1743). De motu animalium. Philosophia de motu animalium. P. Goose, The Hague.
Botezat, E. (1904). Geschmacksorgane und andere nervöse Endapparate im Schnabel der Vögel. *Biol. Zentralbl.* **24,** 722–736.
Botezat, E. (1906). Die Nervenendapparate in den Mundteilen der Vögel und die

einheitliche Endigungsweise der peripheren Nerven bei den Wirbeltieren. *Z. Wiss. Zool.* **84,** 205–360.

Botezat, E. (1910). Ueber Sinnesdrüsenzellen und die Funktion von Sinnesapparaten. *Anat. Anz.* **37,** 513–530.

Bowman, W. C., and Everett, S. D. (1964). An isolated parasympathetically innervated oesophagus preparation from the chick. *J. Pharm. Pharmacol.* **16,** Suppl., 72–79.

Boyden, E. A. (1922). The development of the cloaca in birds with special reference to the origin of the bursa of Fabricius, the formation of the urodaeal sinus, and the regular occurrence of a cloacal fenestra. *Amer. J. Anat.* **30,** 163–201.

Braitmaier, H. (1904). Ein Beitrag zur Physiologie und Histologie der Verdauungsorgane bei Vögeln. Inaugural-Dissertation, University of Tübingen.

Brandes, G. (1896). Über den Einfluss veränderter Ernährung auf die Struktur des Vogelmagens. *Biol. Zentralbl.* **16,** 825–838.

Breteau, J. (1962). Contribution à l'étude anatomique et histologie des glandes salivaires de *Gallus gallus.* Thesis, University of Lyon.

Broussy, J. (1936). Sur un point particulier de l'histophysiologie de la muqueuse du gésier des granivores. *C. R. Soc. Biol.* **121,** 1298–1300.

Brown, E. W. (1904). Digestion experiments with poultry. *U.S. Dep. Agr., Bur. Anim. Ind. Bull.* **96,** 1–112.

Brown, K. M., and Moog, F. (1967). Invertase activity in the intestine of the developing chick. *Biochim. Biophys. Acta* **132,** 185–187.

Browne, T. G. (1922). Some observations on the digestive system of the fowl. *J. Comp. Pathol. Ther.* **35,** 12–32.

Brown-Séquard, C. E. (1850). Existence d'un mouvement rhythmique dans le jabot des oiseaux. *C. R. Soc. Biol.* **2,** 83.

Brücke, E. (1859). Beitrag zur Lehre von der Verdauung. *Sitzungsber. Akad. Wiss. Wien, Math.-Naturwiss. Kl.* **37,** 131–184.

Brüggemann, H. (1931). Die Verdaulichkeit der Rohrfaser bei Hühnern, Tauben, Gänsen, und Enten. *Arch. Tierernaehr. Tierzucht.* **5,** 89–126.

Buckner, G. D., and Martin, J. H. (1922). The function of grit in the gizzard of the chicken. *Poultry Sci.* **1,** 108–113.

Buckner, G. D., Martin, J. H., and Peter, A. M. (1926). Concerning the growth of chickens raised without grit. *Poultry Sci.* **5,** 203–208.

Buckner, G. D., Insko, W. M., Jr., and Henry, A. H. (1944). Does breed, age, sex or laying condition affect the pH of the digestive system of chickens? *Poultry Sci.* **23,** 457–458.

Bujard, E. (1909). Etude des types appendiciels de la muquese en rapport avec les régimes alimentaires. *Int. Monatsschr. Anat. Physiol.* **26,** 1–96.

Burn, J. H. (1968). The development of the andrenergic fibre. *Brit. J. Pharmacol. Chemother.* **32,** 575–582.

Burnstock, G. (1969). Evolution of the autonomic innervation of visceral and cardiovascular systems in vertebrates. *Pharmacol. Rev.* **21,** 247–324.

Burrows, W. H. (1936). The surgical removal of the gizzard from the domestic fowl. *Poultry Sci.* **15,** 290–293.

Cadow, G. (1933). Magen und Darm der Fruchttauben. *J. Ornithol.* **81,** 236–252.

Calhoun, M. H. (1954). "Microscopic Anatomy of the Digestive System." Iowa State Coll. Press, Ames.

Capobianco, F. (1902). Contributo alla costituzione dello strato cuticulo-ventricolare dello stomaco muscoloso degli uccelli. *Boll. Soc. Natur. Napoli* [1] **15,** 160.

Carr, R. H., and James, C. M. (1931). Synthesis of adequate protein in the glands of the pigeon crop. *Amer. J. Physiol.* **97,** 227–231.

Cattaneo, G. (1885). Istologia e sviluppo dell'apparato gastrico degli uccelli. *Atti Soc. Ital. Sci. Natur. Mus. Civ. Stor. Natur. Milano* **27**, 88–175.

Cazin, M. (1886). Observations sur l'anatomie du pétrel géant. *Bibl. Ecole Hautes Etud.* **31**, 27 pp.

Cazin, M. (1887). Recherches anatomiques histologiques et embryologiques sur l'appareil gastrique des oiseaux. *Ann. Sci. Natur. Zool.* [7] **4**, 177–323.

Cazin, M. (1887). Glandes gastriques à mucus et à ferment chez les oiseaux. *C. R. Acad. Sci.* **104**, 590–592.

Champy, C., and Colle, P. (1919). Sur une correlation entre la glande du jabot du pigeon et les glandes genitales. *C. R. Soc. Biol.* **82**, 818–819.

Charbonnel-Sallé, and Phisalix, C. (1886). Sur la sécrétion lactée du jabot des pigeon en incubation. *C. R. Acad. Sci.* **103**, 286–288.

Chauveau, A. (1862). Du nerf pneumogastrique considéré comme agent excitateur et comme agent coordinateur des contractions oesophagiennes. *J. Anat. Physiol. Norm. Pathol. Homme Anim.* **5**, 191–226 and 323–348.

Cheney, G. (1938). Gastric acidity in chicks with gastric ulcers. *Amer. J. Dig. Dis.* **5**, 104–107.

Chitty, D. (1938). A laboratory study of pellet formation in the Short-eared Owl. *Proc. Zool. Soc. London* **108**, 267–287.

Chelbnikow, N. I. (1936). Zur Methodik der Bestimmung der Rohfaserverdauung bei Hühnern. *Biedermann's Zentralbl. Abt. B* **8**, 32–41.

Chodnik, K. S. (1947). A cytological study of the alimentary tract of the domestic fowl (*Gallus domesticus*). *Quart. J. Microsc. Sci.* **88**, 419–443.

Chodnik, K. S. (1948). Cytology of the glands associated with the alimentary tract of domestic fowl (*Gallus domesticus*). *Quart. J. Microsc. Sci.* **89**, 75–87.

Choi, J. K. (1962). Fine structure of the smooth muscle of the chicken's gizzard. *Electron Microsc., Proc. Int. Congr., 5th, 1962* Vol. 2, Art. M-9.

Cholodkowsky, N. (1892). Zur Kenntniss der Speicheldrüsen der Vögel. *Zool. Anz.* **15**, 250–254.

Clara, M. (1924). Das Pankreas der Vögel. *Anat. Anz.* **57**, 257–265.

Clara, M. (1925). Beiträge zur Kenntnis des Vogeldarmes. I. Teil. Mikroskopische Anatomie. *Z. Mikrosk.-Anat. Forsch.* **4**, 346–416.

Clara, M. (1927). Das Problem des Rumpfdarmschleimhautreliefs. *Z. Mikrosk.-Anat. Forsch.* **9**, 1–48.

Clara, M. (1928). Le cellule basigranulose. Un contributo alla conoscenza della composizione dell'epithelio intestinale dei vertebrati superiori (uccelli e mammiferi). *Arch. Ital. Anat. Embriol.* **25**, 1–46.

Clara, M. (1934). Über den Bau des Magendarmkanals bei den Amseln (*Turdidae*). *Z. Anat.* **102**, 718–771.

Clarke, L. F., Rahn, H., and Martin, M. D. (1942). Seasonal and sexual dimorphic variations in the so-called "air sac" region of the sage grouse. *Wyo. Game Fish Dep., Bull.* **2**, 13–27.

Coates, M. E., and Jayne-Williams, D. J. (1966). Current views on the role of the gut flora in nutrition of the chicken. *Physiol. Dome. Fowl, Brit. Egg Mkt. Bd. Symp., 1st, 1964* pp. 181–188.

Coates, M. E., Ford, J. E., and Harrison, G. F. (1968). Intestinal synthesis of vitamins of the B-complex in chicks. *Brit. J. Nutr.* **22**, 493–500.

Cobb, J. L. S., and Bennett, T. (1969). A study of intercellular relationships in developing and mature visceral smooth muscle. *Z. Zellforsch. Mikrosk. Anat.* **100**, 516–526.

Cole, J. R., Jr., and Boyd, F. M. (1967). Fat absorption from the small intestine of gnotobiotic chicks. *Appl. Microbiol.* **15,** 1229–1234.

Collip, J. B. (1922). The activation of the glandular stomach of the fowl. *Amer. J. Physiol.* **59,** 435–438.

Cornselius, C. (1925). Morphologie, Histologie, und Embryologie des Muskelmagens der Vögel. *Gegenbaurs Jahrb.* **54,** 507–559.

Corti, A. (1923). Contributo alla migliore conoscenza dei diverticuli ciechi dell'intestino posteriore degli uccelli. *Ric. Morfol.* **3,** 211–295.

Couch, J. R., German, H. L., Knight, D. R., Parks, P. S., and Pearson, P. B. (1950). The importance of the cecum in intestinal synthesis in the mature domestic fowl. *Poultry Sci.* **29,** 52–58.

Crane, R. K. (1968). A concept of the digestive-absorptive surface of the small intestine. *In* "Handbook of Physiology" (Amer. J. Physiol., J. Field, ed.), Sect. 6, Vol. V, pp. 2535–2542. Williams & Wilkins, Baltimore, Maryland.

Curschmann, H. (1866). Zur Histologie des Muskelmagens der Vögel. *Z. Wiss. Zool.* **16,** 224–235.

Cuvier, G. L. C. F. D. (1805). "Leçons d'anatomie comparée, Vol. III. Cochard, Paris.

Cymborowski, B. (1968). Influence of diet on the histological structure of the gullet and glandular stomach of the Common Tern (*Sterna hirundo* L.). *Zool. Pol.* **18,** 451–468.

Dal Borgo, G., Salman, A. J., Pubols, M. H., and McGinnis, J. (1968). Exocrine function of the chick pancreas as affected by dietary soybean meal and carbohydrate (33448). *Proc. Soc. Exp. Biol. Med.* **129,** 877–881.

Dandrifosse, G. (1970). Mechanism of amylase secretion by the pancreas of the pigeon. *Comp. Biochem. Physiol.* **34,** 229–235.

Danysz, M., and Koskowski, W. (1922). Étude de quelques fonctions digestives chez les pigeons normaux, nourris aux riz poli, et en inanition. *C. R. Acad. Sci.* **175,** 54–56.

Darwin, C. (1896). "The Variation of Animals and Plants Under Domestication," 2nd ed., Vols. 1 and 2. Murray, London.

Davies, W. L. (1939). The composition of the crop milk of pigeons. *Biochem. J.* **33,** 898–901.

Dawson, A. B., and Moyer, S. L. (1948). Histogenesis of the argentophile cells of the proventriculus and gizzard in the chicken. *Anat. Rec.* **100,** 493–507.

Dedič, S. (1930). Über physiologische Formierung und Motilität der Verdauungsorgane bei Habichten (*Astur palumbarius*). *Fortschr. Geb. Roentgenstr.* **43,** 367–371.

de Réaumur, R. A. F. (1756a). Sur la digestion des oiseaux. I. Expériences sur la manière dont se fait la digestion dans les oiseaux vivent principalement de grains et herbes, et dont l'estomac est un gésier. *Mem. Acad. Sci. Inst. Fr.* [1] **1752,** 266–307.

de Réaumur, R. A. F. (1756b). Sur la digestion des oiseaux. II. De la manière dont elle se fait dans l'estomac des oiseaux de proie. *Mem. Acad. Sci. Inst. Fr.* [1] **1752,** 461–495.

de Réaumur, R. A. F. (1756c). Sur la digestion des oiseaux. *Hist. Acad. Roy. Sci. Mém. Math. Phys.* **1752,** 49–71.

De Rycke, P. (1962). Onderzoek over exopeptidasen bij het kuiken. *Natuurwetensch. Tijdschr. (Ghent)* **43,** 82–86.

Desnuelle, P. (1961). Pancreatic lipase. *Advan. Enzymol.* **23,** 129–161.

Desselberger, H. (1931). Der Verdauungskanal der Dicaeiden nach Gestalt und Funktion. *J. Ornithol.* **79,** 353–374.

Diakow, M. J. (1932). Untersuchungen über Verdaulichkeit, Stoff-, und Energiewechsel bei Hühnern. *Wiss. Arch. Landwirt., Abt. B* **7,** 571–637.

Dorst, J. (1952). Contribution à l'étude de la langue des *Méliphagides*. *Oiseaux* **20,** 185–214.

Dow, D. D. (1965). The role of saliva in food storage by the Gray Jay. *Auk* **82,** 139–154.

Doyon, M. (1894a). Contribution a l'étude des phénomènes mécaniques de la digestion gastrique chez les oiseaux. *Arch. Physiol. Norm. Pathol.* [5] **6,** 869–878.

Doyon, M. (1894b). Recherches expérimentales sur l'innervation gastrique des oiseaux. *Arch. Physiol. Norm. Pathol.* [5] **6,** 887–898.

Doyon, M. (1925). Influence des nerfs sur la moitricité de l'estomac chez l'oiseau. *C. R. Soc. Biol.* **93,** 578–580.

Dulzetto, F. (1927a). Sulla funzione digestiva delle ghiandole del gozzo di *Columba livia* (Bonnet). Sulla presenza di amilasi e di saccarasi. Nota I. *Boll. Soc. Ital. Biol. Sper.* **2,** 443–447.

Dulzetto, F. (1927b). Sulla funzione digestiva della ghiandole del gozzo di *Columba livia* (Bonnet). Nota II. *Boll. Soc. Ital. Biol. Sper.* **2,** 575–577.

Dulzetto, F. (1928a). Sulle ghiandole del gozzo di *Columba livia* (Bonnet). *Arch. Biol.* **38,** 173–199.

Dulzetto, F. (1928b). Sulla funzione digestiva delle ghiandole del gozzo di *Columba* mediante la fistola temporanea. *Boll. Soc. Ital. Biol. Sper.* **4,** 950–953.

Dulzetto, F. (1930). La funzione delle ghiandole del gozzo del colombo studiata mediante la fistola temporanea. *Arch. Sci. Biol.* **14,** 430–450.

Dulzetto, F. (1932a). Sui costituenti morfologici del cosidetto latte del gozzo dei colombi. *Boll. Zool.* **3,** 45–51.

Dulzetto, F. (1932b). Struttura e funzione delle ghiandole del gozzo di *Columba livia*. *Arch. Zool. (Ital.) Napoli* **16,** 922–930.

Dulzetto, F. (1938). A propista di una communicazione di Broussy sulla istologia ed istofisiologia dell'apparato gastrico dei Granivori. Ricerche sulla distribuzione e sulla funzione delle "Ghiandole del Gozzo" del colombo e del pollo. *Arch. Biol.* **49,** 369–396.

Dulzetto, F., and di Volsi, N. (1934). Sulla composizione chimica e sui costituenti morfologici del cosidetto latte del gozzo dei colombi. *Atti Congr. Mondi. Pollicolt., 5th,* Vol. 2, pp. 668–674.

Durant, A. J. (1929). The control of blackhead in turkeys by coecal abligation. *N. Amer. Vet.* **10,** 52–55.

Eber, G. (1956). Vergleichende Untersuchungen über die Ernährung einiger Finkenvögel. *Biol. Abh.* **13,** 1–60.

Eberth, C. J. (1867). Untersuchungen über die Leber der Wirbelthiere. *Arch. Mikrosk. Anat.* **3,** 423–440.

Eglitis, I., and Knouff, R. A. (1962). An histological and histochemical analysis of the inner lining and glandular epithelium of the chicken gizzard. *Amer. J. Anat.* **111,** 49–66.

Elias, H. (1955). Embryonic diversity leading to adult identity. The early embryology of the liver of vertebrates. (A preliminary report.) *Anat. Anz.* **101,** 153–167.

Elias, H., and Bengelsdorf, H. (1952). The structure of the liver of vertebrates. *Acta Anat.* **14,** 297–337.

Enemar, A., Falck, B., and Hakanson, R. (1965). Observations on the appearance of norepinephrine in the sympathetic nervous system of the chick embryo. *Develop. Biol.* **11,** 268–283.

Engler, H. (1933). Quantitative Verdauungsversuche am Haushuhn. *Biedermanns Zentralbl., Abt. B* **5,** 329–371.

Everett, S. D. (1966). Pharmacological responses of the isolated oesophagus and crop of the chick. *Physiol. Dom. Fowl, Brit. Egg Mkt. Bd. Symp., 1st, 1964* pp. 261–273.

Everett, S. D., and Mann, S. P. (1968). Catecholamine release by histamine from the isolated intestine of the chick. *Eur. J. Pharmacol.* **1**, 310–320.

Farner, D. S. (1941). Some aspects of the physiology of digestion in birds. A thesis submitted for the degree of Doctor of Philosophy, University of Wisconsin.

Farner, D. S. (1942a). Some aspects of the physiology of digestion in birds. *Summ. Doct. Diss., Univ. Wis.* **6**, 84–85.

Farner, D. S. (1942b). The hydrogen ion concentration in avian digestive tracts. *Poultry Sci.* **21**, 445–450.

Farner, D. S. (1943a). Gastric hydrogen ion concentration and acidity in the domestic fowl. *Poultry Sci.* **22**, 79–82.

Farner, D. S. (1943b). The effect of certain dietary factors on gastric hydrogen ion concentration in the domestic fowl. *Poultry Sci.* **22**, 295–298.

Farner, D. S. (1943c). Biliary amylase in the domestic fowl. *Biol. Bull.* **84**, 240–243.

Farner, D. S. (1960). Digestion and the digestive system. *In* "Biology and Comparative Physiology of Birds" (A. J. Marshall, ed,), Vol. 1, pp. 411–467. Academic Press, New York.

Fedorovskii, N. P. (1951). Zobnoe i zheludochnoe pischevarenie indeek. *Sov. Zootekh.* **1**, 50–58.

Fennell, R. A., and Pearse, A. G. E. (1961). Some histochemical observations on the bursa of Fabricius and thymus of the chicken. *Anat. Rec.* **139**, 93–103.

Fernandes, J. F., and Junqueira, L. C. U. (1955). Protein and ribonucleic acid turnover rates related to activity of digestive enzymes of pigeon pancreas. *Arch. Biochem. Biophys.* **55**, 54–62.

Fischer, E., and Niebel, W. (1896). Über das Verhalten der Polysaccharide gegen einige tierische Sekrete und Organe. *Sitzungsber. Kgl. Preuss. Akad. Wiss.* **1**, 73–82.

Fisher, H., and Dater, E. (1961). Esophageal diverticula in the Redpoll, *Acanthis flammea*. *Auk* **78**, 528–531.

Flower, W. H. (1860). On the structure of the gizzard of the nicobar pigeon, and other granivorous birds. *Proc. Zool. Soc. London* **1860**, 330–334.

Foelix, R. (1970). Vergleichend morphologische Untersuchungen an den Speicheldrüsen körnerfressender Singvögel. *Zool. Jahrb., Abt. Anat. Ontog. Tiere* **87**, 523–587.

Forbes, W. A. (1882). On some points of anatomy of the Indian Darter (*Plotus melanogaster*), and on the mechanism of the neck of the darters (*Plotus*) in connection with their habits. *Proc. Zool. Soc. London* **1882**, 208–212.

Frei, H. J. (1966). Untersuchungen am Pankreas der Fringilliden. Unpublished master's thesis, University of Zürich.

French, G. H. (1898). The glandular stomach of birds. *J. Appl. Microsc.* **1**, 106–107.

French, N. R. (1954). Notes on breeding activities and on gular sacs in the Pine Grosbeak. *Condor* **56**, 83–85.

Friedman, M. H. F. (1939). Gastric secretion in birds. *J. Comp. Cell. Physiol.* **13**, 219–233.

Friedmann, H., and Kern, J. (1956). The problem of cerophagy or wax-eating in the Honey-guides. *Quart. Rev. Biol.* **31**, 19–30.

Fritz, J. C. (1937). The effect of feeding grit on digestibility in the domestic fowl. *Poultry Sci.* **16**, 75–79.

Fritz, J. C., Burrows, W. H., and Titus, H. W. (1935). A comparison of digestibility in gizzardectomized and normal fowls. *Poultry Sci.* **15**, 239–243.

Fruton, J. S., and Bergmann, M. (1940). The specificity of salmon pepsin. *J. Biol. Chem.* **136**, 559–560.

Gadow, H. (1879a). Versuch einer vergleichende Anatomie des Verdauungssystems der Vögel. I. Theil. *Jena. Z. Naturwiss.* **13,** 97–171.

Gadow, H. (1879b). Versuch einer vergleichende Anatomie des Verdauungssystems der Vögel. II. Theil. *Jena. Z. Naturwiss.* **13,** 339–403.

Gadow, H. (1889). On the taxonomic value of the intestinal convolutions in birds. *Proc. Zool. Soc. London* **1889,** 303–315.

Gadow, H. (1891). Vögel. *Bronn's Klassen und Ordnungen des Tierreichs Anat. Theil. Leipzig* **6,** 1–1008.

Galvyaalo, M. Ya., and Goryukhina, T. A. (1937). [Enzyme systems in the developing chick embryo.] *Fiziol. Zh. SSSR im. I. M. Sechenova* **22,** 215–223; from *Chem. Abstr.*

Gardner, L. (1925). The adaptive modifications and the taxonomic value of the tongue of birds. *Proc. U. S. Nat. Mus.* **67,** 1–49.

Garrod, A. H. (1874a). On the "showing off" of the Australian Bustard (*Eupodotis australis*). *Proc. Zool. Soc. London* **1874,** 471–473.

Garrod, A. H. (1874b). On some points in the anatomy of the Columbae. *Proc. Zool. Soc. London* **1874,** 249–259.

Garrod, A. H. (1876). On the anatomy of Chauna derbiana and on the systematic position of the screamers. *Proc. Zool. Soc. London* **1876,** 189–200.

Garrod, A. H. (1878a). Notes on the gizzard and other organs of *Carpophaga latrans*. *Proc. Zool. Soc. London* **1878,** 102–105.

Garrod, A. H. (1878b). On the anatomy of the maleo (*Megacephalon maleo*). *Proc. Zool. Soc. London* **1878,** 629–631.

Garrod, A. H. (1878c). Notes on points in the anatomy of Levaillants darter (*Plotus levaillanti*). *Proc. Zool. Soc. London* **1878,** 679–681.

Giacomini, E. (1890). Sulle glandule salivari degli uccelli. *Monit. Zool. Ital.* **1,** 158–163, 176–188, and 195–210.

Giebel, C. (1857). Bemerkungen zu Vorigem über *Cathartes aura, Falco albicilla, F. lagopus* und *F. buteo*. *Z. Gesamte Naturwiss.* **9,** 426–433.

Glick, B. (1956). Normal growth of the bursa of Fabricius in chickens. *Poultry Sci.* **35,** 843–851.

Glick, B. (1957). Experimental modification of the growth of the bursa of Fabricius. *Poultry Sci.* **36,** 18–23.

Glick, B. (1969). The immunobiological control of the immune response of the fowl. *Poultry Sci.* **68,** 17–22.

Gorham, F. W., and Ivy, A. C. (1938). General function of the gall bladder from the evolutionary standpoint. *Field Mus. Natur. Hist. Publ., Zool. Ser.* **22,** 159–213.

Greschik, E. (1912). Mikroskopische Anatomie des Endarmes der Vögel. *Aquila* **19,** 210–269.

Greschik, E. (1913). Histologische Untersuchungen der Unterkieferdrüse (Glandula mandibularis) der Vögel. Ein Beitrag zur Kenntnis der Mucinbildung. *Aquila* **20,** 331–374.

Greschik, E. (1914a). Die Entstehung der keratinoiden Schicht im Muskelmagen der Vögel. *Aquila* **21,** 99–120.

Greschik, E. (1914b). Histologie des Darmkanales der Saatkrähe (*Corvus frugilegus* L.). *Aquila* **21,** 121–136.

Greschik, E. (1917a). Über den Darmkanal von *Ablepharus pannonicus* Fitz. und *Anguis fragilis* L. *Anat. Anz.* **50,** 70–80.

Greschik, E. (1917b). Der Verdauungskanal der Rotbugamazone (*Androglossa aestiva* Lath.). Ein Beitrag zur Phylogenie der Oesophagealdrüsen der Vögel. *Aquila* **24,** 152–174.

Greschik, E. (1922). Über die Panethschen Zellen und basalgekörnten Zellen im Dünndarm der Vögel. *Aquila* **29**, 149–155.
Grimm, J. D. (1866). Ein Beitrag zur Anatomie des Darmes. Inaugural-Dissertation, University of Dorpat.
Grimm, R. J., and Whitehouse, W. M. (1963). Pellet formation in a Great Horned Owl. *Auk* **80**, 301--306.
Groebbels, F. (1924). Beiträge zur histologischen Physiologie der Verdauungsdrüsen. I. Untersuchungen über die histologische Physiologie der Magenschleimhaut einiger Säugetiere und Vögel. *Z. Biol.* **80**, 1–22.
Groebbels, F. (1929). Über die Farbe der Cuticula im Muskelmagen der Vögel. *Z. Vergl. Physiol.* **10**, 20–25.
Groebbels, F. (1930). Die Verdauung bei der Hausgans untersucht mit der Methode der Dauerkanüle. *Pflueger's Arch. Gesamte Physiol. Menschen Tiere* **224**, 687–701.
Groebbels, F. (1932a). Die Verdauung der Hausente, untersucht mit der Methode der Dauerkanüle. *Wiss. Arch. Landwirt., Abt. B* **8**, 430–435.
Groebbels, F. (1932b). "Der Vogel. Bau, Funktion, Lebenserscheinung, Einpassung," Vol. 1. Borntraeger, Berlin.
Groebbels, F., and Never, H. E. (1932). Die Methodik der Dauerfisteln beim Vogel. *In* "Handbuch der biologischen Arbeitsmethoden," (E. Abderhalden, ed.), Vol. IV, Part 6 II, pp. 1981–1990. Urban & Schwerzenberg, Berlin.
Gross, A. O. (1926). The Heath Hen. *Mem. Boston Soc. Natur. Hist.* **6**, 487–590.
Gross, A. O. (1930). Progress report of the Wisconsin Prairie Chicken Investigation. Wisconsin Conservation Commission, Madison, 112 pp.
Güntherberg, K. (1930). Versuche über die Fettverdauung beim Huhn. *Wiss. Arch. Landwirt., Abt. B* **3**, 339–367.
Habeck, R. (1930). Die Durchgangszeiten verschiedener Futtermittel durch den Verdauungskanal bei Hühnern und Tauben. *Wiss. Arch. Landwirt., Abt. B* **2**, 626–663.
Hafferl, A. (1926). Zur Entwicklungsgeschichte der Kopfarterien beim Kiebitz (*Vanellus cristatus*). *Gegenbauers Jahrb.* **57**, 57–83.
Halnan, E. T. (1928). Digestibility trials with poultry. *J. Agr. Sci.* **18**, 421–431.
Hamilton, T. S., and Card, L. E. (1924). The utilization of lactose by the chicken. *J. Agr. Res.* **27**, 597–604.
Hamilton, T. S., and Mitchell, H. H. (1924). The occurrence of lactose in the alimentary tract of the chicken. *J. Agr. Res.* **27**, 605–608.
Hanke, B., and Niethammer, G. (1955). Zur Morphologie und Histologie des Oesophagus von *Thinocorus orbignyianus*. *Bonn. Zool. Beitr.* **6**, 207–211.
Hanzlik, P. J., and Butt, E. M. (1928). Reactions of the crop (oesophageal) muscles under tension, with a consideration of the anatomical arrangement, innervation and other factors. *Amer. J. Physiol.* **85**, 271–289.
Hassan, T. (1968). A hyoscine-resistant contraction of the isolated chicken esophagus to stimulation of the vagus and descending nerves. *Brit. J. Pharmacol.* **34**, 205–206.
Hassan, T. (1969). A hyoscine-resistant contraction of isolated chicken oesophagus in response to stimulation of parasympathetic nerves. *Brit. J. Pharmacol.* **36**, 268–275.
Hasse, C. (1865). Über den Oesophagus der Tauben und das Verhältnis der Secretion des Kropfes zur Milchsecretion. *Z. Rat. Med.* **23**, 101–132.
Hasse, C. (1866). Beiträge zur Histologie das Vogelmagens. *Z. Rat. Med.* **28**, 1–32.
Haurowitz, F., and Petrou, W. (1925). Über das pH Optimum der Magenlipase verschiedener Tiere. *Hoppe-Seyler's Z. Physiol. Chem.* **144**, 68–79.
Heatley, N. G., McElheny, F., and Lepkovsky, S. (1965). Measurement of the rate of flow of pancreatic secretion in the anesthetized chicken. *Comp. Biochem. Physiol.* **16**, 29–36.

Heidrich, P. K. (1908). Die Mund- und Schlundkopfhöhle der Vögel und ihre Drüsen. *Gegenbaurs Jahrb.* **37,** 10–69.

Heller, V. G., and Penquite, R. (1936). The effect of minerals and fiber on avian intestinal pH. *Poultry Sci.* **15,** 397–399.

Heller, V. G., Morris, L., and Shirley, H. E. (1930). A method of calculating coefficients of digestibility of poultry feed. *Poultry Sci.* **10,** 3–9.

Hellman, B., and Hellerström, C. (1960). The islets of Langerhans in ducks and chickens with special reference to the argyrophil reaction. *Z. Zellforsch. Mikrosk. Anat.* **52,** 278–280.

Henning, H. (1929). Die Verdaulichkeit der Rohfaser beim Huhn. *Landwirt. Vers.-Sta.* **108,** 253–286.

Henry, K. M., MacDonald, A. J., and Magee, H. E. (1933). Observations on the functions of the alimentary canal in fowls. *J. Exp. Biol.* **10,** 153–171.

Herpol, C. (1964). Activité protéolytique de l'appareil gastrique d'oiseaux granivores et carnivores. *Ann. Biol. Anim., Biochim., Biophys.* **4,** 239–244.

Herpol, C. (1965). Onderzoekingen over de spijsvertering bij vogels. Doctoral Dissertation, University of Gent.

Herpol, C. (1966a). Is de voedingswijze bij vogels een determinerende faktor voor de darmlengte? *Gerfaut* **56,** 79–99.

Herpol, C. (1966b). Zuurtegraad en vertering in de maag van vogels. *Natuurwetensch. Tijdschr. (Ghent)* **49,** 201–215.

Herpol, C. (1966c). Influence de l'age sur le pH dans le tube digestif de *Gallus domesticus. Ann. Biol. Anim., Biochim., Biophys.* **6,** 495–502.

Herpol, C. (1967a). Le pH dans le tube digestif de *Columba livia domestica. Z. Vergl. Physiol.* **54,** 415–422.

Herpol, C. (1967b). Étude de l'activité protéolytique des divers organes du système digestif de quelques espèces d'oiseaux en rapport avec leur régime alimentaire. *Z. Vergl. Physiol.* **57,** 209–217.

Herpol, C., and van Grembergen, G. (1961). Le pH dans le tube digestif des oiseaux. *Ann. Biol. Anim., Biochim., Biophys.* **1,** 317–321.

Herpol, C., and van Grembergen, G. (1967a). L'activité protéolytique du système digestif de *Gallus domesticus. Z. Vergl. Physiol.* **57,** 1–6.

Herpol, C., and van Grembergen, G. (1967b). La significance du pH dans le tube digestif de *Gallus domesticus. Ann. Biol. Anim., Biochim., Biophys.* **7,** 33–38.

Herriott, R. M. (1938). Transformation of swine pepsinogen into swine pepsin by chicken pepsin. *J. Gen. Physiol.* **21,** 575–582.

Herriott, R. M. (1941). Isolation, crystallization, and properties of pepsin inhibitor. *J. Gen. Physiol.* **24,** 325–338.

Heupke, W., and Franzen, J. (1957). Ein Vergleich der Verdauungsvorgänge bei Menschen und Vögeln. *Arch. Tierernaehr.* **7,** 48–53.

Heuser, G. F. (1945). The rate of passage of feed from the crop of the hen. *Poultry Sci.* **24,** 20–24.

Hewitt, E. H., and Schelkopf, L. (1955). pH values and enzymatic activity of the digestive tract of chicken. *Amer. J. Vet. Res.* **16,** 576–579.

Hickey, J. J., and Elias, H. (1954). The structure of the liver of birds. *Auk* **71,** 458–462.

Hill, W. C. O. (1926). A comparative study of the pancreas. *Proc. Zool. Soc. London* **1926,** 581–631.

Hillerman, J. P., Kratzer, F. H., and Wilson, W. O. (1953). Food passage through chickens and turkeys and some regulating factors. *Poultry Sci.* **32,** 332–335.

Hirata, G. (1910). Über die Mengenverhältnisse der Diastase in den einzelnen Organen verschiedener Tierarten. *Biochem. Z.* **27,** 385–396.

Hoelting, H. (1912). Ueber den mikroskopischen Bau der Speicheldrüsen einiger Vögel. Inaugural Dissertation, University of Giessen, Hannover.
Hokin, L. E. (1951a). The synthesis and secretion of amylase by pigeon pancreas *in vitro*. *Biochem. J.* **48,** 320–326.
Hokin, L. E. (1951b). Amino-acid requirements of amylase synthesis by pigeon pancreas slices. *Biochem. J.* **50,** 216–220.
Hokin, L. E., and Hokin, M. R. (1956). The actions of pancreaozymin in pancreas slices and the role of phospholipids in enzyme secretion. *J. Physiol. (London)* **132,** 442–453.
Hokin, L. E., and Hokin, M. R. (1961). The synthesis and secretion of digestive enzymes by pancreas tissue *in vitro*. *Exocrine Pancreas; Norm. Abnorm. Funct., Ciba Found. Symp., 1961* pp. 186–207.
Hokin, M. R., and Hokin, L. E. (1953). Enzyme secretion and the incorporation of P^{32} into phospholipides of pancreas slices. *J. Biol. Chem.* **203,** 967–977.
Holdsworth, C. D., and Wilson, T. H. (1966). The development of active sugar and amino acid transport in the yolk sac and intestine of the chicken. *J. Physiol. (London)* **183,** 63P–64P.
Holdsworth, C. D., and Wilson, T. H. (1967). Development of active sugar and amino acid transport in the yolk sac and intestine of the chicken. *Amer. J. Physiol.* **212,** 233–240.
Holmgrén, F. (1867). Physiologiska undersökningar öfver dufvans magar. *Upsala Laekarefoeren. Foerh.* **2:** 631–680; **4:** 691–695.
Holmgrén, F. (1872). Om köttände dufor. *Upsala Laekarefoeren. Foerh.* **7,** 603–614.
Home, E. (1810). On the gizzard of grazing birds. *Phil. Trans. Roy. Soc. London* **100,** 184–189.
Home, E. (1812). On the different structures and situations of the solvent glands in the digestive organs of birds, according to the nature of their food and particular modes of life. *Phil. Trans. Roy. Soc. London* **102,** 394–404.
Home, E. (1814–1828). "Lectures on Comparative Anatomy in which are Explained the Preparations in the Hunterian Collection," 4 vols., Suppl. 2. Nicol & Longman, London.
Honess, R. F., and Allred, W. J. (1942). Structure and function of the neck muscles in inflation and deflation of the esophagus in the sage grouse. *Wyo. Game Fish Dep. Bull.* **2,** 1–12.
Hosizima, T., Takata, H., Uraki, Z., and Sibuya, S. (1930). Über Tauroisolithocholsäure aus Hühnergalle. *J. Biochem. (Tokyo)* **12,** 393–397.
Houssay, F. (1902). Variations organique chez les poules carnivores de seconde génération. *C. R. Acad. Sci.* **135,** 1357–1359.
Houssay, F. (1907). Études sur six générations de poules carnivores. *Arch. Zool. Exp. Gen.* [4] **6,** 137–332.
Howard, W. E. (1958). Food intake and pellet formation of a Horned Owl. *Wilson Bull.* **70,** 145–150.
Hunter, J. (1786). "Observations on Certain Parts of the Animal Economy." Sold at No. 13 Castle Street, Leicester-Square, London.
Ihnen, K. (1927). Bewegungsmechanismus, Fütterungs- und Entleerungszeiten des Kropfes bei Huhn und Taube. *Fortschr. Landwirt.* **2,** 797–798.
Ihnen, K. (1928). Beitrag zur Physiologie des Kropfes beim Huhn und Taube. *Pflueger's Arch. Gesamte Physiol. Menschen Tiere* **218,** 783–796.
Iljin, M. D. (1913). Sur la physiologie du gésier. *C. R. Soc. Biol.* **75,** 293–294.
Immelmann, K. (1964). Zebrafinken (Saugtrinken). *Vogel-Kosmos* **1,** 4–7.

Ivanov, N., and Gotev, R. (1962). Untersuchungen über die aussensekretorische Tätigkeit der Bauchspeicheldrüse bei Hühnen. *Arch. Tierernaehr.* **12,** 65–73.

Ivey, W. E., and Edgar, S. A. (1952). The histogenesis of the esophagus and crop of the chicken, turkey, guinea fowl, and pigeon, with special reference to ciliated epithelium. *Anat. Rec.* **114,** 189–212.

Ivorec-Szylit, O., Mercier, C., Raibaud, P., and Calet, C. (1965). Contribution à l'étude de la dégradation des glucides dans le jabot du Coq. Influence du taux de glucose du régime sur l'utilisation de l'amidon. *C. R. Acad. Sci.* **261,** 3201–3203.

Iwanow, I. F. (1930). Die sympathische Innervation des Verdauungstraktes einiger Vogelarten (*Columba livia* L., *Anser cinereus* L., and *Gallus domesticus*). *Z. Zellforsch. Mikrosk. Anat.* **22,** 469–492.

Jackson, J. T., Mangan, G. F., Machlin, L. J., and Danton, C. A. (1955). Absorption of vitamin B_{12} from the cecum of the hen. *Proc. Soc. Exp. Biol. Med.* **89,** 225–227.

Jaeckel, H. (1925). Über die Bedeutung der Steinchen in Hühnermagen. *Beitr. Physiol.* **3,** 11–38.

Jeuniaux, C. (1962). Digestion de la chitine chez les oiseaux et les mammifères. *Ann. Soc. Zool. Belg.* **92,** 27–45.

Jeuniaux, C. (1963). "Chitine et chitinolyse." Masson, Paris.

Jorbet, C. (1872). Etudes d'anatomie comparée sur les organs du toucher chez divers mammifers, oiseaux, poissons et insectes. *Ann. Sci. Natur.: Bot. Biol. Veg.* [5] **16,** 1–162.

Johnston, D. W. (1958). Sex and age characters and salivary glands of the Chimney Swift. *Condor* **60,** 73–84.

Johnston, D. W. (1961). Salivary glands in the Black Swift. *Condor* **63,** 338.

Jolly, J. (1911). La bourse de Fabricius et les organes lympho-épithéliaux. *C. R. Ass. Anat.* **13,** 164–176.

Jolly, J. (1914). La bourse de Fabricius et les organes lympho-épithéliaux. *Arch. Anat. Microsc.* **16,** 363–547.

Joos, C. (1952). Untersuchungen über die Histogenese der Drüsenschicht des Muskelmagens bei Vögeln. *Rev. Suisse Zool.* **59,** 315–338.

Jung, L., and Pierre, M. (1933). Sur le rôle de la saliva chez les oiseaux granivores. *C. R. Soc. Biol.* **113,** 115–116.

Kaden, L. (1936). Über Epithel und Drüsen des Vogelschlundes. *Zool. Jahrb., Abt. Anat. Ontog. Tiere* **61,** 421–466.

Kallius, E. (1906). Beiträge zur Entwicklung der Zunge. II. Vögel 3. *Melopsittacus undulatus. Anat. Hefte, Abt. 2* **31,** 603–651.

Karpov, L. V. (1919). O perevarivanii nekotorykh rastitelnykh i zhivotnykh belkov gusiiym zheludochnom sokom. *Fiziol. Zh. SSSR im I. M. Sechenova* **2,** 185–196.

Karrer, O., and Parsons, D. S. (1968). Disaccharidase activity in the small intestine of the developing chick. *Life Sci.* **7,** 85–89.

Katayama, T. (1924). Über die Verdaulichkeit der Futtermittel bei Hühnern. *Bull. Imp. Agr. Exp. Sta. Jap.* **3,** 1–78.

Kato, T. (1914). Druckmessungen im Muskelmagen der Vögel. *Pfluegers Arch. Gesamte Physiol. Menschen Tiere* **159,** 6–26.

Kaupp, B. F., and Ivey, J. F. (1923). The digestion coefficients of poultry feeds and rapidity of digestion and fate of grit in the fowl. *N. C., Agr. Exp. Sta., Bull.* **22,** 1–143.

Keeton, R. W., Koch, F. C., and Luckhardt, A. B. (1920). Gastrin studies. III. The response of the stomach mucosa of various animals to gastrin bodies. *Amer. J. Physiol.* **50,** 454–468.

Kerr, W. R., and Common, R. H. (1935). The effect of certain acid treatments for coccidiosis on the hydrogen ion concentration of the fowl intestine. *Vet. J.* **91**, 309–311.

Kerry, K. R. (1969). Intestinal disaccharidase activity in a monotreme and eight species of marsupials (with an added note on the disaccharidases of five species of sea birds). *Comp. Biochem. Physiol.* **29**, 1015–1022.

Klima, M. (1956). Die Entwicklung der Bursa Fabricii und ihre Benützung zur Altersbestimmung der Vögel. *Sb. Prednasek Ornit. Prag* **1**, 60–66.

Klima, M. (1958). Príspevek k morfologii bursy fabriciovy ptáku [Beitrag zur Morphologie der Bursa Fabricii der Vögel.] *Sylvia (Prague)* **15**, 151–170.

Klug, F. (1891). Zur Kenntnis der Verdauung der Vögel, insbesondere der Gänse. *Zentralbl. Physiol.* **5**, 131–135.

Klug, F. (1892). Beitrag Kenntnis der Verdauung bei Vögeln. *Hauptber. Int. Ornithol. Congr., 2nd 1891* Vol. 2, pp. 43–59.

Kokas, E., Phillips, J. L., Jr., and Brunson, W. D., Jr. (1967). The secretory activity of the duodenum in chickens. *Comp. Biochem. Physiol.* **22**, 81–90.

Kolderup, C. F. (1924). Steininholdet i kraasen hos norske liryper. *Bergens Mus. Aarb.* **1**, 1–34.

Kolossow, N. G. (1959). Weitere Beobachtungen am Nervensystem des Darmes. *Z. Mikrosk.-Anat. Forsch.* **65**, 557–573.

Kolossow, N. G. Sabussow, G. H., and Iwanow, J. F. (1932). Zur Innervation des Verdauungskanals der Vögel. *Z. Mikrosk.-Anat. Forsch.* **30**, 257–294.

Koskowski, W. (1922). L'action de l'histamine sur la sécrétion du sac gastrique chez les pigeons. *C. R. Acad. Sci.* **174**, 247–248.

Kostenko, A. S. (1957). Rol nervnoi sistemy v regulatsii sekretornykh zhelydka utok. *Nauchn. Zap. Belotserkov. Sel'skokhoz. Inst.* **6**, 97–105.

Kovshar, A. F., and Nekrasov, B. V. (1967). O podbyazychnykh meshkakh vyurkovykh ptits. *Ornitologiya* **8**, 320–325.

Kratinoff, P. (1929). Beiträge zur Physiologie der Hungertätigkeit des Verdauungsapparates. V. Über den Einfluss des Cholins auf die Hungerbewegungen des Kropfes bei Hühnern. *Z. Gesamte Exp. Med.* **64**, 413–416.

Kratzer, F. H. (1944). Amino acid absorption and utilization in the chick. *J. Biol. Chem.* **153**, 237–247.

Küchler, W. (1936). Anatomische Untersuchungen an *Phytotoma rara* Mol. *J. Ornithol.* **84**, 352–362.

Lack, D. (1956). A review of the genera and nesting habits of Swifts. *Auk* **73**, 1–32.

Lang, E. M. (1963). Flamingos raise their young on a liquid containing blood. *Experientia* **19**, 532.

Lang, E. M., Thiersch, A., Thommen, H., and Wackernagel, H. (1962). Was füttern die Flamingos (*Phoenicopterus ruber*) ihren Jungen? *Ornithol. Beob.* **59**, 173–176.

Langendorff, O. (1879a). Versuche über die Pankreasverdauung der Vögel. *Arch. Anat. Physiol., Physiol. Abt.* **1879**, 1–35.

Langendorff, O. (1879b). Über die Entstehung der Verdauungsfermente beim Embryo. *Arch. Anat. Physiol., Physiol. Abt.* **1879**, 95–112.

Lavrenteva, G. F. (1963). [Bile secretion in chickens.] *Tru. Gor'ko. Sel'skokhoz. Inst.* **13**, 80–85.

Laws, B. M., and Moore, J. H. (1963a). Some observations on the pancreatic amylase and intestinal maltase of the chick. *Can. J. Biochem.* **41**, 2107–2211.

Laws, B. M., and Moore, J. H. (1963b). The lipase and esterase activities of the pancreas and small intestine of the chick. *Biochem. J.* **87**, 632–638.

Leasure, E. E., and Link, R. P. (1940). Studies on the saliva of the hen. *Poultry Sci.* **19,** 131–134.
Lehmann, V. W. (1941). Atwater's Prairie Chicken, its life history and management. *N. Amer. Fauna* **57,** 1–63.
Leiber, A. (1907). Vergleichende Anatomie der Spechtzunge. *Zoologica (New York)* **20,** 1–79.
Lenkeit, W. (1934). Der Einfluss verschiedener Ernährung auf die Grössenverhältnisse des Magen-Darmkanals beim Geflügel. *Arch. Gefluegelk.* **8,** 116–129.
Leopold, A. S. (1953). Intestinal morphology of gallinaceous birds in relation to food habits. *J. Wildl. Manage.* **17,** 197–203.
Lerner, J., and Taylor, M. W. (1967). A common step in the intestinal absorption mechanisms of D- and L-methionine. *Biochim. Biophys. Acta* **135,** 991–999.
Lerner, J., Martin, V., Eddy, C. R., and Taylor, M. W. (1968). Some additional characteristics of methionine transport in the chicken intestine. *Experientia* **24,** 1103–1104.
Leuckart, F. S. (1841). Ueber zusammengesetzte Magenbildungen bei verschiedenen Vögeln. *Zool. Bruchstuecke* **2,** 64–71.
Leydig, F. (1854). Kleinere Mitteilungen zur tierischen Gewebelehre. *Arch. Anat. Physiol., Leipzig* **1854,** 296–348.
Lieberfarb, A. S. (1927). Bewegungen des leeren Kropfes der Hühner bei akuter Hyperthyreose. *Pflueger's Arch. Gesamte Physiol. Menschen Tiere* **216,** 437–447.
Lienhart, R. (1953). Recherches sur le rôle des cailloux contenus dans le gésier des oiseaux granivores. *Bull. Soc. Sci. Nancy* [2] **12,** 5–9.
Linduska, J. P. (1943). A gross study of the bursa of Fabricius and cockspur as age indicates in the Ring-necked Pheasant. *Auk* **60,** 426–437.
Litwer, G. (1926). Die histologischen Veränderungen der Kropfwandungen bei Tauben zur Zeit der Bebrütung und Auffütterung ihrer Jungen. *Z. Zellforsch. Mikrosk. Anat.* **3,** 697–722.
Löffler, H., and Leibetseder, J. (1966). Daten zur Dauer des Darmdurchganges bei Vögeln. *Zool. Anz.* **177,** 334–340.
Lucas, F. A. (1895). The tongues of Woodpeckers. *Bull. U.S. Biol. Surv.* **7,** 35–44.
Lucas, F. A. (1897). The tongues of birds. *Rep. U.S. Nat. Mus.* pp. 1003–1020.
Luppa, H. (1962). Histologie, Histogenese und Topochemie der Drüsen des Sauropsidenmagens. II. Aves. *Acta Histochem.* **13,** 233–300.
McAtee, W. L. (1906). The shedding of the stomach lining of birds. *Auk* **23,** 346.
McBee, R. H., and West, G. C. (1969). Cecal fermentation in the Willow Ptarmigan. *Condor* **77,** 54–58.
McCallion, D. J., and Aitken, H. E. (1953). A cytological study of the anterior submaxillary glands of the fowl, *Gallus domesticus*. *Can. J. Zool.* **31,** 173–178.
Macdonald, A. J., and Bose, S. (1944). Studies on the digestibility coefficients and biological values of the proteins in poultry feeds. *Poultry Sci.* **23,** 135–141.
Magnan, A. (1910a). Sur une certain loi de variation du foie et du pancréas chez les oiseaux. *C. R. Acad. Sci.* **151,** 159–160.
Magnan, A. (1910b). Influence du régime alimentaire sur l'intestin chez les oiseaux. *C. R. Acad. Sci.* **150,** 1706–1707.
Magnan, A. (1911a). La surface digestive du ventricule succenturié et la musculature du gésier chez les oiseaux. *C. R. Acad. Sci.* **153,** 295–297.
Magnan, A. (1911b). Influence du régime alimentaire sur le gros intestin et les caecums des oiseaux. *C. R. Acad. Sci.* **152,** 1506–1508.
Magnan, A. (1911c). Morphologie des caecums chez les oiseaux en fonction du régime alimentaire. *Ann. Sci. Natur. Zool.* [9] **14,** 275–305.

Magnan, A. (1911d). Sur la variation inverse du ventricule succenturié et du gésier chez les oiseaux. *C. R. Acad. Sci.* **152,** 1705–1707.

Magnan, A. (1911e). La surface totale de l'intestin chez les oiseaux. *C. R. Soc. Biol.* **71,** 617–619.

Magnan, A. (1912a). Essai de morphologie stomacale en fonction du régime alimentaire chez les oiseaux. *Ann. Sci. Natur. Zool.* [9] **15,** 1–41.

Magnan, A. (1912b). Variations expérimentales du foie et des reins chez le canards en fonction du régime alimentaire. *C. R. Acad. Sci.* **155,** 182–184.

Magnan, A. (1912c). Variations du ventricule succenturié et du gésier entraînées chez les canards pars divers régimes alimentaires. *C. R. Acad. Sci.* **155,** 1111–1114.

Magnan, A. (1912d). Adaptation fonctionelle de l'intestin chez les canards. *C. R. Acad. Sci.* **155,** 1546–1547.

Magnan, A. (1913). Rapports entre l'alimentation et les dimensions des caecums chez les canards. *C. R. Acad. Sci.* **156,** 85–87.

Malewitz, T. D., and Calhoun, M. L. (1958). The gross and microscopic anatomy of the digestive tract, spleen, kidney, lungs and heart of the turkey. *Poultry Sci.* **37,** 388–398.

Mangold, E. (1906). Der Muskelmagen der körnerfressenden Vögel, seine motorischen Funktionen und ihre Abhängigkeit vom Nervensystem. *Pflueger's Arch. Gesamte Physiol. Menschen Tiere* **111,** 163–240.

Mangold, E. (1911a). Die Magenbewegungen der Krähe und Dohle und ihre Beeinflussung vom Vagus. *Pflueger's Arch. Gesamte Physiol. Menschen Tiere* **138,** 1–13.

Mangold, E. (1911b). Die funktionellen Schwankungen der motorischen Tätigkeit des Raubvogelmagens. *Pflueger's Arch. Gesamte Physiol. Menschen Tiere* **139,** 10–32.

Mangold, E. (1925). Ueber Kohlenhydrat- und Eiweissverdauung bei Tauben und Hühnern, und über das Eindringen von Verdauungsfermenten durch die pflanzliche Zellmembran. *Biochem. Z.* **156,** 4–14.

Mangold, E. (1927a). Über Verdauung und Ernährung der Vögel. *Sitzungsber. Ges. Naturforsch. Freunde Berlin* pp. 49–55.

Mangold, E. (1927b). Über die Bedeutung der Steinchen im Hühnermagen. *Sitzungsber. Ges. Naturforsch. Freunde Berlin* pp. 20–21.

Mangold, E. (1927c). Die Bedeutung von Steinchen und Sand im Hühnermagen. Arch. Gefluegelk. **1,** 145–152.

Mangold, E. (1928). Die Verdaulichkeit der Rohfaser und die Funktion der Blinddärme beim Haushuhn. *Arch. Gefluegelk.* **2,** 312.

Mangold, E. (1929). Die physiologischen Funktionen des Blinddarms allgemein und besonders bei den Vögeln. *Sitzungsber. Ges. Naturforsch. Freunde Berlin* pp. 217–226.

Mangold, E. (1931a). Die Verdauung bei den Vögeln. *Proc. Int. Ornithol. Congr., 7th 1930* pp. 206–214.

Mangold, E. (1931b). Ueber die Verdaulichkeit der Hornsubstanz (Keratin) der Vogelfedern bei Säugetieren und Vögeln. *Sitzungsber. Ges. Naturforsch. Freunde Berlin* pp. 286–291.

Mangold, E. (1934a). Die Physiologie des Ernährungsvorganges beim Geflügel. *Atti Congr. Mond. Pollicolt., 5th 1933* Vol. 2, pp. 14–26.

Mangold, E. (1934b). Die Verwertung der pflanzlichen Rohfaser beim Menschen und den Tieren. *Sitzungsber. Ges. Naturforsch. Freunde Berlin* pp. 345–388.

Mangold, E. (1934c). Ueber Veränderungen des Magen-Darmkanals unter dem Einfluss verschiedener Ernährung. *Sitzungsber. Ges. Naturforsch. Freunde Berlin* pp. 395–399.

Mangold, E. (1943). Die Bedeutung der Magen- und Darmsymbionten der Wirbeltiere. *Ergeb. Biol.* **19,** 1–81.

Mangold, E. (1950). "Die Verdauung bei den Nutztieren." Akademie-Verlag, Berlin.

Mangold, E., and Rüdiger, H. (1934). Über den Einfluss des Grit auf die Nahrungsaufnahme der Hühner. *Arch. Gefluegelk.* **7,** 295–304.

Märk, W. (1944). Die Entwicklung der Bursa Fabricii bei der Ente. *Z. Mikrosk.-Anat. Forsch.* **54,** 1–95.

Markus, M. B. (1964). Intestinal caeca in the South African Columbidae. *Bull. Brit. Ornithol. Club* **84,** 137–138.

Marshall, A. J., and Folley, S. J. (1956). The origin of the nest cement in edible-nest swiftlets (*Collocalia* spp.). *Proc. Zool. Soc. London* **126,** 383–389.

Martin, C. (1970). Vergleichend morphologische Untersuchungen an der Leber von Singvögeln. Unpublished master's thesis, University of Zürich.

Masahito, P., and Fuji, T. (1959). Histology of the pigeon cropsack, with special reference to the early changes evoked by prolactin. *J. Fac. Sci., Univ. Tokyo, Sect. 4* **8,** 401–409.

Maumus, J. (1902). Les caecums des oiseaux. *Ann. Sci. Natur. Zool.* **15,** 1–148.

Maumus, J., and Launoy, L. (1901). La digestion caecale chez les oiseaux. *Bull. Mus. Hist. Natur., Paris* **7,** 361–366.

Mayaud, N. (1950). Téguments et phanères. *In* Traité de Zoologie (P.-P. Grassé, ed.), Vol. XV, pp. 4–77. Masson, Paris.

Mayhew, R. L. (1935). The pH of the digestive tract of the fowl. *J. Amer. Vet. Med. Ass.* **86,** 148–152.

Meckel, J. F. (1829). "System der vergleichenden Anatomie," Thesis IV, Chapter XII, pp. 398–489. Rengersche Buchhandlung, Halle.

Medway, L. (1962). The relation between the reproductive cycle, molt and changes in the sublingual salivary glands of the swiftlet *Collocalia maxima* Hume. *Proc. Zool. Soc. London* **138,** Part 2, 305–315.

Medway, L. (1969). Studies on the biology of the edible swiftlets of Southeast Asia. *Malay. Nature J.* **22,** 57–63.

Mennega, A. M. W. (1938). Waterstofionenconcentratie en vertering in de maag van eenige vertebraten. Dissertation, Rijks Universiteit, Utrecht.

Merkel, F. S. (1878). Die Tastzellen der Ente. *Arch. Mikrosk. Anat.* **15,** 415–427.

Merkel, F. S. (1880). "Über die Endigung der sensiblen Nerven in der Haut der Wirbeltiere." Schmidt, Rostock.

Meyer, H. (1929). Mikroskopische Untersuchungen über den Verlauf der Verdauung von Fleisch und Hühnereiweiss im Magen verschiedener Tiere. *Z. Vergl. Physiol.* **10,** 712–750.

Meyer, W. K. (1927). Vergleichende mikroskopische Untersuchungen über die Verdauung der Kleberzellen verschiedener Zerealien im Magen Darmkanal pflanzenfressender Tiere (Huhn, Taube, Schaf, Kaninchen). *Z. Vergl. Physiol.* **6,** 402–430.

Michalowsky, J. (1909). Zur Frage über funktionelle Aenderungen in den Zellen des Drüsenmagens bei Vögeln. *Anat. Anz.* **34,** 257–275.

Miller, A. H. (1941). The buccal food-carrying pouches in the Rosy Finch. *Condor* **43,** 72–73.

Miller, L. (1962). Stomach stones. *Zoonooz* **35,** 10–12.

Milne-Edwards, H. (1860). "Leçons sur la physiologie et l'anatomie comparée de l'homme et des animaux," Vol. VI. Masson, Paris.

Mitchell, P. C. (1901). On the intestinal tract of birds; with remarks on the valuation and nomenclature of zoological characters. *Trans. Linn. Soc. London, Zool.* **8,** 173–275.

Moller, W. (1931). Ueber die Schnabel- und Zungenmechanik blütenbesuchender Vögel. *Biol. Gen.* **7**, 100–154.

Moore, C. A., and Elliott, R. (1944). Numerical and regional distribution of taste buds on the tongue of the pigeon. *Anat. Rec.* **88**, 449.

Mountfort, G. (1964). Bill. *In* "A New Dictionary of Birds," p. 93–96. Nelson, London.

Müller, I., Runge, W., and Ferner, H. (1956). Cytologie und Gefässverhältnisse des Inselorgans bei der Ente. *Z. Zellforsch. Mikrosk. Anat.* **62**, 165–186.

Müller, S. (1922). Zur Morphologie des Oberflächenreliefs der Rumpfdarmsschleimhaut bei den Vögeln. *Jena. Z. Naturwiss.* **58**, 533–606.

Murie, J. (1868). Observations concerning the presence and function of the gular pouch in *Otis kori* and *Otis australis*. *Proc. Zool. Soc. London* **1868**, 471–477.

Murie, J. (1869). Note on the gular pouch of *Otis tarda*. *Proc. Zool. Soc. London* **1869**, 140–142.

Naik, D. R. (1962). A study of the intestinal caeca of some Indian birds. M.Sc. Thesis, Benares Hindu University.

Naik, D. R., and Dominic, C. J. (1962). The intestinal caeca of some Indian birds in relation to food habits. *Naturwissenschaften* **49**, 287.

Naik, D. R., and Dominic, C. J. (1963). The intestinal caeca as a criterion in avian taxonomy. *Proc. Indian Sci. Congr. 50th, 1962* Part III, p. 533.

Naik, D. R., and Dominic, C. J. (1969a). Intestinal caeca of Owls. *Proc. Indian Sci. Congr., 55th, 1968* Part III, p. 522.

Naik, D. R., and Dominic, C. J. (1969b). A study of the intestinal caeca of some Indian birds. *Proc. Indian Sci. Congr., 56th, 1968* Part III, pp. 473–474.

Nechay, B. R., and Lutherer, B. D. C. (1968). Handling of urine by cloaca and ureter in chickens. *Comp. Biochem. Physiol.* **26**, 1099–1105.

Neergaard, J. W. (1806). "Vergleichende Anatomie und Physiologie der Verdauungswerkzeuge der Säugethiere und Vögel." Realschulbuchhandlung, Berlin.

Nekrasov, B. V. (1958). Funktsionalno-morphologicheskii ocherk chelyustnovo apparata nekotorykh vyrkykh ptits. *Izv. Kazan. Filiala Akad. Nauk SSSR, Ser. Biol. Nauk* **6**, 47–68.

Nekrasov, B. V. (1961). O nekotorykh osobennostyakh stroeniya yazyka i podyazychnovo apparate vyurkovykh ptits. *Tr. Zool. Inst., Akad. Nauk SSSR* **29**, 213–226.

Newton, I. (1967). The adaptive radiation and feeding ecology of some British finches. *Ibis* **109**, 33–98.

Nicolai, J. (1956). Zur Biologie und Ethologie des Gimpels (*Pyrrhula pyrrhula* L.). *Z. Tierpsychol.* **13**, 93–132.

Nicoll, C. S. (1969). Bioassay of prolactin. Analysis of the pigeon crop-sac response to systemic prolactin injection by an improved method of response quantification. *Acta Endocrinol. (Copenhagen)* **60**, 91–100.

Niethammer, G. (1931). Zur Histologie und Physiologie des Taubenkropfes. *Zool. Anz.* **97**, 93–103.

Niethammer, G. (1933). Anatomisch-histologische und physiologische Untersuchungen über die Kropfbildungen der Vögel. *Z. wiss. Zool.* Abt. A, 144:12–101.

Niethammer, G. (1937). Ueber den Kropf der männlichen Grosstrappe. *Ornithol. Monatsber.* **45**, 189–192.

Niethammer, G. (1961). Sonderbildungen an Ösophagus und Trachea beim Weibchen von *Turnix sylvatica lepurana*. *J. Ornithol.* **102**, 75–79.

Niethammer, G. (1966a). Sexualdimorphismus am Oesophagus von *Rostratula*. *J. Ornithol.* **107**, 201–204.

Niethammer, G. (1966b). Ueber die Kehltaschen des Rotflügelgimpels, *Rhodopechys sanguinea*. *J. Ornithol.* **107**, 278–283.

Nikulina, Z. K. (1919). K fiziologii slyunnikh zhelez u gusia. *Fiziol. Zh. SSSR im. I.M. Sechenova* **2,** 199–244.

Nolf, P. (1925a). Influence du vague sur la motricité de l'estomac de l'oiseau. *C. R. Soc. Biol.* **93,** 454–455.

Nolf, P. (1925b). Influence des nerfs sympathiques sur la motricité de l'estomac de l'oiseau. *C. R. Soc. Biol.* **93,** 839.

Nolf, P. (1925c). L'innervation motrice du tube digestif de l'oiseau. *Arch. Int. Physiol.* **25,** 290–341.

Nolf, P. (1927a). Du rôle des nerfs vague et sympathique dans l'innervation motrice de l'estomac de l'oiseau. *Arch. Int. Physiol.* **28,** 309–428.

Nolf, P. (1927b). Le système nerveux entérique de l'oiseau. *Ann. Physiol. Physiochim. Biol.* **3,** 474–476.

Nolf, P. (1928). Le systèm nerveux entérique essai d'analyse par la méthode à la nicotine de Langley. *Arch. Int. Physiol.* **30,** 317–492.

Nolf, P. (1934a). Les nerfs extrinsèques de l'intestin chez l'oiseau. I. Les nerfs vagues. *Arch. Int. Physiol.* **39,** 113–164.

Nolf, P. (1934b). Les nerfs extrinsèques de l'intestin chez l'oiseau. II. Les nerfs coeliaques et mésentériques. *Arch. Int. Physiol.* **39,** 165–226.

Nolf, P. (1934c). Les nerfs extrinsèques de l'intestin chez l'oiseau. III. Le nerf de Remak. *Arch. Int. Physiol.* **39,** 227–256.

Nolf, P. (1935). De l'influence de l'anoxemie et de l'acidose gésier sur la motricité de l'estomac chez l'oiseau. *Arch. Int. Physiol.* **41,** 57–140.

Nolf, P. (1936). De la longue durée des effets chronotrope et inotrope exercés par les nerfs gastro-intestinaux et de la possibilité de les obtenir séparément. *Arch. Int. Physiol.* **44,** 38–111.

Nolf, P. (1938a). Le système nerveux gastro-entérique. *Ann. Physiol. Physiochim. Biol.* **14,** 293–319.

Nolf, P. (1938b). L'appareil nerveux de l'automatisme gastrique de l'oiseau. I. Essai d'analyse par nicotine. *Arch. Int. Physiol.* **46,** 1–85.

Nolf, P. (1938c). L'appareil nerveux de l'automatisme gastrique de l'oiseau. II. Étude des effets causés par une ou plusiers sections de l'anneau nerveux du gésier. *Arch. Int. Physiol.* **46,** 441–559.

Nolf, P. (1938d). Les éléments intrinsèques de l'anneau nerveux du gésier de l'oiseau granivore. I. *Arch. Int. Physiol.* **47,** 453–517.

Nolf, P. (1939a). L'appareil nerveux intrinsèque du gesier de l'oiseau granivore. *Acta Brevia Neer. Physiol., Pharmacol., Microbiol.* **9,** 212–214.

Nolf, P. (1939b). Les éléments intrinsèques de l'anneau nerveux du gésier de l'oiseau granivore. II. *Arch. Int. Physiol.* **48,** 451–542.

Norris, R. A. (1961). Colors of stomach linings of certain passerines. *Wilson Bull.* **73,** 380–383.

Nussbaum, M. (1882). Ueber den Bau und die Tätigkeit der Drüsen. *Arch. Mikrosk. Anat.* **21,** 119–123.

Oakeson, B. B. (1956). Liver and spleen weight cycles in non-migratory White-crowned Sparrows. *Condor* **58,** 45–50.

Ohashi, H., and Ohga, A. (1967). Transmission of excitation from parasympathetic nerve to smooth muscle. *Nature (London)* **216,** 291–292.

Olivo, O. M. (1947). Structure de la membrane kératinoïde de l'estomac musculaire de *Gallus gallus. Acta Anat.* **4,** 213–217.

Olsson, N., Kihlen, G., Ruudvere, A., Wadne, C., and Ånstrand, G. (1950). Smältbarhetsförsök med fjäderfä. *Kgl. Lantbrukshoegsk. Statens Lantbruksfoer., Statens Husdjursfoer. Medd.* **43,** 1–69.

Oppel, A. (1895). Über die Muskelschichten im Drüsenmagen der Vögel. *Anat. Anz.* **11,** 167–172.

Oppel, H. (1900). "Lehrbuch der vergleichenden mikroskopischen Anatomie der Wirbelthiere, Vol. III. Jena.

Paine, C. M., Newman, H. J., and Taylor, M. W. (1959). Intestinal absorption of methionine and histidine by the chicken. *Amer. J. Physiol.* **197,** 9–12.

Paira-Mall, L. (1900). Über die Verdauung bei Vögeln, ein Beitrag zur vergleichenden Physiologie der Verdauung. *Arch. Gesamte Physiol. Menschen Tiere* **80,** 600–627.

Pastea, E., Nicolau, H., Popa, V., Roşca, L., and Căspărin, A. (1967). Observatii morfofiziologice asupra aparatului digestiv la rată. *Lucr. Stiint. Inst. Agron. Nicolae Balcescu* **10C,** 191–202.

Patterson, T. L. (1927). Gastric movements in the pigeon with economy of animal materials. Comparative studies. V. *J. Lab. Clin. Med.* **12,** 1003–1007.

Patterson, T. L. (1933). The comparative physiology of the gastric hunger mechanism. *Ann. N.Y. Acad. Sci.* **34,** 55–272.

Patzelt, V. (1936). Der Darm. *In* "Handbuch der mikroskopischen Anatomie des Menschen" (W. von Möllendorf, ed.), Vol. 5, Part III, pp. 1–448. Springer-Verlag, Berlin and New York.

Pernkopf, E. (1930). Beiträge zur vergleichenden Anatomie des Vertebratenmagens. *Z. Anat. Entwicklungsgesch.* **91,** 329–390.

Pernkopf, E. (1937). Die Vergleichung der verschiedenen Formentypen des Vorderdarmes der Kranioten. *In* "Handbuch der vergleichenden Anatomie der Wirbeltiere" (L. Bolk *et al.,* eds.), Vol. 3, pp. 477–562. Urban & Schwarzenberg, Munich.

Pfleiderer, G., Linke, R., and Reinhardt, G. (1970). On the evolution of endopeptidases. VIII. Cross-reactions of trypsins and chymotrypsins of different species. *Comp. Biochem. Physiol.* **33,** 955–967.

Pilliet, A. H. (1893). Note sur l'appareil salivaire des oiseaux. *C. R. Soc. Biol.* **45,** 349–352.

Plenk, H. (1932). Der Magen. *In* "Handbuch der mikroskopischen Anatomie des *Menschen*" (W. von Möllendorf, ed.), Vol. 5, Part 2, pp. 1–234. Springer-Verlag, Berlin and New York.

Plimmer, R. H. A. (1906). Lactase in the intestine. *J. Physiol. (London)* **35,** 20–31.

Plimmer, R. H. A., and Rosedale, J. L. (1922). Distribution of enzymes in the alimentary canal of the chicken. *Biochem. J.* **16,** 23–26.

Polyakov, I. I. (1958). Nekotorye dannye o podzheludochnom i kishernom soke kur. *Dokl. Mosk. Sel'skokhoz. Akad.* **38,** 238–333.

Pomayer, C. (1902). Kloake und Phallus der Amnioten. III. Die Vögel. *Gegenbauers Jahrb.* **30,** 614–649.

Ponirovskii, N. G. (1915). K voprosu o soderzhanii diastatichemovo fermenta v slyunnikh zhelezach ptits. *Fiziol. Lab. Kharkovsk. Vet. Inst. Imp. Nikolaia I za* 1918. (Original not seen. See Nikulina, 1919.)

Popov, N. A., and Kudriavtsef, A. A. (1931). K voprosy o zhelydochnom i kishechnom pishchevarenii u kur. *Tr. Vses. Inst. Eksp. Vet.* **7,** 48–55.

Popov, N. A., and Kudriavtsef, A. A. (1932). Gizzard- and intestine-digestion by poultry. *Int. Rev. Poultry Sci.* **5,** 38.

Portier, P. (1898). Recherches sur la lactase. *C. R. Soc. Biol.* **50,** 387–389.

Portier, P., and Bierry, H. (1901). Recherches sur l'influence de l'alimentation sur les sécrétions diastatiques. *C. R. Soc. Biol.* **53,** 810–811.

Portmann, A. (1935). Die Ontogenese der Vögel als Evolutionsproblem. *Acta Biotheor.* **1,** 59–90.

Portmann, A. (1938). Beiträge zur postembryonalen Entwicklung der Vögel. I. Vergleichende Untersuchungen über die Ontogenese der Hühner und Sperlingsvögel. *Rev. Suisse Zool.* **45**, 273–348.

Portmann, A. (1942). Die Ontogenese und des Problem der morphologischen Wertigkeit. *Rev. Suisse Zool.* **49**, 169–185.

Portmann, A. (1950). Le tube digestif. *In* "Traité de Zoologie" (P.-P. Grassé, ed.), Vol. 15, pp. 270–289. Masson, Paris.

Postma, G. (1887). Bijdrage tot de kennis van den bouw van het darmkanal der vogels. Proefschrift ter Verkrijging van den Graad van Doctor in de Plant en Dierkunde aan Rijksuniversiteit te Leiden. A. H. Adriani, Leiden.

Prévost, J. (1961). Ecologie du Manchot Empereur (*Aptenodytes forsteri* Gray). *Actual. Sci. Ind.* **1291**, 2.

Prévost, J., and Vilter, V. (1963). Histologie de la sécrétion oesophagienne du manchot Empereur. *Proc. Int. Ornithol. Congr., 13th, 1962* Vol. 2, pp. 1085–1094.

Radeff, T. (1928). Über die Rohfaserverdauung beim Huhn und die hierbei dem Blinddarm zukommende Bedeutung. *Biochem. Z.* **193**, 192–196.

Rand, A. L. (1961). The tongue and nest of certain flower-peckers (Aves: *Dicaeidae*). *Fieldiana: Zool.* **39**, 581–587.

Reed, C. I., and Reed, B. P. (1928). The mechanism of pellet formation in the Great Horned Owl (*Bubo virginianus*). *Science* **68**, 359–360.

Reichel, P. (1883). Beiträge zur Morphologie der Mundhöhlendrüse der Wirbeltiere. *Gegenbaurs Jahrb.* **8**, 1–72.

Retterer, E. (1885). Contribution à l'étude du cloaque et de la bourse de Fabricius chez les oiseaux. *J. Anat. Physiol. Norm. Pathol. Homme Anim.* **21**, 369–559.

Riddle, O. (1963). Prolactin in vertebrate function and organization. *J. Nat. Cancer Inst.* **31**, 1039–1110.

Rogers, F. T. (1916). Contributions to the physiology of the stomach, XXXIX. The hunger mechanism of the pigeon and its relation to the central nervous system. *Amer. J. Physiol.* **41**, 555–570.

Röseler, M. (1929). Die Bedeutung der Blinddärme der Haushuhns für die Resorption der Nahrung und die Verdauung der Rohfaser. *Z. Tierzucht. Zuechtungsbiol.* **13**, 281–310.

Rosenberg, L. E. (1941). Microanatomy of the duodenum of the turkey. *Hilgardia* **13**, 625–643.

Rossi, G. (1904a). Ricerche sulla meccanica dell'apparato digerente del pollo. Dati anatomici e metodo di recerca. *Atti Reale Accad. Naz. Lincei, Rend., Cl. Sci. Fis., Mat. Natur.*, **13**, 356–363.

Rossi, G. (1904b). Ricerche sulla meccanica dell'apparato digerente del pollo. La meccanica della masticazione. *Atti Reale Accad. Naz. Lincei, Rend., Cl. Sci. Fis., Mat. Natur.*, **13**, 473–378.

Rossi, G. (1905). Sulla meccanica dell'apparato digerente del pollo. *Arch. Fisiol.* **2**, 372–383.

Rothlin, D. (1922). Untersuchungen über den Gehalt an diastatischem Ferment des Pankreas bei Beri-beritauben. *Hoppe-Sayler's Z. Physiol. Chem.* **121**, 300–306.

Roux, W. (1906). Über die funktionelle Anpassung des Muskelmagens der Gans. *Wilhelm Roux' Arch. Entwicklungsmech. Organismen* **21**, 461–499.

Ryan, C. A. (1965). Chicken chymotrypsin and turkey trypsin. Part I. Purification. *Arch. Biochem. Biophys.* **110**, 169–174.

Ryan, C. A., Clary, J. J., and Tomimatsu, Y. (1965). Chicken chymotrypsin and turkey trypsin. Part II. Physical and enzymic properties. *Arch. Biochem. Biophys.* **110**, 175–183.

Sahba, M. M., Morisset, J. A., and Webster, P. D. (1970). Synthetic and secretory effects of cholecystokinin-pancreozymin on the pigeon pancreas. *Proc. Soc. Exp. Biol. Med.* **134,** 728–732.

Salman, A. J., Dal Borgo, G., Pubols, M. H., and McGinnis, J. (1967). Changes in pancreatic enzymes as a function of diet in the chick. *Proc. Soc. Exp. Biol. Med.* **126,** 694–698.

Salques, R. (1934). La nature des matières inertes du gésier des granivores. *Oiseaux* **4,** 531–541.

Sandreuter, A. (1951). Vergleichende Untersuchungen über die Blutbildung in der Ontogenese von Haushuhn (*Gallus gallus* L.) und Star (*Sturnus v. vulgaris* L.). *Acta Anat.* **11,** Suppl. 1, 1–72.

Sato, K., Homma, K., and Gotoh, J. (1960a). Studies on the absorption of glucose in the everted small intestine. II. *Jap. J. Vet. Sci.* **22,** 103–109.

Sato, K., Homma, K., and Gotoh, J. (1960b). Studies on the absorption of glucose in chicks. *Jap. J. Vet. Sci.* **22,** 155–158.

Scharnke, H. (1931). Beiträge zur Morphologie und Entwicklungsgeschichte der Zunge der *Trochilidae, Meliphagidae* und *Picidae. J. Ornithol.* **79,** 425–491.

Scharnke, H. (1932). Ueber eine rückgebildete Honigfresserzunge. *J. Ornithol.* **81,** 355–359.

Schepelmann, E. (1906). Ueber die gestaltende Wirkung verschiedener Ernährung auf die Organe der Gans, insbesondere über die funktionelle Anpassung und die Nahrung. I Teil. Allgemeines, Speiseröhre, und Magen. *Wilhelm Roux' Arch. Entwicklungsmech. Organismen* **21,** 500–595.

Schildmacher, H. (1931). Untersuchungen über die Funktion der Herbst'schen Körperchen. *J. Ornithol.* **79,** 374–415.

Schleucher, R., and Hokin, L. E. (1954). The synthesis and secretion of lipase and ribonuclease by pigeon pancreas slices. *J. Biol. Chem.* **210,** 551–557.

Schlotthauer, C. F., Essex, H. E., and Mann, F. C. (1933). Cecal occlusion in the prevention of blackhead (enterhepatitis) in turkeys. *J. Amer. Vet. Ass.* **83,** 218–228.

Schlotthauer, C. F., Mann, F. C., and Essex, H. E. (1935). A study of egg-production, fertility, and hatchability of a flock of turkey with caecums occluded. *J. Amer. Vet. Ass.* **85,** 455–457.

Schmidt, C. R., and Ivy, A. C. (1937). The general function of the gall bladder. Do species lacking a gall bladder possess its functional equivalent? The bile and pigment output of various species of animals. *J. Cell. Comp. Physiol.* **10,** 365–383.

Schmidt-Neilsen, K., Borut, A., Lee, P., and Crawford, E., Jr. (1963). Nasal salt excretion and the possible function of the cloaca in water conservation. *Science* **142,** 1300–1301.

Schorger, A. W. (1960). The crushing of *Carya* nuts in the gizzard of the turkey. *Auk* **77,** 337–340.

Schreiner, K. E. (1900). Beiträge zur Histologie und Embryologie des Vorderdarmes der Vögel. *Z. Wiss. Zool.* **68,** 481–580.

Schumacher, S. (1922). Die Blinddärme der Waldhühner mit besonderer Berücksichtigung eigentümlicher Sekretionserscheinungen in denselben. *Z. Gesamte Anat., Abt. 1* **64,** 76–95.

Schumacher, S. (1925). Der Bau der Blinddärme und des übrigen Darmohres vom Spielhahn (*Lyrurus tetrix* L.). *Z. Gesamte Anat., Abt. 1* **76,** 640–644.

Schumacher, S. (1927a). Die Mundhöhle. *In* "Handbuch der mikroskopischen Anatomie des Menschen" (W. von Möllendorf, ed.), Vol. 1, pp. 1–34. Springer-Verlag, Berlin and New York.

Schumacher, S. (1927b). Die Zunge. *In* "Handbuch der mikroskopischen Anatomie des Menschen" (W. von Möllendorf, ed.), Vol. 1, pp. 35–60. Springer-Verlag, Berlin and New York.

Schumacher, S. (1927c). Die Speiseröhre. *In* "Handbuch der mikroskopischen Anatomie des Menschen" (W. von Möllendorf, ed.), Vol. 1, pp. 301–336. Springer-Verlag, Berlin and New York.

Schwarz, C., and Teller, H. (1924). Beiträge zur Physiologie der Verdauung. VIII. Mitteilung, Über die Kropfverdauung des Haushuhmes. *Fermentforschung* **7,** 254–269.

Selander, U. (1963). Fine structure of the oxyntic cell in the chicken proventriculus. *Acta Anat.* **55,** 299–310.

Semenza, G. (1968). Intestinal oligosaccharidases. *In* "Handbook of Physiology" (Amer. Physiol. Soc., J. Field, ed.), Vol. V, pp. 2543–2566. Williams & Wilkins, Baltimore, Maryland.

Shaw, T. P. (1913). Digestion in the chick. *Amer. J. Physiol.* **31,** 439–446.

Shufeldt, R. W. (1890). Studies of the Macrochires, morphological and otherwise. *J. Linn. Soc.* London **20,** 299–394.

Sick, H. (1954). Zur Biologie des amazonischen Schirmvogels, *Cephalopterus ornatus. J. Ornithol.* **95,** 233–244.

Skadhauge, E. (1967). *In vivo* perfusion studies of the cloacal water and electrolyte resorption in the fowl (*Gallus domesticus*). *Comp. Biochem. Physiol.* **23,** 483–501.

Skadhauge, E. (1968). The cloacal storage of urine in the rooster. *Comp. Biochem. Physiol.* **24,** 7–18.

Smit, H. (1968). Gastric secretion in the lower vertebrates and birds. *In* "Handbook of Physiology" (Amer. Physiol. Soc., J. Field, ed.), Sect. 6, Vol. V, pp. 2791–2805. Williams & Wilkins, Baltimore, Maryland.

Spallanzani, L. (1783). "Expériences sur la digestion de l'homme et de différentes espèces d'animaux." Barthelemi Chirol, Librairie, Geneva.

Spallanzani, L. (1785). "Versuche über das Verdauungsgeschäft des Menschen und verschiedener Tierarten" (Transl. by Michaelis). Leipzig.

Stanković, R., Arnovljević, V., and Matavulj, P. (1929). Enzymatische Hydrolyse des Keratins mit dem Kropfsafte des *Astur palumbarius* (Habicht) und *Vultur monachus* (Kuttengeier). *Hoppe-Seyler's Z. Physiol. Chem.* **181,** 291–299.

Stannius, F. H. (1846). "Lehrbuch der vergleichenden Anatomie der Wirbelthiere," pp. 12–481. Veit & Co., Berlin.

Steinbacher, G. (1934). Zur Kenntnis des Magens blütenbesuchender Papageien. *Ornithol. Monatsber.* **42,** 80–83.

Steinbacher, G. (1935). Der Trinkakt der Colibris. *Ornithol. Monatsber.* **43,** 11–15.

Steinbacher, G. (1951). Die Zungenborsten der Loris. *Zool. Anz.* **146,** 57–65.

Steinbacher, J. (1941). Weitere Untersuchungen über den Zungenapparat afrikanischer Spechte. *Ornithol. Monatsber.* **49,** 126–136.

Steinbacher, J. (1955). Zur Morphologie und Anatomie des Zungenapparates brasilianscher Spechte. *Senckenbergiana Biol.* **36,** 1–8.

Steinbacher, J. (1957). Ueber den Zungenapparat einiger neotropischer Spechte. *Senckenbergiana Biol.* **36,** 259–270.

Steinmetzer, K. (1924). Die zeitlichen Verhältnisse beim Durchwandern von Futter durch den Magendarmkanal des Huhnes. *Pfluegers Arch. Gesamte Physiol. Menschen Tiere* **206,** 500–505.

Stevenson, J. (1933). Experiments on the digestion of birds. *Wilson Bull.* **45,** 155–167.

Stotz, H., and Brüggemann, H. (1933). Untersuchungen über die Verdaulichkeit der Rohfaser verschiedener Gerstensorten bei Hühnern. *Arch. Gefluegelk.* **7**, 202–215.

Stresemann, I. (1934). Aves. *In* "Handbuch der Zoologie" (W. F. Kükenthal and T. Krumbach, eds.), Vol. VII, Part 2, de Gruyter, Berlin.

Studer-Thiersch, A. (1967). Beiträge zur Brutbiologie der Flamingos, Gattung *Phoenicopterus*. *Zool. Garten* **34**, 159–229.

Sunde, M. L., Gravens, W. W., Elvehjem, C. A., and Halpin, J. G. (1950). The effect of diet and cecectomy on the intestinal synthesis of biotin. *Poultry Sci.* **29**, 10–14.

Suomalainen, H., and Arhimo, E. (1945). On the microbial decomposition of cellulose by wild gallinaceous birds (family Tetraonidae). *Ornis Fenn.* **22**, 21–23.

Sutter, E. (1943). Ueber das embryonale und postembryonale Hirnwachstum bei Hühnern und Sperlingsvögeln. *Denkschr. Schweiz. Naturforsch. Ges.* **75**, Abh. 1, 1–110.

Swenander, G. (1899). Beiträge zur Kenntnis des Kropfes der Vögel. *Zool. Anz.* **22**, 140–142.

Swenander, G. (1902). Studien über den Bau des Schlundes und des Magens der Vögel. *Kgl. Nor. Vidensk. Selsk. Skr.* [N.S.] **6**, 1–240.

Szymonowicz, L. (1897). Über den Bau und die Entwicklung der Nervenendigungen im Entenschnabel. *Arch. Mikrosk. Anat.* **48**, 329–358.

Takahashi, K. (1938). Über die Tauro-apochenodesoxycholsäure aus Hühnergalle. *Hoppe-Seyler's Z. Physiol. Chem.* **255**, 277–280.

Tasaki, I., and Takahashi, N. (1966). Absorption of amino acids from the small intestine of domestic fowl. *J. Nutr.* **88**, 359–364.

Teichmann, M. (1889). Der Kropf der Taube. *Arch. Mikrosk. Anat.* **34**, 235–247.

Tichomirov, B. (1925). Eine vergleichende-anatomische Uebersicht der Speicheldrüsen der Vögel. I. Einige Vertreter der Familie Paridae. *Trav. Soc. Natur. Leningrad* **55**, 61–64.

Tiedemann, F. (1810–1814). "Anatomie und Naturgeschichte der Vögel," Vols, II and III. Mohr & Zimmer, Heidelberg.

Tiedemann, F., and Gmelin, L. (1831). "Verdauung nach Versuchung," Vol. II. Gross, Heidelberg-Leipzig.

Tilgner-Peter, A., and Kalms, H. (1957). Glukoseresorption im Verdauungskanal von Vögeln. *Z. Vergl. Physiol.* **40**, 473–478.

Toner, P. G. (1963). The fine structure of resting and active cells in the submucosal glands of the fowl proventriculus. *J. Anat.* **97**, 575–583.

Toner, P. G. (1964). The fine structure of gizzard gland cells in the domestic fowl. *J. Anat.* **98**, 77–86.

Toner, P. G. (1966). Ultrastructure of the developing gizzard epithelium in the chick embryo. *Z. Zellforsch. Mikrosk. Anat.* **73**, 220–233.

Troitzkaja, A. G., Danilowa, A. K., and Palmova, K. J. (1936). Die Verdaulichkeit der Futtermittel bei Enten. *Biedermanns Zentralbl., Abt. B* **8**, 69–86.

Tscherniak, A. (1936). Über die Verdauung der Zellwandbestandteile des Futters (Lignin, Pentosane, Cellulose, und Rohfaser) durch das Haushuhn. *Biedermanns Zentralbl., Abt. B* **8**, 408–462.

Tucker, R. (1958). Taxonomy of the salivary glands of vertebrates. *Syst. Zool.* **7**, 74–83.

Vandeputte-Poma, J. (1968). Quelques données sur la composition du "Lait de pigeon." *Z. Vergl. Physiol.* **58**, 356–363.

van Dobben, W. H. (1952). The food of the cormorants in the Netherlands. *Ardea* **40**, 1–63.

Vigorita, D. (1906). Sulla costituzione e genesi dello strato cuticolare dello stomaco muscoloso degli uccelli. *Boll. Soc. Natur. Napoli* **19,** 193–216.

Vonk, H. J., and Postma, N. (1949). X-ray studies on the movements of the hen's intestine. *Physiol. Comp. Oecol.* **1,** 15–23.

Vonk, H. J., Brink, G., and Postma, N. (1946). Digestion in the stomach of birds. I. The acidity in the stomach of young chickens. *Proc. Kon. Ned. Akad. Wetensch.* **49,** 972–982.

Wackernagel, H. (1964). Was füttern die Flamingos ihren Jungen? *Int. Z. Vitaminforsch.* **34,** 141–143.

Wagh, P. V., and Waibel, P. E. (1967). Alimentary absorption of L-arabinose and D-xylose in chicks. *Proc. Soc. Exp. Biol. Med.* **124,** 421–424.

Waldschmidt-Leitz E., and Shinoda. O. (1928). Vergleich der Aktivierungsleistung von Enterokinase verschiedener Herkunft. *Hoppe-Seyler's Z. Physiol. Chem.* **177,** 301–313.

Walter, W. G. (1939). Bedingte Magensaftsekretion bei der Ente. *Acta Brevia Neer. Physiol., Pharmacol., Microbiol.* **9,** 56–57.

Warner, R. L., McFarland, L. Z., and Wilson, W. O. (1967). Microanatomy of the upper digestive tract of the Japanese quail. *Amer. J. Vet. Res.* **28,** 1537–1548.

Weber, W. (1962). Zur Histologie und Cytologie der Kropfmilchbildung der Taube. *Z. Zellforsch. Mikrosk. Anat.* **56,** 247–276.

Webster, P. D. (1968). Effect of methacholine on pancreatic amylase synthesis. *Gastroenterology* **55,** 375–385.

Webster, P. D. (1969). Effect of stimulation on pancreatic amylase secretion and nuclear RNA synthesis. *Proc. Soc. Exp. Biol. Med.* **132,** 1072–1076.

Webster, P. D., and Tyor, M. P. (1966). Effect of intravenous pancreozymin on amino acid incorporation *in vitro* by pancreatic tissue. *Amer. J. Physiol.* **211,** 157–160.

Weinland, E. (1899). Beiträge zur Frage nach dem Verhalten des Milchzuckers im Körper, besonders im Darm. *Z. Biol.* **38,** 16–62.

Weiser, S., and Zaitschek, A. (1902). Beiträge zur Methodik der Stärkebestimmung und zur Kenntnis der Verdaulichkeit der Kohlenhydraten. *Arch. Gesamte Physiol. Menschen Tiere* **93,** 98–127.

Wetmore, A. (1920). A peculiar feeding habit of grebes. *Condor* **22,** 18–20.

Widmer, B. (1966). Rekonstruktion des Pankreas von *Paroaria coronata.* Unpublished master's thesis, University of Zürich.

Wilczewski, P. (1870). Untersuchungen über den Bau der Magendrüsen der Vögel. Inaugural-Dissertation, University of Breslau.

Willstätter, R., Haurowitz, F., and Memmen, F. (1924). Zur Spezificität der Lipasen aus verschiedenen Organen. *Hoppe-Seyler's Z. Physiol. Chem.* **140,** 203–221.

Windaus, A., and van Schoor, A. (1926a). Über die Bestandteile der Hühnergalle. *Hoppe-Seyler's Z. Physiol. Chem.* **161,** 143–146.

Windaus, A., and van Schoor, A. (1926b). Über die Konstitution der Cheno-desoxycholsäure. III. *Hoppe-Seyler's Z. Physiol. Chem.* **157,** 177–185.

Winokurow, S. I. (1925). Die motorische Funktion des Kropfes bei experimenteller Polyneuritis. *Pfluegers Arch. Gesamte Physiol. Menschen Tiere* **216,** 576–582.

Wiseman, G. (1968). Absorption of amino acids. *In* "Handbook of Physiology" (Amer. Physiol. Soc., J. Field, ed.), Vol. III, pp. 1277–1307. Williams & Wilkins, Baltimore, Maryland.

Wood, C. A. (1924). The polynesian fruit pigeon, *Globicera pacifica,* its food and digestive apparatus. *Auk* **41,** 433–438.

Yamasaki, K. (1933). Vorkommen der Taurocholsäure in der Hühnergalle. *J. Biochem. (Tokyo)* **18,** 323–324.

Yamasaki, K. (1951). Isolierung der Tetraoxynorsterolcholansäure aus der Hühnergalle und über die Gallensäuren der Galle von *Citellus mongolicus ramosus,* Thomas "Hatarisu" und von Schafen. *J. Biochem. (Tokyo)* **38,** 93–98.

Yasukawa, M. (1959). Studies on the movements of the large intestine. VII. Movements of the large intestine of fowls. *Jap. J. Vet. Sci.* **21,** 1–8.

Yonemura, S. (1926). Über die Gallodesoxycholsäure aus Hühnergalle und ihren Einfluss auf Pankreassteapsinwirkung. *J. Biochem. (Tokyo)* **6,** 287–296.

Yonemura, S. (1928). Über die Gallodesoxycholsäure. *J. Biochem. (Tokyo)* **8,** 79–84.

Younathan, E. S., and Frieden, E. (1956). Studies on amylase synthesis by pigeon pancreas slices. *J. Biol. Chem.* **220,** 801–809.

Zaitschek, A. (1904). Zur Physiologie des Muskelmagens der körnerfressenden Vögel. *Arch. Gesamte Physiol. Menschen Tiere* **104,** 608–611.

Zander, R. (1879). Folgen des Vagusdurchschneidens bei Vögeln. *Arch. Gesamte Physiol. Menschen Tiere* **19,** 263–334.

Zeitzschmann, O. (1908). Über eine eigenartige Grenzzone in der Schleimhaut zwischen Muskelmagen und Duodenum bei Vögeln. *Anat. Anz.* **33,** 456–460.

Zietzschmann, O. (1911). Der Verdauungsapparat der Vögel. *In* "Ellenbergers Handbuch der Vergleichenden mikroskopischen Anatomie der Haustiere" Vol. 3, pp. 377–415.

Zietschmann, O., Ackernecht, E., and Grau, H. (1943). (W. Ellenberger and H. Baum, eds.), *In* "Handbuch der vergleichenden Mikroskopischen Anatomie der Haustiere" 18th ed., pp. 1023–1031. Springer-Verlag, Berlin and New York.

Ziswiler, V. (1964). Neue Aspekte zur Systematik körnerfressender Singvögel. *Ver. Schweiz. Naturforsch. Ges.* 133–134.

Ziswiler, V. (1965). Zur Kenntnis des Samenöffnens und der Struktur des hörnernen Gaumens bei körnerfressenden Oscines. *J. Ornithol.* **106,** 1–48.

Ziswiler, V. (1967a). Vergleichend morphologische Untersuchungen am Verdauungstrakt körnerfressender Singvögel zur Abklärung ihrer systematischen Stellung. *Zool. Jahrb., Abt. Syst.* **94,** 427–520.

Ziswiler, V. (1967b). Der Verdauungstrakt körnerfressender Singvögel als taxonomischer Merkmalskomplex. *Rev. Suisse Zool.* **74,** 620–628.

Ziswiler, V. (1967c). Die taxonomische Stellung des Schneefinken, *Montifringilla nivalis* (Linnaeus). *Ornithol. Beob.* **64,** 105–110.

Ziswiler, V. (1968). Die taxonomische Stellung der Gattung *Sporopipes* Cabanis. *Bonn. Zool. Mitt.* **19,** 269–279.

Ziswiler, V. (1969). Adaptive Radiation innerhalb der Prachfinkengattung *Erythrura* Swainson. *Rev. Suisse Zool.* **76,** 1095–1105.

Ziswiler, V., and Trnka, V. (1972). Tastkörperchen im Schlundbereich der Vögel. *Rev. Suisse Zool.* **79,** 307–318.

Zoppi, G., and Shmerling, D. H. (1969). Intestinal Disaccharidase activities in some birds, reptiles, and mammals. *Comp. Biochem. Physiol.* **29,** 289–294.

Chapter 7

THE NUTRITION OF BIRDS

Hans Fisher

I.	Introduction	431
II.	Requirements Represented by Nutrient Expenditures	432
	A. The Maintenance Requirement for Energy	432
	B. Maintenance Requirement for Protein and Amino Acids	434
	C. Water Requirement for Maintenance	437
	D. Mineral Requirement for Maintenance	439
	E. Requirements for the Fat-soluble Vitamins and Vitamin C	442
	F. Requirements for the B Vitamins	445
	G. Nutrient Requirements for Muscular Work (Flight)	446
III.	Requirements Represented by Syntheses and Storages of Nutrients	446
	A. Nutrient Requirements for Growth and During Senescence	446
	B. Storage of Nutrients in the Body	449
	C. Nutrient Requirements for Reproduction	451
IV.	The Net Nutritive Value of Feeds	454
	A. Nutrients Contained in Feed	454
	B. The Wastage of Nutrients in Digestion and Metabolism	455
V.	Production of Nutritive Fluids for the Young	460
References		463

I. Introduction

In writing a chapter under the heading The Nutrition of Birds, I must take refuge behind the recent observation by Evans and Miller (1968) that "the number of species whose nutritional requirements are known with any precision is relatively few. Of the mammals, only about a dozen species have been studied out of a total of over 5,000; the situation with birds is worse." Most of the studies related to nutritional requirements of birds have been carried out with the chicken but every effort will be made to point out possible differences that might be expected in other species of birds. The task is not an easy one but represents a challenge not only to this writer but to all those biologists who may contemplate to add information in this much neglected field.

II. Requirements Represented by Nutrient Expenditures

A. The Maintenance Requirement for Energy

1. Basal Metabolism

The basal energy expenditure of an animal determines the minimum requirement of food energy and is also one of the more constant of biological measurements (Mitchell *et al.*, 1927). Standard or basal metabolic rates (BMR) have been studied for many species of both wild and domestic birds (King and Farner, 1961). In general, the metabolic rate per unit of body weight decreases with increasing body weight of the bird. Although Mitchell (1962) showed a high degree of correlation between surface area and metabolic rate, and on this basis recommended the usefulness of this relationship, it is exceedingly difficult, because of their posture and feather conformation, to obtain accurate measurements of body surface in birds.

Lasiewski and Dawson (1967) have expressed the metabolic rates (M, kcal/day) of different species of nonpasserine birds ranging in body weight (W, kg) from 10 gm to 100 kg by the equation: $\log M = \log 78.3 + 0.723 \log W$. The analogous regression for passerine species is parallel to that for nonpasserines but at a higher level (Y-intercept$=$ log 129).

According to Mitchell *et al.* (1927) female chickens have a lower BMR than males. Castration of the male lowers the rate to that of the female. A similar sex difference has been observed for the Mourning Dove (*Zenaida macroura*) and the House Sparrow (*Passer domesticus*) (King and Farner, 1961).

2. Activity Increment

The energy needed beyond the basal expenditure to secure food, and for all the other activities necessary for effective living, must be considered part of the maintenance requirement. Standing, in comparison with sitting, has been credited with an increased heat production of 42% in chickens (Deighton and Hutchinson, 1940). The total activity increment for the chicken has been calculated at 48% of the basal caloric expenditure. Since chickens normally do not fly much, it is obvious that the activity increment of avian species that do will be appreciably greater.

3. *Energy Requirement and Food Intake*

Birds, like many other animals, eat to satisfy two basic needs: First, the physiological demands of the body, and second, a food volume need to reach satiety. Lehmann (quoted by Scharrer, 1951) has pointed out that ideally both requirements should be satisfied at the same time. If satiety is fulfilled through bulkiness of the diet before the physiological needs are satisfied, the animal will stop eating before it has acquired its nutrient requirements. If the reverse occurs, the animal may overeat physiologically, and inefficiency results.

Hill and Dansky (1954), in a carefully designed and executed experiment (Table I), showed that chickens exercise considerable con-

TABLE I
EFFECT OF DIETARY FIBER CONTENT ON GROWTH, FOOD, AND ENERGY CONSUMPTION OF CHICKENS[a]

Fiber addition[b] (%)	Body weight (gm)	Food consumption per bird: Amount (gm)	Food consumption per bird: productive energy (kcal)
None	645	1354	2908
10	646	1393	2633
20	666	1593	2585
30	648	1664	2283
40	642	1816	2021

[a]Results are averaged for 10 male and 10 female Rhode Island Red × Barred Plymouth Rock crossbred chickens at 6 weeks of age (modified from the data of Hill and Dansky, 1954).

[b]Ground oat hulls were added at the expense of corn and wheat with adjustments in the level of protein concentrate to maintain the level of crude protein constant at 20% of the diet.

trol over energy intake. When corn and wheat were replaced with up to 40% oat hulls, body weight was maintained through a marked increase in food consumption. Nevertheless (last column, Table I), despite the increased food intake, caloric intake decreased with increased levels of oat hulls. Carcass analysis showed a decreased deposition of body fat to account for the equal body weight despite a reduced caloric intake. The authors concluded that "although the rate of feed consumption was determined primarily by the energy level of the diet, the fact that equal amounts of energy were not consumed by the lots receiving rations of different energy levels indicates that some other factor is (also) concerned in the regulation of feed intake." It is interesting to note that H. Fisher and Weiss (1956) expressed the view that non-nutritive fiber per se stimulated feed consumption in chickens beyond its influence on the caloric density of the diet.

Kendeigh *et al.* (1969) have demonstrated that House Sparrows regulate their food intake in proportion to energy stresses occurring both at night and during the day.

4. Energy Sources (Carbohydrate)

Starch, stored mainly in grains, other seeds, and in tubers, serves as the major carbohydrate energy source for birds. Of the simpler sugars, glucose and sucrose are well utilized by chickens (Griminger and Fisher, 1963) while lactose and galactose are not (Rutter *et al.*, 1953). Fructose supports reasonably good growth in chickens but, in comparison with starch or glucose (Griminger and Fisher, 1963), decreases passage time of food through the intestinal tract.

According to Bolton (1955), cellulose and lignin are not digested by the chicken. There is reason to believe that true cellulose is not digested by most species of birds, with the possible exception of those that have relatively large or well-developed ceca (Farner, 1960). Even those with large ceca, however, probably derive little benefit from digestion that takes place within them, since nutrient absorption from and beyond the cecum appears to be limited. The findings of earlier workers that "fiber" may be digested is probably explainable on the basis of the chemical analyses of the materials considered part of the crude fiber content of the diet. Bolton (1955), who has carefully differentiated among the carbohydrate-complex components of cereals, observed considerable digestibility for pentosans, but very little, if any, for true cellulose.

B. Maintenance Requirement for Protein and Amino Acids

The existence of a relatively constant rate of nitrogen (N) catabolism associated with the maintenance of life has been considered to be analogous in its nutritional significance to basal energy expenditure (Mitchell, 1962). This rate of N catabolism has been referred to as the "endogenous" N excretion, which in the chicken is similar in magnitude (2 mg per basal calorie energy expenditure) to that observed in a number of mammalian species (Leveille and Fisher, 1958). The endogenous N excretion level in the adult male chicken is reached after approximately 5 days on a N-free diet. This rate of excretion and the dietary protein required to maintain it is only half the amount of N excreted or required when the birds receive an adequate supply of dietary protein (Leveille and Fisher, 1958).

For maintenance, the adult male chicken requires approximately 7% dietary protein, an amount that compares well with that reported for the Tree Sparrow, *Spizella arborea* (Martin, 1968). Interestingly enough, Martin also suggested that about half of this amount, or 4% of dietary protein, may represent the absolute minimum for continued existence at a reduced body weight. In contrast with the study of Martin, who used a relatively good quality of protein for his investigation, mention must be made of the observation by MacMillen and Snelling (1966) that White-crowned Sparrows, *Zonotrichia leucophrys*, could not maintain body weight on a diet of bird seed (about 12% protein), but remained in excellent condition when given a supplement of a 20% protein starter diet. These disparate results suggest that the White-crowned Sparrows may have been younger than the birds used in Martin's study.

Although the protein requirement thus appears to vary by a factor of two depending upon prior protein intake, from a qualitative point of view the difference is not nearly that striking. In this regard, Shapiro and Fisher (1962) have shown that the difference in amino-acid requirements between adult male chickens previously maintained on a protein-free diet and those maintained on a liberal protein intake (15%) was primarily one of a greater requirement for nonessential amino acids by the latter group. Thus, the maintenance requirement for essential amino acids was nearly the same for both groups of birds. A possible explanation for this observation in terms of activities of catabolic enzymes has been given by Ashley and Fisher (1967).

Table II lists the maintenance requirement, the mimimum maintenance level, and the requirements for growth and for egg production of chickens for essential amino acids. The maintenance requirement has been defined (Leveille *et al.*, 1960) as the lowest level of amino acid that would maintain the N retention observed with the complete starting diet; the minimum maintenance level (Leveille *et al.*, 1960) represents the lowest concentration of an amino acid that would maintain the birds in N equilibrium. The growth requirements (Netke *et al.*, 1969) are based on optimum weight gain of 7-day-old chicks receiving graded levels of each of the essential amino acids. The requirements for egg production (D. Johnson and Fisher, 1958) are

TABLE II
AMINO ACID REQUIREMENT PATTERNS OF CHICKENS FOR GROWTH, MAINTENANCE, AND EGG PRODUCTION WITH THREONINE TAKEN AS UNITY[a]

Amino acid[b]	Growth[c]	Maintenance requirement[d]	Minimum requirement[e]	Egg production
Arginine	1.5	1.6	1.0	1.8[f]
Histidine	0.5	0	0	0.5
Lysine	1.4	0.4	0	1.4
Isoleucine	0.9	1.0	0.9	1.4
Leucine	1.8	1.7	1.0	1.9
Valine	1.3	0.8	1.0	1.5
Tryptophan	0.2	0.2	0.1	0.3
Methionine[g]	0.5	1.0	0.3	0.7
Phenylalanine[h]	0.8	0.4	0.1	1.2
Glycine[i]	0.5	0	0	0
Threonine	1.0	1.0	1.0	1.0

[a]Expressed as ratios of threonine; to convert to absolute requirements multiply as follows: growth, by 0.65% of diet; maintenance requirement, by 74 mg/kg·day; minimum maintenance level, by 55 mg/kg·day; egg production, by 0.36% of diet.

[b]Expressed as the L-isomer except for DL-methionine.

[c]Values from studies of Netke *et al.* (1969).

[d]Defined as the lowest level of amino acid that would maintain the nitrogen retention observed with the complete diet (Leveille *et al.*, 1960).

[e]Defined as the lowest amino acid concentration that would maintain nitrogen equilibrium (Leveille *et al.*, 1960).

[f]Arginine value from Adkins *et al.* (1962); all other values from Johnson and Fisher (1958).

[g]In the presence of cystine.

[h]In the presence of tyrosine.

[i]Can be completely replaced with serine (Baker *et al.*, 1968).

based on N balance and egg production of hens on graded levels of the essential amino acids.

For purposes of comparison, all amino acid requirements are presented in Table II in terms of patterns based on threonine taken as unity. This manner of presentation allows some interesting comparisons which highlight the differences as well as analogies in requirement for the purposes in which the amino acids are utilized for maintenance, growth, or egg production.

Before considering differences in amino acid requirement based on stage of development of the bird, brief mention should be made of differences and similarities in requirements between the bird (chicken) and mammals. For the chicken, arginine is essential during all stages of development; quantitatively, it is almost at the top of the requirement list. Most mammals, including man, rat, and dog (Allison and Fitzpatrick, 1960), do not require arginine, at least for maintenance. An interesting exception among mammals appears to be the rabbit, which requires this amino acid (Gaman *et al.*, 1970). In common with man, the adult chicken does not require histidine; the rat, and probably the dog, do require histidine. The growing chick requires glycine or serine for optimal growth (Baker *et al.*, 1968), a requirement not shared by any mammalian species or by adult chickens, either for maintenance or egg production.

In assessing the differences and similarities in amino acid requirements during the various stages of development in the bird, H. Fisher and Scott (1954) reported a high correlation between the amino acid requirements for growth and the amino acid composition of the chick carcass. In the same vein, Leveille *et al.* (1960) have more recently shown a high correlation ($r = 0.86$) between the minimum maintenance requirement and the amino acid composition of feather protein. A similar, high degree of correlation between the amino acid requirement for egg production and the amino acid composition of egg protein has also been reported (D. Johnson and Fisher, 1958).

Before leaving the subject of amino acid and protein requirements, mention should be made of a peculiar and interesting amino acid deficiency disorder. In the absence of an adequate amount of dietary lysine, turkey poults as well as chicks from such breeds as the Barred Plymouth Rock and the Black Minorca show a distinct achromatosis. Klain *et al.* (1957) have shown that the amino acid composition of achromatose feathers is not different from that of normally pigmented feathers. They did, however, observe that the tyrosinase activity of feather quill as measured by production of pigment in the presence of tyrosine was reduced in achromatosis as compared with normal feathers. There is no information available as to the occurrence of such depigmentation conditions among wild birds that have pigmented feathers. It may be of interest to point out that lysine is one of the two most limiting amino acids in cereal and grain proteins.

C. Water Requirement for Maintenance

The water content of the bird on a fat-free basis is similar (about 70%) to that of small and large mammals (Mitchell, 1962; Johnston, 1964). Body water in chickens (as a percentage of wet weight) varies inversely with age from 72% at 1 week to 57% at 32 weeks of age (Medway and Kare, 1959). This decrease with age is essentially a reflection of an increased body fat content (H. S. Weiss, 1958). Although there is no appreciable difference between the body water of male and female chickens, there is a significant difference in the half-life of the body water pool. Cocks have a water pool half-life of 7.3 days as compared with one of 3.6 days for hens, irrespective of their rate of egg production (Chapman and Black, 1967).

Bartholomew and Cade (1963) have summarized data on the water consumption of terrestrial birds in the absence of temperature stress. They note that water intake was inversely related to body size. A detailed study of water balance in the chicken (Medway and Kare, 1959) also indicated an inverse relationship between water intake and body size. The *ad libitum* daily water consumption of chickens as a percentage of body weight decreased from 43% at 1 week of age to 13% at 16 weeks of age. The values fall within the range listed for terrestrial birds by Bartholomew and Cade. These workers also reviewed a number of studies in which the minimum water consumption necessary to maintain body weight was investigated in wild, terrestrial birds under laboratory conditions. In each case, this minimum was considerably less than the *ad libitum* consumption of water. MacMillen (1962) has suggested that "the possibility remains, however, that the *ad libitum* water intake of captive birds may more nearly reflect the greater water demands of normally active, non-captive birds." Table III summarizes a water balance study in hens which suggests that the *ad libitum* water intake closely reflects water expenditure and loss.[1]

TABLE III
WATER BALANCE OF LAYING HENS[a]

Intake (gm/bird-day)	
Drinking water	230
Water in food	15
	245
Output (gm/bird-day)	
Respiration	118
Excreta	96
Eggs	19
	233
Balance = 245 − 233 = 12 gm/bird-day	

[a]White Leghorns, average weight 1.75 kg (Rauch, 1961); cutaneous evaporative water loss was not measured (Bernstein, 1969), hence the 12-gm positive balance.

[1]The water balance study of Medway and Kare (1959) must be interpreted with caution since the rate of water retention is much greater than the observed body weight gains.

This writer has made numerous observations concerning factors that influence the *ad libitum* water intake of chickens housed in cages. Generally, more water was consumed when it was available in an open trough in close proximity to the feeder as compared with an overhead nipple that had to be activated with the beak. The fineness of the mash also seemed to influence water consumption; the finer the feed particles, the higher the water consumption. It is our opinion that because of the lack of a pliable tongue that can help move food from the front of the mouth to the rear, the bird, when given easy access to water, will use it as a means of propelling the food toward the esophagus.

From a practical viewpoint, the water needs of animals have been expressed in terms of the ratio of water intake to dry matter consumed. Glista and Scott (1949) have reported this ratio to be 2:1 for chicks.

According to Schmidt-Nielsen (1964) there is little evidence that birds that live in areas of low water accessibility or desert regions have developed special mechanisms that permit them to survive on reduced water intake. However, more recently Poulsen (1965) and O. W. Johnson and Mugaas (1970) have shown that birds that concentrate urine will have larger numbers of loops of Henle.

Early symptoms of reduced water intake in birds include atrophy of the gonads and the loss of reproductive activity. In the domestic chicken, a relatively common disorder that appears to result from water deprivation is known as Blue Comb or Pullet disease (Jungherr and Levine, 1941; H. Fisher *et al.*, 1961). This condition has been observed when young pullets at the start of egg production are moved from their growing quarters into egg-laying quarters where the watering system may be different. It has been noted (H. Fisher *et al.*, 1961) that such birds may continue to eat in a normal manner without having adjusted to the new water supply; they develop a cyanotic comb, a profuse diarrhea, and uric nephritis (Jungherr and Levine, 1941). Recovery from this disorder is rapid upon restoring a readily available water supply for such birds.

D. Mineral Requirement for Maintenance

According to Mitchell (1962), the maintenance requirement for a mineral "possesses a quite different significance from that of an organic nutrient. In the latter case, certain of the organic constituents of the body tissues seem inevitably to be drawn into the 'metabolic mixture' from which the organism is deriving its energy. They emerge from this experience as the end product of metabolism differing in constitution from, and generally much simpler in chemical structure than, the nutrients from which they have been derived." For the mineral nutrients, the situation is very different. Although inorganic ions may be liberated from various functional combinations they are not basically changed in character or made unavailable for further use by the tissues. Thus, under conditions of high demand and short supply it is conceivable that the mineral elements may be reused to a large degree. Evidence for such reuse has been provided for iron, which has been shown to be reused for hemoglobin synthesis (Cruz *et al.*, 1942), for iodine (Fawcett and Kirkwood, 1954), and possibly for calcium (Mitchell, 1962). As a consequence of such reuse, it is obvious that an exact maintenance requirement cannot be stated for most, if not all, of the essential mineral elements.

A good review of the mineral requirements for chickens can be found in Chapter 5 of Scott, Nesheim, and Young's "Nutrition of the Chicken" from which Table IV has been reproduced (Scott *et al.*, 1969).

TABLE IV
MINERAL REQUIREMENTS OF THE GROWING CHICKEN AND EGG PRODUCING HEN[a]

	Chick requirement		Hen requirement	
	0–8 wk of age	8–20 wk of age	20 wk of age	after 40 wk of age
Structural elements (%)				
Calcium	1.0	0.6	3.3[b]	3.7[b]
Phosphorus (available)	0.45	0.4	0.55	0.55
Homeostatic elements (%)				
Sodium	0.15	0.15	0.15	0.15
Potassium	0.3	0.3	0.3	0.3
Chlorine	0.15	0.15	0.15	0.15
Trace elements (mg/kg)				
Magnesium	500	500	500	500
Manganese	50	50	33	33
Zinc	50	30	30	30
Iron	80	40	40	40
Copper	5	5	5	5
Molybdenum	2	2	2	2
Selenium	0.15	0.1	0.1	0.1
Iodine	0.35	0.35	0.3	0.3
Cobalt[c]	–	–	–	–
Chromium	?	?	?	?

[a]Taken from Scott *et al.* (1969).
[b]These data represent daily needs in grams; the dietary percentage depends upon feed consumption rate.
[c]Required only as a part of vitamin B_{12} molecule.

Chickens, ducks, and turkeys are very prone to leg disorders, often of nutritional origin. A condition known as perosis is characterized by an enlargement of the tibiometatarsal joint, with twisting and bending of the distal end of the tibia and proximal end of the tarsometatarsus, resulting in shortening of the leg bones with concommitant slippage of the gastrocnemius from its condyles. This disorder has been shown by Wilgus *et al.* (1936) to be the result of a manganese deficiency, an element that seems a particularly important additive to the diet of young, growing chicks, ducks, and turkeys.

A leg weakness known as cage-layer fatigue has been characterized in egg-producing hens that are maintained in cages with wire floors (Bell and Siller, 1962). The disorder appears to involve hormonal balance (Bell and Siller, 1962) as well as calcium (Bell and Siller, 1962) and phosphorus (Simpson *et al.*, 1964) nutrition.

With the exception of the high calcium requirement of the laying hen, which will be discussed later, the requirements for calcium, phosphorus, and most of the other mineral elements are similar to those of mammalian species that have been studied. In the young, growing bird, the dietary ratio of calcium to phosphorus is optimally at or near 2:1 (Scott *et al.*, 1969), while in the adult the ratio is reversed, with phosphorus representing a higher requirement than calcium (Mitchell, 1962). According to Tyler and Willcox (1942), the minimum maintenance requirement for calcium is about 100 mg/day in laying or nonlaying hens.

Since birds void urine and feces jointly it has been of interest to ascertain the route of mineral excretion. Hurwitz and Griminger (1961) showed that colostomized hens on a plant protein–cereal type diet voided 19% of the total excreted calcium and 62% of the total excreted phosphorus in the urine.

Recent evidence has suggested a greater dietary role for zinc, copper, and selenium than previously assumed. Typical zinc deficiency symptoms in chickens include poor feathering, shortened legs, and a dermatitis of the feet. Zinc deficiency in diets containing the element may nevertheless result from the presence of high levels of dietary calcium or phytic acid (Scott *et al.*, 1969).

Copper deficiency in chickens, aside from its known interaction with iron, has recently been shown to produce dissecting aneurysm of the aorta and various bone deformities resembling lathyrism (Scott *et al.*, 1969).

The nutritional importance of selenium has been recognized for the last decade. Although the nutritional role of selenium has been closely related to that of vitamin E, to be discussed in the next section of this chapter, it has recently been shown to be an essential nutrient, even in the presence of high levels of vitamin E, for the prevention of exudative diathesis and to promote adequate growth (Thompson and Scott, 1969). Work from our laboratory (H. Fisher *et al.*, 1969) has indicated, however, that under certain conditions exudative diathesis of chickens may not be completely relieved by dietary selenium and may require vitamin E.

E. Requirements for the Fat-soluble Vitamins and Vitamin C

1. Vitamin A

It is generally held (Mitchell, 1964) that the fat-soluble vitamins and perhaps also vitamin C are required in proportion to body size and that they play a role in maintaining the integrity of body cells rather than in metabolism in which the B vitamins are known to be involved.[2] The vitamin A requirement of many animal species has been studied and expressed on a body weight basis. According to Rubin and De Ritter (1954), the vitamin A requirement of the chicken is considerably higher, per unit of body weight (100 IU/kg), than for other animal species (20 IU/kg). This higher requirement for chickens may be due to the relative instability of vitamin A sources used in chicken feed 20–30 years ago, rather than to a metabolically conditioned higher requirement. Recent studies with more stable preparations of vitamin A point toward this possibility (Donovan, 1965).

Birds usually obtain a considerable portion of their vitamin A through the ingestion of carotene precursors. The chicken is an efficient converter of beta carotene to vitamin A (Tiews *et al.*, 1969). This conversion, for the most part, occurs in the intestinal wall.

Carotenoids are important not only as precursors of vitamin A but also in the pigmentation of feathers. Lutein has been obtained in crystalline form from the yellow feathers of the Golden Oriole (*Oriolus auratus*). Astaxanthin has been isolated from the red feathers of the Crimson-breasted Shrike (*Laniarius atrococcineus*). Rhodoxanthin has been isolated from the red feathers of the Magnificent Fruit-Dove (*Megaloprepia magnifica*), and canthaxanthin has been isolated from the feathers of the Scarlet Ibis, (*Eudocimus ruber*) (Völker, 1963). Wackernagel (1963) has pointed out that birds kept in zoos often lose their pigmentation unless proper precaution is observed in feeding them the necessary carotenoids in their diet.

[2]Requirement values for all vitamins have, however, been reported per unit of diet weight, as in Table V.

The role of vitamin A in the visual cycle is well established, and vision in birds will be discussed in a separate chapter of a later volume of this treatise.

As is true for the other fat-soluble vitamins, vitamin A as well as carotene can be stored in the egg and an increased dietary intake will be reflected in increased deposition in the egg (Hogan, 1950).

2. *Vitamin D*

The vitamin D requirement of chickens is given in Table V. Vitamin D_3 is the more active form of the vitamin for chickens in comparison with D_2 which has equal activity with D_3 for most mammalian species (Scott *et al.*, 1969).

TABLE V
VITAMIN REQUIREMENT OF CHICKENS[a]

Vitamin	Starting chicks (0–8 wk)	Growing chickens (8–18 wk)	Laying hens	Breeding hens
		IU/kg diet		
A	2000	2000	4000	4000
D	200	200	500	500
E	13.6[b]	?	?	?
		mg/kg diet		
K_1	0.53	?	1[c]	?
B_1	1.8	?	?	?
B_2	3.6	1.8	2.2	3.8
B_6	3.0	?	3.0	4.5
B_{12}	0.009	?	?	0.003
Pantothenic acid	10	10	2.2	10
Nicotinic acid	27	11	?	?
Folic acid	1.2	?	0.25	0.35
Biotin	0.09	?	?	0.15
Choline	1300	?	?	?

[a]Taken in part from Nutrient Requirements of Poultry (1966).
[b]From Scott *et al.* (1969).
[c]Not more than 1 mg/kg (Griminger, 1964).

Contrary to earlier suggestions (Rosenberg, 1953), the preen glands of ducks, geese, and chickens do not contain pro-vitamin D. Rosenberg (1953) has shown that a high concentration of pro-vitamin D is present in the skin of the feet of chickens.

There is considerable carry-over of vitamin D from the diet into the egg and from the egg into the developing chick. Griminger (1966) has shown improper bone mineralization and reduced growth rates in chicks from hens given inadequate amounts of vitamin D_3.

3. *Vitamin E*

Table V shows that little is known about the quantitative requirements for this vitamin. This is partly due to the fact that the requirement for this vitamin depends to a large degree upon stress conditions which may be induced through a high intake of unsaturated fatty acids, or a deficiency of selenium as previously indicated (Scott *et al.*, 1969), or both. Some doubt has been expressed whether the tocopherols that supply vitamin E activity should be considered as vitamins or merely as biological antioxidants (Draper *et al.*, 1964; Machlin and Gordon, 1960).

Seed-eating birds generally should have an adequate intake of vitamin E since vegetable oils provide considerable amounts of tocopherol; at the same time, some of these oils may provide fairly high levels of unsaturated fatty acids.

4. *Vitamin K*

There appears to be little synthesis of vitamin K in the digestive tract of the chicken (Griminger, 1965). Thus, in contrast with a number of mammalian species that have been studied, the young of this species on a vitamin K deficient diet will develop a hemorrhagic syndrome typical of a deficiency of this vitamin. Some sulfa drugs act as vitamin K antagonists and thus increase the chick's requirement for this vitamin (Griminger, 1965). A point that has not been satisfactorily answered relative to the vitamin K deficiency symptoms of chickens concerns the hemorrhaging even in the absence of any apparent injury. The role of vitamin K in the clotting process is well established, but whether the vitamin is also involved in maintaining capillary integrity has not been clearly established (Griminger, 1965).

5. *Vitamin C*

Although most animal species other than man, primates, and guinea pigs have no dietary requirement for vitamin C (ascorbic acid) and are able to synthesize it from glucose, there are birds belonging to the Passeriformes, such as the Red-vented Bulbul (*Pycnonotus cafer*), that do require a dietary source of vitamin C (Roy and Guha, 1958). A number of bird species including the chicken, pigeon, and Collared Scops-Owl (*Otus bakkamoena*) synthesize ascorbic acid in the kidney; others such as the Chestnut Mannikin (*Lonchura malacca*), Hill Myna (*Gracula religiosa*), Dyal (*Copsychus saularis*), and Indian Tree-Pie

(*Dendrocitta vagabunda*) can do so only in the liver. Two species, the House Crow (*Corvus splendens*) and the Common Myna (*Acridotheres tristis*), can synthesize ascorbic acid in both liver and kidney (Roy and Guha, 1958).

It has been suggested but not well proven that egg laying performance, and in particular egg shell strength, of hens under conditions of heat stress is improved when the diet is supplemented with vitamin C (Perek and Kendler, 1962).

F. Requirements for the B Vitamins

Vitamin B_1 was first discovered in relation to a polyneuritis in chickens (with symptoms similar to those of beri-beri in humans) by workers in a hospital laboratory in Indonesia (Scott *et al.*, 1969). The birds so afflicted had been fed on a diet of polished rice; the symptoms were prevented when brown rice or rice polishings were added to the diet. Historically, much of our knowledge concerning the B vitamins was made through the use of the chicken as the experimental animal. Laboratory animals such as the rat, although requiring the same B vitamins as chickens, through coprophagy and perhaps also through greater intestinal synthesis do not develop B vitamin deficiencies as readily as do chicks. Coates *et al.* (1968) concluded that although appreciable amounts of all the B vitamins were synthesized by microbial action in the gut of conventionally maintained chickens, they derived little benefit from the synthesized vitamins, folic acid being a possible exception.

It has generally been accepted that the requirement for the B vitamins is related to the caloric intake of the animals. However, ambient temperature may also have a role since Mills *et al.* (1947) have shown that the thiamine requirement of the chicken, when expressed per unit of diet, is higher at 32° than at 21°C. No such temperature effect was observed for other B vitamins (Mills *et al.*, 1947).

Deficiencies of several of the B vitamins cause marked leg disorders. In thiamine deficiency, a muscle dysfunction results in extended legs making it impossible for the chick to stand properly. Riboflavin and biotin deficiencies give rise to a curled toe paralysis in which, again, standing is awkward. In nicotinic and folic acid deficiency, as well as in choline deficiency, there occur perosis-like leg disorders in young chicks and poults reminiscent of manganese deficiency.

The requirements for the B vitamins, insofar as they have been studied, are listed in Table V. Attention is drawn to the higher requirement of breeding hens for certain of the vitamins to produce viable embryos, in comparison with the requirement by hens for egg production only. H. Fisher and Hudson (1956) noted a high mortality among newly hatched chicks which was overcome by administering a mixture of all the B vitamins and was later discovered to be primarily due to a pantothenic acid deficiency in the diet of the hens. Scott *et al.* (1969) give good tables listing sources for the B vitamins.

G. Nutrient Requirements for Muscular Work (Flight)

Although the metabolism of flight will be considered in detail in a later chapter, it is pertinent to conclude the consideration of nutrient expenditures by referring to the energy cost of flight, which may represent an important fraction of total caloric expenditure. In hummingbirds, the effect of flight has been shown to increase heat production by ten to thirty times the resting metabolism (Lasiewski, 1963). However, these measurements are for hovering flight, which is known to involve a substantially greater energy expenditure than straight-away flight. For the latter, the energy requirement is probably about three to fifteen times standard metabolism, depending, of course, on the type of bird, the kind of flight, and flight conditions. (See Farner, 1970, for a brief review.)

III. Requirements Represented by Syntheses and Storages of Nutrients

A. Nutrient Requirements for Growth and During Senescence

1. Growth

During our discussion of the requirements for maintenance, we have been primarily concerned with the replacement of nutrients lost from the body as a result of catabolic processes. In considering growth, reproduction, and egg production we shall be concerned with nutrient requirements for anabolic processes.

Surprisingly little is known concerning changes in body composition of birds throughout their growing period. Mitchell (1962) has described two studies in chickens, one with a heavy breed of White Plymouth Rocks carried out in 1926 and another with the smaller White Leghorn breed reported in 1931. Although great changes have occurred in the management and feeding practices of domestic chickens, this writer considers the growth studies by Mitchell to be still of considerable significance because most birds living under natural conditions will probably grow more nearly at a rate conforming to that of chickens fed diets formulated in the 1920's than would be true for diets based upon the latest nutritional information in the 1970's.

Figure 1 shows the changes in dry-matter, protein, ash, fat, and gross energy of pullets, capons, and cockerels of the White Plymouth Rock

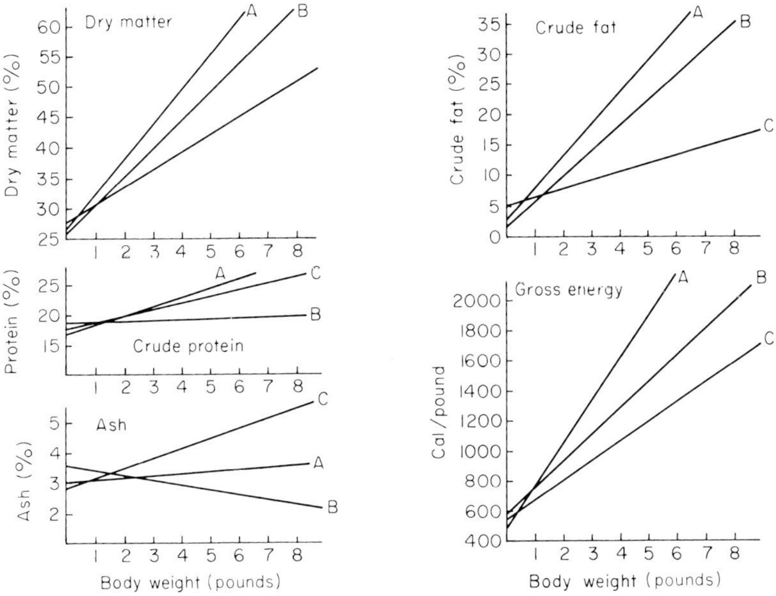

FIG. 1. The change in chemical composition of growth gains in body weight of White Plymouth Rock chickens. Curves A: pullets; B: capons; C: cockerels. (From Mitchell, 1962.)

breed. It can be seen that the pullets fatten considerably more during growth than the cockerels, while caponization of the male largely removes this difference. Also, the water content of the body weight gains varied inversely with the fat content (ether extract) while the gross calorie content varied directly with the fat content.

We have already indicated that the requirement for certain nutrients such as that for the B vitamins does not depend upon a specific physiological body function, but rather depends on food intake since most of the B vitamins are concerned in the metabolism of nutrients. For the fat-soluble vitamins, the requirement appears to depend upon body weight and is, therefore, also not specifically related to rate of growth. For protein and calories, however, the requirement depends strictly on the rate of growth, which can only be determined from an accurate knowledge of the chemical composition of the gains made during the growth period. As Mitchell has pointed out (1962) some tissues continue to grow beyond the age of maturity and, in fact, throughout life. This has been called adult growth. Those tissues that continue to grow include the epidermis and the mucosa of the alimentary tract. We have previously stated that for maintenance the amino acid requirement of birds is influenced considerably by feather formation, which, in a sense, is part of the adult growth.

Since free-living birds frequently do not have a full supply of food that permits maximum growth, it may be of interest to note the effect of mild food restriction on body composition. White Leghorn males were permitted either *ad libitum* access to feed or were daily given an amount of food equal to 80% of the food consumed by the full-fed birds, corrected for body weight (H. Fisher and Griminger, 1963). Table VI shows the body composition of the two groups of birds after

TABLE VI
BODY COMPOSITION OF CHICKENS FULL-FED OR RESTRICTED IN THEIR FOOD INTAKE FOR 3 YEARS

	Food intake[a]	
	Full-fed	Restricted
Final body weight (gm)	2261 ± 132[b]	1832 ± 26
Moisture (%)	63.81 ± 0.34	63.77 ± 0.20
Nitrogen (% dry wt.)	11.98 ± 0.33	12.06 ± 0.11
Lipid extract (% dry wt.)	12.99 ± 1.38	11.70 ± 0.62

[a] 15 full-fed and 24 restricted birds were analyzed (H. Fisher and Griminger, 1963).
[b] Mean ± SE.

37.5 months on the two feeding regimens. Despite a difference in body weight of almost 20%, both groups had essentially the same percentage body composition. The restricted birds were simply smaller in all respects but were not particularly lean or different, at least insofar as gross body components were concerned.

2. *Senescence*

Little is known about changes in nutritional requirements with aging. For man it is well established (Mitchell, 1962) that the basal metabolic rate decreases after the age of 40 by approximately 10%. However, few other important changes in nutritional requirements with age have been documented even for man.

Using the collagen content of certain tissues as a criterion of aging, H. Fisher and Griminger (1963) found a peak in collagen concentration of skin, liver, muscle, comb, and aorta of male chickens between 1 and 3 months of age. In the case of muscle and comb, a further increase occurred between 1 and 3 years of age.

Birds are considered to be very susceptible to atherosclerosis (Ratcliffe and Snyder, 1965), an important disease of aging. It has been reported (Ratcliffe and Snyder, 1965) that the larger psittacines, including macaws, cockatoos, and parrots, will develop massive atheroma of the aorta and brachiocephalic arteries on apparently imbalanced diets made up mainly of sunflower seed. These lesions will disappear when these birds are fed a "high quality mixed ration." Thus, nutrition may be considered to be an important factor in certain debilitating diseases that accompany senescence. In the writer's experience with avian atherosclerosis, caloric restriction is an important factor in retarding the incidence and severity of atherosclerotic lesions (Griminger *et al.*, 1963). The amount of dietary protein and the type of fat and carbohydrate have also been shown to influence the onset and severity of atherosclerosis in birds (H. Fisher *et al.*, 1959, 1964, 1965).

B. Storage of Nutrients in the Body

The storage of nutrients in the body, as distinct from the accretion of nutrients during growth or reproduction, plays an important homeostatic role in liberating the "highly organized forms of life—mammals and birds—from the vicissitudes of the external environment, which tend to disturb . . . the balance existing among its (tissue) nutrient contents" (Mitchell, 1962).

A unique example of this type of storage is represented by the fat stores of migratory birds. Odum and Perkinson (1951) have shown that just before spring migration fat accounted for 17% of total body weight in the White-throated Sparrow (*Zonotrichia albicollis*) and after migration for only 7%. In another study Odum (1958) showed that the fat content of three species of wood-warblers, which have a longer migration flight than the White-thoated Sparrow, averaged 30% of body weight, as compared to 6% in birds that had reached the end of their migration (see chapter by Peter Berthold in a future volume).

Although protein storage can hardly be considered in the same light as fat, good evidence for the presence of so-called "protein reserves" has been obtained in the chicken (H. Fisher, 1967). In our laboratory we observed a greater resistance to Newcastle disease virus in growing chickens fed an excess of good quality protein beyond that needed for maximum growth. Similarly, birds on an excess protein intake were able to cope better with a subsequent regimen in which the dietary amino acid pattern was imbalanced (H. Fisher, 1967). Perhaps the most direct evidence that has as yet been obtained concerning the storage and availability of protein reserves was observed with the laying hen (Leveille *et al.*, 1961). As shown in Table VII, hens given

TABLE VII
INFLUENCE OF PROTEIN LEVEL AND METHIONINE SUPPLEMENTATION ON EGG PRODUCTION AND FEED CONSUMPTION IN THE LAYING HEN[a]

	Dietary variables			
	Protein level (%)		Methionine supplementation	
Week	12	15	−	+
First	3.6 (11)[b]	4.8 (3)	3.2 (6)	5.8 (1)
Second	1.0 (0)	2.8 (0)	0.8 (2)	4.6 (1)
Third	3.6 (2)	4.2 (0)	2.5 (3)	4.8 (2)
Fourth	3.4 (4)	3.4 (0)	−	4.6 (0)

[a]Average weekly egg production for five hens (eggs/hen) (Leveille *et al.*, 1961, © New York Academy of Sciences; Reprinted by permission).

[b]Numerals in parentheses represent the number of hen-days on which sufficient feed had accumulated, due to refusals, obviating the need for daily 100-gm feed allotment.

diets containing either a suboptimal protein level or a suboptimal level of the essential amino acid methionine soon decreased their food intake sharply. Within 1 week egg production ceased, and as soon as it did, the hens returned to a normal food intake. After a period of continuous normal food intake, egg production again commenced, to be accompanied once more by a sharp decrease in voluntary food intake. These results may be explained as follows: The inadequate diets led to a food refusal since they did not provide enough protein or amino acid to maintain proper egg production. As food refusal led to cessation of egg production, the level of protein and amino acids became adequate to meet the maintenance needs only, and the hens reacted by consuming an adequate amount of feed. Protein reserves must have accumulated during the period of cessation in egg production while the hens consumed an adequate amount of diet. This resulted, ultimately, in a renewed onset of egg production.

Another case of nutrient storage peculiar to birds concerns the formation of medullary bone in birds during the egg-laying cycle. This bone, which is an outgrowth from the endosteal lining of the shaft of the long bones, appears to be formed to accommodate calcium for use in the formation of the egg shell (Clavert, 1948; Mitchell, 1962).

Storage of vitamins depends to a considerable extent upon dietary intake, and they are distributed to most body tissues. The adrenal glands are noteworthy for their contents of vitamin C, and the skin of chickens has been reported a good depot for vitamin D (Mitchell, 1964). Kodicek (1954) has divided the vitamins into three groups relative to their storage in the liver: "1. Those that are used by the tissues to be built into co-enzyme systems," [which he calls prosthetins]. "Their level in the liver does not appear to exceed a certain saturation point. . . . 2. The second group contains the vitamins of the fat-soluble group which behave differently in that their concentration . . . . can be increased enormously by dosing. . . ." 3. The third group of vitamins is intermediate between the other two groups.

C. Nutrient Requirements for Reproduction

The nutrient requirements by female birds for egg production are determined to a large extent by the nutrient content of the egg. Romanoff and Romanoff (1949) suggest that the nutrient content of eggs from different types of birds is fairly similar, with the largest variation occurring in the mineral content as a result of species variation in the relative amount of egg shell. The eggs of altricial birds such as passerines have thin shells and, therefore, a much lower percentage of minerals than chicken eggs. The eggs of the Guineafowl appear to have a higher mineral content than those of most other birds.

Everson and Souders (1957) have summarized existing knowledge on the composition of hen eggs in terms of minerals, vitamins, lipids, and amino acids. They have also included the variation in egg composition that may be expected with change in the nutrient content of the hen's diet.

The egg contains considerable quantities of most vitamins, but is devoid of ascorbic acid. The fat-soluble vitamins appear to be greatly influenced by vitamin intake, while there are differences in this regard among the water-soluble vitamins. Adrian (1954) has shown that feeding hens with excess amounts of pantothenic acid leads to a sustained increase in the pantothenic acid content of the egg. On the other hand, the feeding of excess amounts of biotin and of riboflavin leads to an early increase in the egg content of these two vitamins, with a subsequent return to normal levels even though the excess feeding may be continued. The excess feeding of nicotinic acid does not lead to any appreciable increase in nicotinic acid content of the egg.

Feigenbaum and Fisher (1959), among others, have shown that the polyunsaturated fatty acid content of egg fat can be modified considerably by dietary means. As such changes occur, the oleic acid will change in the opposite direction by a similar magnitude. The saturated fatty acids including stearic and palmitic acid are little influenced by dietary change. The cholesterol content of the egg can be increased but appears not to be easily reduced as a result of dietary manipulation of the hen (J. F. Weiss *et al.*, 1967).

The shell from an average-sized chicken egg contains about 2 gm of calcium. Calcium retention efficiency by the Leghorn hen is about 60% (Hurwitz and Griminger, 1960) so that the dietary requirement for calcium approximates 3 gm per egg (see also Table IV).

The efficiency of utilizing dietary nutrients for egg formation is dependent upon the rate of egg production; the higher the rate the greater the efficiency. Table VIII lists some efficiency values for the hen in converting organic nutrients of the feed into egg nutrients.

TABLE VIII
EFFICIENCY OF THE HEN IN CONVERTING ORGANIC NUTRIENTS OF THE FEED INTO EGG NUTRIENTS[a]

	Nutrients in 135 gm of feed	Nutrients in 51.6 gm of egg contents	Conversion of nutrients (%)
Total organic matter, dry (gm)	126.0	13.2	10.5
Protein (gm)	21.6	6.6	30.5
Lipids (gm)	6.7	6.1	91.1
Vitamins			
Vitamin A (units)	1080.0	200–800	46.3
Vitamin B_1 (units)	135.0	20–40	22.2
Vitamin D (units)	108.0	10–50	27.8
Riboflavin (μg)	337.0	100–200	44.5
Pantothenic acid (μg)	1890.0	600–1200	47.6
Total energy (kcal)	396.0	82	20.7

[a]From Romanoff and Romanoff (1949).

Since these figures were compiled at least 20 years ago (Romanoff and Romanoff, 1949), it may be assumed that they represent only a modest rate of egg production by hens. According to Leitch and Godden (as reported by Bolton (1958)), the energetic as well as the protein efficiency of the hen in relatively good egg production (300 eggs per year) is of a similar order of magnitude as that of a dairy cow in good milk production.

Kendeigh *et al.* (1956) and El-Wailly (1966) have shown that the energetic efficiency of egg production in the House Wren (*Troglo-*

dytes aedon) and in the Zebra Finch (*Poephila guttata castanotis*), at high environmental temperatures (29°–34°C), are of a similar magnitude as for the hen. These authors state a value of 77% for the energetic efficiency of egg formation above the energy cost of existence.

Figure 2 shows the relationship between protein intake and nitrogen retention in the egg-laying hen. Maximum nitrogen retention

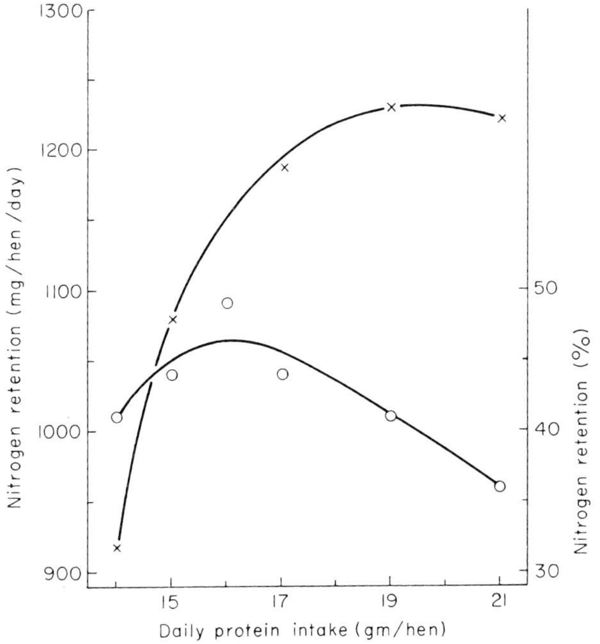

FIG. 2. The effect of protein intake on nitrogen retention and utilization; × represents absolute daily N retention (mg/hen), and O represents percentage of ingested N which is retained. (From Shapiro and Fisher, 1965.)

efficiency was obtained on a daily protein intake of about 16 gm egg protein/day (Shapiro and Fisher, 1965), although absolute nitrogen retention increased beyond the level of maximum efficiency. This additionally retained nitrogen may be beneficial, as was suggested earlier, under conditions of stress such as an inadequate intake of total protein or the ingestion of proteins deficient in essential amino acids.

During the last few years, reports have appeared on the nutritional requirement for male fertility. Wilson *et al.* (1968) fed dietary protein levels ranging from 5 to 16% to male chickens from 7½ to 22 weeks of age and observed a retardation in sexual maturity on levels of dietary protein below 16%. However, semen volume, concentration, and percentage of dead sperm cells were not adversely affected by the low protein intake in comparison with the 16% protein intake. A level of 9% protein fed during the growing period resulted in highest fertility among the mature birds during their subsequent mating year. Varying the protein intake between 7 and 17% during the mating year was found (Arscott and Parker, 1963) not to affect fertilizing capacity of semen in 8-month-old White Leghorn cockerels. It should be recalled that 7% protein meets the protein maintenance needs of the adult cockerel.

IV. The Net Nutritive Value of Feeds

A. Nutrients Contained in Feed

In order to implement the principles of nutrition and to formulate balanced diets, tables of chemical composition for foods and feedstuffs are of considerable importance. Since the listing of such tables is beyond the scope of this chapter, attention is drawn to the composition tables given by Scott *et al.* (1969), as well as to the food composition tables of Watt and Merrill (1963). The value of tables of feed or food composition depends, of course, upon the accuracy of the analytical procedures with which given nutrients have been assayed, as well as upon the deviation of a given item from the average for that particular food class.

Mitchell (1964) has defined "conutrients" as the many chemical compounds in food products that may exert a favorable or unfavorable effect upon nutrient requirements, nutritive status, and nutrient utilization. These substances may modify the effect of nutrients without having any inherent nutritive value. Conutrients that have been studied in connection with birds, usually chickens, include toxins, enzymes–antienzymes, antivitamins, saponins, tannins, and fluorine. The roles of several conutrients have recently been reviewed by Mickelsen and Yang (1966). These include the growth-depressing properties of raw soybeans that contain antitrypsin activity as well as other, as yet incompletely identified, protein fractions that inhibit growth and cause pancreatic hypertrophy in the chicken. The cotton seed contains a toxic yellow pigment known as gossypol. Recently, some gossypol-free varieties of cotton have been grown. The toxicity of gossypol can be reduced by adding iron salts to diets containing cottonseed meal.

In addition to other symptoms of gossypol intoxication, hens eating cottonseed meal produce eggs which, on storage, develop abnormal coloring in the yolk as well as in the albumen. The cyclopropene fatty acids present in cottonseed oil cause a pink color in egg white, while gossypol changes the color of the yolk (Mickelsen and Yang, 1966).

Linseed meal contains a growth-depressing substance that has recently been shown to be a potent pyridoxine antagonist (Mickelsen and Yang, 1966). The castor bean also contains several growth depressants (Bolley and Holmes, 1958). Some, if not all, of the depressant activity has been removed in our laboratory by repetitive hot-water extraction of the defatted meal.

Another group of compounds that affect nutrient utilization are the saponins, a class of glycosides widely distributed in plants. In chickens, saponins extracted from alfalfa have been shown to impair growth at levels of 0.1% of diet and also to impair egg production (Anderson, 1957). The growth depression caused by saponin may be counteracted by cholesterol, which complexes with saponin. Thus, saponins may also be beneficial in reducing dietary cholesterol absorption (Newman *et al.*, 1958). The deleterious properties of saponin may, therefore, be in balance with the more beneficial properties for birds under natural conditions. They may consume quantities of saponin insufficient to cause severe adverse effects but sufficient to provide some protection against dietary cholesterol, or even against the reabsorption of endogenous cholesterol, particularly under conditions of suboptimal protein intake (Griminger and Fisher, 1958).

The tannins, less well defined chemically than the saponins, but also widely distributed in plant materials, also cause growth depression in chicks (Vohra *et al.*, 1966). Chang and Fuller (1964) have demonstrated that chick rations in which the cereal component included certain varieties of sorghum supplied enough tannic acid to cause appreciable growth depression. Hens fed 2% tannic acid showed a significant loss in egg production and the eggs produced had an olive-green discoloration of the yolk (Porter *et al.*, 1967).

A major food contaminant that has caused widespread death in turkeys and ducklings (Wogan, 1966) has been isolated from peanuts. The active toxicant, a mycotoxin, is called aflatoxin. It is quite likely that wild birds may also be affected by this material.

Another conutrient, fluorine, which is relatively toxic to mammals, has been shown to be tolerated better by chickens and turkeys (Mitchell, 1964).

B. The Wastage of Nutrients in Digestion and Metabolism

Since birds excrete urine and feces jointly, there are few meaningful digestion studies available upon which to base information concerning the loss of nutrients during digestion. Consequently, losses during digestion and metabolism are considered together in this section.

1. Energy

In recent years, metabolizable energy has been used as a standard for expressing the available energy content of feeds for chickens (Scott *et al.*, 1969). Metabolizable energy is that fraction of the gross energy of a feed ingredient that undergoes metabolism after subtraction of the energy content of indigestible components (which appear in the feces) and subtraction of non-metabolizable urinary energy. The procedure that has been successfully applied to the determination of the metabolizable energy content of many feed ingredients consumed by chickens is given by Potter and Matterson (1960). Selected data from their report are shown in Table IX.

Caution should be exercised in applying the values determined for chickens to other birds. Martin (1968) found experimentally determined metabolizable energy values for Tree Sparrows generally higher than those calculated from values determined for chickens.

2. Protein

Essentially no information of a comparative nature was available until recently concerning the losses of dietary protein in digestion and metabolism in birds. In 1961 a carcass analysis procedure, based on a method developed with the rat by Bender and Miller (1953), was developed for the chicken in the writer's laboratory (Summers and Fisher, 1961). The procedure determined the *net protein utilization value* (NPU) which is defined as the efficiency (percentage of retention) with which the ingested dietary nitrogen (protein) is retained in the body of the bird. The values are highly reproducible and provide a meaningful comparison of protein quality for the chicken (Table X).

3. Fat

The utilization of fats depends primarily upon the absorption of the fatty acids from the digestive tract. Table XI gives absorption values in chickens for various fatty acids and fats that can be used to estimate the metabolizable energy value of the fat. This is done by multiplying the percentage of absorption by the gross energy value of the fat (approximately 9.4 kcal/gm for most fats (Scott *et al.*, 1969)). This estimation of metabolizable energy for fats is accurate because fatty acids are generally not excreted in the urine.

4. Minerals

The utilization of minerals is more difficult to estimate accurately than that of other nutrients because of re-use and because of re-

TABLE IX
METABOLIZABLE ENERGY OF FEED INGREDIENTS FOR CHICKENS

Feed ingredient	Number of samples	Percent of gross energy metabolized, average	Metabolizable kcal per pound[a] Range	Metabolizable kcal per pound[a] Average
Alfalfa meal (15% protein)	1	16		290
Alfalfa meal (18.5–19.0% protein)	2	26	480–510	490
Barley, ground	1	62		1130
Bread meal	1	85		1640
Glucose monohydrate	2	99		1470
Corn, ground yellow	10	83	1450–1600	1530
Corn, ground white, degermed, dehulled	1	91		1630
Corn gluten meal	1	68		1440
Corn oil	2	94	3980–4030	4000
Distillers dried grains, corn	1	34		740
Distillers dried grains with solubles, corn	3	54	990–1180	1110
Distillers dried solubles, corn	1	62		1340
Fermentation product, antibiotic	2	22	370–420	400
Fish meal, menhaden (62–64% protein)	2	60	1110–1360	1240
Hominy feed	2	71	1250–1510	1380
Meat and bone scraps (44% protein)	1	56		810
Milo, ground	1	82		1530
Oats, ground	2	60	1090–1190	1140
Soybean oil meal (44% protein)	6	54	970–1080	1020
Soybean oil meal (50% protein)	8	56	1030–1140	1090
Starch	1	98		1680
Sucrose	1	92		1670
Tallow, fancy no. 1 beef	2	74	3170–3210	3190
Wheat, ground	1	72		1320
Wheat flour	1	76		1390
Wheat middlings, standard	3	50	780–1140	960
Whey, dried	1	54		860
Yeast, brewers', dried	1	44		840

[a]Metabolizable energy was measured on an oven-dry basis but the values are reported on an air-dry basis, assuming 10% moisture in all ingredients except corn oil, sucrose, and tallow, which are assumed to be moisture-free (from Potter and Matterson, 1960).

TABLE X

NET PROTEIN UTILIZATION (NPU) VALUES OF DIFFERENT PROTEINS FOR THE GROWING CHICKEN[a]

Protein source	NPU (%)
Soybean meal (dehulled)	55
+ methionine	69
Isolated soy protein	46
+ methionine	62
Linseed meal	36
+ methionine + lysine	39
Cottonseed meal	38
+ amino acids[b]	50
Safflower meal	30
+ methionine + lysine	46
Peanut meal	39
+ amino acids[c]	63
Isolated peanut protein	32
+ methionine + lysine	54
Rapeseed meal + lysine	51
Dried beef	76
Meat scrap meal	32
Egg powder	79
Egg white	66
Menhaden fish meal	63
Corn gluten	40
Zein	19
Gelatin	22
Casein	42
+ amino acids[d]	60
Sesame meal	56
+ lysine	63

[a] Taken from various reports by H. Fisher *et al.* (1962); H. Fisher (1964, 1965); Summers and Fisher (1962).

[b] Methionine, lysine, threonine, leucine, and isoleucine.

[c] Methionine, lysine, and threonine.

[d] Arginine, methionine, and glycine.

secretion into the digestive tract for purposes of elimination. Of the many factors known to influence the utilization of minerals by birds as well as mammals, only a few will be mentioned here. Calcium utilization has been shown in mammals (Bendano-Brown and Lin, 1959) to be closely related to the oxalate content of plant foods. Common vegetables such as lettuce, onion, and pepper are fairly high in oxalate, and calcium retention from such sources is correspondingly impaired. It is likely that wild birds that rely for their calcium upon

TABLE XI
ABSORBABILITY VALUES OF VARIOUS FATTY ACIDS, MONOGLYCERIDES, TRIGLYCERIDES, AND HYDROLYZED TRIGLYCERIDES AS DETERMINED IN THE CHICKEN[a]

		Absorption (%)	
		Chicks (3–4 wk)	Chickens (over 8 wk)
Fatty acids			
Lauric	12:0	65	–
Myristic	14:0	25	29
Palmitic	16:0	2	12
Stearic	18:0	0	4
Oleic	18:1	88	94
Linoleic	18:2	91	95
Monoglycerides			
Caprylic	8:0		100
Caprin	10:0		93
Laurin	12:0		89
Myristin	14:0		67
Palmitin	16:0		55
Stearin	18:0		41
Elaidin	18:1 (*trans*)		93
Olein	18:1 (*cis*)		98
Linolein	18:2		96
Triglycerides			
Soybean oil		96	96
Corn oil		94	95
Lard		92	93
Beef tallow		70	76
Menhaden oil		88	–
Hydrolyzed triglycerides			
Soybean oil fatty acids		88	93
Corn oil fatty acids		90	92
Lard fatty acids		82	83
Beef tallow fatty acids		61	67

[a] From Scott *et al.* (1969).

plant sources encounter a low utilization of dietary calcium. Phytin phosphorus, a constituent also of plant foods, increases the calcium requirement of the growing chicken (Nelson *et al.*, 1968b).

Gillis *et al.* (1954) have published a useful table of the biological availability of different inorganic phosphates for the chicken. The orthophosphates, dicalcium phosphates, and defluorinated phosphates were found to have a high availability, whereas pyrophosphates and metaphosphates were essentially unavailable for utilization by the growing chicken. Phytin or organic phosphorus represents, according to Nelson *et al.* (1968a), between 60 and 80% of the total phosphorus of common grains and other plant feed ingredients. Depending upon the level and source of all of the phosphates in the diet, the phytin phosphate is only about one-tenth as effectively utilized by chicks as is inorganic phosphate. Young turkey poults utilize phytin phosphorus even less well than chickens.

5. *Vitamins*

The utilization of the fat-soluble vitamins depends upon a number of factors that may prevail in the food or in the digestive tract of the bird. Scott *et al.* (1969) have summarized the case for vitamin A: (1) destruction of the vitamin in feed through oxidation, high temperature, catalytic effect of trace minerals and peroxidizing effects of polyunsaturated fatty acids; (2) destruction in the digestive tract by prooxidants, coccidia, capillaria, or bacteria; (3) impaired absorption through possible parasitic damage to intestine, low fat level in diet, or lack of sufficient bile salt for micelle formation.

One of the B vitamins, nicotinic acid, is known to be poorly available to the chicken from plant sources. Most of the nicotinic acid present in the common cereals, including rice, wheat, corn, and barley, is apparently in a bound form that is not readily available (Ghosh *et al.*, 1963). Even for the adult hen the nicotinic acid availability from corn and wheat middlings was found to be only 30 and 36%, respectively (Manoukas *et al.*, 1968).

V. Production of Nutritive Fluids for the Young

It is well known that columbiform species feed their young on a holocrine secretion of the crop. Apparently first described by Hunter (1786), who noted that it was produced by both sexes, the formation of this secretion was described in some detail by Claude Bernard (1859) who noted that, unlike mammalian milk, it contained cells; he also noted that it contained about 23% protein, about 10% fat, and no sugar. The formation of crop milk by the domestic pigeon has been investigated extensively from aspects of histology, histophysiology, and cytophysiology (e.g., Beams and Meyer, 1931; Dulzetto, 1938; Litwer, 1926; Niethammer, 1933; Weber, 1962). It is well known that the proliferation of the crop mucosa and the formation of crop milk are caused by prolactin. (See Riddle 1963, for review; also see Chapter 6).

The crop milk of the domestic pigeon bears certain similarities to mammalian milks (Table XII). In protein content and absence of carbohydrate it compares closely to the composition of the milk from marine mammals (Pilson and Kelly, 1962), while the water and fat content resembles that of dog milk (Mitchell, 1962). Most of the unsaturated lipids are not bound as lipoprotein, and are, therefore, readily available to the young bird. The milk proteins contain, at least qualitatively, most of the amino acids essential to the growing chicken (Vandeputte-Poma, 1968).

TABLE XII
COMPOSITION OF PIGEON CROP MILK AS FED TO SQUABS[a]

Item	Percentage
Water	74.00
Ash	1.37
Protein	12.40
Other nitrogen compounds	1.32[b]
Carbohydrate	trace[c]
Lipids	8.61

[a] From Vandeputte-Poma (1968).
[b] Includes free amino acids.
[c] Traces of ribose and deoxyribose from nucleic acids.

The writer has attempted to hand-feed Ring Dove chicks diets of varied composition ranging from infant pediatric formulas to slurries of chicken starting feed. The results indicated remarkable tolerance to differences in nutrient composition, while feeding frequency (once per hour) was a most important factor in obtaining reasonable growth and development.

Greater Flamingos (*Phoenicopterus ruber*) have been shown (Lang *et al.*, 1962; Lang, 1963; Wackernagel, 1964) to produce a fluid of esophageal origin that is initially the sole source of nutrition for the chicks. This fluid contains 8% protein, 0.2% glucose, and about 18% fat; it also contains canthaxanthin, xanthophyll, and traces of β-carotene. Histologic studies indicate that "flamingo fluid" is produced seasonally by merocrine secretion of acinar glands in the esophagus.

The male Emperor Penguin (*Aptenodytes forsteri*), which incubates the single egg and performs the initial brooding of the chick, produces a nutritive fluid in the "crop" at the end of the incubation period and during the brooding period (Prévost, 1961; Prévost and Vilter, 1963). The chick increases in weight from 300 gm at hatching to about 600 gm during the period in which it is nourished by the brooding male. This is a remarkable phenomenon when one bears in mind that the male performs the entire incubation (62–66 days) without food intake, while his body weight declines 12–15 kg from an initial weight of about 35 kg. The dry content of the nutritive fluid contains about 29% lipid, 59% protein, and 5.5% carbohydrate.

Among the Procellariiformes many—perhaps all—species have a much enlarged proventriculus densely packed with glands (Matthews, 1949; Kuroda, 1960) that has been suggested as the source of the waxy oil characteristic of the digestive tract. The function of this fluid has been the subject of much discussion, the most probable functions being nutritive and protective. J. Fisher (1952) has argued that the latter is the case in the Northern Fulmar (*Fulmarus glacialis*) which regurgitates the fluid (both through the mouth and nostrils) free from solid pieces of food that may be present. Kritzler (1948) for the same species argues that the fluid is a digestive by-product and not secretory, primarily because its composition could be altered by changing the dietary fat composition. This, however, is not a conclusive argument! Rice and Kenyon (1962) have offered very strong evidence that this "stomach oil," regardless of its origin, is used by adult Laysan Albatrosses (*Diomedea immutabilis*) and Black-footed Albatrosses (*D. nigripes*) to nourish their chicks. Ashmole and Ashmole (1967) suggest that this fluid may be used widely among the Procellariiformes in feeding their young. They suggest that in these species, which often collect food at great distances from the nest, it represents an adaptation for maximizing the food calories carried per unit weight of nutrient carried; as yet undigested food items constitute the source of protein for the chicks (see Chapter 6, Volume I). This hypothesis, of course, does not exclude the further use of the fluid for protective purposes.

Lewis (1966, 1969) has analyzed in some detail the stomach oils of four procellariiform species. The composition varies widely, even within a species. In samples from Leach's Storm-Petrel (*Oceanodroma leucorhoa*) he found glycerylether ester, largely with C-16 and C-18 side chains, to predominate. In one sample from *Puffinus carneipes* he found 11% squalene, 51% diacyl glycerol ethers, and 16.6% triglycerides, whereas in another he found 78% triglycerides. Lewis therefore proposes that the oils are digestive residues and rejects the hypothesis of Matthews that they are secreted. If Lewis is correct, the role of the specialized procellariiform proventriculus is without explanation. The proposed nutritive role of the stomach oil then must be re-examined in terms of the metabolic capabilities of the young.

REFERENCES

Adkins, J. S., Harper, A. E., and Sunde, M. L. (1962). The L-arginine requirement of the laying pullet. *Poultry Sci.* **41**, 657–663.

Adrian, J. (1954). Etude de quelques vitamines du complexe B2 dans l'oeuf de poule. *World's Poultry Congr. Proc., 10th, 1954* pp. 148–149.

Allison, J. B., and Fitzpatrick, W. H. (1960). "Dietary Proteins in Health and Disease." Thomas, Springfield, Illinois.

Anderson, J. O. (1957). Effect of alfalfa saponin on the performance of chicks and laying hens. *Poultry Sci.* **36**, 877–880.

Arscott, G. H., and Parker, J. E. (1963). Dietary protein and fertility of male chickens. *J. Nutr.* **80**, 311–314.

Ashley, J. H., and Fisher, H. (1967). Protein reserves and muscle constituents of protein-depleted and repleted cocks. *Brit. J. Nutr.* **21**, 661–670.

Ashmole, N. P., and Ashmole, M. J. (1967). Comparative feeding ecology of sea birds of a tropical oceanic island. *Peabody Mus. Natur. Hist., Yale Univ. Bull.* **24**, 1–131.

Baker, D. H., Sugahara, M., and Scott, H. M. (1968). The glycine-serine interrelationship of chick nutrition. *Poultry Sci.* **47**, 1376–1377.

Bartholomew, G. A., and Cade, T. J. (1963). The water economy of land birds. *Auk* **80**, 504–539.

Beams, H. W., and Meyer, R. K. (1931). The formation of pigeon milk. *Physiol. Zool.* **4**, 486–500.

Bell, D. J., and Siller, W. G. (1962). Cage layer fatigue in Brown Leghorns. *Res. Vet. Sci.* **3**, 219–230.

Bendano-Brown, A., and Lin, C. Y. (1959). Availability of calcium in some Philippine vegetables. *J. Nutr.* **67**, 461–468.

Bender, A. E., and Miller, D. S. (1953). A new brief method of estimating net protein value. *Biochem. J.* **53**, vii.

Bernard, C. (1859). "Les propriétés physiologiques et les altérations pathologiques des liquides de l'organisme," pp. 220–238. Baillière et Fils, Paris.

Bernstein, M. H. (1969). Cutaneous and respiratory evaporation in Painted Quail, *Excalfactoria chinensis*. *Amer. Zool.* **9**, 1099.

Bolley, D. S., and Holmes, R. L. (1958). Inedible oilseed meals. *In* "Processed Plant Protein Foodstuffs" (A. M. Altschul, ed.), pp. 829–857. Academic Press, New York.

Bolton, W. (1955). The digestibility of the carbohydrate complex by birds of different ages. *J. Agr. Sci.* **46**, 420–424.

Bolton, W. (1958). The efficiency of food utilization for egg production by pullets. *J. Agr. Sci.* **50**, 97–101.

Chang, S. I., and Fuller, H. L. (1964). Effect of tannin content of grain sorghums on their feeding value for growing chicks. *Poultry Sci.* **43**, 30–36.

Chapman, T. E., and Black, A. L. (1967). Water turnover in chickens. *Poultry Sci.* **41**, 761–765.

Clavert, J. (1948). Contribution à l'étude de la formation des oeufs télolécithiques des oiseaux. Mécanismes de l'édification de la coquille. *Bull. Biol. Fr. Belg.* **82**, 290–330.

Coates, M. E., Ford, J. E., and Harrison, G. F. (1968). Intestinal synthesis of vitamins of the B complex in chicks. *Brit. J. Nutr.* **22**, 493–500.

Cruz, W. O., Hahn, P. F., and Bale, W. F. (1942). Hemoglobin radioactive iron liberated by erythrocyte destruction (acetylphenylhydrazine) promptly reutilized to form new hemoglobin. *Amer. J. Physiol.* **135**, 595–599.

Deighton, T., and Hutchinson, J. C. D. (1940). Studies on the metabolism of fowls. II. The effect of activity on metabolism. *J. Agr. Sci.* **30**, 141–157.

Donovan, G. A. (1965). Vitamin A requirement of growing birds. *Poultry Sci.* **44**, 1292–1298.

Draper, H. H., Bergan, J. G., Chiu, M., Csallany, A. S., and Boaro, A. V. (1964). A further study of the specificity of the vitamin E requirement for reproduction. *J. Nutr.* **84**, 395–400.

Dulzetto, F. (1938). A propista di una communicazione di Broussy sulla istologia ed istofisiologia dell'apparato gastrico dei Granivori. Richerche sull distribuzione e sulla funzione delle "Ghiandole del Gozzo" del colombo e del pollo. *Arch. Biol.* **49**, 369–396.

El-Wailly, A. J. (1966). Energy requirements for egg-laying and incubation in the Zebra Finch, *Taeniopygia castanotis*. *Condor* **68**, 582–594.

Evans, E., and Miller, D. S. (1968). Comparative nutrition, growth and longevity. *Proc. Nutr. Soc.* **47**, 121.

Everson, G. J., and Souders, H. J. (1957). Composition and nutritive importance of eggs. *J. Amer. Diet. Ass.* **33**, 1244–1254.

Farner, D. S. (1960). Digestion and the digestive system. *In* "Biology and Comparative Physiology of Birds" (A. J. Marshall, ed.), Vol. I, pp. 411–467. Academic Press, New York.

Farner, D. S. (1970). Some glimpses of comparative avian physiology. *Fed. Proc., Fed. Amer. Soc. Exp. Biol.* **29**, 1649–1663.

Fawcett, D. M., and Kirkwood, S. (1954). Tyrosine iodinase. *J. Biol. Chem.* **209**, 249–256.

Feigenbaum, A. S., and Fisher, H. (1959). The influence of dietary fat on the incorporation of fatty acids into body and egg fat of the hen. *Arch. Biochem. Biophys.* **79**, 302–306.

Fisher, H. (1964). The limiting amino-acids in groundnut meal provided at optimal and suboptimal protein levels to growing chickens. *J. Sci. Food Agr.* **15**, 539–542.

Fisher, H. (1965). Unrecognized amino acid deficiencies of cottonseed protein for the chick. *J. Nutr.* **87**, 9–12.

Fisher, H. (1967). Nutritional aspects of protein reserves. *Newer Methods Nutr. Biochem.* **3**, 101–124.

Fisher, H., and Griminger, P. (1963). Aging and food restriction: Changes in body composition and hydroxyproline content of selected tissues. *J. Nutr.* **80**, 350–354.

Fisher, H., and Hudson, C. B. (1956). Chick viability and pantothenic acid deficiency in the breeding diet—a case report. *Poultry Sci.* **35**, 487–488.

Fisher, H., and Scott, H. M. (1954). The essential amino acid requirements of chicks as related to their proportional occurrence in the fat free carcass. *Arch. Biochem. Biophys.* **51**, 517–519.

Fisher, H., and Weiss, H. S. (1956). Feed consumption in relation to dietary bulk and energy level: The effect of surgical removal of the crop. *Poultry Sci.* **35**, 418–423.

Fisher, H., Feigenbaum, A., Leveille, G. A., Weiss, H. S., and Griminger, P. (1959). Biochemical observations on aortas of chickens. Effect of different fats and varying levels of protein, fat and cholesterol. *J. Nutr.* **69**, 163–171.

Fisher, H., Griminger, P., Weiss, H. S., and Hudson, C. B. (1961). Observations on water deprivation and blue comb disease. *Poultry Sci.* **40**, 813–814.

Fisher, H., Summers, J. D., Wessels, J. P. H., and Shapiro, R. (1962). Further evaluation of proteins for the growing chicken by the carcass retention method. *J. Sci. Food Agr.* **13,** 658–662.

Fisher, H., Griminger, P., Weiss, H. S., and Siller, W. G. (1964). Avian atherosclerosis: Retardation by pectin. *Science* **146,** 1063.

Fisher, H., Siller, W. G., and Griminger, P. (1965). Restricted protein intake and avian atherosclerosis. *Nature (London)* **207,** 329–330.

Fisher, H., Griminger, P., and Budowski, P. (1969). Anti-vitamin E activity of isolated soy protein for the chick. *Z. Ernaehrungswiss.* **9,** 271–278.

Fisher, J. (1952). "The Fulmar." Collins, London.

Gaman, E., Fisher, H., and Feigenbaum, A. S. (1970). An adequate purified diet for rabbits of all ages. *Nutr. Rep. Int.* **1,** 35–48.

Ghosh, H. P., Sarkor, P. K., and Guha, B. C. (1963). Distribution of the bound form of nicotinic acid in natural materials. *J. Nutr.* **79,** 451–453.

Gillis, M. B., Norris, L. C., and Heuser, G. F. (1954). Studies on the biological value of inorganic phosphates. *J. Nutr.* **52,** 115–126.

Glista, W. A., and Scott, H. M. (1949). Growth and feed efficiency studies with broilers. II. Gross water consumption in relation to the level of soybean oil meal in the ration. *Poultry Sci.* **28,** 747–749.

Griminger, P. (1964). Effect of vitamin K nutrition of the dam on hatchability and prothrombin levels in the offspring. *Poultry Sci.* **43,** 1289–1296.

Griminger, P. (1965). Blood coagulation. *In* "Avian Physiology" (P. D. Sturkie, ed.), 2nd ed., pp. 21–26. Cornell Univ. Press, Ithaca, New York.

Griminger, P. (1966). Influence of maternal vitamin D intake on growth and bone ash of offspring. *Poultry Sci.* **45,** 849–850.

Griminger, P., and Fisher, H. (1958). Dietary saponin and plasma cholesterol in the chicken. *Proc. Soc. Exp. Biol. Med.* **99,** 424–426.

Griminger, P., and Fisher, H. (1963). Fructose utilization in the growing chicken. *Poultry Sci.* **42,** 1471–1473.

Griminger, P., Fisher, H., and Weiss, H. S. (1963). Food restriction and spontaneous avian atherosclerosis. *Life Sci.* **6,** 410–414.

Hill, F. W., and Dansky, L. M. (1954). Studies of the energy requirements of chickens. 1. The effect of dietary energy level on growth and feed consumption. *Poultry Sci.* **33,** 112–119.

Hogan, A. G. (1950). The vitamin requirements of poultry. *Nutr. Abstr. Rev.* **19,** 751–791.

Hunter, J. (1786). "Observations on Certain Parts of the Animal Economy." Sold at No. 13 Castle Street, Leicester-Square, London.

Hurwitz, S., and Griminger, P. (1960). Observations on the calcium balance of laying hens. *J. Agr. Sci.* **54,** 373–377.

Hurwitz, S., and Griminger, P. (1961). Partition of calcium and phosphorus excretion in the laying hen. *Nature (London)* **189,** 759–760.

Johnson, D., Jr., and Fisher, H. (1958). The amino-acid requirement of laying hens. 3. Minimal requirement levels of essential amino-acids: Techniques and development of diet. *Brit. J. Nutr.* **12,** 276–285.

Johnson, O. W., and Mugaas, J. N. (1970). Quantitative and organizational features of the avian renal medulla. *Condor* **72,** 288–292.

Johnston, D. W. (1964). Ecologic aspects of lipid deposition in some postbreeding arctic birds. *Ecology* **45,** 848–852.

Jungherr, E., and Levine, J. M. (1941). The pathology of so-called pullet disease. *Amer. J. Vet. Res.* **2,** 261–271.

Kendeigh, S. C., Kramer, T. C., and Hamerstrom, F. (1956). Variations in egg characteristics of the House Wren. *Auk* **73,** 42–65.

Kendeigh, S. C., Kontogiannis, J. E., Mazac, A., and Roth, R. R. (1969). Environmental regulation of food intake by birds. *Comp. Biochem. Physiol.* **31,** 941–957.

King, J. R., and Farner, D. S. (1961). Energy metabolism, thermoregulation, and body temperature. *In* "Biology and Comparative Physiology of Birds" (A. J. Marshall, ed.), Vol. 2, pp. 215–288. Academic Press, New York.

Klain, G. J., Hill, D. C., Gray, J. A., and Branion, H. D. (1957). Achromatosis in the feathers of chicks fed lysine-deficient diets. *J. Nutr.* **61,** 317–328.

Kodicek, E. (1954). Storage of vitamins in liver. *Proc. Nutr. Soc.* **13,** 125–135.

Kritzler, H. (1948). Observations on behavior in captive fulmars. *Condor* **50,** 1–15.

Kuroda, N. (1960). Histological observations of the proventriculus of *Diomedea. Misc. Rep. Yamashina Inst. Ornithol. Zool.* **3,** 227–238.

Lang, E. M. (1963). Flamingoes raise their young on a liquid containing blood. *Experientia* **19,** 532.

Lang, E. M., Thiersch, A., Thommen, H., and Wackernagel, H. (1962). Was füttern die Flamingos *(Phoenicopterus ruber)* ihren Jungen? *Ornithol. Beobachter* **59,** 173–176.

Lasiewski, R. C. (1963). Oxygen consumption of torpid, resting, active, and flying hummingbirds. *Physiol. Zool.* **36,** 122–140.

Lasiewski, R. C., and Dawson, W. R. (1967). A re-examination of the relation between standard metabolic rate and body weight in birds. *Condor* **69,** 13–23.

Leveille, G. A., and Fisher, H. (1958). The amino acid requirements for maintenance in the adult rooster. 1. Nitrogen and energy requirements in normal and protein-depleted animals receiving whole egg protein and amino acid diets. *J. Nutr.* **66,** 441–453.

Leveille, G. A., Shapiro, R., and Fisher, H. (1960). Amino acid requirements for maintenance in the adult rooster. IV. The requirements for methionine, cystine, phenylalanine, tyrosine and tryptophan; the adequacy of the determined requirements. *J. Nutr.* **72,** 8–15.

Leveille, G. A., Fisher, H., and Feigenbaum, A. S. (1961). Dietary protein and its effects on the serum proteins of the chicken. *Ann. N. Y. Acad. Sci.* **94,** 265–271.

Lewis, R. W. (1966). Studies of the glyceryl ethers of the stomach oil of Leach's Petrel *Oceanodroma leucorhoa (Viellot). Comp. Biochem. Physiol.* **19,** 363–377.

Lewis, R. W. (1969). Studies on the stomach oils of marine animals-II. Oils of some procellariiform birds. *Comp. Biochem. Physiol.* **31,** 725–731.

Litwer, G. (1926). Die histologischen Veränderungen der Kropfwandung bei Tauben zur Zeit der Bebrütung und Ausfütterung ihrer Jungen. *Z. Zellforsch. Mikrosk. Anat.* **3,** 695–722.

Machlin, L. J., and Gordon, R. S. (1960). *Nutr. Rev.* **18,** 95–96.

MacMillen, R. E. (1962). The minimum water requirements of Mourning Doves. *Condor* **64,** 165–166.

MacMillen, R. E., and Snelling, J. C. (1966). Water economy of the White-crowned Sparrow and its use of saline water. *Condor* **68,** 388–395.

Manoukas, A. G., Ringrose, R. C., and Teeri, A. E. (1968). The availability of niacin in corn, soybean meal and wheat middlings for the hen. *Poultry Sci.* **47,** 1836–1842.

Martin, E. W. (1968). The effects of dietary protein on the energy and nitrogen balance of the Tree Sparrow. (*Spizella arborea arborea*). *Physiol. Zool.* **41,** 313–331.

Matthews, L. H. (1949). The origin of stomach oil in the petrels, with comparative observations on the avian proventriculus. *Ibis* **91,** 373–393.

Medway, W., and Kare, M. R. (1959). Water metabolism of the growing domestic fowl with specific reference to water balance. *Poultry Sci.* **38,** 631–637.

Mickelsen, O., and Yang, M. G. (1966). Naturally occurring toxicants in foods. *Fed. Proc., Fed. Amer. Soc. Exp. Biol.* **25,** 104–123.

Mills, C. A., Cottingham, E., and Taylor, E. (1947). The influence of environmental temperature on dietary requirement for thiamine, pyridoxine, nicotinic acid, folic acid and choline in chicks. *Amer. J. Physiol.* **149,** 376–382.

Mitchell, H. H. (1962). "Comparative Nutrition of Man and Domestic Animals," Vol. 1, Academic Press, New York.

Mitchell, H. H. (1964). "Comparative Nutrition of Man and Domestic Animals," Vol. 2. Academic Press, New York.

Mitchell, H. H., Card, L. E., and Haines, W. T. (1927). The effect of age, sex, and castration on the basal heat production of chickens. *J. Agr. Res.* **34,** 945–960.

Nelson, T. S., Ferrara, L. W., and Storer, N. L. (1968a). Phytate phosphorus content of feed ingredients derived from plants. *Poultry Sci.* **47,** 1372–1374.

Nelson, T. S., McGillivray, J. J., Shieh, T. R., Wodzinski, R. J., and Ware, J. H. (1968b). Effect of phytate on the calcium requirement of chicks. *Poultry Sci.* **47,** 1985–1989.

Netke, S. P., Scott, H. M., and Allee, G. L. (1969). Effect of excess amino acids on the utilization of the first limiting amino acid in chick diets. *J. Nutr.* **99,** 75–81.

Newman, H. A. I., Kummerow, F. A., and Scott, H. M. (1958). Dietary saponin, a factor which may reduce liver and serum cholesterol levels. *Poultry Sci.* **37,** 42–46.

Niethammer, G. (1933). Anatomisch-histologische und physiologische Untersuchungen über die Kropfbildungen der Vögel. *Z. Wiss. Zool.* **144,** 12–101.

Nutrient Requirements of Poultry. (1966). *Nat. Acad. Sci.—Nat. Res. Counc., Publ.* **1345.**

Odum, E. P. (1958). The fat deposition picture in the White-throated Sparrow in comparison with that in long-range migrants. *Bird-Banding* **29,** 105–108.

Odum, E. P., and Perkinson, J. D. (1951). Relation of lipid metabolism to migration in birds. Seasonal variation in body lipids of the migratory White-throated Sparrow. *Physiol. Zool.* **24,** 216–230.

Perek, M., and Kendler, J. (1962). Vitamin C supplementation to hens' diets in a hot climate. *Poultry Sci.* **41,** 677–678.

Pilson, M. E. Q., and Kelly, A. L. (1962). Composition of the milk from *Zalophus californianus,* the California Sea Lion. *Science* **135,** 104–105.

Porter, D. K., Fuller, H. L., and Blackshear, C. D. (1967). Effect of tannic acid on egg production and egg yolk mottling. *Poultry Sci.* **46,** 1508–1512.

Potter, L. M., and Matterson, L. D. (1960). The metabolizable energy of feed ingredients for chickens. *Conn., Agr. Exp. Sta., Progr. Rep.* **39.**

Poulson, T. L. (1965). Countercurrent multipliers in avian kidneys. *Science* **148,** 389–391.

Prévost, J. (1961). Ecologie du Manchot Empereur (*Aptenodytes forsteri* Gray). *Actual. Sci. Indu.* **222.**

Prévost, J., and Vilter, V. (1963). Histologie de la sécrétion oesophagienne du Manchot empereur. *Proc. Int. Ornithol. Congr., 13th, 1962,* Vol. 2, pp. 1085–1094.

Ratcliffe, H. L., and Snyder, R. L. (1965). Atherosclerosis in mammals and birds in the Philadelphia zoo. *In* "Comparative Atherosclerosis" (J. C. Roberts, Jr., R. Straus, and M. C. Cooper, eds.), pp. 127–128. Harper, New York.

Rauch, W. (1961). Versuche zur Beurteilung des Wasserhaushaltes von Legehennen. *Arch. Gefluegelk.* **25,** 76–81.

Rice, D. W., and Kenyon, K. W. (1962). Breeding cycles and behavior of Laysan and Black-footed Albatrosses. *Auk.* **79,** 517–567.

Riddle, O. (1963). Prolactin in vertebrate function and organization. *J. Nat. Cancer Inst.* **31,** 1039–1110.

Romanoff, A. L., and Romanoff, A. J. (1949). "The Avian Egg." Wiley, New York.
Rosenberg, H. R. (1953). The site and nature of provitamin D in birds. *Arch. Biochem. Biophys.* **42,** 7–11.
Roy, R. N., and Guha, B. C. (1958). Species difference in regard to the biosynthesis of ascorbic acid. *Nature (London)* **182,** 319–320.
Rubin, S. H., and De Ritter, E. (1954). Vitamin A requirements of animal species. *Vitam. Horm. (New York)* **12,** 101–135.
Rutter, W. J., Krichevsky, P., Scott, H. M., and Hansen, R. G. (1953). The metabolism of lactose and galactose in the chick. *Poultry Sci.* **32,** 706–715.
Scharrer, K. (1951). Die energetische Bewertung der Futter und Nahrungsmittel. *Ber. Oberhess. Ges. Natur- Heilk. Giessen, Naturwiss. Abt.* **24,** 240–255.
Schmidt-Nielsen, K. (1964). "Desert Animals." Oxford Univ. Press, London and New York.
Scott, M. L., Nesheim, M. C., and Young, R. J. (1969). "Nutrition of the Chicken." Humphrey Press, Geneva, New York.
Shapiro, R., and Fisher, H. (1962). Protein reserves: Relationship of dietary essential and nonessential amino acids to formation and maintenance in the fowl. *J. Nutr.* **76,** 106–112.
Shapiro, R., and Fisher, H. (1965). The amino acid requirement of laying hens. 6. The absolute daily protein requirement for peak production. *Poultry Sci.* **44,** 198–205.
Simpson, C. F., Waldroup, P. W., Ammerman, C. B., and Harms, R. H. (1964). Relationship of dietary calcium and phosphorus levels to the cage layer fatigue syndrome. *Avian Dis.* **8,** 92–100.
Summers, J. D., and Fisher, H. (1961). Net protein values for the growing chicken as determined by carcass analysis: Exploration of the method. *J. Nutr.* **75,** 435–442.
Summers, J. D., and Fisher, H. (1962). Net protein values for the growing chicken from carcass analysis with special reference to animal protein sources. *J. Sci. Food Agr.* **13,** 496–500.
Thompson, J. N., and Scott, M. L. (1969). Role of selenium in the nutrition of the chick. *J. Nutr.* **97,** 335–342.
Tiews, J., Zucker, H., and Wendel, E. (1969). Zur Vitamin-A-Biosynthese des Kückens aus β-Carotin. *Int. J. Vitamin Res.* **39,** 141–156.
Tyler, C., and Willcox, J. S. (1942). The calcium requirements of poultry with particular reference to their needs for maintenance. *J. Agr. Sci.* **32,** 62–69.
Vandeputte-Poma, J. (1968). Quelques données sur la composition du "Lait de Pigeon." *Z. Vergl. Physiol.* **58,** 356–363.
Vohra, P., Kratzer, F. H., and Joslyn, M. A. (1966). The growth depressing and toxic effects of tannins to chicks. *Poultry Sci.* **45,** 135–142.
Völker, O. (1963). Die Bedeutung der Carotinoide für das tierische Farbkleid, erläutert an einigen Beispielen. *Wiss. Veroeff. Deut. Ges. Ernaehr.* **9,** 282–291.
Wackernagel, H. (1963). Praktische Erfahrungen in der Carotinoidversorgung von Vögeln. *Wiss. Veroeff. Deut. Ges. Ernaehr.* **9,** 294–299.
Wackernagel, H. (1964). Was füttern die Flamingos ihren Jungen? *Int. Z. Vitaminforsch.* **34,** 141–143.
Watt, B. K., and Merrill, A. L. (1963). Composition of foods. *U.S., Dep. Agr., Agr. Handb.* **8.**
Weber, W. (1962). Zur Histologie und Cytologie der Kropfmilchbildung der Taube. *Z. Zellforsch. Mikrosk. Anat.* **56,** 247–276.
Weiss, H. S. (1958). Application to the fowl of the antipyrine dilution technique for the estimation of body composition. *Poultry Sci.* **37,** 484–489.

Weiss, J. F., Johnson, R. M., and Naber, E. C. (1967). Effect of some dietary factors and drugs on cholesterol concentration in the egg and plasma of the hen. *J. Nutr.* **91,** 119–128.

Whittow, G. C. (1965). Energy metabolism. *In* "Avian Physiology" (P. D. Sturkie, ed.), 2nd ed., pp. 239–271. Cornell Univ. Press, Ithaca, New York.

Wilgus, H. S., Jr., Norris, L. C., and Heuser, G. F. (1936). The role of certain inorganic elements in the cause and prevention of perosis. *Science* **84,** 252–253.

Wilson, H. R., Rowland, L. O., and Harms, R. H. (1968). Reproduction in males fed various grower diets. *Poultry Sci.* **47,** 1733.

Wogan, G. N. (1966). Physiologically significant food contaminants. *Fed. Proc., Fed. Amer. Soc. Exp. Biol.* **25,** 124–129.

Chapter 8

THE INTERMEDIARY METABOLISM OF BIRDS[1]

Robert L. Hazelwood

I. General Considerations 472
A. Carbohydrate Metabolism 472
B. Lipid Metabolism 475
C. Protein Metabolism 476
D. Water Metabolism 477
II. Tissue Metabolism 478
A. Liver 478
B. Skeletal Muscle 479
C. Heart 480
D. Central Nervous System 481
E. Adipose Tissue 482
F. Glycogen Body 484
G. Red Blood Cells 485
III. Influence of Diet 486
A. Embryo Metabolism 486
B. Postnatal Metabolism 489
C. Experimental Manipulation 491
IV. Endocrine Control of Metabolism 496
A. Pancreas 496
B. Thyroid and Parathyroid 501
C. Adenohypophysis 503
V. Influence of Reproductive (Laying) Cycle 507

[1]This chapter as well as unpublished data contained herein was supported in part by NSF:GB-8457.

VI. Climatic Effects on Metabolism ... 510
A. Heat ... 510
B. Cold ... 511
C. Diurnal Rhythms and Photoperiodicity ... 512
VII. Influence of Migration on Metabolism ... 514
A. Short-bout Exercise ... 514
B. Prolonged Flight (Migration) ... 515
VIII. General Summary and Conclusions ... 518
References ... 519

I. General Considerations

By title the topic at hand is all-encompassing and therein a difficult task to complete. Added to the vastness of this topic is the disturbing dearth of information pertinent to large areas of avian metabolic processes, an abundant volume of data obtained from studies employing indirect approaches, and the fact that most studies have been carried out only in the "white feathered rat," *Gallus "domesticus."* Thus, the approach that follows dwells on selective topics and phases of intermediary metabolism of Aves with concerted effort made to draw from studies conducted on nondomestic forms. Many excellent texts on biochemical aspects of metabolism are available to the reader and he is referred to them for the specific details of mammalian pathways of anabolism and catabolism, most of which also occur in birds (Herrmann and Tootle, 1964; Deucher, 1965; Romanoff, 1967; Drummond, 1967).

A. Carbohydrate Metabolism

As in mammals, blood glucose circulates in the form of *d*-glucose in birds and is available to anabolic and catabolic reactions by intracellular translocation followed by phosphorylation at the carbon-6 position. Other sugars (fructose and galactose) enter the metabolic scheme by ATP and specific enzyme reactions to become phosphorylated sugars (fructose-1-phosphate or fructose-6-phosphate and glucose-1-phosphate, respectively). Synthesis of glycogen proceeds in avian tissues along pathways identical to those described in mammals, namely, glucose-1-phosphate is converted to uridine diphosphoglucose (liver, muscle, red blood cells), then catalyzed by glycogen synthetase to form successive α-1,4-linked glucose units in straight chains. Under the aegis of a branching enzyme (amylotransglucosidase) chains of seven to ten α-1,4-glucose units are transferred irreversibly to an α-1,6 linkage on the same chain (or another near by) forming highly branched, dendritic structures. Thus, the glycogen molecule can be conceptualized as a spherical molecule with a radius less than 180 Å and with outer terminal branches arborized as are the peripheral branches of a tree. These outer tiers (maltose fractions), therefore, represent the most recent portion of the glycogen molecule synthesized and are available for subsequent transfer to the inner core of the molecule (limit dextrin) or for degradation as the first step in glycogenolysis.

Alpha-1,4-glucosyl units are mobilized repetitively from the outer tiers of the glycogen branches by the direct action of phosphophosphorylase. Recent emphasis on the degradative steps has indicated the need for adenyl cyclase, an enzyme which cyclizes ATP into 3′,5′-AMP. The latter participates in the activation of phosphorylase to initiate the glycogenolytic step at the outer branches of the molecule. Successive units are removed down to the α-1,6 linkages at which point a special debranching enzyme is required. Splitting of the α-1,6 links results in free glucose liberation and then phosphorylase continues to split the α-1,4 links of glucosyl units once again. Thus, the end result of glycogenolysis in mammals and birds is glucose-1-phosphate and a small amount of nonphosphorylated (free) glucose. Through a mutase reaction glucose-1-phosphate is converted to glucose-6-phosphate.

The significance of the direct oxidative pathway of carbohydrate in avian tissues is yet to be firmly established. This pathway, frequently referred to as the pentose phosphate or hexosemonophosphate shunt, is an alternative pathway for glucose-6-phosphate to reach the three-carbon phosphate state, CO_2 and water by circumventing the classical anaerobic glycolytic pathway of Embden–Meyerhof and the citric acid cycle. The importance of this pathway in mammals has been established in liver, adipose, and mammary tissue but in birds equivalent studies are sparse, and available significant evidence appears to center on brain and intestinal tissue (see below). In effect three glucose molecules are degradated (oxidized by dehydrogenation with NADP as the acceptor) resulting in three CO_2 molecules and three pentoses. The latter are rearranged to form two glucose-6-phosphate molecules and one glyceraldehyde-3-phosphate. Complete oxidation occurs when two of the three carbon fragments recombine and essentially by a reversal of glycolysis form another glucose-6-phosphate molecule. Characteristic of this pentose shunt pathway is the production of pentoses (for nucleotide and nucleic acid synthesis), reduced NADP (essential for lipogenesis and steroid synthesis), and CO_2. Invariably, those tissues in which lipogenesis or steroidogenesis is prominent also depend heavily on the pentose phosphate pathway for availability of NADPH. Thus, while mammalian (but not avian) liver and adipose tissue have high pentose oxidation activity, muscle tissue does not.

From the above, it is obvious that glucose-6-phosphate plays an important (critical!) role in carbohydrate metabolism since this molecule can lead to glycogen synthesis, to CO_2, triosephosphates, and pentoses, or can be dephosphorylated to free glucose and released to the blood stream. Removal of the high-energy phosphate from carbon-6 requires presence of the enzyme glucose-6-phosphatase which is abundant in liver and kidney tissue. Absence of this enzyme in muscle tissue indicates that muscle glycogen does not contribute directly to blood glucose levels; rather, glucose-6-phosphate thereby formed must proceed down the anaerobic glycolytic path. Anaerobic glycolysis occurs in the extramitochondrial cytoplasmic compartment by proceeding through isomerase and kinase reactions to form the fructose-1,6-diphosphate intermediate which in turn is split into two three-carbon fragments. Through successive steps these triose-phosphates are converted into pyruvate which is subsequently decarboxylated to acetyl-CoA ("active acetate") and which in turn condenses with oxaloacetate to form citrate. This last step is the first of many steps involved in the aerobic catabolism of glucose and produces 30 of the total 38 net high-energy bonds obtained from a molecule of glucose. The detailed steps involved in the citric acid cycle are given in standard textbooks and will not be considered at this point. Nonetheless, the end result is the formation of CO_2 and water and the release of energy from the original glucose-6-phosphate molecule.

While glucose appears to be a major immediate source of energy for avian cells, extensive fatty acid utilization (especially during migration) and gluconeogenesis must constantly be borne in mind. Furthermore, while glucose and glycogen are the major forms of ingested carbohydrate, possible conversion of other sugars such as fructose and galactose to glycolytic intermediates should not be overlooked. The major route of phosphorylation of fructose appears to be under the influence of the specific enzyme fructokinase, the end product being fructose-1-phosphate. Fructose-6-phosphate may occur to a small extent by action of hexokinase (which also phosphorylates glucose and mannose). The former enzyme is unaffected by nutritional or hormonal manipulation and therefore differs from glucokinase and hexokinase.

Hepatic tissue rapidly converts galactose to glucose, first by phosphorylation at the carbon-1 position and then by reaction with uridine diphosphate glucose to form UDP-galactose and glucose-1-phosphate.

The glucose moiety thus formed is then available to enter the glycogenic scheme as described earlier or be transformed to glucose-6-phosphate. The galactonucleotide, however, is converted to UDP-glucose and thereby is also available for glycogenic reactions. Since the latter conversion is freely reversible, galactose is not essential in the diet even though the sugar is necessary for cerebrosides of nervous tissue, mucoproteins of connective tissue, and for milk synthesis in mammals.

Important to the fast growing embryo and neonatal organism is an alternative oxidative pathway for glucose-6-phosphate, namely, the uronic acid pathway which incorporates the early glycogenic sequence of reactions. However, at the UDP-glucose step carbon-6 is oxidized to glucuronic acid and thus the final product is UDP-glucuronic acid. This product is an extremely important reactant in the synthesis of chondroitin sulfate and in steroid hormone conjugation. In contrast to mammals, birds utilize a portion of the UDP-glucuronic acid as precursor substrate for ascorbic acid synthesis. Once again, the important central role of glucose-6-phosphate initially in these many reactions is apparent, for truly this intermediate appears to be critical in the regulation of carboydrate, fat, and protein metabolism. Thus, glucose-6-phosphate can be said to be at the metabolic crossroads of glycolysis, glycogenesis, the direct oxidative pathway, gluconeogenesis, and glycogenolysis.

B. Lipid Metabolism

Most authorities feel that a considerable portion of the carbohydrate absorbed by the avian intestine (glucose, fructose, galactose, and mannose) is converted to fat prior to its use as an energy source. Fat is then released from depots (usually under the influence of a lipase) in the form of nonesterified (free) fatty acids which enter the bloodstream and circulate loosely complexed with plasma albumin. However, free fatty acids (FFA) represent only a small portion of the total lipid found in plasma ($< 6\%$), the majority existing as triglyceride, phospholipid, and cholesterol components. Triglycerides are hydrolyzed in adipose tissue to their constituent fatty acids and glycerol prior to release to the plasma. Here a lipoprotein complex is formed and subsequent uptake and oxidation occur in cells of liver, heart, muscle, brain, and kidney. Once inside tissue cells the fatty acids are converted to an "active" intermediate, a step requiring ATP and a thiokinase enzyme, and then undergo β-oxidation, namely removal of the first two (α and β) carbon atoms from the carboxyl end of the molecule. The oxidation proceeds in the mitochondrion as described, two carbon atoms at a time, until the final degraded form exists as acetyl-CoA which then condenses with oxaloacetate to form citric acid. During starvation or other situations of high fatty acid oxidation excessive build-up of acetyl-CoA occurs in the liver of birds and subsequently results in formation and release to the plasma of ketone bodies (acetoacetate, β-hydroxybutyrate, and acetone). Formation of acetyl-CoA molecules in the above manner is attended by energy release in the form of high-energy bonds. Additional ATP bonds are formed as the citric acid oxidation proceeds. Thus, fat potentially is a most convenient and efficient storage form of energy since it contains at least twice the caloric value (9.3 kcal/gm) of either carbohydrate or protein and because it is deposited with a minimum of water content. Total catabolism of common dietary fatty acids leads to an efficiency of 41% in trapping potential energy as high-energy bonds, a figure which agrees well with that obtained from total combustion of carbohydrate.

The principal site of lipogenesis (fatty acid synthesis) apparently is in the cell cytoplasm although elongation of the fatty acid chain occurs within the mitochondrion and possibly as well in microsomes of liver cells. The rate-limiting step in lipid formation appears to be the conversion of acetyl-CoA to malonyl-CoA by the action of acetyl-CoA carboxylase. Subsequently, sources of NADPH are required at least at two different reductase steps. Generators of reduced NADP are most likely the pentose phosphate pathway, and malic enzyme and isocitrate dehydrogenase activities. This series of reactions occurs in most tissues as long as the diet provides adequate potential sources of glucose-6-phosphate or pyruvate, or both, and thereby makes available the acetyl-CoA necessary to react with acetyl-CoA carboxylase. Diets that contain high levels of carbohydrate accelerate these reactions; any form of dietary restriction, as in fasting or situations of high fatty acid intake or insulin deficiency, depresses lipogenesis markedly.

C. Protein Metabolism

Proteins enter the metabolic scheme as amino acids and therefore, as in the case of lipid, are degradated in a schema distinct from glycolysis preparatory to entering the citric acid cycle. A discussion of the metabolism of each amino acid is beyond the scope of this treatise; however, the well-documented interconversion of amino acids from one to another renders it conceivable that but five amino acids need be considered in terms of ultimate degradation and energy release. Thus, under certain (normal) metabolic circumstances, and due to the aforementioned interconversion, alanine typifies such amino acids as cysteine, methionine, serine, glycine, tryptophan, and hydroxyproline in that all (through devious paths) enter the degradative scheme at the pyruvate level. [It will be recalled that pyruvate is subsequently decarboxylated to acetyl-CoA (active acetate).] Singularly, aspartate enters the citric acid scheme by conversion to oxaloacetate. Phenylalanine, leucine, tyrosine, and isoleucine can, through various intermediates, result in acetyl-CoA formation (and thereby a juncture is made with fatty acid metabolism) before condensation with oxaloacetate to form citrate. Lysine, histidine, glutamine, proline, hydroxyproline, arginine, and ornithine can enter the metabolic scheme at the α-ketoglutarate step, while methionine and valine may enter at the succinyl-CoA step of the cycle. Finally, phenylalanine and tyrosine may be converted to fumarate and thereby contribute to the energy-releasing activity of the oxidative scheme.

The critical role that oxaloacetate plays in intermediate metabolism is worthy of note because it serves as an intermediate in both the citric acid cycle and in forming carbohydrate from noncarbohydrate sources. Thus, labeled fatty acids as well as certain amino acids can be found in glucose and glycogen depending upon the prevailing physiological conditions. Certain reaction "barriers" have to be overcome for glucose (or glycogen) to be produced from lower intermediates of metabolism. Thus, the best evidence at hand allows that such anabolic processes occur by circumventing several key steps of glycolysis, not merely by reversing the Embden–Meyerhof reactions. In both birds and mammals liver glycogen may result from injection of lower intermediates (Krebs *et al.*, 1964) but only by complete randomization of the original carbon atoms. Contrarily, in avian and mammalian muscle, glycogenesis may occur by a simple reversal of glycolysis. Gluconeogenesis apparently occurs during times of "energy starvation" in the liver and to a lesser extent in the kidney.

D. Water Metabolism

Virtually ignored has been quantitative study of water ingestion and turnover in birds other than those studies emphasizing production-economical aspects. Evidence is at hand, however, that sex differences exist as far as water ingestion is concerned and that the rate of intake is closely related to rhythmic physiological functions. Hens of the domestic fowl ingest more water than cocks, and the laying cycle (see below) along with the large amount of water contained in the fresh egg (two thirds by weight) are of prime significance in accounting for these sex differences (Lifschitz *et al.*, 1967). Water turnover studies in chickens have been carried out using tritiated water, and the results reported indicate once again sex differences in the half-life of the body water pool (Chapman and Black, 1967). Kinetic studies on the total body water pool indicate a half-life of 3.6 ± 0.33 days for hens and 7.3 ± 0.32 days for cocks. While variations occur in hens, the half-life for body water is always significantly below that of comparable age males regardless of the rate of egg production. The size of the water pool, however, as a percentage of body weight is not significantly different in the two sexes (Chapman and Black, 1967).

Respiratory and cutaneous body water loss is very high on the first day after hatching at a time when the basal metabolic rate is quite low. The loss of water in the first 2 weeks decreases greatly in young chicks, then rises markedly as the weight-specific metabolic rate increases sharply to a maximum by 3–4 weeks of age (Medway and Kare, 1957). Both water loss and metabolic rate slowly decline as the chicken matures. Deprivation of water in poults leads to reduction in body temperature, the magnitude of which corresponds closely with that of the length of time water is withheld (Haller and Sunde, 1966). Replacement of cool water corrects the observed depression of food intake and growth rate during the water-deprivation period and also assists return of the suppressed body temperature toward normal. Estimates of the tolerance of chickens of different ages to loss of body water have been reported, indicating that the major loss of body weight during such periods is attributed to loss of water and that death ensues when approximately 45% of total body water is lost (Mulkey and Huston, 1967). Evidently, muscle fibers, particularly the white (outer) fibers retain the majority of this tissue's water (Froning and Norman, 1966).

II. Tissue Metabolism

A. Liver

Hepatic tissue of birds has been employed frequently in both *in vitro* and *in vivo* studies of intermediary metabolism. Suffice it to say that most—if not all—the pathways discussed above in Sections I,A,B, and C have been found in avian hepatic tissue, though pentose shunt oxidation of glucose appears to be minimal (O'Hea and Leveille, 1968). Animals of experimental choice have been primarily the domestic chicken, pigeon, duck, and, on occasion, the goose. Once absorbed through the gut, foodstuffs are carried to the liver via the portal vein and thence are either stored, altered to other potential substrate forms, or degradated to CO_2, water, and energy. Daily rhythms of carbohydrate and fat concentration have been observed in embryos and also in adult chickens, Red-winged Blackbirds (*Agelaius phoeniceus*), and European Starlings (*Sturnus vulgaris*). Unlike the avian embryo, which exhibits up to three cycles daily, adult forms show only one diurnal cycle, which is related to photoperiodicity. Thus, Starlings and Red-winged Blackbirds have been observed to have highest levels of both liver fat and glycogen during early evening hours and lowest levels in early morning; altering light patterns alters the substrate levels correspondingly (H. I. Fisher and Bartlett, 1957).

Avian liver, unlike that of mammals, metabolizes fructose with relative ease and readily forms triose-phosphates from the labeled keto-sugar (Heald, 1963). Despite the ease with which fructose is metabolized by the liver, birds do not tolerate increased galactose levels well. The UDP-glycogen pathway appears early during embryonic development and persists in the adult bird, yet dietary intake of galactose (10% or greater) results in high blood galactose, low liver glycogen, normal blood glucose levels, and eventually epileptiform convulsions and death. High lactose intake does not lead to similar sequelae and therefore the avian liver probably does not degradate lactose to galactose and glucose. The fact that galactose readily penetrates intracellular compartments indicates that the deleterious effects of induced galactosemia in birds may well be due to suppression of glucose utilization. Avian liver UDP-glucose levels are markedly reduced in galactose poisoning (Nordin *et al.*, 1960).

Hepatic tolerance of dietary fat is in contrast to that of galactose; well over 90% of the nonprotein caloric intake can be supplied by fat without an untoward effect on growth of chicks. Also well documented is the role of the avian liver in lipid biosynthesis. Evidence has been supplied by several laboratories indicating that hepatic lipid synthesis accounts for 90–95% of *de novo* fatty acid synthesis in chicks, the remaining synthesis occurring in various adipose sites and skin. Since the activity of pentose shunt dehydrogenase enzymes is quite low in avian liver, while that of malic enzyme is high, the former pathway probably plays a small role in providing reducing equivalents for lipid synthesis (O'Hea and Leveille, 1968). Once formed in liver, lipids are transported in the blood mainly as β or low-density lipoproteins to various depots in the avian organism (O'Hea and Leveille, 1969). Further details of avian lipid metabolism and endocrine control thereof are presented in Sections IV,A and C.

B. Skeletal Muscle

Studies in birds have been limited largely to the pectoral muscles because of their active metabolic patterns and use in flight. Of particular importance are studies on pigeon breast muscle, which has been shown histochemically and biochemically to possess an abundance of active and inactive phosphorylase (Vallyathan *et al.*, 1964). Particularly, the white skeletal muscle fibers tend to lie to the outside of the muscle fasciculus, receive sparse vascular elements, contain very high phosphorylase and glycogen levels but very few lipid inclusions, and have a relative paucity of mitochondria and myoglobin (George and Telesara, 1961; George and Berger, 1966). Red fibers appear to be narrower and to have sparse glycogen granulations but in all other respects have considerably greater levels of substrates, intracellular inclusions, and enzymes. Red fibers contain an abundance of the citric acid cycle enzymes. It would appear, therefore, that white muscle fibers are geared metabolically for fast bursts of energy employing anaerobic glycolysis (as in take-off activity) while red fibers are poorly equipped for anaerobiosis and therefore depend on aerobic degradation of noncarbohydrate foodstuffs. Since there is no evidence of protein stores serving as fuel for metabolic activity in birds normally, sustained muscular effort would appear to be an attribute of red fibers wherein fatty acid metabolism predominates (Drummond, 1967). Inactivity of muscle invariably leads to disappearance of both phosphorylase and glycogen inclusions from white fibers in particular (Vallyathan *et al.*, 1964).

C. Heart

Cardiac muscle of birds has not been studied as extensively as avian skeletal muscle despite normal heart rates which are three to seven times faster than in man and cardiac outputs which average up to seven times greater than man on a unit body weight basis. Generally, it is agreed that protein depots contribute little or nothing to cardiac metabolism in birds and mammals, that carbohydrate contributes a definite, though relatively small, amount to the fuel supply, and that fatty acids, pyruvate, and lactate degradation dominate heart tissue energetics. In addition to the bioenergetics of flight muscles and the acute demands of the laying cycle, cardiac metabolism also depends on ability of this tissue to extract nonesterified fatty acids, triglycerides, ketone bodies, pyruvate, and lactate from the circulation for subsequent intracellular utilization. Dietary triglycerides are available for cardiac fuel as well as for glyceride synthesis.

Utilization of lipid for metabolic fuel requires the presence of specific enzymes, lipases, of which the myocardium has an abundance and which release fatty acids for subsequent β-oxidation reactions to the heart. Studies on mammalian perfused hearts or *in vitro* heart muscle strips indicate that over 85% of the energy needs of cardiac tissue are normally met by fatty acid metabolism and that of the total amount extracted from perfusing fluids a small portion (< 15%) is stored within the tissue for subsequent use. Comparable studies in birds are yet to be reported. Uptake and utilization of noncarbohydrate substrates appear to be directly related to the amount presented to the myocardium per unit time and in doing so reduces simultaneous utilization of carbohydrate. This interplay between carbohydrate and fat utilization by heart muscle appears to be virtually independent of hormone control, yet readily responds to nutritional, hormonal, and humoral influences. Most of these influences act by making available more cardiac lipase to split triglyceride substrate during situations where increased cardiac metabolism is required. Thus, exercise, starvation, catecholamine release, and insulin deficiency all encourage greater fatty acid utilization. It may be assumed that fatty acid metabolism is preferred by cardiac tissue of all warm-blooded animals, a point emphasized by the trebling of cardiac glycogen depots in chickens after 48 hours of fasting. Gluconeogenesis—and fat utilization—evidently spares cardiac glycogen stores during nutritional hypoglycemia while other tissue glycogen depots are markedly depleted (Hazelwood and Lorenz, 1959).

D. Central Nervous System

Nervous tissue of birds largely has been neglected by workers on avian tissue metabolism. Aside from a few reports on the role of galactose in formation of cerebroside components of nervous elements there appears to be a dearth of information as to the nature and metabolism of nerve tissue and cerebrospinal fluid of birds. Phosphopentose shunt activity appears to increase in the spinal cord of developing chick embryos, reaching a peak at 4¾ days, and then subsiding to very low levels for the subsequent 17 days (Burt, 1965). This observation implies that mitosis and early neuroblastic differentiation may be dependent upon pentose cycle activity and NADPH production. Noting the temporal fluctuations in 6-phosphofructokinase and aldolase activities in neural tissue of chick embryos, Burt (1967) suggests that glycolysis is low during early neurogenesis. As pentose-shunt activity declines rapidly after day 5, glycolytic metabolism increases and remains elevated in spinal cord tissue while functional maturation proceeds.

Malhorta *et al.* (1967) reported that the distribution and concentration of fatty acids in the chicken brain agree, in general, with those found in avian cerebrospinal fluid. Thus, there is an ample distribution of C-16 through C-22 fatty acids in both brain tissue and fluid. Stearic and oleic acids are the predominant FFA in adult chicken cerebrospinal fluid (Hunzicker, 1969).

Other studies on avian cerebrospinal fluid are few, have been limited to the chicken, and indicate glucose levels double and protein levels four to five times higher than those found in mammalian cerebrospinal fluid. Na, K, Cl, and amino-N_2 levels are comparable to those found in mammalian species (Anderson and Hazelwood, 1969). The absence of a blood–brain barrier in chicks, a poorly developed barrier in adult chickens, and the presence of large amounts of protein in avian cerebrospinal fluid suggest a porous "barrier" indeed, one which conceivably allows large molecules to pass into brain tissue with relative ease. Presence of insulin-like material in chicken cerebrospinal fluid has been reported and strengthens this suggestion (Hunzicker and Hazelwood, 1970), though the metabolic significance of such observations is yet to be established.

E. Adipose Tissue

Fat deposits recently have been the result of vigorous investigation in mammals because of the growing body of evidence indicating inherent metabolic activity and responsivity to both humoral and hormonal agents. Major depots of lipid are found predominantly in the abdomen of birds as well as in perigonadal regions, and under the microscope the tissue appears much like mammalian adipose tissue. An exception is found in the ankle joint of chickens where lipid droplets are associated with cytoplasmic filamentous meshwork 70–100 Å wide (Luckenbill and Cohen, 1966). Variation in source of dietary fat apparently has little or no effect on the fatty acid composition of the birds' depot. Rather, studies on Common Redpolls (*Acanthis flammea*) indicate that environmental (photoperiod, temperature, stress) and physiological (migration, breeding, laying) conditions exert a more significant effect on adipose tissue composition than does diet per se (West and Meng, 1968). Similar studies on other avian species appear to agree with these observations and there is no evidence that the rapidity or frequency, or both, of food ingestion influences avian tissue lipogenesis, an effect which differs considerably from that of liver carbohydrate metabolism (Lepkovsky *et al.*, 1960).

Pyruvate kinase of chicken adipose tissue is remarkably sensitive to the presence of fructose diphosphate and in this manner differs considerably from mammalian tissues (Leveille, 1969b). Thus, competent anaerobic glucose degradation could very well control pyruvate formation and adipose tissue lipogenesis by providing adequate acetyl-CoA (as a result of fructose-1,6-phosphate influence on depressing the K_m of the enzyme for phosphoenolpyruvate). As shown by several laboratories the regulation of lipolysis in avian adipose tissue *in vitro* and probably *in vivo* differs markedly from that of most mammals studied. Reports from Langslow and Hales (1969), Goodridge (1968a), Goodridge and Ball (1965, 1966, 1967c), and Carlson *et al.* (1964) are pertinent to this topic. Lipogenesis in adipose tissue of chick embryos, chicks, adult chickens, and pigeons has been studied *in vitro* and *in vivo*, and while acetyl-CoA carboxylase, citrate-cleavage enzyme, and malic enzyme are present their activities are considerably less than that found in hepatic tissue (O'Hea and Leveille, 1968). The well-known effect of starvation and refeeding on these enzyme activities in mammals is not observed in birds, indicating a minor role of adipose tissue in fatty acid synthesis (Goodridge and Ball, 1966). Fatty acid synthesis by 7–8-day neonatal chick adipose tissue is high compared to younger and older chicks but still is less than 5% of that observed in adipose tissue of young rats (Goodridge, 1968a). Further relative inactivity of adipose tissue in lipogenesis in birds is indicated by glucose-U-^{14}C incorporation studies *in vivo* in the pigeon (Goodridge and Ball, 1967c). From these and other studies it appears that less than 4% of total body lipogenesis from glucose can be attributed to avian adipose tissue, the vast majority occurring in the liver (Goodridge and Ball, 1967c; O'Hea and Leveille, 1968, 1969; Leveille, 1969a). Evidently, fat depots derive glyceride-fatty acids from circulating triglycerides and their glyceride-glycerol from plasma glucose.

Regulation of lipolysis in birds also appears to differ from that reported in mammals, primarily in the (in)sensitivity of adipose tissue to hormone action. Carlson *et al.* (1964) found that mesenteric adipose tissue from both hens and cocks did not respond *in vitro* to doses of norepinephrine 20 times higher than effective doses on rat adipose tissue. No effect was observed *in vivo* either to the catecholamine or *in vitro* to porcine ACTH (Langslow and Hales, 1969). While the spontaneous lipolytic rate in adipose tissue obtained from fasted pigeons was increased concomitant with depressed oxygen consumption, addition of epinephrine did not alter activity of the tissue (Goodridge and Ball, 1965). Finally, that the pattern of response of avian adipose tissue differs from that of other experimental animals is indicated further by data indicating a remarkable sensitivity of intact chicken adipose tissue and isolated fat cells to glucagon-induced lipolysis, a sensitivity which is further enhanced by addition of insulin (Langslow and Hales, 1969; also see below, Section IV,A).

F. Glycogen Body

Enigmatic to students of avian carbohydrate metabolism is a structure originally described as the "corpora sciatica" and which subsequently has been the object of investigation because of a superabundance of polysaccharide stored within. This structure, the glycogen body, is an opaque, semigelatinous-appearing mass lying within the vertebral column in juxtaposition with the dorsal spinal cord at the level of emergence of the sciatic nerve plexus. The embryogenesis, early histology, glycogen accumulation, and vascularity of this structure have been carefully detailed by Watterson and his colleagues (Doyle and Watterson, 1949; Watterson, 1949). Electron micrographs of the glycogen body indicate a peripherally placed nucleus which, along with the juxtanuclear material and sparse mitochondria, is crowded by massive amounts of glycogen granules (Revel *et al.*, 1960). Glycogen-body cells are first identified on days 7–8 of incubation; however, the greatest increase in polysaccharide appears to occur during the last week of embryonic development, the increase possibly being dependent on the presence of an intact adenohypophysis (Watterson *et al.*, 1954). Glycogen levels continue to increase during the first 3 weeks after hatching only to decline gradually, though very slowly, thereafter. Approximately 60–80% of the lipid-free weight is represented by glycogen and, therefore, this structure contains up to 25% of total hepatic glycogen and 5–10% of the total body glycogen of nonfasted chicks.

Careful studies by Snedecor *et al.* (1963) on the chemical nature and turnover of the glycogen moiety indicate that glycogenesis and glycogenolysis occur *in situ* at about the same slow rate, resulting in constant polysaccharide levels. Injected ^{14}C-glucose appears uniformally in the outer tiers of the glycogen molecule (maltose) which at all times have a higher specific activity than the inner core (limit dextrin). A transfer of labeled glucosyl units from the outer branches of the glycogen molecule into the inner cores is evidenced with elapsed time. These observations of carbohydrate activity in the glycogen body along with evidence of phosphorylase and acid and alkaline phosphatases (Hazelwood *et al.*, 1962; Snedecor *et al.*, 1963) raise the question of why the polysaccharide moiety is unaltered by dietary, metabolic, or hormone action (Hazelwood, 1965).

The fact that the glycogen body is in direct physical contact with the central nervous system and that both tissues share the same vascular distribution has prompted speculation that the glycogen body is important in CNS metabolism. Suggestions that the physiological role of the glycogen body is to provide energy substrate to the CNS during sustained flight, that it protects the CNS from the impact of peripheral hypoglycemic crises by providing substrates to cerebrospinal fluid, or that it responds only to hormones of avian origin are not supported by a growing body of negative data. The slow turnover of glucose in this structure appears to obviate any important role it may play during sustained flight; other data indicate that CSF glucose levels fall sharply in response to i.v. or intracisternal insulin while glycogen–body glycogen levels remain unaltered (Anderson and Hazelwood, 1969). Also, isolation and characterization of chicken insulin allow for reappraisal of the suggestion of species-specific action of hormones on this tissue. Chicken insulin *in vitro* and *in vivo* is without effect on the polysaccharide levels of the glycogen body (Hazelwood and Barksdale, 1970).

Thus, the significance of this tissue, the presence of which is restricted to Aves, is yet to be established. In part, its relative glycogenolytic lethargy may be due to limited availability of some enzyme(s) associated with the glycolytic pathway, since the addition of liver or muscle homogenates to the glycogen body has been shown to increase glycogenolysis markedly (Lervold and Szepsenwohl, 1961). At this time, however, it appears that the glycogen body lies outside the avian metabolic sphere.

G. Red Blood Cells

Erythrocytes have been studied in respect to their metabolism and with respect to their contribution to the overall metabolism of the bird. While the nonnucleated mammalian RBC metabolizes glucose readily via the pentose phosphate shunt it is incapable of metabolizing acetate and the various organic intermediate components of the citric acid cycle. The nucleated RBC of birds possesses the enzymatic machinery necessary to carry out aerobic glycolysis and thereby aid porphyrin synthesis during early hemoglobin formation stages. Workers have demonstrated different hemoglobins in RBC's of chickens; the synthesis of one type is found in the primitive erythroid cells of the 5-day embryo while a different type is synthesized in the definitive cell type of the 12-day or older embryo (D'Amelio and Constantino, 1968). Energy release via the citric acid cycle probably waxes and wanes throughout the incubation period, the increases and decreases approximating globin and hemoglobin synthesis in the primitive and definitive cell lines. Fraser (1966) has estimated that the 5-day chick embryo possesses RBC's in which a ribosome can complete the synthesis of a single polypeptide chain (required in hemoglobin structure) within 1.6 minutes.

III. Influence of Diet

Dietary composition varies widely with the species according to its stage of development, availability of food, season of the year, and its geographical distribution and habitat.

A. Embryo Metabolism

Masterful treatises on the biochemical patterns associated with enzyme development and embryo morphogenesis exist, rendering further treatment of this subject here unnecessary (see Herrmann and Tootle, 1964; Deuchar, 1965; Moog, 1965; Romanoff, 1967; Grau, 1968). Once formed, the embryo develops fully within its own nutritive environment isolated from all maternal influences. Delicacy and sparseness of embryo tissue have limited most avian studies to that of the entire embryo and almost exclusively to that of the domestic fowl. Thus, respiratory quotients *in ovo* during the first week of incubation approach unity indicating that carbohydrate utilization predominates at this time. Of the available carbohydrate about 60% resides in the albumen, mostly in combination with proteins, and 40% can be found in the yolk, most of which is uncombined. As albumen carbohydrate levels gradually decrease (due to utilization) to barely detectable amounts on day 15, yolk carbohydrate levels increase to a maximum concomitantly. These converse alterations are not proportionate, however, and despite the gradual shift of carbohydrate from albumen to subembryonic fluid and membranes, and thence to yolk, the total carbohydrate content decreases rapidly throughout the 21-day incubation period (Romanoff, 1967). Gradually, the total carbohydrate levels of the embryo increase as does the level of circulating *d*-glucose. Glycogen levels in albumen and yolk are low and apparently are not too important as metabolic support for the developing embryo. Extra-embryonic sources of glucose have been indicated as essential for prolongation of embryo survival time in yolk-sac perfusion studies (Austic *et al.*, 1966) though the early embryo has been shown to utilize several naturally occurring sugars, namely, *d*-glucose, *d*-mannose, *d*-galactose, and *d*-maltose. Additionally, possible importance of the yolk-sac membrane in providing certain substrates has been indicated by observation of a sodium-dependent, phlorizin-sensitive active-transport system for sugars on day 11 of incubation (Holdsworth and Wilson, 1967).

Virtually all intermediates as well as enzymes of the classical Embden–Meyerhof anaerobic glycolytic pathway have been isolated from homogenates of whole embryos (Stumpf, 1947). These observations, together with the finding of pentose-shunt intermediates (Wenger *et al.*, 1967) and essential enzymes for the uronic acid pathway in chick embryo liver, indicate that the fast growing organism has multiple pathways to metabolize carbohydrate for differentiation and morphogenesis (Kilsheimer *et al.*, 1960; Okuno *et al.*, 1964; Conklin, 1966; Krompecher *et al.*, 1966). The uronic acid pathway may contribute considerably to the polysaccharide levels found in embryos since it affords a source of depot carbohydrate in the absence of insulin. Finding a hepatic UDP-glucuronyl-transferase prior to the onset of β-cell granulation in chick embryos further documents this postulate. Endocrine regulation of carbohydrate metabolism in the embryo has been ably reviewed by Leibson (1965–1966).

The fact that the total carbohydrate and plasma glucose of the embryo increases over the entire incubation period is not to imply that fluctuations do not occur throughout this 21-day period. To the contrary, during days 7–14 there is a definite plateau (or decrease) in circulating glucose levels at a time when the RQ decreases toward 0.85. Thus, energy sources other than carbohydrate must support morphogenetic processes distinct from the very early differentiative processes. Different levels of energy requirements exist from one 7-day period of incubation to another as well as from one tissue to another within the same 7-day period. Protein uptake and utilization appear to predominate during the second week of incubation, decreasing yolk ovovitellin, ovolivetin, and vitellomucoid (Romanoff, 1967). In contrast to yolk, protein represents 85% of the total solids of fresh egg albumen, which is characterized (in decreasing order of amounts) by ovalbumin, ovomucoid, and ovoconalbumin. These proteins remain at high levels during the first 13 days of embryonic life and thereafter are transferred rapidly into the amniotic cavity and sac (Romanoff, 1967). Since collectively (but not individually) the levels of 16 amino acids identified in yolk and albumen do not change to any great extent during the first 14 days of development, investigators have diverted their efforts recently to the possibility of whole (intact) protein uptake and utilization by the developing embryo to support rapid growth. Growth and development of chick embryos have been followed in careful studies by Klein (1968) on explants cultured *in vitro* on media supplemented with protein from various sources. (*In ovo* the chick embryo appears to utilize mainly yolk proteins for structural component lay-down and after the first week appears preferentially to draw proteins from the albumen depots. However, whether these proteins are degraded to the component amino acids prior to uptake by the embryo, or whether they are absorbed intact as protein has been a matter of controversy.) These explant studies emphasize the importance of intact protein, particularly of yolk origin during days 1–6, in contributing to structural development of the embryo. Maximum growth potential is reached when intact protein is provided as the nitrogenous substrate to cultured embryos. Since the profile of hen plasma protein is quite similar to that found in egg yolk, and since most yolk proteins are synthesized in the liver of hens, it may be surprising to find poor support of embryonic growth with use of sera from laying hens (Klein, 1968). Such findings may indicate chemical transformations during transport of the plasma protein to the developing egg.

Contributions of albumen protein to the embryo have been quantitated in explant studies, and the results indicate, again, that addition of ovalbumin to the media doubles protein levels of the embryo within 48 hours over levels achieved by embryos cultured on free amino acid media (Klein, 1968). Such results are not to be interpreted to mean that free amino acids are not taken up and used by the developing embryo. To the contrary, they are used as shown by utilization studies and single amino acid deficiency studies (Taussig, 1965; Austic *et al.*, 1966). Still, it remains that developing embryos preferentially employ intact proteins, first from yolk and later from albumen, as sources of amino acid precursors to support growth. Certain evidence (Coffey *et al.*, 1966; Klein, 1968) also indicates that different structural development *in ovo* may be supported by different protein precursors (extra-embryonic membranes vs. embryo per se, etc.).

It is generally agreed that lipid utilization mainly supports terminal chick embryo growth and metabolism during the final 7 days of incubation (Sova *et al.*, 1968). RQ values (*en toto*) approaching 0.70 during this time are supported by observations that up to 90% of the total caloric requirement of the embryo can be traced to fatty acid utilization (Hazelwood, 1965). The vast majority of egg lipids are confined to the yolk compartment where at least 65% of the total solids are represented as lipid components. Transport of yolk lipid to the yolk-sac membrane occurs rapidly between days 13 and 17 of incubation and by the latter date the membrane contains as much lipid as the egg yolk. All this lipid is in the form of triglycerides, no monoglyceride, diglyceride, or FFA fractions being detectable in either yolk material or in sac membranes. Evidence has been presented indicating that the major proportion of yolk triglyceride and phospholipid is absorbed unchanged by the yolk-sac membrane and that extensive esterification of yolk cholesterol (mainly with oleic acid) is involved in the transport of lipid from yolk to the membraneous structures (Noble and Moore, 1967a,b).

In addition to the lipid sources discussed above, plasma cholesterol is available to the developing embryo for nutritive purposes. Evidence has been presented suggesting that circulating cholesterol levels are under a pituitary–adrenal hormonal influence, an effect that may also facilitate transport of lipid from the sac-membranes for subsequent utilization (Thommes and Shulman, 1967). Studies on adipose tissue (as an indication of viability in support of embryonic growth) have been carried out by many investigators. Accord is generally found in that embryonic adipose tissue responds poorly to most hormonal treatments. A possible exception is adipose tissue from 17–18 day embryos, which is sensitive both to the anti-lipogenic and lipolytic effects of glucagon (Goodridge, 1968a,b). Yet, what sensitivity to such hormone action is observed in the embryo in no degree compares with the sharp increase in response of this tissue to glucagon 6–8 days after hatching (Goodridge, 1968a,b; for details, see Section IV,A). Thus, even though glucagon granules and secretion appear early in incubation the role of this hormone in making available FFA from embryonic adipose tissue is probably small, if not inconsequential. What alterations occur at the receptor sites causing insensitive adipose tissue to become sensitive several days after hatching remains an unanswered question.

In conclusion, it is interesting to note that the sequence of predominant substrate utilization to support differentiation and morphogenesis (carbohydrate–protein–lipid) is not unique to avian embryos, but has been described also in embryos of fish, insects, crustaceans, nematods, and oliogochaets as well.

Residual ("secondary") yolk is about one third of the original yolk volume and is described as that yolk remaining at hatch-time. This yolk material is drawn into the body cavity and remains as an outpocketing of the intestine. Carbohydrate and triglycerides are transported from this yolk reservoir into the plasma very rapidly during the first 4–5 days of post-hatch existence. In this manner, these endogenous spare yolk substrates appear to supplement the dietary intake the first few days after hatching. Plasma FFA, cholesterol, and phospholipid levels decrease markedly (50–67%) after day 6 (Sova *et al.*, 1968) and by 7–8 days after hatching virtually all the nutritive value of the spare yolk is exhausted.

B. Postnatal Metabolism

The role of lipid metabolism during the terminal days of incubation is firmly established (Sova *et al.*, 1968). Probably no better example of use of this substrate is established than by the "hatching muscle," a muscle that extends from the posterior skull bones to the upper cervical vertebrae. Within 48 hours of hatching, gross water inbibition occurs within this muscle as glycogen levels therein decrease by at least 80% (Ramachandran *et al.*, 1969). Concomitantly, gross depletion of triglycerides within the hatching muscle occurs as the muscle contracts repeatedly in forcing the egg tooth against the shell during the pipping process (Hsiao and Ungar, 1969). Thus, earliest among active contractile elements are these skeletal muscle fibers which employ lipid—and to a lesser extent carbohydrate—to meet the metabolic needs of this first performed work effort. The significance of the large water uptake may be related to the equally marked increase in acid mucopolysaccharide levels in this muscle on day 21, forming a spongy mass that cushions the chick's skull as force is applied to the shell (Klicka *et al.*, 1969).

Coincident with the lipid, water, and mucopolysaccharide changes in the hatching muscle, glycogen levels remain virtually static at approximately 0.3% in all other skeletal muscles. These levels gradually increase to 0.4–0.5% (except in heart) as full endocrine–nutritional regulation and expression is achieved in the young chicken. Contrarily, and in parallel with the rapid decrease of lipid levels in hatching muscle, a marked decrease in hepatic glycogen occurs (to the extent of 66%) during the pipping process and extending through at least 24 hours postnatal (Ramachandran *et al.*, 1969). Subsequently, liver glycogen levels rise quickly to approach levels identical with those of adults as the rapid change-over from prenatal lipid metabolism is replaced with the high carbohydrate diets characteristic of postnatal life.

Undoubtedly, dietary regimen, hormonal influence, and physical activity collectively dictate deposition of carbohydrate and lipid in tissues of the fast-growing chick (when protein anabolism predominates). Since nutrients must cross the intestinal wall to enter portal vein blood prior to hepatic uptake, storage, conversion, and utilization, a few words are appropriate concerning intestinal absorption in the early postnatal organism. Generally, it may be stated that as early as 2–3 days of age the chick gut demonstrates selective absorptive ability for sugars, most carrier actions being energy-dependent and some being sodium-dependent (Alvarado, 1967; Wagh and Waibel, 1967). Glucose and galactose transport across intestinal segments occurs by active (ATP-dependent) processes in the 18-day embryo as well as in chicks 48 or 96 hours old. Absorption coefficients for gut transport of various sugars in newly hatched chicks are greatest for galactose, intermediate for glucose and xylose, and least for fructose. While regional differences exist in active transport of galactose by the avian gut (upper and lower third of small intestine are greatest) *d*-glucose appears to be actively transported equally well along the entire length of small intestine (Fearon and Bird, 1968). Comparable quantitative studies on amino acid and lipid transport by the gut of newly hatched chicks appear lacking in the literature.

C. Experimental Manipulation

1. *Starvation*

There have been numerous studies of fasting in birds, owing largely to interest in the many peculiarities of carbohydrate metabolism in this class (see Section IV,A). Newly hatched chicks withstand fasting very well, blood glucose levels decreasing only slightly throughout a 6-day period of observation. Residual yolk utilization is of paramount importance to these postnatal fasted chicks and efficiency of utilization is apparent by the fact that blood glucose levels decrease less than 25% over a 6-day fasting period. These decrements in blood glucose appear to be correlated with a decreasing body temperature. The probability exists that liver glycogen depots assist in maintenance of circulating levels of glucose since hepatic glycogen is decreased sixfold during the same 6-day period, and pectoralis muscle glycogen levels decrease about 25%, from over 1000 mg% to approximately 750 mg%. Recall, however, that muscle glycogenolysis does not contribute directly to plasma glucose, due to the absence of glucose-6-phosphatase, and can support these levels only by hepatic reconstitution of glycogen from circulating lactate (the "Cori cycle").

Fasting is endured by adult birds (pigeons, chickens, turkeys, pheasants, etc.) for 8–15 days, apparently with relative metabolic ease. After early decreases in liver glycogen and blood glucose, there is a progressive rise in both with only the glucose levels surpassing prefasting levels. The role of protein degradation and subsequent gluconeogenesis in supporting avian metabolism during prolonged starvation has been indicated and probably accounts for the temporary elevation of plasma glucose levels (Hazelwood and Lorenz, 1959).

Of peculiar interest in both mammals and birds is the observation that total fasting invariably depletes glycogen depots of every tissue in the organism with the exception of cardiac tissue, where glycogen concentration is elevated! Thus, as in mammals, cardiac glycogen (chickens) levels increase during fasting from the relatively low levels of 0.1 to 0.25% in 12 hours and remain elevated between 0.20 and

0.35% for 2 days or more. These data are in accord with those obtained in rats and the essentiality of circulating growth hormone to effect these alterations in cardiac muscle has been established. The lipolytic effect of growth hormone in conjunction with preferential use of fatty acids as fuel by cardiac muscle allows for a sparing and accumulation of glycogen within. Fasting periods in excess of 3–4 days invariably lead to a relative depletion of glycogen in cardiac muscle long before the bird becomes moribund.

Entrance of glucose into tissue cells is decreased in starvation, possibly as a coincident result of decreased insulin secretion or possibly as a result of the lowered plasma glucose levels shortly after commencement of the starvation period. However, contribution of plasma glucose to the overall oxidative metabolism of chickens decreases and becomes relatively steady at 8–10% after 48 hours of food deprivation, values that are in good agreement with those reported for the fasting dog, rat, and man (Annison *et al.*, 1966). Contrary to observations in mammals, FFA levels increase in fasting birds (chickens) without attendant glucose changes, and glucose infusion during the fasting period does not lower these fatty acid levels (Heald *et al.*, 1965; Lepkovsky *et al.*, 1967). Apparently, insulin levels remain fairly stable during fasting periods in chickens despite fluctuating FFA, α-amino nitrogen, and glucose levels (Langslow *et al.*, 1970). Such observations further emphasize the growing body of evidence indicative of relative independence of carbohydrate metabolism from insulin control in birds (also, see Section IV,A).

Birds differ from mammals also in the response of amino-acid metabolism (as a reflection of overall protein metabolism) to starvation. Over a 36-hour fasting period lysine levels increase almost 500% in chickens and, unlike the condition in mammals, plasma threonine and methionine levels also increase though to a lesser extent than that observed for lysine (Boomgaardt and McDonald, 1969). Fasting also markedly increases plasma uric acid levels, approaching, after 3–10 days of starvation, 10–40 times initial values (Okumura and Tasaki, 1969). Re-feeding corrects these excessive levels of uric acid to normal within 6 hours; in general, both uric acid and ammonia levels of plasma are proportionately correlated with dietary protein intake (Okumura and Tasaki, 1969). Thus, while the contribution of glucose to satisfy metabolic needs appears to be quite small during starvation, the utilization of fatty acids (cardiac muscle) and the mobilization of protein reserves (general tissue demands) loom as major homeostatic mechanisms by which domestic fowl respond to dietary crises. Unfortunately, virtually no information relative to tissue metabolic patterns of nondomestic species is available.

2. Synthetic Diets

Considerably more investigative effort has been directed toward determining the effects of abnormal (exotic in many cases!) diets on the domestic fowl than toward characterization of normal metabolic–biochemical patterns in tissues of avian species, a statement readily apparent from surveying the literature of the last 50 years. The reader is referred to the many texts dealing with the nutrition of the domestic fowl for this information, or to the recent review by Bird (1968). Non-pelagic birds are primarily "nibblers," that is, food intake occurs more or less uniformly (when food is abundant) rather than being confined to brief periods of intensive feeding. Attempts to train birds to eat to capacity within a prescribed time each day have been reported, and indications are that such birds (chickens) adapt at the crop level rather than at the tissue level. Such "trained" birds, therefore, hold more food in the crop sac, releasing little by little to the proventricular cavity over a long period of time, and therefore require little metabolic adaptation by the tissues (Lepkovsky *et al.*, 1960). While no difference can be detected in the lipogenic potential of tissues obtained from "trained" vs. untrained chickens, those birds allowed to eat *ad libitum* (untrained) digest food particles more rapidly, have greater flow of proteolytic enzymes, and greater and faster deposition of liver glycogen than their experimental counterparts (Lepkovsky *et al.*, 1960).

Inclusion of the pentoses *l*-arabanose and *d*-xylose in diets of chicks leads to a severe depression of growth, feed-efficiency, liver and gut size, as well as depletion of hepatic and skeletal muscle glycogen (Wagh and Waibel, 1966). Concomitant with these dramatic effects are observations that uric acid levels are increased greatly as are blood glucose and cholesterol levels (Wagh and Waibel, 1966). Absorption of these pentoses, therefore, encourages protein and triglyceride catabolism to meet metabolic needs; assistance from carbohydrate depots occurs also but the intracellular transport of *d*-glucose for utilization appears to be hindered.

Low levels (2%) of various forms of cellulose, gums, and pectin also depress growth of birds. This depression of protein anabolism is not overcome by high-fat or high-protein diets despite the fact that negative nitrogen balance exists along with decreases in absorption of fat and metabolizable energy (Kratzer *et al.*, 1967).

As discussed in Section VI,C, the fatty acid composition of avian depot fat appears to be far more dependent upon the physiological state of the bird and its surrounding environmental conditions than on the composition of the diet. This is true of domestic as well as of wild species when allowed to eat *ad libitum*. However, if synthetic diets or natural diets with added fats from various sources are evaluated in chicks (or adult hens) definite changes are observed. Oleic and linoleic acid levels of egg yolk lipid are increased greatly by increased soybean oil supplementation; adipose tissue levels of these two fatty acids are increased significantly (Sell *et al.*, 1968), also. Rapeseed oil supplementation of the diet results in substantial levels of erucic acid in both egg yolk and in adipose depots indicating that a portion of the ingested fatty acids are deposited directly in the yolk and tissues of hens (Sell *et al.*, 1968). Oleic acid supplementation of the diet of laying hens results in a greater degree of unsaturation in the fatty acids of the yolk and embryo; irrespective of the maternal diet embryo triglycerides contain higher levels of saturated fatty acids than those found in yolk throughout the entire incubation period (Donaldson, 1967). While feeding of cholesterol increases concentrations of oleic acid in the cholesteryl ester fractions in the yolk, liver, and plasma of hens, no alteration of the lipids of abdominal or thigh tissue is observed (Chung *et al.*, 1967).

Literally hundreds of studies relative to the effects of specific dietary deficiencies on hatchability and subsequent growth of newly hatched chicks have been reported. These will not be reviewed here except for one example that relates dietary intake specifically with avian morphogenesis. The linoleic acid requirement of chicks and adult domestic fowl is well established; also, hatchability of eggs obtained from hens maintained on linoleic acid deficient diets (though iso-caloric) is reduced almost to zero. Omission of this fatty acid from the maternal diet leads to reduced egg size, a 13% reduction in caloric content of egg yolk, lowered yolk volume, dry matter, and lipid content. Interesting, however, is the finding that maternal deficiency of linoleic acid leads to a 60% mortality in embryos therefrom, the vast majority of which die between days 0 and 4 or days 20 and 22 (Calvert, 1967). Essentiality of this fatty acid for both differentiation and rapid growth appears to be indicated, especially since the yolk levels of linoleic and arachidonic acids are barely detectable in these eggs. Direct evidence of involvement of linoleic acid in growth and morphogenesis appears from the observation that among the embryos that survive to the final days of incubation a very large percentage (at least 75%) are in an abnormal body position with their heads tucked over the right wing instead of under the wing. Thus, improper head position would be instrumental to the failure to pip through at hatching time (Calvert, 1967). Here, therefore, is one of many examples whereby deficiency of a single dietary ingredient leads to abnormal growth and possibly morphogenesis; the biochemical pathway (or lesion) connecting linoleic acid with such diverse effects as proper head–wing placement *in ovo* deserves more attention.

Fasting increases free amino acids in the plasma of chicks, especially the levels of lysine and threonine (Shao and Hill, 1967). Feeding a nonprotein diet high in carbohydrate or fat suppresses this increase in amino acids in fasted chicks as does administration of insulin (Shao and Hill, 1967). Thus, there appears to be a relationship between intracellular transport (and therefore true availability) of glucose and the utilization of endogenous protein to support day-to-day metabolism. During fasting, glucose availability is low, and plasma amino acids increase. Feeding other caloric substrates or making existing glucose levels available to peripheral tissue cells via insulin depresses protein catabolism (and therefore free amino acids in the plasma) to meet metabolic demands of the fasting organism. These observations are supported further by the fact that chickens fed diets containing excessive protein levels survive subsequent fasting periods better than normal animals (H. Fisher and Ashley, 1967).

Gradual reduction of dietary protein intake (as would be expected to occur in restrictive environmental conditions) over a 14-day period results in concomitant decreases in urinary excretion of ammonia, uric acid, and urea while creatine and amino acid nitrogen levels appear to remain constant (Teekell *et al.*, 1968). The source of the "unknown" nitrogen excretion via urine presumably is of endogenous tissue origin since there is no relationship between this excretion and the decrease in dietary protein. Strangely, carbohydrate (anthrone-reacting) compounds in the urine increase as dietary protein levels decrease, implying the possibility that protein catabolism is efficient enough to meet metabolic needs and thereby increase gluconeogenesis to such an extent that circulating glucose levels surpass the renal threshold. Alternatively, the renal threshold for glucose may be reduced with a reduction in protein intake. In any event, it appears that the hen adapts well to decreases in dietary protein to a certain critical maintenance level below which endogenous protein catabolism increases and body weight decreases (Teekell *et al.*, 1968).

Mineral metabolism studies in adult avian species have been restricted mainly to those on domestic fowl and have been oriented toward observations on early embryo development or toward efficiency of productivity. Many singular dietary element deficiencies result in the same or similar physiological defects or dysfunctions possibly indicating a common basis or biochemical step for the defects observed. That this may not be the case has been emphasized by studies truly characterizing lesions beyond superficial or obvious defects. Thus, while folic acid, choline, biotin, and niacin deficiencies individually produce perosis-like abnormalities of the chicken leg similar to those of manganese deficiency, there is no basis for assuming that a common metabolic defect embraces this group. To the contrary, while there is generalized reduction in endochronal maturation of chicken bones in all the aforementioned deficiencies, only a lack of manganese decreases the width of ipiphyseal cartilage by an impairment in production of extracellular matrix distinct from a deleterious effect on cell proliferation (Leach, 1968). The biochemical step at which manganese plays a regulating role in the formation of bone matrix is unknown in chickens and, along with the other similar, yet different, dietary deficiencies, merits further investigation to ascertain what metabolic paths are inoperable or have gone awry as a result of the incomplete diet.

Calcium deficiency has been studied in domestic fowl for a number of years, largely because of its role in eggshell development. However, more fundamental and antecedent to formation of shell is the role that this metal plays in the laying cycle. It appears that the level of blood calcium controls the release of adenohypophyseal gonadotrophins preparatory to ovulation and subsequent shell formation in the oviduct. Thus, chickens placed on Ca-deficient diets (less than 1%) cease to lay eggs (with or without shells) within 2 weeks time, a period during which the skeleton may be depleted of 40% of its calcium content (Taylor, 1970). Yet, if these chickens are injected with extracts of whole avian pituitary glands egg laying resumes despite continuous ingestion of the Ca-deficient diet (Taylor, 1970). Whether or not plasma calcium acts via hypothalamic releasing factors or directly on the adenohypophysis is unknown at this time, though almost certainly the mechanism resides in a membrane-release phenomenon.

A wide variety of deficiency states has been described for other major elements (almost always in the domestic fowl) such as Zn, Fe, Cu, Mg, and P, details of which may be found in any standard textbook on nutrition (see also Chapter 7).

IV. Endocrine Control of Metabolism

A. Pancreas

The avian pancreas has been adequately described anatomically, cytologically, and histochemically (e.g., Mikami and Ono, 1962; Epple, 1965; Przybylski, 1967). The avian pancreas produces at least two endocrine secretions, glucagon and insulin, from the α-islet cells and β-islet cells, respectively. The possibility of a third pancreatic "hormone" with effects on lipid metabolism has been suggested by several workers (e.g., Epple, 1965). Recently a linear polypeptide containing 36 amino acid residues (MW 4244) distinct from glucagon has been identified in chicken pancreas (Kimmel *et al.*, 1968). This polypeptide exerts a powerful secretogogic action on the chicken proventriculus (Hazelwood, unpublished).

1. *Insulin*

Early observations that most birds do not become permanently diabetic following surgical pancreatectomy or alloxan injection, as well as the more recent observation that sodium tolbutamide is effective in producing hypoglycemia in depancreatized and enterectomized fowl, have prompted many studies on the role of the pancreas in regulating avian carbohydrate metabolism (for review, see Hazelwood, 1965). With rare exception, studies of insulin in avian intermediary metabolism have been made employing insulin obtained from mammalian sources, usually porcine, bovine, or ovine. From the available data there appear no qualitative differences between the response of birds to these or to chicken insulin. The only reported difference is quantitative; that is, chickens are more sensitive to the metabolic effects of chicken insulin than to equal amounts of the nonhomologous hormone (Hazelwood *et al.*, 1968; Hazelwood and Barksdale, 1970). Such "resistance" of birds to hypoglycemic effects of mammalian insulin has been reported many times and in part is attributed to avian plasma factor(s) that interfere(s) with full expression of the hormone at the cellular level. Nonetheless, it is well documented that insulin in birds causes hypoglycemia, hepatic glycogenesis, and ketonemia while depressing gluconeogenesis.

Granulations in the insulin-secreting β-cells of the embryonic chick pancreas appear by day 5 (Przybylski, 1967). Insulin has been detected in the pancreas by day 12 and in the plasma by day 13, and the presence of protein-bound sulfhydryl bonds has been reported in β-cells by day 14. Thus, the growing embryo and the extra-embryonic membranes are under the influence of the anabolic pancreatic hormone during the last 10 days of development. The rapid recovery of hepatic glycogenesis from the nadir on day 12, therefore, is probably linked in some way to the response of hepatic hexokinase to the endogenous pancreatic hormone.

At the embryonic cardiac tissue level bovine and chicken insulins appear to have the same effect, namely, altering the K_m more than the transport maximum for glucose in intracellular distribution of sugar. The effect of insulin in increasing distribution of glucose occurs at approximately the same time that pancreatic islet tissue is most active in insulogenesis, namely, after the tenth day (Guidotti *et al.*, 1966).

Insulin is without effect, however, on glucose transport in the 7-day embryo heart, indicating development (between days 7 and 10 of incubation) of an insulin-sensitive "membrane" which limits glucose transport. Data have been reported indicating that insulin applied to the blastoderm on day 3 increases yolk-sac membrane glycogen over a subsequent 7-day period (Thommes and Mathew, 1969). Thus the insulin-sensitive UDPGlycogen synthetase system may be present as early as day 4 in the chick. Certain direct relationships are seen with pancreatic β-cell granulation and yolk-sac glycogen appearance; also, fluctuations between the 13th and 15th day of incubation in yolk-sac membrane glycogen are similar to those observed in liver, both of which coincide with α-cell activity.

Both chicken and beef insulin increase the maximal velocity of transport of a nonmetabolizeable amino acid (AIBA) in the 5-day chick embryo heart (Guidotti *et al.*, 1968a), an action that is independent of simultaneous glucose transport or *de novo* protein synthesis. Also, it appears that metabolism of certain amino acids to CO_2 and water by embryonic cardiac tissue is accelerated whereas incorporation into protein is uninfluenced by insulin (Guidotti *et al.*, 1968b). Thus, the requirements of cardiac tissue energy substrates during middle incubation may be met, through hormonal control, by catabolism of amino acids and/or by fatty acids; early needs can be met by free diffusion of glucose across insulin-insensitive cardiac membranes.

It is well established that neither beef nor chicken insulin has any effect on FFA and glycerol release from adipose tissue obtained from 15–18 day chicks (Goodridge, 1968a,b). Furthermore, insulin slightly increases glucose oxidation, glycogen deposition, glyceride-glycerol and FFA synthesis in adipose tissue in late stages of incubation, observations which are increased 14-fold when evaluated 7–8 days after hatching (Goodridge, 1968a). Such alterations may well be related to the abrupt change in dietary intake which occurs after hatching.

Investigations of adipose tissue, *in vitro* and *in vivo*, from early postnatal and adult chickens indicate that regulation of lipid metabolism in birds is grossly different from that of mammals. Thus, insulin appears to have no anti-lipolytic effect, *in vivo* or *in vitro*, in adipose tissue or in isolated fat cells; this is in general agreement with the lack of effect of catecholamines and ACTH on similar systems (Langslow and Hales, 1969). This evidence is confirmed further by the lack of an effect of insulin on glycolysis or the release of fatty acids and glycerol from pigeon adipose tissue *in vitro* (Goodridge and Ball, 1965). From an anabolic point of view lipogenesis from pyruvate or glucose substrates proceeds at a very slow rate in pigeon adipose tissue and is remarkably unaffected by addition of insulin (Goodridge and Ball, 1966). Presence of several critical citric acid cycle and pentose phosphate shunt enzymes have been established in chicken and pigeon adipose tissue but activity of these enzymes is considerably below that found in liver tissue (Goodridge and Ball, 1966; O'Hea and Leveille, 1968). It may be concluded, therefore, that *de novo* lipogenesis in these species occurs mainly in the liver, leaving to the adipose tissue the role of depositing lipid synthesized elsewhere.

It is well established in mammals that the availability of plasma glucose and caloric requirements to a large extent dictate circulating levels (and therefore endogenous availability) of free fatty acids. Generally, there is an inverse relationship between circulating glucose and FFA levels in mammals. Contrary to this body of evidence are observations in birds that the intravenous infusion of glucose does not alter FFA while injections of amounts of beef insulin (glucagon-free), sufficiently large to cause severe hypoglycemia, increase FFA levels in chickens (Heald *et al.*, 1965). The plasma fatty acid response to exogenous insulin is immediate (as is the hypoglycemia), occurring in immature, mature, laying, and depancreatized chickens (Heald *et al.*, 1965; Lepkovsky *et al.*, 1967). Thus, increased FFA release cannot be attributed to insulin-induced glucagon release (data from depancreatized birds), a fact further confirmed by the insulin augmentation (permissive?) of glucagon action on adipose tissue *in vitro* (Goodridge, 1968b; Langslow and Hales, 1969).

Mialhe's early work (1958) on pancreatic control of blood glucose levels in ducks was extended to chickens by Mikami and Ono (1962) and subsequently, to geese by Sitbon (1967). Collectively, these investigators have demonstrated that *total* removal of the avian pancreas invariably leads to a profound hypoglycemia, convulsions, and death unless glucose or glucagon is administered. Such observations differ from most reports on the effects of surgical ablation of the pancreas in fowl. Consequently, some authors have attributed these differences to removal of the splenic pancreatic lobe in those experiments where *total* removal was achieved. This isthmus of pancreatic tissue is very difficult to remove due to its elusive position beneath hepatic and splenic tissue and may well have remained intact in earlier studies on "depancreatized" birds. Conversely, transitory diabetic effects have been reported to occur only in carnivorous forms (falcon, some owls, buzzard, raven) after pancreatic extirpation (total?), the metabolic disarrangements disappearing within a week after surgery. However, Mirsky and Gitelson (1958) reported that the herbivorous goose responds like mammals to pancreatectomy; furthermore, operated geese have diabetic glucose tolerance curves and do not respond to tolbutamide injection. Thus, it appears that of all avian species investigated, only in geese does true altered carbohydrate metabolism occur as a result of insulin deficiency. Contrarily, pancreatectomy adversely alters lipid metabolism (at least in chickens), leading to ketonemia, lipemia, and spontaneous atherosclerosis (Hazelwood, 1965).

Studies on the effects of insulin on intermediary metabolism of birds other than pigeons or chickens are sparse and the most significant emanate from the investigations of Grande on owls, ducks, and geese. Insulin injection into owls and geese leads to a prompt hypoglycemia without any effect on plasma FFA levels (Grande, 1969, 1970). These observations in geese are of considerable interest because of the aforementioned fact that this species stands out "mammalian-like" as an exception among birds in responding to pancreatectomy and other manipulations of carbohydrate metabolism. Yet, confronted with insulin hypoglycemia geese do not respond with depressed free fatty acid levels (as do mammals) and thereby appear more "avian-like." Certainly, additional work is merited by the dichotomy of metabolic responses observed in this species.

2. *Glucagon*

Pancreatic α-cells are readily identifiable on day 6 of incubation and secretory (glucagon) granules on day 14; a hyperglycemic response to exogenous glucagon can be detected in embryos as early as day 6 and is an effect dependent upon adequate hepatic glycogen stores (Thommes and Firling, 1964). The action of glucagon in increasing blood glucose in birds appears to be due to the effect of the hormone in increasing levels of adenyl cyclase, increasing active phosphorylase, and thereby initiating splitting of α-1,4 linkages of liver glycogen. Work with glucagon extracted and crystallized from avian pancreas has not been reported but evidence from Kimmel's laboratory indicates that the avian hormone is a linear polypeptide with a molecular weight slightly greater than that of mammalian glucagon (MW 3485). Phosphorylase activity of the chick embryo liver is detectable early in incubation but rises to a maximum level abruptly between days 16 and 18 at a time when liver glycogen and UDP-Glycogen transferase livers decrease markedly (Okuno *et al.*, 1964).

Glucagon greatly inhibits synthesis of glycogen, glyceride-glycerol, and FFA in adipose tissue obtained from chicks 7–8 days old, effects which are considerably greater than those observed in 15–18-day embryos (Goodridge, 1968a). Insulin increases these glucagon effects even further. On the other hand, glucagon accelerates lipolysis (release of FFA and glycerol) to a small extent in embryonic adipose tissue of chicks; chicks 7–8 days old respond tenfold greater and addition of insulin increases sensitivity of the tissue even more (Goodridge, 1968b). Such evidence emphasizes even further the co-ordination of hormone regulation of metabolism as the *in ovo* organism shifts from heavy fat metabolism to the high carbohydrate intake so characteristic of the growing chick after hatching. Studies on the alterations in sensitivity of receptor sites would be most helpful in furthering our understanding of tissue maturation and regulatory processes.

Production of FFA and glycerol are accelerated when glucagon is incubated with abdominal fat tissue obtained from either fasted or normally fed pigeons and chicks (Goodridge and Ball, 1965; Langslow and Hales, 1969). The hormone also increases plasma free fatty acids immediately and markedly, probably by a direct action on adipose tissue since infusion of glucose (to simulate glucagon action) is without an effect on these levels (Heald *et al.*, 1965). Both FFA and blood glucose are rapidly elevated in geese, ducks, owls, and turkeys injected with minute amounts of crystalline beef glucagon (Grande, 1968, 1970). Actually one observes in these species better correlation of log-dose response to glucagon with fatty acid levels than with blood glucose levels (Grande, 1968). Contrary to effects observed in the studies of adipose tissue *in vitro*, insulin does not appear to augment this plasma response to glucagon in the aforementioned species (Grande, 1969). A comparison of the effects of epinephrine with those of glucagon in ducks and geese indicates that, unlike observations in mammals, epinephrine has no effect on hepatic triglycerides and plasma FFA in ducks. Yet, glucagon has a marked adipokinetic effect in ducks, as a result of increase in plasma triglyceride and subsequent deposition in the liver (Grande and Prigge, 1970). The changes in fatty acid composition (C-16 to C-18:2) in liver lipid are consistent with the suggestion that glucagon mobilizes adipose tissue lipid to increase fatty acid and triglyceride levels in plasma and thereby encourages hepatic uptake and deposition of newly synthesized lipid (Grande and Prigge, 1970). Precisely at what biochemical step glucagon exerts its effect on adipose tissue in birds deserves further investigative attention.

B. Thyroid and Parathyroid

Snedecor's laboratory has contributed much to our understanding of the physiological significance of the avian thyroid gland in regulating tissue metabolism of embryos and adults (Snedecor and King, 1964; D. B. King and Snedecor, 1966; Singh *et al.*, 1968; Snedecor, 1968).

Early reports from the work of Oscar Riddle in pigeons verified previous observations in mammals indicating that thyroidectomy results in low blood glucose levels. Contrarily, injection of thyroxine into birds depletes hepatic glycogen by increasing glucose-6-phosphatase activity and thereby inducing a mild hyperglycemia. Chicks that are radiothyroidectomized (by ^{131}I injection) have greatly increased liver weights, increased liver glycogen (50 times) and lipid levels, along with decreased testes, comb, and spleen weights (Snedecor and King, 1964; Snedecor, 1968). No effect of hypothyroidism is observed on blood glucose or glycogen level in the glycogen body and skeletal muscles. Hepatic carbohydrate changes are probably not due to thyroxine absence alone but rather due to an extrathyroidal effect of pituitary TSH since hypophysectomized chicks receiving TSH and PTU (propylthiouracil) have enlarged livers and supranormal heptatic glycogen levels (D. B. King and Snedecor, 1966). Effects of thyroid hormone at the cellular level are slow to occur in chicks—as is true in mammals—regardless of whether the hormone is in T_3 or T_4 form; increases in basal metabolic rate occur only after a lag of several hours (Singh *et al.*, 1968). Proteinaceous effects of thyroid hormone have not been reported in birds other than effects on "body growth" and a direct relationship between the hormone availability and plasma free amino acid levels of chicks. Lysine, threonine, tryptophan, methionine, and leucine levels are reduced significantly in plasma of PTU treated chicks, changes which do not appear to be related to alterations of dietary protein intake (Shao and Hill, 1968).

The possibility that thyroxine plays a permissive role in maintaining the level of glycogen in the yolk-sac membrane in chick embryos has been suggested, since inhibition of thyroxine by PTU between days 8 and 18 of incubation results in decreased membrane glycogen depots. These inhibitory effects are counteracted by cortisone acetate (but not thyroxine), indicating the possible regulatory action of the hormone in providing potential energy substrates to the embryo (Thommes *et al.*, 1968).

Obviously missing from our knowledge are the regulatory patterns of metabolism directed by thyroxine in species other than domestic fowl as well as the cellular reactions involved in meeting the needs of the organism.

Parathyroidal activity in birds has been neglected largely until recent years when the importance of calcitonin and parathormone in regulating calcium metabolism in mammals has been recognized. No definitive studies have been reported relative to metabolic effects of purified parathormone in birds, most studies dealing with localization of Ca^{++} in the shell gland (Hohman and Schraer, 1966), Ca^{++} requirements during reproductive cycles (Hurwitz and Bar, 1966, 1967; Dean *et al.*, 1967), and Ca^{++} turnover and shift to the skeleton of the developing embryo (De Vincentiis and Marmo, 1966; Creger and Colvin, 1967). The action of parathormone in elevating plasma Ca^{++} levels in mammals stems from its action in promoting osteoclastic activity in bones leading to bone resorption and subsequent release of Ca^{++}. This activity is somewhat counterbalanced by the action of calcitonin (thyrocalcitonin), a hormone distinct from parathormone and released by the parafollicular cells of the thyroid, which lowers plasma Ca^{++} levels by depressing bone resorption. The ultimobranchial body secretes calcitonin in birds. What little evidence exists at the present time does not indicate that the ultimobranchial tissue plays a dominant role in skeletal growth in chicks (Dent *et al.*, 1969). Such a statement is based on the fact that even though this tissue is largest (per unit body weight) 3 days after hatching when bone growth is rapid, chicks that are chronically ultimobranchialectomized show no alterations in bone structure, growth, or serum calcium, phosphorus, and alkaline phosphatase levels (Dent *et al.*, 1969). It would appear unlikely that either parathormone or calcitonin is responsible for differences in thickness of egg shells since under standard dietary and environmental conditions Ca^{++} turnover and bone mass of hens are virtually equal and constant (Hurwitz and Bar, 1967).

C. Adenohypophysis

With the exception of prolactin very little work has been reported on the effects of pituitary gland secretions on avian metabolism. To the contrary, what little is known of the possible effects of these hormones on embryogenesis and metabolism is derived from studies on birds that were hypophysectomized before experimentation. Hypophysectomy is accomplished in the chick embryo by surgical removal of the upper beak and three fourths of the cranial contents and cranium, resulting in a true decapitate–hypophysectomized preparation. These operated embryos survive most of the remaining incubation period and thereby provide the investigator with a pituoprivic tool for experimentation. Hyperglycemia (marked) and hepatic glycogenesis attend hypophysectomy at day 8 in chick embryos and persist until day 12, after which time these two variables slowly return to normal levels by day 16. Concomitant with these tissue changes are observations that the level of glycogen in the yolk-sac membrane is significantly reduced and remains so at least through day 18 (Thommes and Aglinskas, 1966). Despite these carbohydrate alterations, which imply early hypophyseal control over tissue metabolism in the chick embryo, some data indicate that hypophysectomy possibly is without effect on glycogen body glycogen levels (Thommes and Just, 1966). Since removal of the pituitary gland deprives the embryo of all hypophyseal secretions it remains to be established precisely which hormone(s) is responsible for the aforementioned carbohydrate changes. At this time avian hypophyseal hormones have not been isolated, purified, and characterized.

Implication of the hypophysis (or a hypophyseal–target tissue axis) in control of carbohydrate and lipid metabolism in young and adult chickens has been made. An impairment in lipid metabolism of hypophysectomized adult chickens is suggested by the observation that such birds become markedly obese; however, no data are available as to the effects of selective hormone replacement in these pituitary-deprived chickens and therefore it is uncertain as to which adenohypophyseal hormone is responsible for the maintenance of proper lipid balance. It appears that this obesity may be due to a defect in lipolysis (Gibson and Nalbandov, 1966). As observed in the embryo liver, glycogen levels are greatly increased in cockerels hypophysectomized at 3 weeks of age and observed 12 days postoperatively (D. B. King, 1969).

1. Growth Hormone

While "growth promoting activity" exists in crude extracts of chicken pituitary tissue no information is available as to the effects, especially metabolic, of avian growth hormone (GH). Generally speaking, very little effect is observed in birds injected with mammalian growth hormone: namely, little growth augmentation, slightly decreased egg production, and a tendency to molt. The only tissue response to mammalian GH reported is the strong glycogenesis in cardiac tissue of nonfasted chickens. Such data are in accord with those obtained previously in mammals (Hazelwood and Lorenz, 1959). The fact that heterologous GH is without striking effects in birds is not to be interpreted as indicative of the normal avian metabolic pattern. Suggestive that such is not the case are observations of King, who reported grossly decreased body weights and bone lengths (shank and middle toe) of hypophysectomized cockerels, changes which could not be attributed to decreased food intake (D. B. King, 1969).

2. Gonadotrophins

The activity of pituitary gonadotrophins on tissue metabolism has been evaluated only by action through the respective target gonadal tissue. Thus, testosterone administered to adult chickens increases circulating FFA of males but not females even though blood glucose levels remain stable (Lepkovsky *et al.*, 1967). These sex differences are similar to those found with another protein-sparing hormone, insulin, in birds. Furthermore, the effect of androgens on increasing the uptake of ^{14}C-acetate and incorporation into triglyceride fraction of liver lipid is established. Therefore it appears that while this hormone may not affect hepatic lipid composition, it definitely exerts a significant influence on the rates of hepatic synthetic and degradative metabolic processes (Balnave, 1968).

Probably because of the rapid lipid mobilization and utilization by the laying hen, estrogens and progesterone effects have been studied more fully than those of other steroids. Estrogen rapidly increases total weight and dry weight as well as total lipid levels of the liver, observations that are supported further by studies of incorporation of acetate into triglycerides in this tissue (Balnave, 1968, 1969).

3. *Prolactin*

The role that this hormone plays in initiating broodiness and nesting activities in birds was well established by Riddle's classic observations in the early 1930s. However, the metabolic effect of prolactin in birds has been the object of intense investigation only in recent years (for earlier work, see Riddle, 1963). Despite the sensitivity that adipose tissue demonstrates to carbohydrate metabolism within, studies on abdominal fat pads of migratory and nonmigratory finches indicate that prolactin does not encourage conversion of labeled glucose or acetate into fatty acids in adipose tissue *in vitro* or *in vivo* (Goodridge, 1964). Contrary to the insensitivity of adipose tissue to prolactin, hepatic lipogenesis is markedly stimulated by this hormone. In fact, prolactin injection in pigeons causes an immediate increase in liver weight, increased concentrations and activities of malic enzyme, citrate cleavage enzyme, and malate dehydrogenase, and increased rate of incorporation of glucose, pyruvate, and acetate into hepatic fatty acids (Goodridge and Ball, 1967b). Failure to observe any concomitant changes in glycogen synthesis and hexokinase, glucokinase, pyruvate kinase, hexose monophosphate shunt dehydrogenase, aconitase, or isocitrate dehydrogenase activities indicates that probably the aforementioned dehydrogenases play a minor role in hepatic lipogenesis (Goodridge and Ball, 1967b). Prolactin *in vitro* increases the rate of hepatic uptake of acetate, pyruvate, and glucose, and incorporation into fatty acids two to six times but is without effect on glycogen levels (Goodridge and Ball, 1967a,b). Time-course studies in pigeons treated previously with prolactin indicate that labeled substrate appears promptly in the liver as labeled fatty acids and subsequently in plasma and adipose tissue (Goodridge and Ball,

1967a). Thus, once again, it appears that these data are consistent with the hypothesis that the major site of lipogenesis in birds is the liver where fatty acids are both synthesized and processed, then released to the plasma and carried subsequently to peripheral adipose tissue depots for uptake and storage.

Extensive work by Meier and colleagues on the diurnal and seasonal responsivity of migrant and nonmigrant forms to exogenous prolactin has been most revealing (also, see Section VI,C). The antigonadal effects of prolactin are well known, although only recently has it been demonstrated that these "inhibitory" effects on gonadal development are probably a result of a blocking action at the target (gonad) tissue receptor site in nonmigrant or weak migrant avian species (Meier and Dusseau, 1968; Meier, 1969). Strong migrants, such as the White-throated Sparrow (*Zonotrichia albicollis*), are unresponsive to the antigonadal effects of daily prolactin (LTH) injections. Metabolically, prolactin may cause either fat catabolism or fat anabolism, depending upon time of day hormone injections are made. Injections of prolactin made up to 5 hours after the start of a 16-hour photoperiod invariably suppresses body weight and lipid reserves; however, injections made 5 or 10 hours after commencement of the photoperiod increase both body weight and lipid depots (Meier and Davis, 1967). Endogenous prolactin release from the anterior pituitary gland is known to be at a maximum shortly after the mid-photoperiod in photosensitive species. Prolactin's anabolic effect on lipid reserves, appetite, and body weight is augmented by light in migratory forms. Such observations are in accord with those indicating that the hormone encourages profound catabolism when injected during dark periods, data which are consistent with metabolic patterns expected of nocturnal migrant forms (Meier and Farner, 1964). Collectively, these observations indicate that there exists a prolactin dominance over depot lipid accumulation, hyperphagia, increased nocturnal activity (requiring corticosterone also), and body weight preceding vernal migration. The most convincing evidence at hand indicates that seasonal variations in peak hypophyseal levels of prolactin as well as diurnal rhythms (e.g., May, Noon vs. August, midnight in *Z. albicollis*) dictate to some degree the extent and direction which metabolic gears will be shifted (Meier *et al.*, 1969). The beneficial results of prolactin release in temperate zone vernal migrant species, therefore, may be considered as a dual action to retard gonadal development until migration to breeding grounds has been completed concomitant with induction of metabolic and migratory behavior (Meier, 1969).

Though prolactin has been reported to be "diabetogenic" in certain mammalian preparations, similar observations are yet to be made in birds. Early reports that prolactin is a hyperglycemic agent in birds are now open to question because only impure preparations were available at that time. Apparently, the only evidence of an effect of prolactin on protein metabolism is the *de novo* synthesis of RNA and protein that is a prerequisite for expression of the crop-sac (bioassay) response to the hormone (Sherry and Nicoll, 1967). Both puromycin and actinomycin D block this effect of prolactin on dermal tissue in pigeons.

4. Other

Adrenocorticotropic hormone (ACTH) and thyroid-stimulating hormone (TSH) are the remaining adenohypophyseal secretions to be considered. Neither of these hormones has been studied thoroughly as a regulator of avian intermediary metabolism. Reference has already been made to what little is known about TSH effects in birds, namely, increase in liver size and glycogen levels in chicks (D. B. King and Snedecor, 1966). The cellular basis for such TSH activity is yet to be established, as is also the hormone's effect on other tissues, metabolic pathways, and in other avian species.

Studies on ACTH effects in birds have been restricted to chickens, and what evidence is at hand indicates that only supraphysiological doses of mammalian (usually porcine) ACTH cause any metabolic effect whatsoever. The hyperglycemia, hepatic (but not muscle) glycogenesis, hypercholesterolemia, and retarded growth observed with ACTH in chickens result secondarily from the subsequent adrenal secretion of corticosterone and hydrocortisone. While glucose-elevating effects of ACTH in birds can be traced to an increase in gluconeogenesis largely at the expense of protein depots, the concomitant hepatic glycogenesis appears to be due to an inhibitory action of the cortex steroids (in response to ACTH) on glycogenolytic enzymes in the liver. Associated with this hyperglycemia is an increase in FFA levels (Heald *et al.*, 1965). Data obtained *in vitro* from chicken adipose tissue and isolated fat cells indicate that large doses of ACTH are required to demonstrate any effect, the usual response being increased lipolysis (Langslow and Hales, 1969). Until avian ACTH is isolated and purified further critical evaluation of metabolic effects of this hormone in birds cannot be made with certainty.

V. Influence of Reproductive (Laying) Cycle

The intermediary metabolic effects associated with the laying cycle of birds are associated with dietary variables as well as with demands of egg formation. Chapters in future volumes of this treatise deal with various detailed aspects of this topic, leaving the following paragraphs to treat what little is known about biochemical adaptations during laying periods.

Release of gonadotrophins associated with egg laying influences lipid metabolism, in particular via the subsequent secretion of estrogen(s) and progesterone. Laying hens have considerably higher plasma lipid levels than nonlayers or male chickens; deposition of lipids occurs in the liver and intra-abdominal storage sites. Gonadal hormone treatment enlarges the liver and increases fatty acid components mainly as an increase in triglyceride concentration relative to the phospholipid moiety (Balnave and Brown, 1967; Balnave, 1969). Obviously, these changes are adaptations for ovogenesis.

The biological half-life of plasma cholesterol in laying hens is approximately 36 hours. Cholesterol apparently is derived almost totally from available dietary and liver sources, and the contribution of cholesterol by the ovary to egg yolk appears to be very small indeed (Andrews *et al.*, 1968). Since plasma provides the major immediate source of cholesterol to the egg, egg laying can be regarded in birds as a major physiological excretory route for derived lipids.

That protein metabolism is in some way involved in successful laying cycles is evident from observations that reduction in available dietary protein invariably leads to diminished clutch size and total egg production. Any adverse environmental (inadequate photostimulation, temperature, humidity, etc.) or physiological (longevity of production, intestinal infection, dietary amino acid imbalance, etc.) condition that leads to a decreased rate and quality of laying also increases the protein requirements of such hens. As in the case of lipid metabolism, however, the specific metabolic steps or enzymatic reactions involved during these physiological adaptations are unknown.

Mineral metabolism in laying birds has received considerable attention over the years (largely for economic reasons) and has centered mainly on the roles played by calcium, phosphate, and copper. The crystalline calcium carbonate of eggshell is laid down in the shellgland (uterus) after essential salts and water are transported through the shell membranes surrounding the albumen layer. About 40% of the eggshell weight is calcium, which is laid down as a "shell" during the last 15–16 hours of eggshell formation and which is transported largely by the shellgland mucosa (Hohman and Schraer, 1966; Taylor, 1970). The mitochondrial fraction of shellgland mucosa, and to a lesser extent the microsomal fraction, appears to be most active in calcium transport from plasma to the shell structure (Hohman and Schraer, 1966).

The major sources of calcium for eggshell are obviously diet and bone; however, the former cannot provide sufficient Ca^{++} for shell formation to support the removal of five times the total circulating Ca^{++} levels per hour, as required during calcification! Apparently medullary ("secondary") bone of the marrow cavities supplies most of the calcium to the circulation for successful shellgland activity, and under duress can release up to 10–12% of total (body) bone substance per 24 hours. Hormonal control of this decalcification process undoubtedly involves both ovarian steroid (under gonadotrophin influence) and parathormone activity. The former is involved with deposition and the latter with osteocytic dissolution of medullary bone, releasing Ca^{++} to plasma for shellgland use. [The reader is referred to the excellent review by Taylor (1970) for details of the entire eggshell formation process.]

Also under parathyroidal influence is mobilization of phosphate (along with Ca^{++}) from bone, absorption by the gut, and renal excretion during the laying cycle. Laying chickens reabsorb 41% less of the filtered phosphate (at renal tubules) than do nonlayers, indicating a high phosphate excretion as medullary bone is destroyed. Excretion of phosphate by the kidney appears to be a reflection of both plasma phosphate levels and parathyroidal inhibition of proximal tubular reabsorption (Martindale, 1969).

Implication of copper metabolism in shell membrane formation has been suggested by evidence that the mucosa of the isthmus contains much more copper than any other portion of the oviduct (Moo-Young *et al.*, 1970). Such a heavy endowment of Cu^{++} reflects the activity of the isthmus mucosa in forming the keratin-like components of the shell membranes. Disulfide linkages are characteristic of keratin and the oxidative closure of sulfhydral groups is regulated by cuproenzymes, much as elastin synthesis may be mediated by amine oxidase. Further work on this aspect appears warranted.

Studies of mineral interrelationships and of enzyme–mineral metabolic functions are sorely needed and offer promise of improving our understanding of complex metabolic regulatory patterns concerned with avian growth and reproduction. Savage (1968) has reviewed the role of trace (dietary) minerals on reproductive performance of avian species.

The temporary metabolic respite following clutch laying is reflected largely by reversal of the metabolic patterns shifted into high gear during egg-laying periods. Such reversals, therefore, are undoubtedly the result of waxing and waning of gonadotrophic, estrogenic, and parathyroidal secretions which in turn have cellular effects opposite to those of the previous period. The longer hiatus in the laying cycle brought about by molting appears to be very complex and poorly understood indeed with respect to metabolic patterns. Considerable information is needed on the biochemical basis of molting, particularly in relation to the effects mediated by thyroxine and the metabolic alterations occurring in the integument.

VI. Climatic Effects on Metabolism

A. Heat

While there have been many studies on the effects of hot environments on birds, there have been few relating to effects on specific avian metabolic processes. Short-term or continuous exposure to 40°C during incubation decidedly increases growth of the liver and total growth of the embryo. Furthermore, evidence of decreased hepatic glycogenesis or increased glycogenolysis, or both, between days 7 and 8 has been presented and is attributed to the higher incubation temperature (Delphia *et al.*, 1967).

Abrupt exposure of adult chickens to elevated temperatures decreases the heart rate rapidly, decreases blood pressure modestly, increases respiratory rate and muscle activity, and increases body temperature. Panting can be observed periodically at 35°C, indicating a possible lowering of the threshold of the panting center in birds acclimated to high environmental temperature (Harrison and Biellier, 1969). The occurrence of heat-induced metabolic disarrangement is implied by observations that oxygen consumption increases concurrently with a rapid decrease in egg content specific gravity when birds are subjected to 35°C (from 5 or 21°C ambient). Thus, the increased respiratory rate may not compensate completely for the lowered cardiovascular function, and a consequent acid–base imbalance directly affects the availability of shell and egg contents derived from such affected blood (Harrison and Biellier, 1969). Acclimation to increased temperatures usually occurs within 12–24 hours following exposure as long as the ambient heat remains within the tolerable range. Apparently, "summer" birds tolerate heat better than do "winter" birds subjected to the same elevated temperature, which is a response attributable both to a slower rise in core temperature and to a higher respiratory response of the "summer" birds (Weiss and Borbely, 1957). The concurrent effects of water and feed deprivation with alterations in environmental temperature have been discussed above (Section I,D).

Changes in tissue catecholamine levels associated with exposure to heat have been studied. Exposure of adult chickens to 32°C for 4–20 weeks increases adrenomedullary epinephrine content but does not affect circulating levels (Lin and Sturkie, 1968). This may indicate either an inhibition of release of the amines from the adrenal tissue or a more rapid than normal destruction of the circulating forms, the latter a result of the increased body temperature. Since avian adipose tissue (in contrast to that of mammals) is insensitive to the lipolytic effects of catecholamines (see Section II,E) little metabolic significance can be placed on the adrenal vs. plasma catecholamine levels of birds other than increased body temperatures (induced by ambient conditions) reduce the "need" for release of calorigenic substances.

B. Cold

Metabolism studies carried out in advanced embryos and newly hatched chicks (less than 30 minutes old) indicate that the RQ of most 19-day embryos rises during exposure to cold (20°C); however, 56% of the chicks have reduced RQ values at hatching time and at 1-day post hatching all chicks have marked reductions in RQ. The disproportionate increase in O_2 consumption relative to that of CO_2 production in these cold-exposed chicks is indicative of rapid mobilization and utilization of lipid (Freeman, 1967). Significant decreases in hepatic glycogen and blood glucose levels attend exposure of late embryos to cold temperature though there are no demonstrable alterations in cardiac or skeletal muscle glycogen levels or plasma lactate levels (Freeman, 1967). Contrarily, exposure to cold increases the plasma levels of free fatty acids, especially in the 1-day-old chick, indicating thermogenic compensation for the cold environment because lipid metabolism releases more heat (per unit weight) than either carbohydrate or protein. Since no shivering is observed in the 1-day neonate chick exposed to cold, and since avian adipose cells are resistant to the lipolytic action of sympathomimetic amines, these neonate organisms must rely upon thermogenic mechanisms other than shivering or catecholamine release to increase their metabolic rates (Freeman, 1966, 1967). The possibility that glucagon plays a major role via adipose tissue or hepatic lipid depots, or both, is strengthened by the absence of brown fat in the neonate chick.

Sudden exposure and subsequent acclimitization of domestic fowl to cold environments cause a rapid increase in heart rate, decreased blood pressure, decreased respiratory muscle activity, and an increased O_2 consumption (Harrison and Biellier, 1969). While exposure to cold probably induces alterations in thyroxine release and utilization (which in turn affect all the aforementioned cardiorespiratory functions) the major hormonal response to low ambient temperature appears to be that of the adrenomedullary hormones (Hendrich and Turner, 1967; Lin and Sturkie, 1968). Circulating catecholamines increase as a result of cold exposure though adrenal tissue levels remain normal, thereby indicating increased biosynthesis (Lin and Sturkie, 1968). The biochemical role that the amines play, other than as general calorigenic agents in regulating intermediary metabolism in birds exposed to cold, is yet to be established. Of equal or greater interest and need are temperature–metabolic studies in birds indigenous to frigid zones.

C. Diurnal Rhythms and Photoperiodicity

Metabolic variations (maxima and minima) have been detected in early chick embryos but it is not until the 7th or 8th day of incubation that a definite diurnal pattern evolves which is temperature or light dependent (Johnson, 1966). Such observations are based on continuous O_2 consumption data and, in part, support earlier reports of daily cyclic changes in hepatic fat and glycogen levels of various avian species. Actually, seasonal differences in diurnal patterns of chick embryos resemble those reported for nonavian forms, and therein add support to the suggestion that a "receptor system" exists in avian embryos which is responsive to exteroceptual cues (Johnson, 1966).

Diurnal rhythms have been noted in adult birds also, some details of which have been described above in Section II. Circulating plasma glucose levels in chickens appear to reach a nadir and subsequent plateau during night hours approximately 30 mg% below the maximal levels approached during light hours (Twiest and Smith, 1970). This response may not be wholly unexpected since chickens are primarily daylight eaters while nocturnal eaters (such as rodents) have maximal plasma glucose levels shortly after midnight. However, of particular interest is the fact that on a rigidly controlled lighting system the most abrupt increment in blood glucose levels in chickens occurs approximately 30 minutes *prior* to commencement of the light phase and prior to normal eating activity (Twiest and Smith, 1970). Such observations, therefore, are suggestive of a true circadian fluctuation, one that "anticipates" metabolic demands. The signal(s) for such glucose diurnal cycles has not been elucidated.

Studies of protein metabolism have not been undertaken (other than those related to egg production) relevant to the impact of photoperiod on metabolic variables. Evidence exists indicating that inadequate photostimulation (and consequent declining production) of laying pullets is attended by higher requirements of dietary protein to maintain "good health" of the hen (Bray, 1968).

Lipid deposition in birds as a result of hormonal or photoperiod manipulation has been of continuing interest to investigators and has been reviewed succinctly by Farner and Follett (1966) and Farner *et al.* (1968). Prolactin causes profound carcass lipid deposition in birds if given at about the middle of the photoperiod. Thus, studies on the Golden Topminnow and on the White-throated Sparrow (*Zonotrichia albicollis*) indicate that prolactin injected during the 8th hour of a 16-hour photoperiod doubles carcass lipid content (Meier *et al.*, 1966). Yet, if the hypophyseal hormone is injected near the beginning or near the end of the photoperiod, carcass lipid deposition is depressed and body weight decreases. Certain gonadotrophins augment the diurnal sensitivity to lipid-inducing hormones in birds and presumably act by assisting the effect of prolactin on liver to synthesize triglycerides and release fatty acids to peripheral depots. These hormonal influences on lipid composition and deposition appear to be much more critical than dietary intake as far as the quality of fat laid down is concerned.

When White-crowned Sparrows (*Zonotrichia leucophrys*) are maintained for periods exceeding a year in a constant refractory photoperiod (LD 20:4), cycles of molt and lipid storage persist with normal frequency and intensity. It appears, therefore, that the direct (?) effects of photoperiod constancy are mainly focused on the frequency and phasing of physiological periodicities (J. R. King, 1968). Attempts have been made to correlate fattening processes with photoperiodicity, and both of these with hormonal signal regulators, but for the most part a firm picture does not emerge. Epple (1963) surveyed pancreatic islet cytology of many small bird species in an attempt to relate temporary adiposity with insulinogenesis and release. Only in the European Blackbird (*Turdus merula*) is diminished β-islet activity correlated with vernal fattening. Increased β-cell activity is observed during summer months in this species as hyperglycemia, and decreased lipid stores occur. While such metabolic alterations can be explained as being the response of islet cells to circulating glucose levels (which in turn are controlled to some extent by anti-insulin agents such as adrenosteroids) there appears now to be a more formidable explanation. Though gonadal secretions play a large role in the premigratory adiposity and adrenocortical hormones act in concert, the role of prolactin and glucagon in providing fatty acid substrates should not be underemphasized. Thus, during the hyperphagic season increase in prolactin and glucagon release would lead to a profound rise in fatty acid synthesis and ultimate lipid deposition at various storage sites in the avian organism. It is well known that increased lipid levels (especially those attended by hyperphagia) "arrest" β-cell activity in mammals, and one may be observing a similar physiological mechanism in *Turdus merula*.

The fact that distinct annual cyclic alterations of pancreatic islet activity are not found in all species of birds is clearly demonstrated by more recent work on the White-crowned Sparrow (Epple and Farner, 1967). Thorough histologic study of insulin-secreting cells in this species during breeding and wintering periods has been made and the results indicate that while islet activity is generally high throughout the year there are no distinct seasonal changes in the activity of the endocrine pancreas of these free-living forms. Neither is there any correlation between β-cell activity and degree of lipid deposition regardless of whether the sparrows were studied in their natural habitat or in captivity under controlled long or short photoperiods (Epple and Farner, 1967). Thus, there appears to exist no definitive pattern relating annual cycles, photoperiodicity, and insulin secretion in most small avian species. Again, the roles of prolactin or lipid mobilization, or both, in response to external (cyclic) cues merit thorough investigation.

VII. Influence of Migration on Metabolism

Since of necessity migration involves considerable muscular, cardiovascular, and respiratory effort a treatment of exercise appears appropriate at this point.

A. Short-bout Exercise

Availability of O_2 is a prime requisite for muscular activity because of the need for an H^+ acceptor in the preferential use of fatty acids as fuel. Thus, the effective interplay of the respiratory system to provide adequate O_2, of the cardiovascular system to transport the O_2 to tissue cells, and the "signals" (glucagon?) to liberate fatty acids from lipid depots to be transported to cardiac and skeletal muscle fibers is required. Free fatty acids enter muscle cells rapidly and therefore do not require previous hepatic modification to smaller, degradative intermediates. Intracellular translocation and subsequent oxidation of fatty acids rise from 22% at rest to approximately 86% during treadmill exercise in dogs. Comparable quantitative studies in birds are lacking but what exercise (flight) studies are available indicate that similar cardiorespiratory responses occur which are directed to meet the metabolic needs of flight, respiratory, and cardiac muscles. Heart rates of 165 at rest increase to 565/minute during stable flight in pigeons, pulmonary ventilation increases 20 times over basal levels mainly as a result of the abrupt respiratory rate increase, and pulmonary ventilation during flight is at least 2.5 times greater than that required to meet metabolic needs (Hart and Roy, 1966). The latter is apt testimony of the efficiency of the avian respiratory system, indeed, as the evaporative heat loss via exhaled air accounts for 17% of the heat production during flight. Of particular interest is the post-flight recovery changes in metabolism as indicated by cardiopulmonary variables. Respiratory rates and volumes decrease quickly during recovery from flight exercise while cardiac rates decrease more slowly (Hart and Roy, 1966). Thus, it appears that short durations of exercise lead to a reasonable degree of O_2 debt being incurred by the pigeon in flight and, since the pulmonary ventilation is considerably above that required by the tissues, the incurred debt is probably a result of cardiovascular limitations. Oxygen uptake during recovery from exercise in birds, therefore, is not associated as much with fatty acid utilization to restore ATP levels as it is needed as an H^+ acceptor in removing lactic acid formed from anaerobic glycolysis.

B. Prolonged Flight (Migration)

Since separate chapters are devoted to the energetics of flight and physiological aspects of migration only selected remarks will be made here concerning the metabolic adaptation (and hormonal control thereof) associated with sustained flights. It is well documented that the broad (white) pectoralis muscle fibers of birds are poorly endowed with mitochondria, lipid inclusions, myoglobin, and vascular elements. These fibers are rich in glycogen depots (over three times those of other skeletal muscles) and therefore are suited best for rapid, explosive action such as sudden take-off or braking action in landing. Such actions would be carried out under anaerobic muscle conditions resulting in rapid glycogenolysis.

The inner, thin, red pectoralis muscle fibers are poorly endowed with glycogen inclusions but have dense accumulations of myoglobin, mitochondria, lipid inclusions, and vascular elements, not to omit the essential enzymatic machinery for β-oxidation of fatty acids. Truly, these red muscle fibers appear to require an abundance of O_2 during flight to meet the aerobic needs of fatty acid degradation and, as a result, are the contractile structures of choice for sustained flight such as occurs in migration. The migratory and powerful flying non-migratory bird, therefore, can "switch" muscle fiber activities to meet the airborne activity desired, whether it be short bursts (white fibers) of anaerobic work or prolonged steady (red fibers) aerobic effort. Each activity is supported by different myometabolic reactions enabling the bird to switch from predominantly anaerobic to aerobic and then back to anaerobic contractile activity as it lifts off, maintains a steady flight pattern, and ultimately descends to alight on a surface, respectively. [Definitive treatments of this topic are presented in the excellent monographs of George and Berger (1966), Drummond (1967), and Tucker (1969).]

It has been noted for many years that preparatory to any sustained flight migratory birds undergo certain metabolic alterations which predispose rapid tissue deposition of lipid. This topic has been reviewed expertly by Farner *et al.* (1968) and Helms (1968). Premigratory fat storage has received considerable investigative attention and accord has been reached that generally it is associated with, or immediately follows, a marked hyperphagia in the bird. This temporary hyperphagia represents an altered physiologic state and not merely an increase in time available for feeding (J. R. King and Farner, 1956; J. R. King, 1961). Thus, vernal hyperphagia may be a result of increased photostimulation of the ventrolateral nuclei of the avian hypothalamus, the so-called "hunger center," and increased substrate intake occurs without an increase in energy expenditure. The result of this greater food intake is tissue deposition of C-14, C-16, and C-18 saturated fatty acids and, to a much larger extent, C-18 unsaturated fatty acids. Premigratory fat deposition readily increases body weight by at least 25% in some species, and by up to 50% in others (e.g., Bobolink, *Dolichonyx oryzivorus*) that migrate across oceans. (See Chapter by Peter Berthold in a future volume.) The fact that this transitory fat storage occurs concomitantly with decreased fat utilization is suggested by data indicating that pectoralis muscle homogenates demonstrate decreased mitochondrial lipase activity simultaneously with increased phosphorylase and glycogen levels when examined in the Rose-colored Starling (*Sturnus roseus*) prior to migration. These data (which tend to reverse in postmigratory periods) indicate that some control is exerted over activity of the key lipase enzyme during hyperphagic fat deposition, enabling the bird to draw on its carbohydrate stores and thereby sparing fatty acid utilization while encouraging lipid deposition (Drummond, 1967). Such an adaptive process would ready the bird for migration by providing fuel for the journey (of which some are 1000 miles or more in distance) and may in some way be the result of photoperiodic hypothalamic stimulation and ensuing endocrine expression at the muscle cell level. The fact that this interplay is not simply a hypothalamic "driving" force, but has many physiological interactions, has been observed by the increased physical activity of mammals prior to eating or during starvation and is seen in migratory birds as increased nocturnal activity ("Zugunruhe") which usually occurs several days after fat deposition commences. Apparently, Zugunruhe and lipid deposition are independent processes, yet both are dependent upon similar physiological or exteroceptive cues (J. R. King and Farner, 1963; Lofts *et al.*, 1963).

Hormonal control of fat deposition may well be a reflection of hypothalamic response to light, since a major hypothalamic efferent pathway leads to the avian adenohypophysis. Gonadotrophin secretion is increased greatly during increased vernal photoperiods, but, as indicated above, these hormones per se have little *direct* effect on adipose depots unless prolactin is available simultaneously. Avian prolactin, however, is very much under direct (stimulating) hypothalamic control and has known lipid effects in liver tissue (but not fat depots), encouraging fatty acid synthesis and deposition. Subsequently, these newly formed lipid substrates are made available to peripheral adipose storage sites via hepatic release to the plasma, and fat deposition occurs rapidly, usually within 10 days. Though removed from direct hypothalamic–hypophyseal control, the role of glucagon in the pre-migratory and migratory metabolic changes must be considered. (Insulin, it appears, can be eliminated as a potent regulator of lipogenesis/lipolysis in avian tissue.) Goodridge (1964), working on migratory and nonmigratory finches, suggests that high levels of glucagon during nonmigratory periods of the year inhibit lipogenesis by increasing FFA and associated coenzyme A esters resulting from the hormonal effect on adipose tissue enzymes. Thus, lipogenesis is suppressed and fatty acid utilization is encouraged. During premigratory periods, glucagon levels fall, allowing lipogenesis to increase as a result of less inhibitory "feedback"; fat deposition ensues. Supportive data for this concept come from the very low pancreatic α-cell activity observed in European Blackbirds during fattening periods, and also by the depressed blood glucose observed during fattening periods in White-crowned Sparrows. Such indirect evidence, however, needs to be replaced by observations on circulating glucagon as well as glucagon effects on mitochondrial lipases during various annual seasons.

Equally attractive is the possibility that glucagon levels are high *during* the migration period per se, since the red skeletal muscle fibers are drawing heavily on fatty acid substrates to sustain flight. Thus, glucagon may provide the essential fuel by stimulating mobilization of fat pad triglycerides and FFA, an action that is enhanced by the concomitantly high prolactin levels. Critical evidence is yet to be presented on this possibility but certainly is required in an effort to relate the environmental trigger via the hypothalamus and pituitary gland to the tissue level where control over lipase is registered. Precisely how glucagon fits in this attractive concept is yet to be determined.

VIII. General Summary and Conclusions

A condensation of the foregoing is inappropriate at this point; space would be better devoted to what was not stated above due to lack of information. Many aspects of avian intermediary metabolism merit investigation in order to crystallize our picture of metabolic homeostasis under a wide variety of conditions. Among those problems deserving early and critical investigation are

1. What are normal metabolic patterns for species residing continuously in arctic regions, how do these differ from those of more temperate climatic zones, and how do these metabolic patterns adapt the bird to its surroundings?
2. The dearth of information pertinent to normal metabolism, particularly protein metabolism, in pelagic species is as obvious from the above paragraphs as it is disturbing. Particularly from a nutritional approach, but also due to expected altered hormonal patterns associated with water availability and saline abundance, this area justifies enthusiastic exploration and exposition.
3. What precise role does photostimulation play in advancing the vernal premigratory fattening response in some species? Is it an action *directly* on the "hunger" center which in turn predisposes extraprehension of food? Or is the hyperphagic adiposity syndrome an *indirect* result of photostimulation of an hypothalamic–hypophyseal–target tissue axis which increases tissue utilization of substrate and thereby induces hunger reactions?
4. Recent evidence indicates that thyroxine may not regulate enzyme activity as sluggishly as previously thought, yet to this date virtually unknown are the cellular activities controlled by this hormone in Aves. How does thyroxine regulate molting sequelae; i.e., at what biochemical step does the hormone induce changes that lead to the loss of feathers, integument alteration, and the temporary hiatus in laying?
5. Of fascinating interest is the alteration(s) in tissue sensitivity that

occurs in many tissues between the last week of embryonation and the first postnatal week. Thus, why is embryonic adipose tissue virtually insensitive to hormonal presence only to become very sensitive to the same hormone a few days after hatching? Are tissue receptor sites "exposed" and freed from "covers" during this period?

The aforementioned are, of course, a mere tasty selection of fascinating problems available for enthusiastic investigation. These and many more unmentioned questions await our experimental efforts and should be rewarded with exciting results.

REFERENCES

Alvarado, F. (1967). D-xylose transport in the chicken small intestine. *Comp. Biochem. Physiol.* **20,** 461–470.

Anderson, D. K., and Hazelwood, R. L. (1969). Chicken cerebrospinal fluid: Normal composition and response to insulin administration. *J. Physiol. (London)* **202,** 83–95.

Andrews, J. W., Jr., Wagstaff, R. K., and Edwards, H. M., Jr. (1968). Cholesterol metabolism in the laying fowl. *Amer. J. Physiol.* **214,** 1078–1083.

Annison, E. F., Hill, K. J., Shrimpton, D. H., Shringer, D. A., and West, C. E. (1966). Glucose turnover and oxidation in the domestic fowl. *Brit. Poultry Sci.* **7,** 319–320.

Austic, R. E., Grau, C. R., and Matteson, G. C. (1966). Requirements for glucose and amino acids in defined media for chick embryos. *J. Nutr.* **90,** 175–182.

Balnave, D. (1968). The influence of gonadal hormones on the uptake of ^{14}C-acetate by liver lipid fractions in the immature male chick. *J. Endocrinol.* **42,** 119–127.

Balnave, D. (1969). The effects of certain gonadal hormones on the content and composition of lipids in the blood and liver of immature male chicks. *Comp. Biochem. Physiol.* **28,** 709–716.

Balnave, D., and Brown, W. O. (1967). Liver fatty acid composition of normal and gonadal hormone-treated chicks. *Comp. Biochem. Physiol.* **22,** 313–317.

Bird, F. H. (1968). Role of the avian small intestine in amino acid metabolism. *Fed. Proc., Fed. Amer. Soc. Exp. Biol.* **27,** 1194–1198.

Boomgaardt, J., and McDonald, B. E. (1969). Comparison of fasting plasma amino acid patterns in the pig, rat, and chicken. *Can. J. Physiol. Pharmacol.* **47,** 392–395.

Bray, D. J. (1968). Photoperiodism and age as factors affecting the protein requirements of laying pullets. *Poultry Sci.* **47,** 1005–1013.

Burt, A. M. (1965). Glucose metabolism and chick neurogenesis. I. Glucose-6-phosphate dehydrogenase activity in the embryonic brachial spinal cord. *Develop. Biol.* **12,** 213–232.

Burt, A. M. (1967). Glucose metabolism and chick neurogenesis. II. 6-Phosphofructokinase and aldolase activity in the embryonic brachial spinal cord. *J. Exp. Zool.* **165,** 317–323.

Calvert, C. C. (1967). Studies on hatchability of fertile eggs from hens receiving a linoleic acid deficient diet. *Poultry Sci.* **46,** 967–973.

Carlson, L. A., Liljedahl, S., Verdy, M., and Wirsen, C. (1964). Unresponsiveness to the lipid mobilizing action of catecholamines *in vivo* and *in vitro* in the domestic fowl. *Metabo., Clin. Exp.* **13,** 227–231.

Chapman, T. E., and Black, A. L. (1967). Water turnover in chickens. *Poultry Sci.* **46,** 761–765.

Coffey, R. G., Morse, H., and Newburgh, R. W. (1966). The synthesis of nucleic acid constituents in the early chick embryo. *Biochim. Biophys. Acta* **114,** 547–558.

Conklin, J. L. (1966). Histochemical localization of enzymes in the embryonic chick liver. *J. Exp. Zool.* **161,** 251–270.

Creger, C. R., and Colvin, L. B. (1967). The transference of ^{89}Sr from the shell to the developing chick embryo. *Radiat. Res.* **32,** 131–137.

Chung, R. A., Davis, E. Y., Munday, R. A., Tsao, Y. C., and More, A. (1967). Effect of cholesterol with different dietary fats on the fatty acid composition of egg yolk and various body tissues. *Poultry Sci.* **46,** 133–141.

D'Amelio, V., and Constantino, E. (1968). The globins and their synthesis in the erythroid cells of the chick embryo. *Biochim. Biophys. Acta* **155,** 614–615.

Dean, W. F., Scott, M. L., Young, R. J., and Ash, W. J. (1967). Calcium requirement of ducklings. *Poultry Sci.* **46,** 1496–1499.

Delphia, J. M., Singh, S., and Baskin, H. (1967). The relationship of hyperthermia to liver growth and liver glycogen in the chick embryo. *Poultry Sci.* **46,** 1454–1459.

Dent, P. B., Brown, D. M., and Good, R. A. (1969). Ultimobranchial calcitonin in the developing chicken. *Endocrinology* **85,** 582–585.

Deucher, E. M. (1965). Biochemical patterns in early developmental stages of vertebrates. *In* "The Biochemistry of Animal Development" (R. Weber, ed.), Vol. 1, pp. 246–306. Academic Press, New York.

De Vincentiis, M., and Marmo, F. (1966). The ^{45}Ca turnover in the membranous labyrinth of chick embryos during development. *J. Embryol. Exp. Morphol.* **15,** 349–354.

Donaldson, W. E. (1967). Lipid composition of chick embryo and yolk as affected by stage of incubation and maternal diet. *Poultry Sci.* **46,** 693–697.

Doyle, W. L., and Watterson, R. L. (1949). The accumulation of glycogen in the "glycogen body" of the nerve cord of the developing chick. *J. Morphol.* **85,** 391–404.

Drummond, G. I. (1967). Muscle metabolism. *Fortschr. Zool.* **18,** 359–429.

Epple, A. (1963). Pancreatic islets and annual cycle in some avian species. *Proc. Int. Ornithol. Congr., 13th, 1962* pp. 974–982.

Epple, A. (1965). Investigations on a third pancreatic hormone. *Gen. Comp. Endocrinol.* **5,** 674.

Epple, A., and Farner, D. S. (1967). The pancreatic islets of the White-crowned Sparrow, *Zonotrichia leucophrys gambelii,* during its annual cycle and under experimental conditions. *Z. Zellforsch. Mikrosk. Anat.* **79,** 185–197.

Farner, D. S., and Follett, B. K. (1966). Light and other environmental factors affecting avian reproduction. *J. Anim. Sci.* **25,** Suppl., 90–118.

Farner, D. S., King, J. R., and Stetson, M. H. (1968). The control of fat metabolism in migratory birds. *Proc. Int. Congr. Endocrinol., 3rd, 1968* pp. 152–157.

Fearon, J. R., and Bird, F. H. (1968). Site and rate of active transport of *d*-glucose in the intestine of the fowl at various initial glucose concentrations. *Poultry Sci.* **47,** 1412–1416.

Fisher, H., and Ashley, J. H. (1967). Protein reserves and survival of cocks on a protein-free diet. *Poultry Sci.* **46,** 991–994.

Fisher, H. I., and Bartlett, L. M. (1957). Diurnal cycles in liver weights in birds. *Condor* **59,** 364–372.

Fraser, R. C. (1966). The rate of hemoglobin chain formation in developing chick embryos. *Exp. Cell Res.* **44,** 195–200.

Freeman, B. M. (1966). The effects of cold, noradrenaline and adrenaline upon the oxygen consumption and carbohydrate metabolism of the young fowl (*Gallus domesticus*). *Comp. Biochem. Physiol.* **18,** 369–382.

Freeman, B. M. (1967). Some effects of cold on the metabolism of the fowl during the perinatal period. *Comp. Biochem. Physiol.* **20,** 179–193.

Froning, G. W., and Norman, G. (1966). Binding and water retention properties of light and dark chicken meat. *Poultry Sci.* **45,** 797–800.

George, J. C., and Berger, A. J. (1966). "Avian Myology." Academic Press, New York.

George, J. C., and Telesara, C. L. (1961). Histochemical observations in the succinic dehydrogenase and cytochrome oxidase activity in pigeon breast muscle. *Quart. J. Microsc. Sci.* **102,** 131–141.

Gibson, W. R., and Nalbandov, A. V. (1966). Lipolysis and lipogenesis in liver and adipose tissue of hypophysectomized cockerels. *Amer. J. Physiol.* **211,** 1352–1356.

Goodridge, A. G. (1964). The effect of insulin, glucagon and prolactin on lipid synthesis and related metabolic activity in migratory and non-migratory finches. *Comp. Biochem. Physiol.* **13,** 1–26.

Goodridge, A. G. (1968a). Metabolism of glucose-U-^{14}C *in vitro* in adipose tissue from embryonic and growing chicks. *Amer. J. Physiol.* **214,** 897–901.

Goodridge, A. G. (1968b). Lipolysis *in vitro* in adipose tissue from embryonic and growing chicks. *Amer. J. Physiol.* **214,** 902–907.

Goodridge, A. G., and Ball, E. G. (1965). Studies on the metabolism of adipose tissue. XVIII. *In vitro* effects of insulin, epinephrine and glucagon on lipolysis and glycolysis in pigeon adipose tissue. *Comp. Biochem. Physiol.* **16,** 367–381.

Goodridge, A. G., and Ball, E. G. (1966). Lipogenesis in the pigeon: *In vitro* studies. *Amer. J. Physiol.* **211,** 803–808.

Goodridge, A. G., and Ball, E. G. (1967a). The effect of prolactin on lipogenesis in the pigeon: *In vivo* studies. *Biochemistry* **6,** 1676–1682.

Goodridge, A. G., and Ball, E. G. (1967b). The effect of prolactin on lipogenesis in the pigeon: *In vitro* studies. *Biochemistry* **6,** 2335–2343.

Goodridge, A. G., and Ball, E. G. (1967c). Lipogenesis in the pigeon: *In vivo* studies. *Amer. J. Physiol.* **213,** 245–249.

Grande, F. (1968). Effect of glucagon on plasma free fatty acids and blood sugar in birds. *Proc. Soc. Exp. Biol. Med.* **128,** 532–536.

Grande, F. (1969). Lack of insulin effect on free fatty acid mobilization produced by glucagon in birds. *Proc. Soc. Exp. Biol. Med.* **130,** 711–713.

Grande, F. (1970). Effects of glucagon and insulin on plasma free fatty acids and blood sugar in owls. *Proc. Soc. Exp. Biol. Med.* **133,** 540–543.

Grande, F., and Prigge, W. F. (1970). Glucagon infusion, plasma FFA and triglycerides, blood sugar, and liver lipids in birds. *Amer. J. Physiol.* **218,** 1406–1411.

Grau, C. R. (1968). Avian embryo nutrition. *Fed. Proc., Fed. Amer. Soc. Exp. Biol.* **27,** 185–192.

Guidotti, G. G., Loreti, L., Gaja, G., and Foá, P. P. (1966). Glucose uptake in the developing chick embryo heart. *Amer. J. Physiol.* **211,** 981–987.

Guidotti, G. G., Borghetti, A. F., Gaja, G., Loreti, L., Ragnotti, G., and Foá, P. P. (1968a). Amino acid uptake in the developing chick embryo heart. The effect of insulin on α-aminoisobutyric acid accumulation. *Biochem. J.* **107,** 565–574.

Guidotti, G. G., Gaja, G., Loreti, L., Ragnotti, G., Rottenbert, D. A., and Borghetti, F. 1968b). Amino acid uptake in the developing chick embryo heart. The effect of insulin on glycine and leucine accumulation. *Biochem. J.* **107,** 575–580.

Haller, R. W., and Sunde, M. L. (1966). The effects of withholding water on the body temperature of poults. *Poultry Sci.* **45,** 991–997.

Harrison, P. C., and Biellier, H. V. (1969). Physiological response of domestic fowl to abrupt changes of ambient air temperature. *Poultry Sci.* **48,** 1034–1045.

Hart, J. S., and Roy, O. Z. (1966). Respiratory and cardiac responses to flight in pigeons. *Physiol. Zool.* **39,** 291–306.

Hazelwood, R. L. (1965). Carbohydrate metabolism. *In* "Avian Physiology" (P. D. Sturkie, ed.), pp. 313–357. Cornell Univ. Press, Ithaca, New York.

Hazelwood, R. L., and Barksdale, B. K. (1970). Failure of chicken insulin to alter polysaccharide levels of the avian glycogen body. *Comp. Biochem. Physiol.* **36,** 823–827.

Hazelwood, R. L., and Lorenz, F. W. (1959). Effects of fasting and insulin on carbohydrate metabolism of the domestic fowl. *Amer. J. Physiol.* **197,** 47–51.

Hazelwood, R. L., Hazelwood, B., and McNary, W. (1962). Possible hypophyseal control over glycogenesis in the avian glycogen body. *Endocrinology* **71,** 334–336.

Hazelwood, R. L., Kimmel, J. R., and Pollock, H. G. (1968). Biological characterization of chicken insulin activity in rats and domestic fowl. *Endocrinology* **83,** 1331–1337.

Heald, P. J. (1963). The metabolism of carbohydrate by liver of the domestic fowl. *Biochem. J.* **86,** 103–110.

Heald, P. J., McLachlan, P. M., and Rookledge, K. A. (1965). The effects of insulin, glucagon and adrenocorticotrophic hormone on the plasma glucose and free fatty acids of the domestic fowl. *J. Endocrinol.* **33,** 83–95.

Helms, C. W. (1968). Food, fat, and feathers. *Amer. Zool.* **8,** 151–167.

Hendrich, C. E., and Turner, C. W. (1967). A comparison of the effects of environmental temperature changes and 4.4°C. cold on the biological half-life ($t_{1/2}$) of thyroxine-I^{131} in fowls. *Poultry Sci.* **46,** 3–5.

Herrmann, H., and Tootle, M. L. (1964). Specific and general aspects of the development of enzymes and metabolic pathways. *Physiol. Rev.* **44,** 289–371.

Hohman, W., and Schraer, H. (1966). The intracellular distribution of calcium in the mucosa of the avian shell gland. *J. Cell Biol.* **30,** 317–331.

Holdsworth, C. D., and Wilson, H. T. (1967). Development of active sugar and amino acid transport in the yolk sac and intestine of the chicken. *Amer. J. Physiol.* **212,** 233–240.

Hsiao, C. Y. Y., and Ungar, F. (1969). Lipid changes in the chick hatching muscle. *Proc. Soc. Exp. Biol. Med.* **132,** 1047–1051.

Hunzicker, M. E. (1969). Fatty acid composition and insulin-like activity of chicken cerebrospinal fluid. Master's Thesis, University of Houston.

Hunzicker, M. E., and Hazelwood, R. L. (1970). Chicken cerebrospinal spinal fluid: Insulin-like activity. *Comp. Biochem. Physiol.* **36,** 795–801.

Hurwitz, S., and Bar, A. (1966). Calcium depletion and repletion in laying hens. 2. The effect on radiocalcium and radiostrontium retention in bone and deposition in the egg shell. *Poultry Sci.* **45,** 352–358.

Hurwitz, S., and Bar, A. (1967). Calcium metabolism of hens secreting heavy or light egg shells. *Poultry Sci.* **46,** 1522–1527.

Johnson, L. G. (1966). Diurnal patterns of metabolic variations in chick embryos. *Biol. Bull.* **131,** 308–322.

Kilsheimer, G. S., Weber, D. R., and Ashmore, J. (1960). Hepatic glucose production in developing chicken embryo. *Proc. Soc. Exp. Biol. Med.* **104,** 515–518.

Kimmel, J. R., Pollock, H. G., and Hazelwood, R. L. (1968). Isolation and characterization of chicken insulin. *Endocrinology* **83,** 1323–1330.

King, D. B. (1969). Effect of hypophysectomy of young cockerels, with particular reference to body growth, liver weight, and liver glycogen level. *Gen. Comp. Endocrinol.* **12,** 242–255.

King, D. B., and Snedecor, J. G. (1966). Effect of TSH on the liver glycogen level of the hypophysectomized chick. *Amer. Zool.* **6,** 561.

King, J. R. (1961). On the regulation of vernal pre-migratory fattening in the White-crowned Sparrow. *Physiol. Zool.* **34,** 145–157.

King, J. R. (1968). Cycles of fat disposition and molt in White-crowned Sparrows in constant environmental conditions. *Comp. Biochem. Physiol.* **24,** 827–837.

King, J. R., and Farner, D. S. (1956). Bioenergetic basis of light-induced fat deposition in the White-crowned Sparrow. *Proc. Soc. Exp. Biol. Med.* **93,** 354–359.

King, J. R., and Farner, D. S. (1963). The relationship of fat deposition to Zugunruhe and migration. *Condor* **65,** 200–223.

Klein, N. W. (1968). Growth and development of chick embryo explants on various protein substrates. *J. Exp. Zool.* **168,** 239–256.

Klicka, J., Edstrom, R., and Ungar, F. (1969). Acid mucopolysaccharide changes in chick hatching muscle. *J. Exp. Zool.* **171,** 249–252.

Kratzer, F. H., Rajaguru, R. W. A. S. B., and Vohra, P. (1967). The effect of polysaccharides on energy utilization, nitrogen retention and fat absorption in chickens. *Poultry Sci.* **46,** 1489–1493.

Krebs, H. A., Dierks, C., and Gascoyne, T. (1964). Carbohydrate synthesis from lactate in pigeon-liver homogenate. *Biochem. J.* **93,** 112–121.

Krompecher, S., Laszlo, M. B., and Ladanyi, P. (1966). Some metabolic changes during the chick's ontogenesis. *Acta Morphol. Acad. Sci. Hung.* **14,** 267–268.

Langslow, D. R., and Hales, C. N. (1969). Lipolysis in chicken adipose tissue *in vitro*. *J. Endocrinol.* **43,** 285–294.

Langslow, D. R., Butler, E. J., Hales, C. N., and Pearson, A. W. (1970). The response of plasma insulin, glucose and nonesterified fatty acids to various hormones, nutrients and drugs in the domestic fowl. *J. Endocrinol.* **46,** 243–260.

Leach, R. M., Jr. (1968). Effect of manganese upon the epiphyseal growth plate in the young chick. *Poultry Sci.* **47,** 828–830.

Leibson, L. (1965–1966). The endocrine factors in the regulation of carbohydrate metabolism in the developing chick embryo. *Bio. Neonatorum* **9,** 249–262.

Lepkovsky, S., Chari-Bitron, A., Lemmon, R. M., Ostwald, R. C., and Dimick, M. K. (1960). Metabolic and anatomic adaptations in chickens "trained" to eat their daily food in two hours. *Poultry Sci.* **39,** 385–389.

Lepkovsky, S., Dimick, M. K., Furuta, F., Snapir, N., Park, R., Narita, N., and Komatsu, K. (1967). Response of blood glucose and plasma free fatty acids to fasting and to injection of insulin and testosterone in chickens. *Endocrinology* **81,** 1001–1006.

Lervold, A. M., and Szepsenwohl, J. (1961). Glycogenolysis in aliquots of glycogen bodies of the chick under the influence of various tissues. *Fed. Proc., Fed. Amer. Soc. Exp. Biol.* **20,** 77.

Leveille, G. A. (1969a). *In vitro* hepatic lipogenesis in the hen and chick. *Comp. Biochem. Physiol.* **28,** 431–435.

Leveille, G. A. (1969b). The influence of fructose diphosphate on pyruvate kinase activity in liver, muscle and adipose tissue of the rat, mouse, pig and chicken. *Comp. Biochem. Physiol.* **28,** 733–740.

Lifschitz, E., German, O., Favret, E. A., and Manso, F. (1967). Difference in water ingestion associated with sex in poultry. *Poultry Sci.* **46,** 1021–1023.

Lin, Y., and Sturkie, P. D. (1968). Effect of environmental temperatures on the catecholamines of chickens. *Amer. J. Physiol.* **214,** 237–240.

Lofts, B., Marshall, A. J., and Wolfson, A. (1963). The experimental demonstration of pre-migratory activity in the absence of fat deposition in birds. *Ibis* **105,** 99–104.

Luckenbill, L. M., and Cohen, A. S. (1966). The association of lipid droplets with cytoplasmic filaments in avian sub-synovial adipose cells. *J. Cell Biol.* **31,** 195–199.

Malhorta, H. C., Khorana, S., Misra, U. K., and Venkitasubramian, T. A. (1967). Brain lipids in experimental tuberculosis in chickens. *Indian J. Biochem.* **4,** 219–222.

Martindale, L. (1969). Phosphate excretion in the laying hen. *J. Physiol. (London)* **203,** 83P.

Medway, W., and Kare, M. R. (1957). Water metabolism of the domestic fowl from hatching to maturity. *Amer. J. Physiol.* **190,** 139–141.

Meier, A. H. (1969). Antigonadal effects of prolactin in the White-throated Sparrow, *Z. albicollis. Gen. Comp. Endocrinol.* **13,** 222–225.

Meier, A. H., and Davis, K. B. (1967). Diurnal variations of the fattening response to prolactin in the White-throated Sparrow, *Z. albicollis. Gen. Comp. Endocrinol.* **8,** 110–114.

Meier, A. H., and Dusseau, J. W. (1968). Prolactin and the photoperiodic gonadal response in several avian species. *Physiol. Zool.* **41,** 95–103.

Meier, A. H., and Farner, D. S. (1964). A possible endocrine basis for premigratory fattening in the White-crowned Sparrow, *Z. leucophrys gambelii. Gen. Comp. Endocrinol.* **4,** 584–595.

Meier, A. H., Burns, J. T., and Dusseau, J. W. (1969). Seasonal variations in the diurnal rhythm of pituitary prolactin content in the White-throated Sparrow, *Z. albicollis. Gen. Comp. Endocrinol.* **12,** 282–289.

Meier, A. H., Davis, K. B., and Lee, R. (1966). The diurnal rhythm of the fattening response to prolactin in the Golden Topminnow and the White-throated Sparrow. *Amer. Zool.* **6,** 519.

Mialhe, P. (1958). Glucagon, insuline et regulation endocrine de la glycemie chez le canard. *Acta Endocrinol. Suppl.* **36,** 9–134.

Mikami, S. I., and Ono, K. (1962). Glucagon deficiency induced by extirpation of alpha islets of the fowl pancreas. *Endocrinology* **71,** 464–473.

Mirsky, I. A., and Gitelson, S. (1958). The diabetic response of geese to pancreatectomy. *Endocrinology* **63,** 345–348.

Moog, F. (1965). Enzyme development in relation to functional differentiation. *In* "The Biochemistry of Animal Development" (R. Weber, ed.), Vol. 1, pp. 307–367. Academic Press, New York.

Moo-Young, A. J., Schraer, H., and Schraer, R. (1970). The copper content of the isthmus mucosa and certain organs of the domestic fowl. *Proc. Soc. Exp. Biol. Med.* **133,** 497–499.

Mulkey, G. J., and Huston, T. M. (1967). The tolerance of different ages of domestic fowl to body water loss. *Poultry Sci.* **46,** 1564–1569.

Noble, R. C., and Moore, J. H. (1967a). The partition of lipids between the yolk and yolk-sac membrane during the development of the chick embryo. *Can. J. Biochem.* **45,** 949–958.

Noble, R. C., and Moore, J. H. (1967b). The transport of phospholipids from the yolk to the yolk-sac membrane during the development of the chick embryo. *Can. J. Biochem.* **45,** 1125–1133.

Nordin, J. H., Wilkin, D. R., Bretthauer, R. K., Hansen, R. G., and Scott, H. M. (1960). A consideration of galactose toxicity in male and female chicks. *Poultry Sci.* **39,** 802–812.

O'Hea, E. K., and Leveille, G. A. (1968). Lipogenesis in isolated adipose tissue of the domestic chick (*Gallus domesticus*). *Comp. Biochem. Physiol.* **26,** 111–120.

O'Hea, E. K., and Leveille, G. A. (1969). Lipid biosynthesis and transport in the domestic chick (*Gallus domesticus*). *Comp. Biochem. Physiol.* **30,** 149–159.

Okumura, L., and Tasaki, I. (1969). Effect of fasting, refeeding and dietary protein level on uric acid and ammonia content of blood, liver and kidney in chickens. *J. Nutr.* **97,** 316–320.

Okuno, G., Grillo, T. A. I., Price, S., and Foá, P. P. (1964). Development of hepatic phosphorylase in the chick embryo. *Proc. Soc. Exp. Biol. Med.* **117,** 524–526.

Przybylski, R. J. (1967). Cytodifferentiation of the chick pancreas. I. Ultrastructure of the islet cells and the initiation of granule formation. *Gen. Comp. Endocrinol.* **8,** 115–128.

Ramachandran, S., Klicka, J., and Ungar, F. (1969). Biochemical changes in the musculus complexus of the chick (*Gallus domesticus*). *Comp. Biochem. Physiol.* **30,** 631–640.

Revel, J., Napolitano, L., and Fawcett, D. (1960). Identification of glycogen in electron micrographs of thin tissue sections. *J. Biophys. Biochem. Cytol.* **8,** 575–589.

Riddle, O. (1963). Prolactin in vertebrate function and organization. *J. Nat. Cancer Inst.* **31,** 1039–1110.

Romanoff, A. L. (1967). Chemistry of the non-embryonic portions of the egg. *In* "Biochemistry of the Avian Embryo," pp. 179–232. Wiley, New York.

Savage, J. E. (1968). Trace minerals and avian reproduction. *Fed. Proc., Fed. Amer. Soc. Exp. Biol.* **27,** 927–931.

Sell, J. L., Choo, S. H., and Kondra, P. A. (1968). Fatty acid composition of egg yolk and adipose tissue as influenced by dietary fat and strain of hen. *Poultry Sci.* **47,** 1296–1302.

Shao, T.-C., and Hill, D. C. (1967). A comparison of the effect of dietary fat and carbohydrate on free amino acids in the blood plasma of chicks. *Can. J. Physiol. Pharmacol.* **45,** 225–234.

Shao, T.-C., and Hill, D. C. (1968). Effect of thiouracil on free amino acids in the blood plasma of chicks. *Poultry Sci.* **47,** 1806–1810.

Sherry, W. E., and Nicoll, C. S. (1967). RNA and protein synthesis in the response of pigeon crop-sac to prolactin. *Proc. Soc. Exp. Biol. Med.* **126,** 824–829.

Singh, A., Reineke, E. P., and Ringer, R. K. (1968). Influence of thyroid status of the chick on growth and metabolism, with observations on several parameters of thyroid function. *Poultry Sci.* **47,** 212–219.

Sitbon, G. (1967). La pancreatectomie totale chez l'oie. *Diabetologia* **3,** 427–434.

Snedecor, J. G. (1968). Liver hypertrophy, liver glycogen accumulation, and organ-weight changes in radiothyroidectomized and goitrogen-treated chicks. *Gen. Comp. Endocrinol.* **10,** 277–291.

Snedecor, J. G., and King, D. B. (1964). Effect of radiothyroidectomy in chicks with emphasis on glycogen body and liver. *Gen. Comp. Endocrinol.* **4,** 144–154.

Snedecor, J. G., King, D. B., and Henrikson, R. (1963). Studies on the chick glycogen body: Effects of hormones and normal glycogen turnover. *Gen. Comp. Endocrinol.* **3,** 176–183.

Sova, Z., Jicha, J., Koudela, K., Houska, J., and Vrbenska, A. (1968). The values of cholesterol, phospholipids and esterified fatty acids in wl chickens in the first month of life. *Deut. Tieraerztl. Wochenschr.* **75,** 169–175.

Stumpf, P. K. (1947). Phosphorylated carbohydrate compounds in developing chick embryo. *Fed. Proc., Fed. Amer. Soc. Exp. Biol.* **6,** 296.

Taussig, M. P. (1965). Incorporation of amino acids into chick proteins during embryonic growth. *Can. J. Biochem.* **43,** 1099–1110.

Taylor, T. G. (1970). How an eggshell is made. *Sci. Amer.* **222,** 89–94.

Teekell, R. A., Richardson, C. E., and Watts, A. B. (1968). Dietary protein effects on urinary nitrogen components of the hen. *Poultry Sci.* **47**, 1260–1266.

Thommes, R. C., and Aglinskas, A. S. (1966). Endocrine control of yolk sac membrane glycogen in the developing chick embryo. II. Effects of hypophysectomy. *Gen. Comp. Endocrinol.* **7**, 179–185.

Thommes, R. C., and Firling, C. E. (1964). Blood glucose and liver glycogen levels in glucagon-treated chick embryos. *Gen. Comp. Endocrinol.* **4**, 1–8.

Thommes, R. C., and Just, J. J. (1966). A re-evaluation of the effects of "hypophysectomy" by surgical decapitation on the glycogen content of the glycogen body of the developing chick embryo. *Endocrinology* **79**, 1021–1022.

Thommes, R. C., and Mathew, G. (1969). Endocrine control of yolk-sac membrane glycogen in the developing chick embryo. IV. effects of insulin addition. *Physiol. Zool.* **42**, 311–319.

Thommes, R. C., and Shulman, R. W. (1967). Endocrine control of lipid metabolism in the developing chick embryo. I. blood cholesterol. *Gen. Comp. Endocrinol.* **8**, 54–60.

Thommes, R. C., McCarter, C. F., and Nguyen, L. H. (1968). Endocrine control of yolk-sac-membrane glycogen in the developing chick embryo. III. effects of thiourea addition. *Physiol. Zool.* **41**, 491–499.

Tucker, V. A. (1969). The energetics of bird flight. *Sci. Amer.* **220**, 70–78.

Twiest, G., and Smith, C. J. (1970). Circadian rhythm in blood glucose level of chickens. *Comp. Biochem. Physiol.* **32**, 371–375.

Vallyathan, N. V., Cherian, K. M., and George, J. C. (1964). Histochemical and quantitative changes in glycogen and phosphorylase during disease atrophy of the pigeon pectoralis. *J. Histochem. Cytochem.* **12**, 721–728.

Wagh, P. V., and Waibel, P. E. (1966). Metabolizability and nutritional implications of l-arabinose and d-xylose for chicks. *J. Nutr.* **90**, 207–211.

Wagh, P. V., and Waibel, P. E. (1967). Alimentary absorption of l-arabinose and d-xylose in chicks. *Proc. Soc. Exp. Biol. Med.* **124**, 421–424.

Watterson, R. L. (1949). Development of the glycogen body of the chick spinal cord. *J. Morphol.* **85**, 337–339.

Watterson, R. L., Veneziano, R., and Brown, D. (1954). Development of the glycogen body of the chick spinal cord: Effects of hypophysectomy on its glycogen content. *Physiol. Zool.* **31**, 49–59.

Weiss, H. S., and Borbely, E. (1957). Seasonal changes in the resistance of the hen to thermal stress. *Poultry Sci.* **36**, 1383–1384.

Wenger, E., Wenger, B. S., and Kitos, P. A. (1967). Pentose phosphate pathway activity of the chick embryo *in ovo*. *J. Exp. Zool.* **166**, 263–270.

West, G. C., and Meng, M. S. (1968). The effect of diet and captivity on the fatty acid composition of redpoll (*Acanthis flammea*) depot fats. *Comp. Biochem. Physiol.* **25**, 535–540.

Chapter 9

OSMOREGULATION AND EXCRETION IN BIRDS

Vaughan H. Shoemaker

I. Introduction 527
II. The Avian Kidney 528
A. Renal Structure 528
B. Renal Function 533
III. Postrenal Modification of the Urine—Role of the Cloaca and Intestine 546
IV. The Avian Salt Gland 551
A. Morphology of the Salt Gland 551
B. Function of the Salt Gland 556
References 566

I. Introduction

It is generally assumed that the excretory organs together with their control mechanisms are responsible for maintenance of the internal environment. As stated by Homer Smith (1953), "the composition of the body fluids is determined not by what the mouth takes in but by what the kidneys keep." Indeed, it is the organs of excretion to which this chapter is primarily devoted. However, decisive osmoregulatory roles may be played by other factors, since homeostasis requires only that input and output are equalized.

Maintenance of water balance in birds means that water ingested by drinking and eating, plus that formed in oxidative metabolism, must equal the amount that is evaporated as well as that lost via urine, feces, and glandular secretions. Similarly, electrolytes in the drinking water and food must be balanced by electrolytes in the excreta. Each of these quantities is variable and to some extent under the bird's control by behavioral and physiological means. Many are also profoundly influenced by the environment and life-style of the bird.

For example, evaporative water loss and metabolic water production depend on body size, activity, and environmental conditions. Food habits also play an important role in osmoregulation, as does the quality and availability of water for drinking. Thus, the great diversity in size, habitat, behavior, and physiological capacities among birds means that a similar diversity must exist in their water and electrolyte budgets. Information on these important aspects of avian osmoregulation are contained elsewhere in this treatise and the reader is referred to Bartholomew and Cade (1963) for a lucid discussion of these factors and their interrelations. Some information on these aspects of the water and electrolyte economies of captive birds is available (e.g., K. Schmidt-Nielsen *et al.*, 1958; Cade and Greenwald, 1966; Macmillen and Trost, 1966; Willoughby, 1968; McNabb, 1969a,b). A prototype for studies of osmoregulation in free-ranging birds is provided by Smyth and Bartholomew (1966a).

II. The Avian Kidney

A. Renal Structure

1. Gross Morphology

The kidneys of birds are paired structures lying in pelvic skeletal depressions along the backbone. They comprise about 1% of the body weight, being somewhat larger in small birds (O. W. Johnson, 1968) and in species having functional salt glands (Hughes, 1970a). Each kidney is usually composed of three lobes (cranial, middle, and caudal), although this lobation is more or less distinct and the relative size of the lobes varies considerably among species (O. W. Johnson, 1968). The ureter lies along the ventral surface of the kidney and receives major branches from each lobe. Major circulatory connections have been studied primarily in chickens and are shown in Figs. 47 and 48 of Chapter 4, this volume. The arterial blood supply is from the renal artery which supplies the cranial lobe, and branches of the sciatic artery to the middle and caudal lobes (Siller and Hindle, 1969). The femoral artery may also branch to the kidney (see Sturkie, 1965). Avian kidneys receive venous blood from the legs, tail, and mesenteries via the renal portal system, and renal veins provide for drainage to the vena cava.

2. *Nephrons*

Birds possess two basic types of nephrons, including cortical nephrons composed of a glomerulus and proximal and distal convoluted tubules, and less numerous juxtamedullary or looped nephrons in which the proximal and distal tubules are separated by a loop of Henle which extends into the medulla (Fig. 1c). There is surprisingly little to add to Sperber's (1960) account of the microanatomy of avian nephrons. The glomeruli are small and contain few capillary loops. The cells of the proximal tubule have a brush border on the luminal surface (Sperber, 1960) similar to those of mammals and vertebrates in general. The basal (capillary-facing) surfaces of these cells exhibit numerous small projections similar to those reported for a gecko (Roberts and Schmidt-Nielsen, 1966), rather than the extensive folding into mitochondria-containing compartments typical of mammals. The cells of the avian distal tubule also differ from those of mammals in that they have relatively flat basal membranes (Sperber, 1960; Poulson, 1968).

Henle's loop in birds also differs in several respects from that of mammals. There are no abrupt transitions between thin and thick segments. Instead the proximal tubule gradually narrows, loses its brush border, and continues into the medulla with no change in staining properties (Poulson, 1968). The cells of the thin limb of birds are higher than the squamous epithelium of the mammalian thin limb (Sperber, 1960). As in the short-looped nephrons of mammals, the turn of the loop occurs in the thick (ascending) limb. The thick limb resembles that of mammals in that the cells are characterized by striations due to parallel mitochondria (Sperber, 1960), and the general association of this cell type with active sodium transport is consistent with the probable role of the thick limb as a component of a countercurrent multiplier system (see Section II,B,3).

The nephrons of birds, like those of mammals, have a region of attachment of the afferent glomerular arteriole and a loop of the distal convoluted tubule at Bowman's capsule. Cells in the arteriole, called "juxtaglomerular cells," exhibit characteristic granules thought to contain renin. The adjacent cells of the distal convoluted tubule possess large prominent nuclei and form the "macula densa." Juxtaglomerular cells and macula densa together form the "juxtaglomerular apparatus" (see Taylor *et al.*, 1970). Maculae densae have been observed in a wide variety of avian species (O. W. Johnson and Mugaas, 1970b).

3. *Organization of the Avian Kidney*

The division of the avian kidney into cortical and medullary regions is somewhat more complex than in mammals. The medulla of the avian kidney is actually a number of separate units or lobules within each lobe of the kidney. The cortex is also organized into lobules, several of which are associated with each medullary lobule (Fig. 1a).

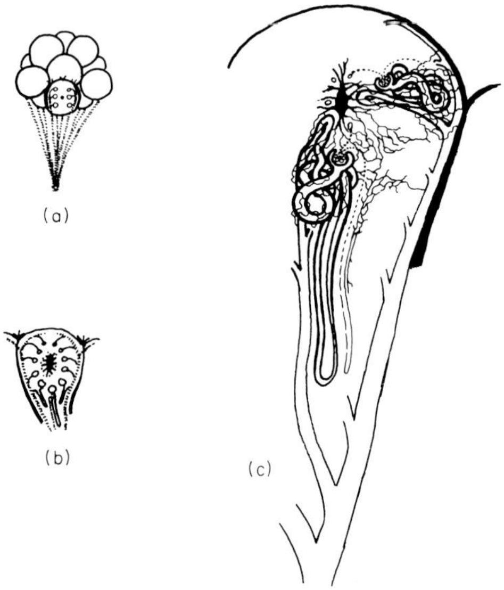

FIG. 1. (a) Diagrammatic representation of cortical lobules showing association with a single medullary lobule or "cone." (b) Cortical lobule showing arrangement of glomeruli around central vein with peripheral collecting tubules (dotted lines) and interlobular veins (solid lines). (c) Cortical and juxtamedullary (looped) nephrons. Note origin of afferent glomerular arterioles from arteries near central vein. Efferent glomerular arterioles of cortical nephrons run peripherally to join capillary net from interlobular vein. Efferent glomerular arterioles from looped nephrons form vasa recta which subsequently join peritubular capillary network in cortex. (From Poulson, 1968.)

a. Renal Cortex. Each cortical lobule consists of a group of nephrons arranged in a roughly spherical manner around a central artery and a branch of the renal vein (Fig. 1b). The artery branches to provide the afferent arterioles to the glomeruli which lie about halfway between the center and the outer limits of the lobule (Fig. 1c). From the glomerulus, the proximal convoluted tubule courses peripherally to the edge of lobule, turns back to the central vein, and then toward the periphery again. Cortical nephrons, at this point, narrow to a short segment which, according to Sperber (1960), corresponds to the thick segment of Henle's loop. This segment then attaches to Bowman's capsule forming the juxtaglomerular apparatus from which point it becomes the distal convoluted tubule. The situation in looped nephrons is similar except that the proximal tubule finally turns toward the medulla to give rise to the loop of Henle. As in cortical nephrons, the thick limb returns to the juxtaglomerular apparatus (Fig. 1c) and becomes the distal convoluted tubule. The distal tubule, after coursing toward the central vein, returns after several convolutions to become the initial collecting tubule and joins peripherally with other collecting tubules. The collecting tubules continue to fuse as they progress through the medulla to the ureter.

Peritubular circulation in the renal cortex of birds consists of a capillary network that has inputs from renal portal veins as well as from the efferent glomerular arteriole. Veins of renal portal origin (interlobular veins) run between the cortical lobules and send numerous small branches into each lobule. These subdivide to form a capillary network around each nephron, and the efferent glomerular arteriole joins the network near the periphery. Occasionally arterioles course directly to the capillary network without glomerular involvement (Siller and Hindle, 1969). The capillary blood flows toward the central or intralobular vein (Figs. 1c and 2).

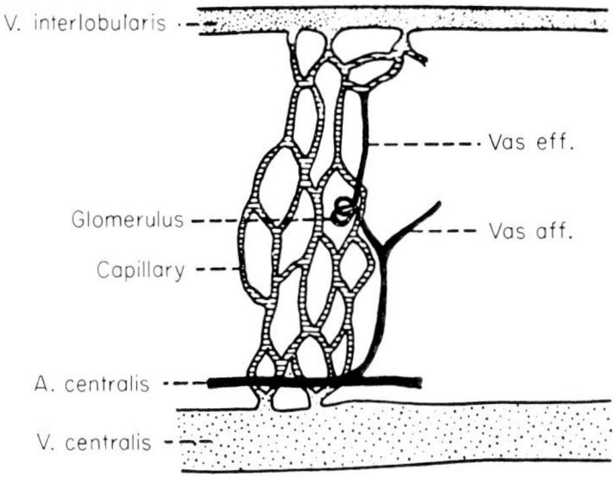

FIG. 2. Scheme of the peritubular circulation in the avian renal cortex. The portal blood flows in through the interlobular vein and mixes with blood from the efferent glomerular arteriole. (From Sperber, 1948.)

The prominence of the renal portal connections to the peritubular circulation is particularly intriguing in view of the complexity of the venous anastomoses which make up the renal portal input. The presence of a strategically placed sphincter, the renal portal "valve," has long led to speculation that special control mechanisms govern the flow of portal blood to the kidney. Akester (1967), using radiographic techniques, has recently shown that numerous patterns of flow may occur through these vessels, with varying proportions of the portal blood being shunted away from the kidney. However, the physiological significance of these observations remains obscure (see Sections II,B,1 and 4).

b. Renal Medulla. Medullary lobules are essentially cones surrounded by a connective tissue sheath open to the cortex at the large end and attached to a major ureteral branch at the other (Fig. 3).

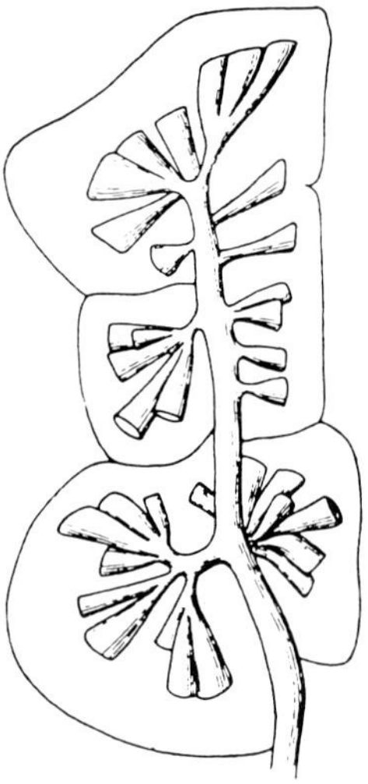

FIG. 3. Ureter with medullary lobules (cones) of pigeon. (From Sperber, 1960.)

Each medullary lobule or cone receives the collecting tubules, loops of Henle, and vasa recta from several cortical lobules (Fig. 1a). There is a progressive diminution of each medullary lobule toward the ureter because of the varying lengths of the loops and the fusion of collecting tubules, but neighboring lobules frequently fuse toward the base to be enclosed within the same connective tissue sheath (Fig. 4).

Comparative studies of medullary characteristics have been made by Poulson (1965) and O. W. Johnson and Mugaas (1970a,b). Poulson noted a correlation between the concentrating capacity of the kidney and the number of medullary lobules in three passerine forms. Johnson and Mugaas also found that the length of the medullary units was not correlated with the ability to conserve water, whereas "water conservers" tended to have a higher proportion of the kidney composed of medulla (see Section II,B,3), and they reported considerable interspecific variation in the configuration and distribution of the medullary units.

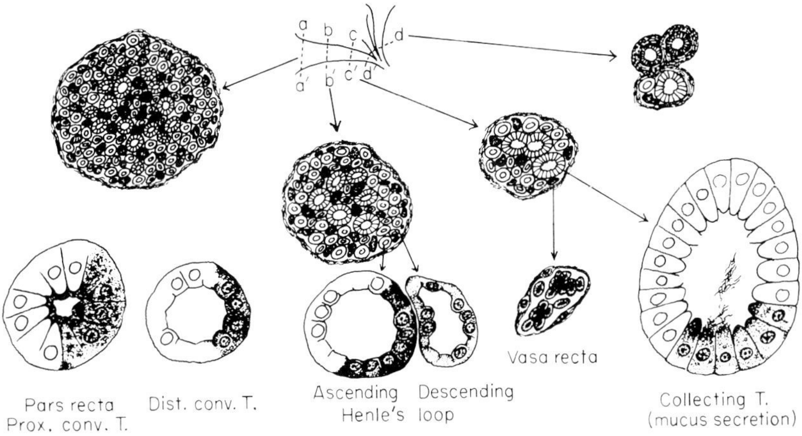

FIG. 4. Cross sections of medullary lobule progressing toward the ureter showing arrangement and appearance of its components. (From Poulson, 1968.)

B. RENAL FUNCTION

In keeping with the variety of osmoregulatory problems encountered by birds, their kidneys show considerable versatility in the volume and composition of the urine produced (Table I). In birds the kidneys are primarily responsible for the excretion of nitrogen and all electrolytes (except sodium and chloride in the case of forms possessing salt glands; see Section IV,B). Additionally, the kidneys may be called upon to excrete excess water or to conserve it. The renal mechanisms by which this flexibility is achieved involve regulation of glomerular filtration rate, transtubular transport of solutes and water, and renal plasma flow.

1. Control of Rate of Urine Production

Birds are capable of varying rates of urine production greatly to meet demands imposed by other factors in the water and electrolyte budgets. There are basically two ways in which this can be accomplished by the kidneys of vertebrates; through variation in the rate at which an ultrafiltrate of the plasma enters the kidney tubules, and through variation in the amount of water subsequently resorbed across the tubules. Both mechanisms are employed by vertebrates,

TABLE I

URINARY CONCENTRATIONS AND RATES OF URINE PRODUCTION IN REPRESENTATIVE BIRDS UNDER VARIOUS CONDITIONS[a]

Species	Treatment	Electrolyte concentrations (mEq liter^{-1})			Osmolarity (osmmoles liter^{-1})	Rate of flow (ml min^{-1} kg^{-1})	Reference
		Na^+	K^+	Cl^-			
Chicken[b]	Water load	38[c]	6[c]	27[c]	115	0.298	Skadhauge and Schmidt-Nielsen (1967a)
	NaCl load	161[c]	25[c]	140[c]	362	0.181	Skadhauge and Schmidt-Nielsen (1967a)
	Dehydrated	134[c]	40[c]	70[c]	538	0.018	Skadhauge and Schmidt-Nielsen (1967a)
Domestic Duck	NaCl load	133	117	193	462		Skadhauge and Schmidt-Nielsen (1967a)
	Drinking fresh water	10[c]	34[c]	12[c]	205[c]	0.023	Holmes *et al.* (1968)
	Drinking saline	91[c]	28[c]	110[c]	413[c]	0.025	Holmes *et al.* (1968)
Glaucous-winged Gull (*Larus glaucescens*)	Seawater acclimated	45[d]	55[d]	30[d]			Hughes (1970b)
	Seawater acclimated and NaCl load	140[d]	20[d]	145[d]			Hughes (1970b)
White Pelican (*Pelecanus erythrorhynchos*)	"Normal"	35 (7–78)	16 (4–43)		231 (52–434)		Calder and Bentley (1967)
	No food or water	18 (8–33)	114 (64–187)		580 (292–734)		Calder and Bentley (1967)
Greater Roadrunner (*Geococcyx californianus*)	"Normal"	70 (16–165)	43 (20–60)		453 (156–718)		Calder and Bentley (1967)
	No food or water	70 (46–89)	75 (32–159)		593 (454–736)		Calder and Bentley (1967)

Mourning Dove (*Zenaida macroura*)	Drinking fresh water			23	512		Smyth and Bartholomew (1966b)
	Drinking 0.2 *M* NaCl			189	544		Smyth and Bartholomew (1966b)
	NaCl load			327			Smyth and Bartholomew (1966b)
Red Crossbill (*Loxia curvirostra*)	Drinking fresh water	20[c] (2–140)				0.045[c] (0.004–0.19)	Dawson *et al.* (1965)
	Drinking 0.2 *M* NaCl	240[c] (168–430)				0.41[c] (0.07–1.2)	Dawson *et al.* (1965)
Black-throated Sparrow (*Amphispiza bilineata*)	NaCl load			ca. 500 (350–703)			Smyth and Bartholomew (1966a)
Salt-marsh race of Savannah Sparrow (*Passerculus sandwichensis beldingi*)	Drinking fresh water			50	300		Poulson and Bartholomew (1962)
	Drinking 0.6 *M* NaCl			960	2000[d]		Poulson and Bartholomew (1962)
Gambel's Quail (*Lophortyx gambelii*)	Drinking fresh water	50			154		McNabb (1969b)
	Drinking 0.27 *M* NaCl	266			669		McNabb (1969b)
Red-tailed Hawk (*Buteo jamaicensis*)	Fed beef heart	38	61		840		I. M. Johnson (1969)

[a]Values shown are means with ranges in parentheses and represent spontaneously voided urine except as noted.
[b]Ureteral urine
[c]Calculated from data given by authors.
[d]Estimated from graph.

but there is considerable variation in their relative importance. Glomerular filtration rate (GFR) has been shown to be an important variable in the control of renal water excretion in fishes (Holmes and McBean, 1963), amphibians (B. Schmidt-Nielsen and Forster, 1954), and reptiles (LeBrie and Sutherland, 1962; Shoemaker *et al.*, 1966). In mammals, GFR does not normally vary with the state of hydration, although is may increase following salt-loading (B. Schmidt-Nielsen, 1964). The fraction of the filtered water that is resorbed appears to be facultatively variable in vertebrates of all classes.

a. Glomerular Filtration. Rates of glomerular filtration in domestic fowl appear to be somewhat less variable than in lower vertebrates. Korr (1939) found filtration rates in water-loaded chickens to be about four times those in dehydrated individuals, and concluded that reduction in GFR is an important factor in conservation of water by birds. However, Shannon (1938) found that GFR was essentially constant over a tenfold range in rate of urine output of "normally hydrated" fowl. More recently, Skadhauge and Schmidt-Nielsen (1967a) compared filtration rates in dehydrated and water-loaded roosters and observed only modest differences in GFR (1.7 vs. 2.1 ml kg^{-1} min^{-1}) despite marked differences in ureteral urine flow (0.018 vs. 0.30 ml kg^{-1} min^{-1}). Filtration rate in Mourning Doves (*Zenaida macroura*) has been measured at rates of urine production ranging from 0.017 to 0.6 ml kg^{-1} min^{-1} (Shoemaker, 1967). GFR was constant (ca. 2.6 ml kg^{-1} min^{-1}) at moderate to high rates of urine flow. However, at rates of urine flow below 0.15 ml kg^{-1} min^{-1} GFR declined with urine production. Dantzler (1966) found GFR to vary directly with urine production in salt-loaded chickens. Data of Holmes *et al.* (1968) also indicate that GFR is a significant variable related to rate of urine production in ducks, although mean rates of glomerular filtration are similar for ducks drinking seawater or fresh water (2.1 and 2.5 ml kg^{-1} min^{-1}, respectively). Thus, weight-relative filtration rates for unanesthetized chickens, ducks, and doves are similar and are about half those for mammals of similar size (see Florey, 1966). Modifications of GFR appear to be significant in the regulation of urine volume in birds, but less so than in lower vertebrates.

The mechanism for control of GFR in birds is not well understood. Injection of arginine vasotocin has not been shown to influence GFR, although it does elicit an antidiuretic response by enhancing tubular water resorption (Skadhauge, 1964; Dantzler, 1966). It may be that the renin–angiotensin system (see Section II,B,6) regulates GFR through control of constriction of the efferent glomerular arteriole

(Sokabe, 1968). Dantzler (1966) suggests that changes in GFR are related to changes in the number of functioning glomeruli. This possibility is intriguing since it could be of particular advantage for birds to utilize juxtamedullary nephrons preferentially during water stress (see Section II,B,3).

b. Tubular Resorption of Water. As in mammals, the major control of urine production in birds is achieved via tubular water resorption. Skadhauge and Schmidt-Nielsen (1967a) observed urine-to-plasma ratios for inulin ranging from 3 to about 100, and a virtually identical range was found in Mourning Doves (Shoemaker, 1973). Thus as little as two thirds of the filtered water may be resorbed during water diuresis and as much as 99% may be resorbed at low rates of urine production. This provides for about a thirty-fold variation in the rate of urine production based on resorptive changes alone. This, coupled with relatively modest variation in GFR, could easily provide for a 100-fold range in rate of urine production. Holmes *et al.* (1968) found less variation in ducks (ca. 90–99% of filtrate resorbed), but they did not induce extreme diuresis in these animals. The range of water resorption in birds is similar to that of reptiles, whereas mammals do not resorb less than 90% of the filtered water even when maximally diuretic (see Skadhauge and Schmidt-Nielsen, 1967a). Moreover, birds apparently excrete at least 1% of the filtered water whereas mammals may excrete as little as 0.1%.

2. *Excretion of Electrolytes*

Electrolyte concentrations in the urine of birds may range from near zero up to several times body fluid levels in response to a variety of regulatory demands (see Table I). Net transtubular movements of water and solutes interact to produce this flexibility. The urinary concentration (C_U) of a solute depends on its concentration in the plasma ultrafiltrate (C_P), the fraction of the filtered solute which is excreted (F_S) and the fraction of the filtered water which is excreted (F_W) as follows:

$$C_U = \frac{F_S \cdot C_P}{F_W}$$

The production of urine dilute with respect to sodium and chloride obviously involves considerable tubular resorption of these ions since they are the major solutes in the plasma. Water-loaded roosters produce urine that is only one fourth as concentrated as plasma with respect to both sodium and chloride by excreting 2.4% of the filtered sodium and chloride and about 10% of the filtered water (Skadhauge and Schmidt-Nielsen, 1967a). Interestingly, dehydrated birds excreted a smaller fraction of the filtered sodium and chloride than water-loaded birds (ca. 1%), but at higher urinary concentrations since only about 1% of the filtered water was excreted with it. Skadhauge and Schmidt-Nielsen found that salt-loaded roosters excreted 7% of the filtered sodium and chloride along with an equivalent fraction of the filtered water, resulting in urinary concentrations equal to those of the plasma. Dantzler (1966) also observed the production of isosmotic urine by salt-loaded chickens, but in this case about 20–30% of the filtered sodium, chloride, and water was excreted.

Holmes *et al.* (1968) found that ducks maintained on fresh water frequently excreted less than 0.5% of the filtered sodium along with 3–5% of the filtered water to produce urine containing ca. 10 mEq Na^+ per liter. Ducks maintained on hyperosmotic saline excreted about 1.5% of the filtered sodium with 3% of the filtered water, resulting in urinary concentrations of 80 mEq $liter^{-1}$, or about half that of the plasma. This concentration is considerably below that of the saline provided for drinking, and reflects the fact that ducks may rely primarily on the salt gland for the excretion of sodium. Mourning Doves can also excrete sodium and chloride at very low levels (<10 mEq $liter^{-1}$) when given a water load, and this increases to about 150 mEq $liter^{-1}$ when isosmotic saline is provided. This change is accomplished primarily through reduction of the fractional tubular resorption of sodium and chloride (Shoemaker, 1967).

Evidently birds have considerable control over tubular resorption of sodium and chloride, and the fraction of filtered sodium and chloride that is excreted may vary from <0.5 to 30%. Similarly the fraction of filtered water that is excreted has been observed to vary from ca. 1 to 30%. From the urinary sodium and chloride concentrations reported for birds, most appear to be limited to combinations within this range, resulting in urinary sodium and chloride levels between about twice and one-twentieth that of the plasma.

Relatively little attention has been given to the excretion of potassium by birds, and even less to transtubular movements of this ion. Skadhauge and Schmidt-Nielsen (1967a) observed net tubular resorption of potassium resulting in renal excretion of 16, 13, and 36% of filtered potassium in dehydrated, water-loaded, and NaCl-loaded chickens, respectively. Corresponding urinary potassium concentrations approximated 40, 6, and 25 mEq $liter^{-1}$. Holmes *et al.* (1968) also found that less than half of the filtered potassium was excreted by ducks maintained on either fresh water or saline solution. Mourning Doves given 100 mM KCl attained urinary potassium levels of ca. 150 mEq liter^{-1} – roughly thirty times plasma levels. This usually resulted from excretion of about 100% of the filtered potassium along with 3–5% of the water. Net tubular secretion of potassium was indicated on several occasions, with the excretion of potassium reaching nearly two times the amount filtered (Shoemaker, 1967).

Holmes *et al.* (1968) obtained data on the excretion of calcium and phosphate indicating that about half of the filtered amounts of these ions were excreted by ducks maintained on either fresh water or saline.

3. *The Renal Concentrating Mechanism*

The ability of birds and mammals to produce urine hyperosmotic to their body fluids has a morphological correlate in the presence of looped nephrons and their associated vasa recta. Compelling evidence has been accumulated to show that, in mammals, these loops function as a countercurrent multiplier system to produce a gradient of concentration within the medulla. The evidence for this system as well as its mechanism of action has been frequently reviewed (Pitts, 1968; Berliner and Bennett, 1967; B. Schmidt-Nielsen, 1964).

Briefly, the mechanism relies on the active transport of solutes out of the water-impermeable ascending limb of Henle's loop. The solute then diffuses into descending limbs and vasa recta and is thus carried toward the tip of the medulla. The net effect is the accumulation of solutes in the medulla with the establishment of progressively greater concentrations toward its tip. When urine passes through this medullary osmotic gradient in water-permeable collecting tubules, water flows out of these tubules leaving hyperosmotic urine to pass into the ureters. The maximum concentration that the urine can attain is thus identical to the concentration of interstitial fluids at the tip of the medulla. Of course the urine may be much more dilute if the collecting tubules are relatively impermeable to water, and this permeability is under endocrine control (see Section II,B,6).

The establishment of a medullary concentration gradient by the countercurrent system depends upon parallel orientation of loops as is clearly the case throughout the medulla of a typical mammalian kidney. In birds, adjacent medullary lobules are frequently far from parallel. Moreover, medullary lobules may be surrounded by cortical tissue. Poulson (1965) suggested that each avian medullary lobule could contain its own concentration gradient if the connective tissue sheath surrounding it serves as a barrier to the movement of water and solutes. Microcryoscopic examination of kidney sections from dehydrated Budgerigars *(Melopsittacus undulatus)* revealed a medullary concentration gradient of considerable magnitude in all elements of the medulla (Emery and Kinter, 1967). Skadhauge and Schmidt-Nielsen (1967b) demonstrated relatively small but convincing osmotic and ionic differences between medullary and cortical tissue in dehydrated or salt-loaded chickens. They also found, using turkeys, that medullary lobules become more concentrated toward the tip. In both species the increase in osmolarity of medullary tissue was due to sodium chloride, suggesting that sodium is pumped from the ascending limbs of the loops of Henle. Unlike the situation in mammals, urea apparently does not contribute significantly to the medullary osmotic gradient.

Interspecific differences in the renal concentrating capacity of mammals correlate well with the length of the loops of Henle (B. Schmidt-Nielsen and O'Dell, 1961), and this is consistent with the theory of countercurrent multiplication. Similar correlation has not been found in birds. However, birds show interspecific variation of several-fold in the proportion of the kidney composed of medullary tissue (Poulson, 1965; O. W. Johnson and Mugaas, 1970a,b). This presumably reflects variation in the proportion of the nephrons that are looped, whereas mammals possess only looped nephrons. The lack of correlation in birds between loop length and concentrating capacity is therefore not surprising. For example, a kidney in which only 10% of the nephrons contribute to the active concentrating process would not be as effective as one in which 40% of the nephrons are so involved, even if the former has considerably longer loops. In fact, Poulson (1965) and O. W. Johnson and Mugaas (1970a) report good correlations between the relative amount of medullary tissue and the concentrating capacity of bird kidneys (see Section II,A,3,b). Moreover, the concentrating capacity of the avian kidney should be enhanced if only looped nephrons functioned, and Braun and Dantzler (1971) provide evidence for this occurrence in salt-loaded birds.

4. *Renal Plasma Flow*

The total rate of flow of plasma through the kidneys of birds is of interest because it is this rate rather than GFR that limits the rate of excretion of substances, such as uric acid, actively secreted into the renal tubules. Moreover, the circulatory arrangements of the avian kidney suggest that renal plasma flow may be controlled independently of the arterial circulation to the glomeruli. Not only do birds possess a functional renal portal system, but also they possess a valve that apparently regulates the amount of portal blood that is actually shunted through the peritubular circulation of the kidney (see II,A,3,a; also Chapter 4, Section X, this volume).

Renal plasma flows of birds have been measured as clearances of *p*-aminohippurate by several investigators. Skadhauge (1964) found a value of 40 ml min^{-1} kg^{-1} for hens, and Holmes *et al.* (1968) report a value of ca. 27 ml min^{-1} kg^{-1} for ducks maintained on fresh water or saline. Mourning Doves exhibit a similar renal plasma flow (ca. 25–30 ml kg^{-1} min^{-1}) over a wide range of urinary flow rates (Shoemaker, 1967). At low rates of urine production in doves the renal plasma flow declines in proportion to the filtration rate. The filtration fraction (inulin clearance/PAH clearance) thus remains constant, and is about 13%. The filtration fraction of chickens is ca. 15% (Skadhauge, 1964) and that of ducks averaged 11–13% (Holmes *et al.*, 1968) and 14–15% (Stewart *et al.*, 1969). Thus, values for filtration fraction reported for birds of several species are consistent and are lower than comparable values for mammals. It seems likely that the lower filtration fraction is due to the presence of the renal portal system in birds, but the relative contribution of blood from this source to the total peritubular circulation has not been measured. However, experiments in which PAH has been infused into the saphenous vein indicated that about 65% of the saphenous blood passes through the kidney (May and Carter, 1967).

5. *Excretion of Nitrogen*

Although birds are generally regarded as uricotelic, significant quantities of other nitrogenous compounds, particularly ammonia, have been reported in the urine of domestic fowl and ducks (Table II). Urea, creatine, and amino acids are also present in minor amounts. The distribution of urinary nitrogen among these compounds is essentially independent of the protein level of the diet (Teekell *et al.*, 1968) and the source of dietary protein (O'Dell *et al.*, 1960) in chickens, and is the same in ducks maintained on saline or fresh water (Stewart *et al.*, 1969). However, Stewart *et al.* observed that the rate of ammonia excretion was quite dependent on the rate of urine production, suggesting that uric acid may become more predominant when water is in short supply. Uric acid "concentrations" up to 170 gm $liter^{-1}$ have been reported in ureteral urine of the pigeon (McNabb and Poulson, 1970). If, in this case, ammonia nitrogen equivalent to 10% of the uric acid nitrogen were to be excreted as soluble ammonium salts, the concentrating capacity of the kidney would be exceeded.

The relative amounts of uric acid and various urate salts excreted

TABLE II
DISTRIBUTION OF NITROGEN IN THE URINE OF CHICKENS AND DUCKS[a]

Species	Uric acid	Ammonia	Urea	Creatine	Amino acids	Reference
Chicken	75–80	10–15	2–6	1	2	O'Dell *et al.* (1960)
Chicken	82	6	–	–	–	Katayama (1924) (in O'Dell *et al.*, 1960)
Chicken	66	8	6	5	–	Coulson and Hughes (1930) (in O'Dell *et al.*, 1960)
Chicken	63	17	10	8	–	Davis (1927) (in O'Dell *et al.*, 1960)
Chicken[b]	60	23	6	4	2	Teekell *et al.* (1968)
Duck	54	29	1.5	–	–	Stewart *et al.* (1969)

[a]Expressed as percentage of total urinary nitrogen.
[b]Values approximated from graphs.

by birds are not known. Examination of the precipitate by polarized light microscopy and X-ray analysis has revealed it to be composed of minute (2–8 μm) birefringent spheres of varying appearance and solubility (Folk, 1969). Despite Folk's claim that uricotelism in birds is folklore, he has apparently presented evidence that uric acid and urate salts may form diverse and complex crystals and crystal aggregates in bird urine (see Poulson and McNabb, 1970; Lonsdale and Sutor, 1971), and even hints at some interesting differences between species. In the discussion to follow, "uric acid" will be used to designate urate salts as well as the acid unless specified otherwise.

Uric acid clearances approximate the renal plasma flow in ducks (Stewart *et al.*, 1969) and Mourning Doves (Shoemaker, 1967). About 85–90% of the uric acid is excreted via tubular secretion, since the filtration fraction in birds is about 10–15% (see Section II,B,4). Thus it appears that the ability of birds to eliminate nitrogen as uric acid depends primarily upon the renal plasma flow and the concentration of uric acid in the plasma.

The site of tubular secretion of uric acid is not known. However, it appears in precipitated form in the collecting ducts and ureters, and precipitation may be virtually complete before urine leaves the ureters (McNabb and Poulson, 1970). At low rates of urine flow the urine is viscous with mucus which often binds the precipitated uric acid in strands. Urine is propelled through the ureters by peristalsis (Gibbs, 1929), and the ureters and ureteral branches are lined with mucus-secreting goblet cells which presumably serve to lubricate the precipitated uric acid. Uric acid may also appear in colloidal form (Porter, 1966), but McNabb and Poulson (1970) found this to be a minor fraction of the total.

The role of urinary ammonia in birds, presumably formed by deamination in the tubule cells, is not clear. It could be involved in acid–base balance or in the conservation of cations through an exchange mechanism (Stewart *et al.*, 1969). Alternatively, it may simply be an economical vehicle for nitrogen excretion when water intake is not restricted.

A comprehensive review of avian nitrogen metabolism, including discussion of pathways and enzymes in formation of nitrogenous wastes, is provided by Brown (1970).

6. *Hormonal Control of Renal Function*

The volume and osmolarity of the urine of terrestrial vertebrates have long been known to be influenced by octapeptides from the neurohypophysis known as antidiuretic hormones. Arginine vasotocin (AVT) is thought to be the antidiuretic hormone (ADH) in birds (Acher, 1963; Munsick *et al.*, 1960) as in other nonmammalian vertebrates. Skadhauge (1964) infused small (7–50 ng) doses of AVT into a leg vein of hens made diuretic through intravenous infusion of a hypoosmotic saline. The result was an immediate reduction in the rate of urine production with a concomitant rise in urinary concentration (Fig. 5). The response was consistently greater on the infused

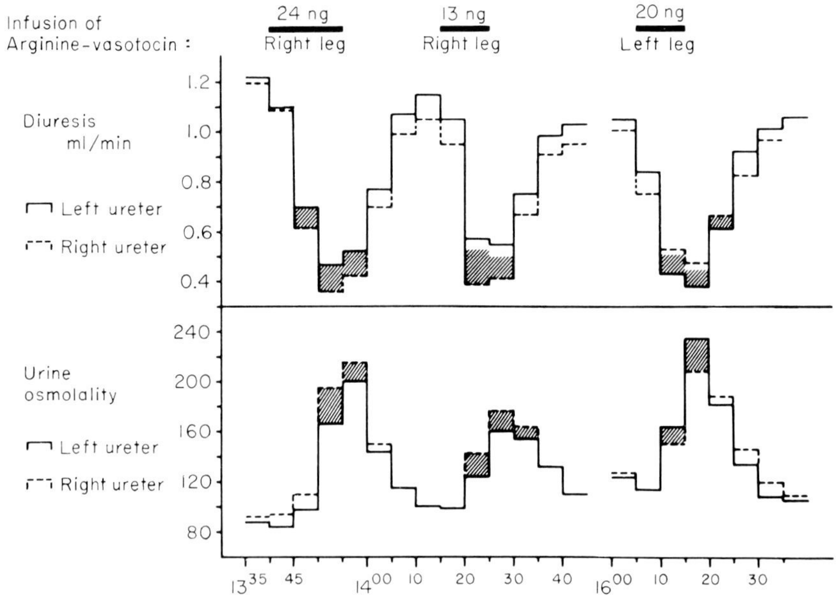

FIG. 5. Effect of arginine vasotocin on the volume and concentration of urine in a 2.8-kg diuretic hen. Hatched areas represent the extent to which the response on the infused side exceeds that of the contralateral kidney. (From Skadhauge, 1964.)

side, indicating that hormone reaching the kidney through the renal portal system (and hence reaching only the nonluminal side of the tubules) is effective. No significant changes in glomerular filtration rate or renal plasma flow accompanied the antidiuretic response, even when the hormone dosage was increased 100-fold. Dantzler (1966) also reported no consistent depression of GFR with massive doses of AVT. It thus appears likely that the renal response to ADH in birds, like that of mammals, consists primarily of an increased permeability of the distal convoluted tubules and collecting tubules to water. This permits the urine to reach osmotic equilibrium with the interstitial fluid surrounding the tubules, thus becoming isosmotic with the plasma in the distal convoluted tubule and hyperosmotic to the plasma as the collecting tubules pass through the medulla (see Section II,B,3). ADH is also thought to stimulate tubular resorption of sodium, which enhances its antidiuretic effect.

The observed effects of arginine vasotocin are consistent with the finding that the rate of urine production in birds can vary considerably while GFR remains constant. GFR can apparently be varied in a controlled fashion (see Section II,B,1), but it seems doubtful that AVT alone is responsible for this.

Control of the release of ADH from the neurohypophysis appears to be similar in a variety of terrestrial vertebrates. The primary stimulus is probably an increase in the concentration of nonpermeating solutes in the extracellular fluids, but a reduction of blood volume is also effective (Moses *et al.*, 1967). Thus, dehydration may stimulate ADH release via both pathways. The diuresis noted during intravenous infusion of hyperosmotic saline (Dantzler, 1966) suggests that expansion of plasma volume may override the osmotic effect. Oksche *et al.* (1963) observed increased neurohypophyseal activity in dehydrated birds and in birds drinking hyperosmotic saline. Von Lawzewitsh and Sarrat (1970) found dehydration or NaCl-loading caused depletion of the neurosecretory depot in the neurohypophysis.

Steroid hormones from the adrenal glands are presumably involved in the control of electrolyte excretion by the avian kidney, as they are in mammals (Pitts, 1968) and lower vertebrates (Maetz, 1969). However, there is little direct evidence bearing on this point (see reviews by deRoos, 1963; Wright *et al.*, 1968). The effects of adrenalectomy (Phillips *et al.*, 1961) and adenohypophysectomy (Wright *et al.*, 1966) on the renal response of ducks to saline loading are slight. In neither case was there appreciable change in the rate of urine production or in the urinary concentrations of sodium and potassium. Diuresis was prolonged in adrenalectomized ducks, but this could have resulted secondarily from impairment of the salt gland response (see Section II,B,4). Donaldson and Holmes (1965) found no differences in patterns of corticosteroidogenesis between ducks maintained on fresh water or saline solution. Similarly, sodium depletion did not affect concentrations of corticosterone or aldosterone in the renal vein of chickens.

Holmes and Adams (1963) tested the effects of corticosterone, cortisol, and aldosterone on the renal response of intact ducks to water loading. Each caused depression of urinary sodium levels, and aldosterone reduced potassium as well. The response to aldosterone differs from that of mammals, where potassium excretion is generally enhanced (see Sawin, 1969). Hollander (1969) found *in vitro* aldosterone production by adrenals from birds given NaCl was depressed, whereas KCl-loaded birds did not differ appreciably from controls given water.

Corticosterone is the primary steroid secreted by the avian adrenal gland, which also produces significant amounts of aldosterone but little or no cortisol (deRoos, 1961, 1963). Mammalian ACTH and extracts of avian adenohypophysis stimulate adrenal secretion of corticosterone and, to a much lesser extent, aldosterone *in vitro* (deRoos and deRoos, 1964; Donaldson *et al.*, 1965). *In vivo* administration of ACTH markedly increased corticosterone concentrations in the adrenal effluent of chickens, but corresponding levels of aldosterone did not change significantly (Taylor *et al.*, 1970). Frankel *et al.* (1967a) and Taylor *et al.* (1970) found that adenohypophysectomy reduced, but did not abolish, corticosterone in the adrenal vein, and Taylor *et al.* also noted a significant drop in aldosterone levels. It thus appears that secretion of both corticosterone and aldosterone is controlled to some extent via the pituitary. Hypothalamic control of adrenal secretion, possibly through a corticotropin releasing factor, is suggested by lesion experiments (Frankel *et al.*, 1967b) and by *in vitro* studies (Stainer and Holmes, 1969).

Synthesis of steroids, especially aldosterone, by mammalian adrenals is influenced by the renin–angiotensin system (see Davis, 1967). Birds apparently possess such a system since angiotensinogen and renin have been found in chicken plasma and kidneys, respectively (Schaffenburg *et al.*, 1960). However, deRoos and deRoos (1963) found no stimulation of steroidogenesis when chicken adrenals were incubated with mammalian angiotensin II. Similarly, Taylor *et al.* (1970) found that infusion of chicken kidney extracts into hypophysectomized chickens did not significantly increase the concentration of aldosterone or corticosterone in blood from the adrenal vein, although a typical pressor response was observed. However, they found marked increases in juxtaglomerular granulation and renin content in the kidneys of sodium-depleted birds. Studies on pigeons have implicated ACTH in the control of renin release (Chan and Holmes, 1971). The renin–angiotensin system is not yet well understood, but its potential influence on renal hemodynamics and electrolyte transport in birds should not be ignored.

III. Postrenal Modification of the Urine—Role of the Cloaca and Intestine

The ureters of birds feed into the cloaca, and the urine is often voided mixed with fecal material. This raises the possibility of significant modification of the urine in the posterior portion of the digestive tract. Despite the plausibility of such a mechanism (see K. Schmidt-Nielsen *et al.*, 1963), its role in osmoregulation is still unclear.

When the digestive and urinary systems of domestic birds are separated, either by exteriorization or cannulation of the ureters, or by rectal fistula or obstruction, large increases in rates of urinary water loss have been reported (Korr, 1939; Dicker and Haslam, 1966; Scheiber and Dziuk, 1966). Other investigators have found these procedures to be without significant effect (Dixon, 1958; Hart and Essex, 1942; Hester *et al.*, 1940). Adequate controls for experiments of this kind are elusive, and manipulations of the cloacal region may have effects unrelated to its normal role (Sturkie and Joiner, 1959a,b).

A number of recent studies have clearly demonstrated that urine may move into the large intestine from the cloaca in chickens, despite the assertion of Weyrauch and Roland (1958) that such movement is prevented by folds (Houston's valves) at the intestinal–cloacal junction. Akester *et al.* (1967) observed that radioopaque urine formed within a few minutes after intravenous injection of sodium diacetamidotricodoacetate. The urine accumulated in the coprodeum and then was propelled into the colon by retroperistaltic waves originating in the coprodeum and terminating at the ileo-ceco-colic junction. Other radiographic studies indicate that the urine begins to move into the large intestine as soon as it reaches the cloaca (Nechay *et al.*, 1968), that retrograde movement of materials into the small intestine is prevented (Koike and McFarland, 1966), and that this phenomenon is not unique to chickens (Ohmart *et al.*, 1970a). It is not clear to what extent refluxing of urine into the large intestine may be controlled. Akester *et al.* (1967) observed that the coprodeum acted at times like a reservoir from which retrograde movement did not occur and suggested the presence of a control mechanism for the initiation of retroperistaltic activity. However, Skadhauge (1968) observed its occurrence (evidenced by the retrograde movement of both inulin and uric acid) during dehydration, water-loading, and salt-loading.

Even if urine were not refluxed into the large intestine, some opportunity for modification of the urine exists in the cloaca itself. The cloaca is apparently not impermeable to water, as shown by the rapid equilibration of tritiated water introduced into the cloaca of the domestic duck (Peaker *et al.*, 1968) following obstruction of the large intestine. They observed no differences in the rate of equilibration related to the state of the animal's water balance. However, since the rate of exchange diffusion far exceeds the net flux, such differences should not be expected (see Potts and Parry, 1963). Of more interest is their observation that 0.7% saline introduced into the cloacas of saline-loaded and water-deprived birds quickly disappeared. Labeled ions have also been shown to appear in the body fluids after cloacal instillation, but at a much slower rate than similar exchanges occur across the large intestine in dogs and man (Weyrauch and Roland, 1958).

Recent comparisons of ureteral urine with urine voided normally in chickens and ducks show only limited differences. Skadhauge (1968) collected ureteral urine through funnels sewn over the ureteral openings of birds subjected to water-loading, salt-loading, and dehydration. He found no difference in the maximum urinary concentrations obtained following salt-loading or dehydration, or in minimal concentrations seen following water-loading, when compared to normally voided excreta from similarly treated birds. Comparisons of total salt and water output were made only for dehydrated birds, and these suggest a resorption of about half of the water and sodium leaving the ureters. Skadhauge points out that this comparison is not particularly compelling because rates of ureteral urine output were measured for about 1 hour and compared to 24-hour collections of normally voided excreta.

Another promising approach is the collection of urine directly from one kidney while allowing the other to drain into the cloaca. Nechay and Lutherer (1968) used this method to study cloacal resorption in chickens given hyperosmotic (2%) and hypoosmotic (0.45%) NaCl intravenously. Ureteral and cloacal urine were virtually identical with respect to both volume and composition during infusion of 2% NaCl. When 0.45% NaCl was given, the volume of urine collected via the cloaca was 81% of that collected through the cannula and averaged 18% greater in total solute concentration. This appears to reflect some osmotic flux of water from hypoosmotic urine across the cloaca. In these experiments urine was prevented from refluxing into the large intestine and the cloaca was emptied frequently.

In vivo perfusion experiments (Skadhauge, 1967) have shed light on the permeability and transport characteristics of the cloaca and large intestine. In these experiments the intestine was ligated just caudal to the ceca, and about 5 ml of perfusate was circulated through the coprodeum and large intestine. The perfusate contained inulin to indicate net water fluxes, and its composition with respect to specific solutes and total osmolarity was varied. When hyperosmotic solutions of nonpermeating solutes were used, water moved into the lumen. The osmotic permeability coefficient, calculated from this water flux, the osmotic gradient, and the area of membrane perfused, was 19.6 ×

10^{-4} cm sec^{-1}. Interestingly, a somewhat higher osmotic permeability coefficient (31×10^{-4} cm sec^{-1}) was found to describe the movement of water out of the lumen from a hypoosmotic perfusate, suggesting some rectification of osmotic flux by the perfused membrane favoring the resorption of water. It would be interesting to test the effect of arginine vasotocin on these permeabilities. The diffusional permeability determined from the rate of disappearance of tritiated water from the perfusate was considerably lower (8.3×10^{-5} cm sec^{-1}). A discrepancy in this direction is to be expected, but it is somewhat larger than that for frog skin, which has an osmotic permeability coefficient of 3.3×10^{-4} cm sec^{-1} and a diffusional permeability coefficient of 6.5×10^{-5} cm sec^{-1} (Dainty and House, 1966). Sodium was found to move out of the lumen when the sodium concentration of the perfusate was above 30 mEq $liter^{-1}$, and into the lumen at lower concentrations. The rate of sodium transport was relatively constant (80–90 μEq kg^{-1} hr^{-1}) at luminal sodium concentrations ranging from 80 to 240 mEq $liter^{-1}$. High concentrations of potassium in the perfusate or prior sodium-loading of the bird reduced the rate of sodium resorption as well as the magnitude of the transmembrane potential (lumen negative) presumably generated by the sodium pump. Resorption of sodium was generally accompanied by a secretion of potassium of lesser magnitude, and the movement of chloride appeared to balance the charge differentials of the sodium and potassium movement. The interrelations of water and electrolyte movement across the coprodeum and intestine following introduction of hyperosmotic NaCl are shown in Fig. 6. Sodium is resorbed and potassium secreted at relatively constant rates. Water initially moves into the lumen along an osmotic gradient. This movement, and the resorption of sodium, dilutes the perfusate and the direction of net water movement is subsequently reversed. This reversal occurs while the perfusate is still slightly hyperosmotic to the plasma, and Skadhauge interprets this in terms of solute-linked water transport (see Schultz and Curran, 1968).

Observations of Hughes (1970b) on Glaucous-winged Gulls (*Larus glaucescens*) are generally consistent with the ionic fluxes observed by Skadhauge in perfusion experiments. Urine collected directly from the cloaca tended to be higher in sodium and chloride than spontaneously voided urine and lower in potassium.

Despite a number of contradictory findings relevent to postrenal modification of the urine, some aspects seem reasonably clear if generalizations from domestic fowl are valid. There appears to be no means of keeping urine from exposure to membranes (coprodeum and

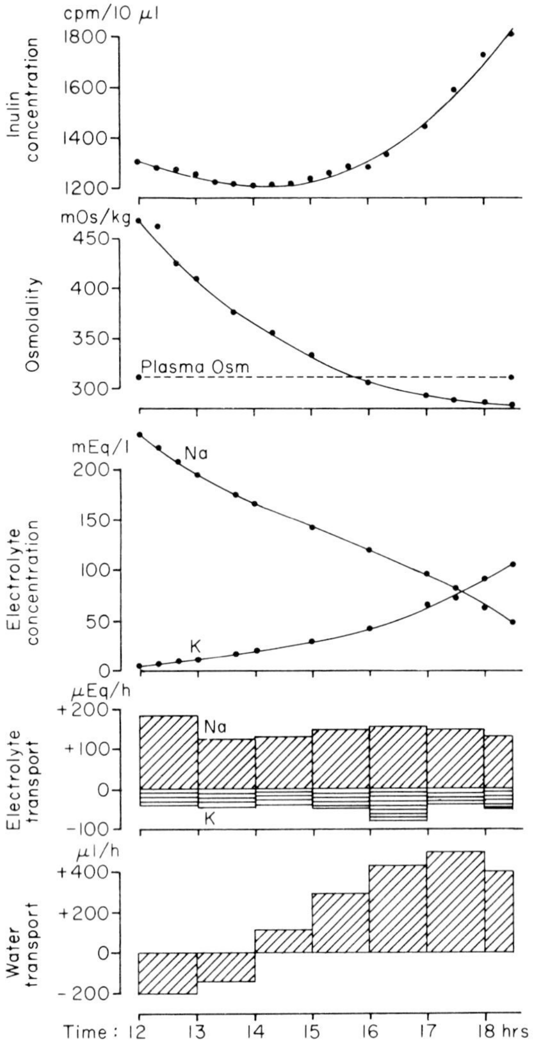

FIG. 6. Cloacal modification (*in vivo*) of initially hyperosmotic perfusate. Note initial osmotic influx of water into the cloaca, followed by subsequent resorption of both water and sodium. (From Skadhauge, 1967.)

large intestine) across which exchange of water and solutes occurs. Nevertheless, these rates of exchange are sufficiently low so that they can have only minor effects when urine is produced at moderate to high rates (see Skadhauge, 1967). The situation at very low rates of urine flow is less clear. Present evidence makes it appear highly unlikely that the concentrating capacity of the kidney can be further enhanced by postrenal mechanisms. Indeed, it is almost certain that a large osmotic gradient cannot be maintained across these membranes for prolonged periods. The resorption of both salts and water can be significant at low rates of renal output but it has not been demonstrated that postrenal modifications of the urine accomplish anything that is beyond the capabilities of the kidneys themselves. It may be that refluxing of liquid urine, as well as precipitated uric acid, into the large intestine allows for resorption of more of the water from the combined mass of fecal material and urates than would be possible if they were handled separately.

IV. The Avian Salt Gland

All birds possess glands in the head, known as nasal glands, which are distinct both anatomically and embryologically from the lacrimal glands and Harderian glands. In some species, particularly those of marine habit, these glands are well developed and are capable of elaborating copious secretions containing high concentrations of sodium chloride. To emphasize this osmoregulatory function, the term "salt gland" was coined (see K. Schmidt-Nielsen and Fänge, 1958) and will be employed here. Functional salt glands have been reported in numerous representatives of thirteen orders of birds as summarized in Table III.

A. Morphology of the Salt Gland

Although the salt glands of birds show variation in size, shape, and location (see Technau, 1936), they have a number of properties in common. Avian salt glands are derived embryologically from invaginations in the nasal epithelium which persist as the main ducts of the gland. In many species there are two such ducts per gland, each giving rise to distinct but closely applied parts (Marples, 1932). All are paired (Fig. 7a), discharge through main *ducts* into the anterior nasal cavity, and appear similar in basic structure, even in species in which the gland has no apparent function (McLelland *et al.*, 1968). Avian salt glands are composed of tubular *lobes* which are surrounded by a thick layer of connective tissue (Fig. 7b). Most of these are parallel and run the length of the gland, although some shorter lobes oriented in various directions have been observed. Each lobe has a *central canal* that is continuous with a duct of the gland. The secretion of the gland is elaborated in *secretory tubules* which are arranged radially around the central canal of each lobe and are continuous with it (Fig.

TABLE III

BIRDS THAT HAVE BEEN REPORTED TO POSSESS FUNCTIONAL SALT GLANDS[a]

Order Family	Common name	Reference
Struthioniformes		
Struthionidae	Ostrich	K. Schmidt-Nielsen *et al.* (1963)
Podicipediformes		
Podicipedidae	Grebes	Cooch (1964)
Sphenisciformes		
Spheniscidae	Penguins	K. Schmidt-Nielsen and Slayden (1958); Douglas (1968)
Procellariiformes		
Diomedeidae	Albatrosses	Frings and Frings (1959)
Hydrobatidae	Storm-petrels	Schmidt-Nielsen (1960)
Pelecaniformes		
Sulidae	Boobies and gannets	McFarland (1960)
Phalacrocoracidae	Cormorants	McFarland (1960)
Pelecanidae	Pelicans	K. Schmidt-Nielsen and Fänge (1958)
Fregatidae	Frigatebirds	McFarland (1960)
Anseriformes		
Anatidae	Ducks	Scothorne (1959); Cooch (1964)
	Geese	Gill and Burford (1968); Hanwell *et al.* (1970)
Phoenicopteriformes		
Phoenicopteridae	Flamingo	McFarland (1959)
Ciconiiformes		
Ardeidae	Herons	Lange and Staaland (1966)
Falconiformes		
Accipitridae	Hawks and eagles	Cade and Greenwald (1966)
Falconidae	Falcons	Cade and Greenwald (1966)
Galliformes		
Phasianidae	Desert Partridge	K. Schmidt-Nielsen *et al.* (1963)
Gruiformes		
Rallidae	Rails, coot	Cooch (1964); Carpenter and Stafford (1970)
Charadriiformes		
Charadriidae	Plovers	Staaland (1967a)
Scolopacidae	Sandpipers, snipe	Staaland (1967a)
Laridae	Gulls	Fänge *et al.* (1958a)
	Terns	Hughes (1968)
Alcidae	Auks, guillemots, puffins	Staaland (1967a)
Cuculiformes		
Cuculidae	Roadrunner	Ohmart *et al.* (1970b)

[a] The salt gland has not been confirmed as the source of secretion in all cases.

7c,d). The secretory tubules branch toward the periphery of the lobes where they are closely packed and roughly parallel. The tubules are closed distally, and there appears to be no means for ultrafiltration into the tubules. Thus the single layer of wedge-shaped epithelial cells that make up the tubules is responsible for the secretion process.

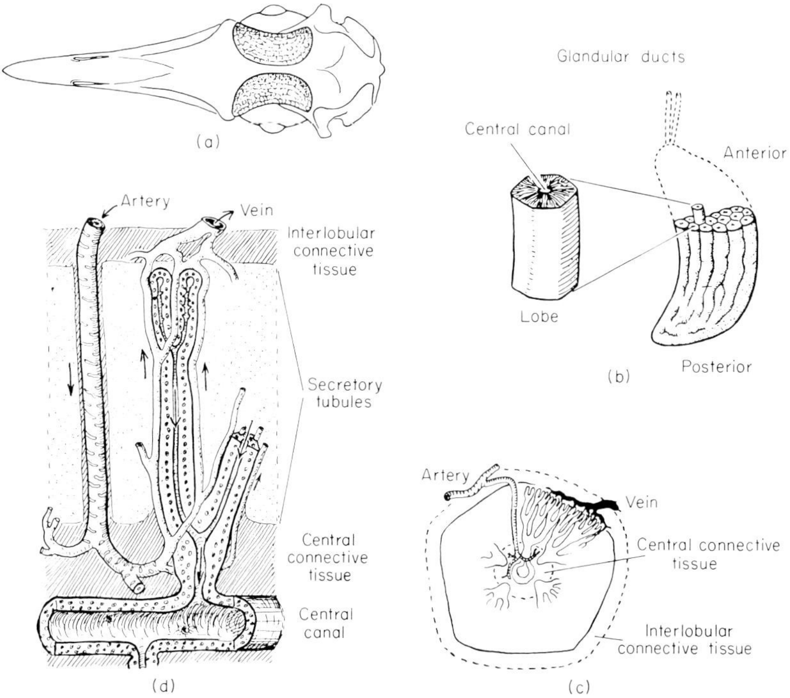

FIG. 7. (a) Skull of the Herring Gull (*Larus argentatus*) from above showing the position of the salt glands. (b) Gross structure of the salt gland. (c) Transverse section through a lobe of the salt gland. (d) Circulation in the salt gland showing opposing directions of flow in the secretory tubules and in the capillaries. The tubules branch repeatedly, but only two ramifications are shown. (Diagrams from Fänge *et al.*, 1958b.)

Blood flow to the gland is primarily via branches of the internal and external ophthalmic arteries. These form interlobular arteries which in turn send branches toward the central canal of each lobe. There they subdivide forming a network of capillaries coursing along the tubules to the periphery of the lobes. Interlobular veins finally collect the blood near the surface of the lobes (Fig. 7c,d). The flow of capillary blood along the secretory tubules is thus countercurrent to the tubular secretion, but the functional significance of this fact is obscure.

Peripheral cells, comprising the distal portion of each tubule, provide for growth of the tubules and frequently exhibit mitotic figures (Ellis, 1965). The lateral and basal surfaces of these cells are relatively flat, and the cells contain few mitochondria. Differentiation into the more specialized *principal secretory cells* involves elaboration of folds on the lateral and basal surfaces as well as proliferation of the mitochondria. This sequence is illustrated in Fig. 8. Fully developed

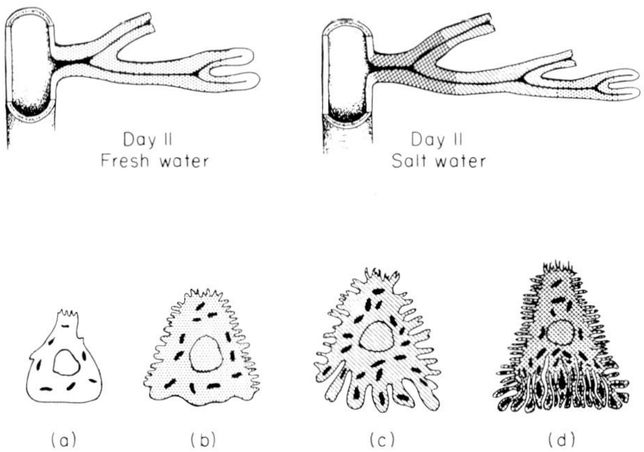

FIG. 8. Effect of salt and freshwater regimens on the development of the secretory epithelium of the salt gland of ducklings. Peripheral cells (A) exhibit little surface specialization and contain few mitochondria. Partially specialized cells (B) have short lateral folds and flat basal plasma membranes. In the next stage (C), the cells exhibit some folding at the basal surface in addition to the lateral plications. Mitochondria are more numerous and are quite evenly distributed in the cytoplasm. In the fully developed secretory cell (D) both the lateral and the basal plasma membranes are extensively folded, forming complex intracellular compartments and extracellular spaces. The mitochondria are increased dramatically in number and pack the basal labyrinths. The distribution of these cell types in the secretory tubules of freshwater- and salt-water-adapted glands on day 11 of the regimen is shown. (Ernst and Ellis, 1969.)

principal cells contain very deep basal infoldings forming a complex interdigitation of intracellular and extracellular compartments, with the intracellular compartments being densely packed with mitochondria (Fig. 9). According to Ernst and Ellis (1969), the luminal surfaces of the secretory cells are very small and have only a few short microvilli extending into the small (ca. 1 μ) lumen. Adjacent cells are joined near the lumen by tight junctions which apparently prevent direct communication between the lumen and the intercellular spaces (Fig. 10). Komnick (1965) and Burock *et al.* (1969) provide additional information on the histology of the gland.

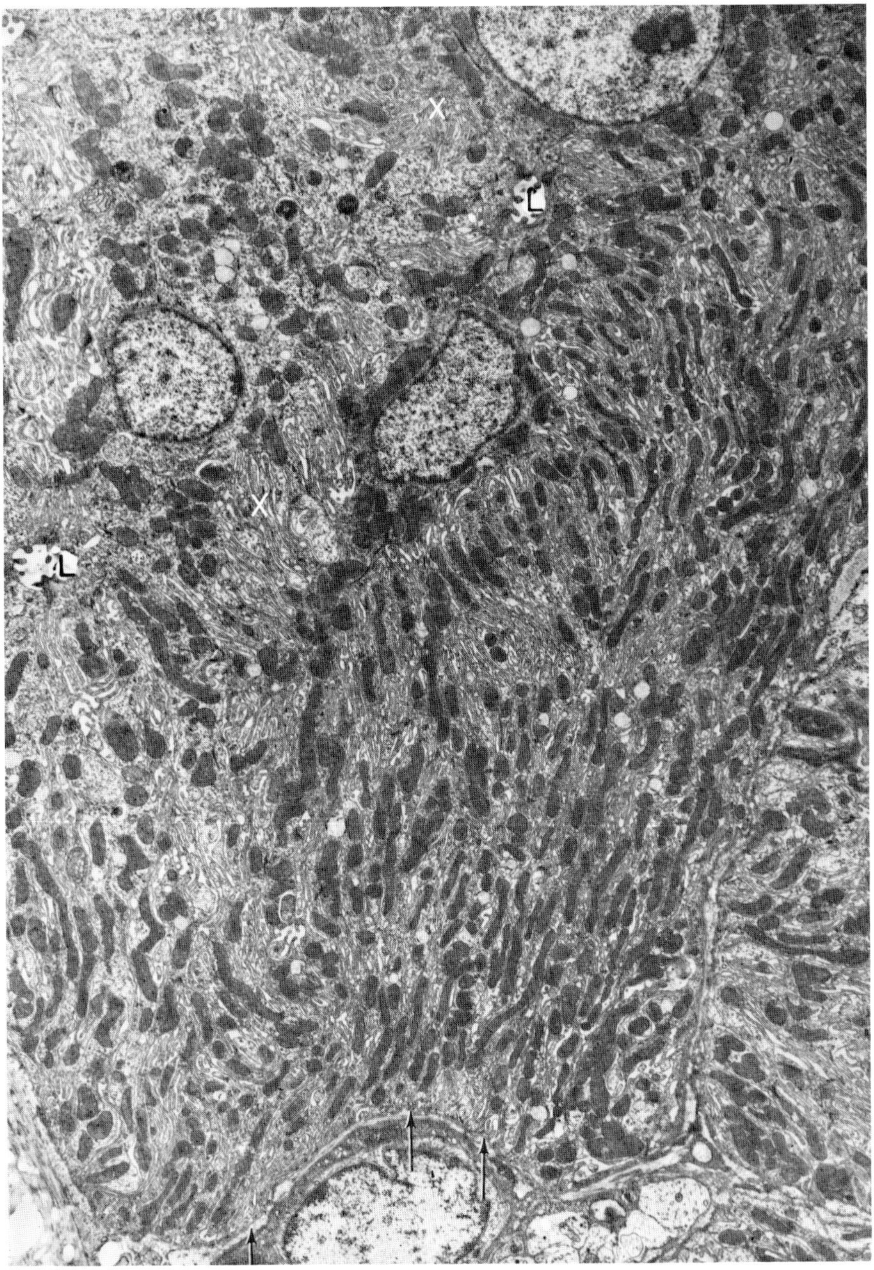

FIG. 9. Micrograph of fully specialized secretory cells from the salt gland of duckling showing the vast increase in surface area resulting from the complex folding at the lateral (X) and basal (arrows) cell surfaces. The area of the apical interface is very small, and few microvilli protrude into the lumen (L). ×6600. (Ernst and Ellis, 1969.)

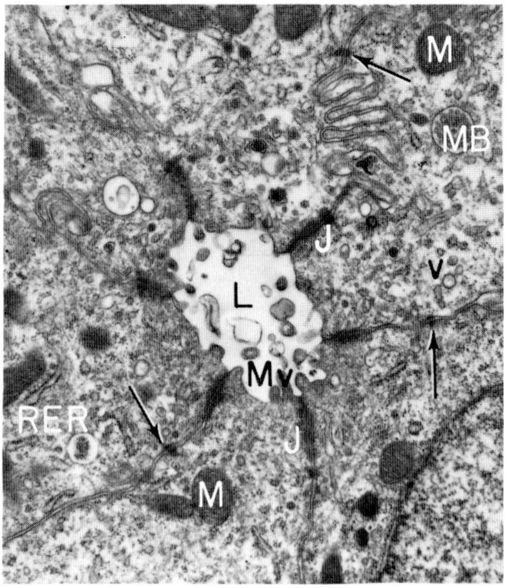

FIG. 10. The apical surfaces of the secretory cells from the salt gland of a duckling showing short microvilli (Mv) extending into the lumen (L) of the tubule. The cytoplasm contains abundant free ribosomes and a few mitochondria (M). The opposed lateral surfaces of the cells are usually flat, but some bear short folds. Intermittently, they are joined by desmosomes (arrows). Junctional complexes (J) block direct passage from the intercellular spaces to the lumen. Multivesicular bodies (MB) are found near the apical surface, and rough surface endoplasmic reticulum (RER) and smooth vesicles (v) are scattered throughout the cytoplasm. × 12,400. (Ernst and Ellis, 1969.)

B. Function of the Salt Gland

1. *Composition of Salt Gland Secretion*

The secretion of the avian salt gland is typically high in both sodium and chloride and contains small amounts of potassium and other ions (see Table IV). These ions apparently account for the total osmotic potential of the secreted fluid (Lange and Staaland, 1966). There is some interspecific variation in the concentrating capacity of the salt gland but these concentrations generally fall between one and two times those of seawater (Table V). These concentrations are quite characteristic in any given species, being relatively independent of the rate of secretion and plasma levels of sodium or potassium (see K. Schmidt-Nielsen, 1965; Ash, 1969). Fractions of arterial plasma components extracted by the salt gland in the domestic goose are sodium, 15%; potassium, 35%; chloride, 21%; and water, 5.8% (Hanwell *et al.*, 1971).

Some ecological correlates of the concentrating capacity of the avian salt gland are apparent (see K. Schmidt-Nielsen, 1960; Zaks and Sokolova, 1961). Concentrations generally are greatest in marine species, particularly in oceanic forms that feed on invertebrates, such as petrels and albatrosses. Staaland (1967a), in an extensive comparative survey of Charadriiformes, found a convincing correlation between the concentration of the salt gland secretion and the degree to which a species is associated with the marine environment. Interestingly, Staaland also demonstrated a good correlation between the concentrating capacity of the salt gland and the length of the secretory tubules within this group, both interspecifically and intraspecifically. Although there are obviously genetic adaptations relating to the concentrating capacity of the salt gland, acclimation cannot be ignored. Ducks provided with saline solution produce a somewhat more concentrated secretion than those maintained on fresh water, and the rate of secretion is also increased (see Table V). Salt glands of Herring Gulls (*Larus argentatus*) show a reduction in the length of the secretory tubules when the birds are given little salt (Komnick, 1965). However, Hughes (1970b) found no significant differences in secretory capacity of the salt glands of Glaucous-winged Gulls acclimated to fresh water and seawater.

TABLE IV
TYPICAL COMPOSITION OF THE FLUID FROM THE SALT GLAND OF HERRING GULLS[a]

Cations	mEq liter^{-1}	Anions	mEq liter^{-1}
Na^+	718	Cl^-	720
K^+	24	HCO_3^-	13
$Ca^{++} + Mg^{++}$	2.0	SO_4^{--}	0.68

[a] K. Schmidt-Nielsen (1960).

Concentrations of nasal secretions collected from birds not loaded with NaCl are more variable and difficult to interpret. Hughes (1970b) reports highly variable concentrations and ratios of Na^+, K^+, and Cl^- in Glaucous-winged Gulls, and K. Schmidt-Nielsen *et al.* (1963) found that K^+ could greatly exceed Na^+ in heat-stressed Ostriches (*Struthio camelus*). Cade and Greenwald (1966) also found considerable intraspecific variation among Falconiformes, although sodium and chloride always predominated. It should be pointed out that rates of secretion may be rather low in these situations, and small samples of fluid collected from the nares may not represent the true concentration of

TABLE V
CONCENTRATIONS AND RATES OF SALT GLAND SECRETION IN VARIOUS BIRDS[a]

Species	Concentration (mEq liter^{-1})		Maximum flow rate (ml min^{-1} kg^{-1})	Maximum excretory rate (mmoles NaCl min^{-1} kg^{-1})	Reference
	Na^+	Cl^-			
Adélie Penguin (*Pygoscelis adeliae*)		788 (700–900)	0.074[b]	0.058[b]	Douglas (1968)
Humboldt Penguin (*Spheniscus humboldti*)	726–840	635–805	0.058[b]	0.045[b]	K. Schmidt-Nielsen and Sladen (1958)
Brown Pelican (*Pelecanus occidentalis*)	698	722	0.12[b]	0.085[b]	K. Schmidt-Nielsen and Fänge (1958)
Double-crested Cormorant (*Phalacrocorax auritus*)	529	517	0.13[b]	0.07[b]	K. Schmidt-Nielsen *et al.* (1958).
Greater Frigatebird (*Fregata minor*)	768	889			McFarland (1960)
Great Black-backed Gull (*Larus marinus*)	756–800		0.38[b]	0.30[b]	K. Schmidt-Nielsen (1960)
Herring Gull (*Larus argentatus*)	600–800		0.50	0.35[b]	K. Schmidt-Nielsen (1960)
Leach's Storm-Petrel (*Oceanodroma leucorhoa*)	900–1100				K. Schmidt-Nielsen (1960)
Black-footed Albatross (*Diomedea nigripes*)	826				Frings *et al.* (1958)

Common Tern (*Sterna hirundo*)	500		0.16[b]	0.08[b]	Hughes (1968)
Glaucous-winged Gull (*Larus glaucescens*) freshwater acclimated	706	767	0.12[b]	0.09[b]	Hughes (1970b)
Glaucous-winged Gull sea water acclimated	785	872	0.14[b]	0.11[b]	Hughes (1970b)
Domestic Duck raised on fresh water	435 (270–523)		0.064	0.028[b]	K. Schmidt-Nielsen and Kim (1964)
Domestic Duck raised on salt water	525 (372–599)		0.113	0.059[b]	K. Schmidt-Nielsen and Kim (1964)
Gray Heron (*Ardea cinerea*)	559 (522–600)	662 (638–697)			Lange and Staaland (1966)
American Coot (*Fulica americana*)		500–600	0.133	0.073	Carpenter and Stafford (1970)
Savanna Hawk (*Heterospizias meridionalis*) fed mouse, no salt load	433 (255–890)	455 (263–949)	0.1 (transient)	0.045[b]	Cade and Greenwald (1966)
Savanna Hawk (fed mouse, 0.5 gm NaCl)	1010	1040			Cade and Greenwald (1966)
Greater Roadrunner (*Geococcyx californianus*)	776				Ohmart *et al.* (1970b)

[a]All measurements were made following administration of about 1 gm NaCl/kg body wt unless otherwise noted. Values shown are means or "typical values" with ranges in parentheses.

[b]Values calculated from data given by authors.

nasal gland secretions, particularly since a hyperosmotic fluid may appear in the nares of salt-loaded gulls following surgical removal of the salt glands (see K. Schmidt-Nielsen, 1965). The only two terrestrial birds shown to respond to salt-loading, the Greater Roadrunner (*Geococcyx californianus*) (Ohmart *et al.*, 1970b) and the Desert Partridge (*Ammoperdix heyi*) (K. Schmidt-Nielsen *et al.*, 1963), have produced NaCl solutions similar to those of marine forms. Functional salt glands have not been found in passerines, despite investigation of some of the more likely prospects (Rounsevell, 1970; Paynter, 1971).

2. *Rate of Salt Gland Secretion*

The rate of salt gland secretion is much more variable (individually, intraspecifically, and interspecifically) than the concentration. Following oral administration of a salt load, secretion does not reach maximum rates for about 1½ hours and begins to fall when most of the load has been eliminated (K. Schmidt-Nielsen, 1960; Ash, 1969). Intravenous salt-loading elicits a similar pattern of excretion except that the response begins within a few minutes. Because of the difficulties of quantitative collection, data on rates of secretion are less numerous than for concentrations. However, some approximate maximum rates of secretion for various species are included in Table V. Many marine species can produce more than 0.1 ml of fluid kg^{-1} min^{-1} which is equivalent to about 15% of the body weight per day. However, it is by no means clear that these birds attain these rates in the wild. The size of the gland appears to determine the rate of secretion, but not the concentration (Douglas, 1968). Within the Charadriiformes the size of the gland is directly related to the number of lobes, which range from two in the freshwater-inhabiting Green Sandpiper (*Tringa ochropus*) to more than 20 in several alcids and gulls. Similarly, the weight of the gland ranges from about 0.1 to 1 gm per kg body wt (Staaland, 1967a).

Whereas marine birds can maintain high rates of secretion for protracted periods, thereby excreting physiologically significant amounts of salt, the situation in fresh water waders and terrestrial species is less clear. For example, the Gray Heron (*Ardea cinerea*) can secrete hyperosmotic fluid but is still incapable of survival when given 2 or 3% NaCl for drinking (Lange and Staaland, 1966), and waders excrete most of a salt load via the cloaca with considerable expenditure of water (Staaland, 1968). Similarly, rates of secretion in Falconiformes may approach those of marine species for short periods, but subside as soon as feeding is completed (Cade and Greenwald, 1966). The accumulation of encrustations around the nares of hawks following salt-loading implies that the salt gland may play a physiologically significant role in some of these species. However, I. M. Johnson (1969) found that the salt gland of the Red-tailed Hawk (*Buteo jamaicensis*) accounted for only 3% of the sodium excreted in 24 hours after feeding.

3. *Stimulus for Salt Gland Secretion*

The usual and best-studied stimulus for salt gland secretion in birds is the intake of sodium chloride. Most investigators have administered about 1 gm (17 mmoles) of NaCl per kilogram of body weight either intravenously or orally, although smaller doses may be effective (McFarland, 1964). Domestic geese are particularly sensitive in this regard, responding to about 2 mmoles of NaCl per kilogram (Hanwell *et al.*, 1970). Effects of other solutes on the secretory activity of salt glands have been investigated. Lanthier and Sandor (1967) found $NaHCO_3$, mannitol, and sucrose stimulated typical salt gland secretion in ducks, but NH_4Cl and urea did not. Similarly, Ash (1969) found that intravenous injections of hyperosmotic sucrose or mannitol solutions were effective whereas urea was not. Sucrose also stimulates nasal secretion in gulls (Fänge *et al.*, 1958a), and Carpenter and Stafford (1970) found $NaHCO_3$ to be as effective as NaCl in American Coots (*Fulica americana*) and Guam Rails (*Rallus owstoni*) while $MgCl_2$ and urea had no effect. These results are consistent with the hypothesis that "osmoreceptors" are involved with the response to a salt load. However, it seems likely that increases in sodium levels have additional effects that are not shared by other solutes, including chloride. Effects of KCl on nasal gland secretion have not been elucidated since birds do not tolerate potassium-loading well (Ash, 1969; Carpenter and Stafford, 1970).

Nonosmotic stimuli have been reported to stimulate salt gland secretion in birds of several species. Photic and auditory stimulation can enhance the rate of secretion in gulls (McFarland, 1965). The act of feeding is frequently observed to initiate salt secretion in birds of prey and, in some cases, presentation of food or heat stress may also initiate secretion (Cade and Greenwald, 1966). Handling and similar "stresses" elicit secretion in albatrosses (Frings and Frings, 1959) and ducks (Ash, 1969). However, excitement inhibits secretion in Adélie Penguins (*Pygoscelis adeliae*) (Douglas, 1968), and captive raptors may not secrete until they become "tame" (Cade and Greenwald, 1966). Dehydration has not been reported to stimulate salt gland secretion, and Douglas and Neely (1969) found it to reduce the volume secreted by Herring Gulls (*Larus argentatus*).

4. *Control of Salt Gland Secretion*

a. Neural Control. The nervous system is definitely implicated in the regulation of the activity of the avian salt glands. The primary innervation of the gland is via fine nonmyelinated fibers arising from a ganglion (ganglion ethmoidale) in the anterior region of the orbit (Fig. 11). The ganglion receives fibers from the ophthalmic division

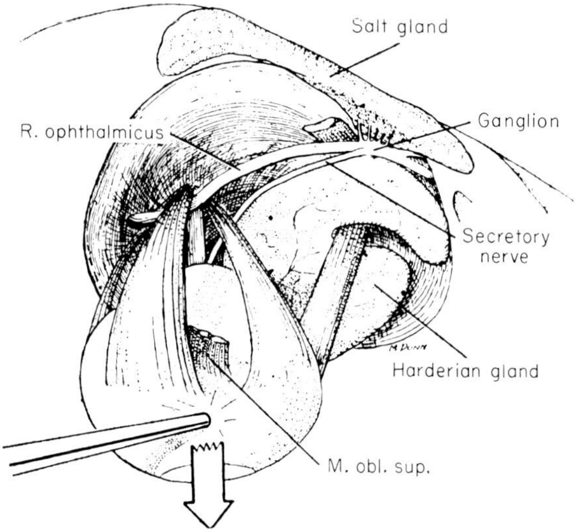

FIG. 11. Exposure of the nerves to the salt gland on the right side of the head of the Herring Gull (*Larus argentatus*). (Fänge *et al.*, 1958a.)

of the trigeminal nerve, a small branch of the facial nerve, and the sympathetic system (Fänge *et al.*, 1958a; Ash *et al.*, 1969). Also, another branch of the trigeminal nerve passes through the posterior end of the gland. Fänge *et al.* (1958a) found stimulation of only the intraorbital branch of the facial nerve to elicit secretion by the salt gland, and termed it the "secretory nerve." Ash *et al.* (1969) and Håkansson and Malcus (1969) obtained similar results, except Ash *et al.* also observed secretion following stimulation of the ramus ophthalmicus. Synapses within the ganglion and "ganglion cells" within the gland are described by Cottle and Pearce (1970).

Ash *et al.* (1969) performed another series of experiments in which the response to a salt load was tested after various nerves were sectioned or excised. Complete denervation of the gland abolished the response, but bilateral section of the secretory nerve, ramus ophthalmicus, and posterior nerve did not. The implication of these experiments is that the gland will respond normally if the ganglion and its connections to the salt gland remain intact, and Ash *et al.* suggested that the osmotic stimulus acts directly on ganglion cells rather than on central "osmoreceptors." It is difficult to reconcile these findings with the report of Gill and Burford (1968) that chronic denervation leads to an enhanced response of the salt gland to an oral salt load.

Pharmacological and histochemical evidence demonstrate the predominantly parasympathetic nature of the innervation of the salt gland. Secretion is depressed by atropine, and stimulated by acetylcholine (especially in the presence of eserine) and a variety of parasympathomimetics (Fänge *et al.*, 1958a; Ash *et al.*, 1969). Acetylcholinesterase-containing fibers are prominent in the salt gland, where they ramify extensively among the secretory tubules. Ellis *et al.* (1963) observed more rapid and pronounced development of cholinesterase-containing fibers in ducklings raised on saline solution. Moreover, denervation of the gland results in the disappearance of acetylcholinesterase staining. The cell bodies presumably reside in the ethmoidal ganglion since sectioning the "secretory nerve" has no effect on acetylcholinesterase staining in the gland (Ash *et al.*, 1969). Blood flow to the salt gland is under autonomic control, being enhanced parasympathetically whereas sympathetic stimulation is vasoconstrictory (Fänge *et al.*, 1963). It is generally assumed that the parasympathetic innervation is secretomotor as well as vasomotor (Ash *et al.*, 1969), but this requires further clarification. Hanwell *et al.* (1971) found denervation abolished both secretion and blood flow increase in the salt glands of geese following salt-loading, whereas atropine blocked secretion but not the increase in blood flow. In normal geese salt-loading elevated mean blood flow through the salt gland from 0.83 to 12 ml per gm of tissue per minute.

b. Endocrine Control. The role of hormones in salt gland activity has been investigated primarily in ducks, and this subject has been recently reviewed by Holmes *et al.* (1969, 1970). Although no hormones have been observed to initiate secretion directly, adrenal cortical hormones are apparently necessary for the secretory response of the salt gland. Adrenalectomy abolishes the response, and cortisol administration repairs the defect (Phillips *et al.*, 1961). Secretion by intact ducks is enhanced if cortisol, cortexone, corticosterone, or mammalian ACTH is administered with a salt load (Phillips and Bellamy, 1962). Moreover, endogenous rates of secretion of corticosterone are increased by either acute or chronic salt-loading (Donaldson and Holmes, 1965). The adrenal influence may be exerted through glucocorticoid effects. ACTH treatment elevates blood glucose levels, and secretion is also enhanced by glucose administration and decreased during insulin hypoglycemia (data of Peaker, Peaker, Phillips, and Wright, given by Holmes *et al.*, 1969). Enhanced intracellular accumulation by the salt gland of labeled cortisol (Phillips and Bellamy, 1967) and corticosterone (Bellamy and Phillips, 1966) following salt-loading argues for a direct mineralocorticoid effect. However, increased blood flow to the gland may be responsible for this observation. Despite its effects on a variety of sodium transport systems, the role of aldosterone in avian salt gland function is not clear (Phillips and Bellamy, 1962).

Wright *et al.* (1966) adenohypophysectomized ducks and virtually abolished the ability of the salt gland to secrete. However, the sham operative procedure also produced a marked depression of secretory capacity which has not been adequately explained. There is some evidence for the involvement of prolactin in salt gland secretion. Peaker *et al.* (1970) found intravenous injection of prolactin to salt-loaded ducks enhanced output of fluid within 5 minutes. Ensor and Phillips (1970) observed reduced pituitary levels of prolactin in ducks and gulls maintained on hyperosmotic saline. The presence of a hypothalamic corticocotropin releasing factor in ducks has also been suggested, but the data are somewhat inconclusive (Stainer and Holmes, 1969) and its relation to salt gland activity has not been investigated.

The neurohypophysis has not been implicated in control of salt gland activity. Wright *et al.* (1968) found no change in salt gland response following removal of the posterior lobe, but this does not necessarily preclude the synthesis and release of neurohypophyseal hormones in the hypothalamus. However, Lanthier and Sandor (1967) found that antidiuretic doses of Pitressin neither elicited salt gland activity nor affected the response of the gland following salt-loading.

5. *Mechanism of Salt Gland Secretion*

The mechanism of secretion in salt glands has been approached from both cytological and physiological standpoints. Of particular interest in this regard is the presence of Na^{+}–K^{+} activated ATPase in the salt gland and other tissues involved with the active transport of cations (Skou, 1965). Hokin (1963) and Bonting *et al.* (1964) reported high levels of this enzyme in avian salt glands and a number of observations link this enzyme with the secretory activity of the gland. The enzyme is characteristically inhibited by ouabain (Bonting *et al.*, 1964; Ernst *et al.*, 1967), and ouabain also inhibits the secretory function of the salt gland when applied via retrograde injection through the duct (Thesleff and Schmidt-Nielsen, 1962). The respiratory response of salt gland slices to acetylcholine is also abolished by ouabain (Hokin, 1963). Bonting *et al.* (1964) found that salt glands of Herring Gulls maintained on fresh water contained much less Na^+–K^+ ATPase per gram than did those of wild individuals, although Hokin (1963) found no difference in ATPase activity between gulls raised on saline or fresh water.

Experiments with ducks, which show dramatic changes in the size and activity of the salt gland depending on salt load, have yielded more conclusive results. Fletcher *et al.* (1967) found that adaptation to saline resulted in a three- to fourfold increase in the specific activity of ATPase which was accompanied by a simultaneous increase in weight-specific sodium-excretory capacity of similar magnitude. Within the saline adapted group, a positive correlation was found between the excretory capacity of the salt gland and its ATPase activity, particularly for the ouabain-sensitive component. These changes were reversed when the ducks were returned to fresh water. Similar results have been obtained by Ernst *et al.* (1967) and Ballantyne and Wood (1968).

Attempts at cytochemical localization of the membrane-bound Na^+–K^+ ATPase in salt glands have been fraught with difficulties (see Ernst and Ellis, 1969). However, Abel (1969), using Herring Gulls, has recently developed a technique that shows localization of ATPase along the inner cytoplasmic surface of the basal foldings of the principal cells, within mitochondria, and on the centriole tubules, but not on the nuclear and apical cell surfaces. Ouabain reduced the intensity of staining at the base of the principal cells, as did maintaining birds on fresh water. Additionally, no reaction product was formed along plasma membranes of peripheral cells, in capillary endothelial cells, or in the central canal cells. Despite hazards of interpretation discussed by Abel, this evidence supports the view that there is a well-developed, ATP-hydrolyzing, enzymatic transport system within the deeply plicated basal surface of the principal cells in avian salt glands.

Active solute uptake across these basal folds could lead to the establishment of a standing osmotic gradient within the extracellular channels, as theoretically described by Diamond and Bossert (1968). Depending on the geometry and the transport and permeability characteristics of the membrane, this could result in the net uptake by the cells of fluid hyperosmotic to the plasma. Several characteristics of avian salt glands fit predictions of this "standing gradient" model. Such a system should work most effectively if the transported solutes are also the major osmotic constituents of the extracellular fluid, as is the case with avian salt glands. Also, theoretically the greatest concentration of transported fluid will result if the active transport mechanism resides along the entire length of the channel. Finally, the model predicts that the osmolarity of the transported fluid will be virtually independent of the rate of solute transport. This is particularly intriguing in view of the fact that the concentration of nasal secretions is normally independent of flow rate (K. Schmidt-Nielsen, 1960; K. Schmidt-Nielsen and Kim, 1964) and remains so even when the gland is cooled locally (Staaland, 1967b). Even if this model should prove to be correct, movement of the secretion into the lumen of the tubule remains to be explained. Obviously the apical cell surface is very small relative to the basal surface, and the intracellular channels between folded lateral membranes appear to be isolated from the lumen by tight junctions (Ernst and Ellis, 1969). Hokin (1967) estimated concentrations of sodium within the principal cells to be about the same as those of the salt gland secretion; thus transport across the luminal surface may be passive. However, Peaker (1971) reports very much lower intracellular sodium concentrations and suggests that active sodium transport across the apical membrane is responsible for secretion.

Many studies of the biochemical and cytochemical properties of salt glands have been made. A series of papers by Van Rossum (see Van Rossum, 1968) deals with electron transfer and its relation to the transport mechanism, and the metabolic requirements of the system have been considered (Borut and Schmidt-Nielsen, 1963; Chance *et al.*, 1964; McFarland *et al.*, 1965). Activities of a number of other enzymes, including cholinesterases, succinic dehydrogenase, carbonic anhydrase, alkaline phosphatase, and glycolytic enzymes have been examined as correlates of salt gland development and activity (see Ballantyne and Fourman, 1966; Wood and Ballantyne, 1968; Ellis *et al.*, 1963; Spanhoff and Jürss, 1967; Stainer *et al.*, 1970; D. P. Smith *et al.*, 1971a,b). Also, the effects of acclimation on the size of the gland and its content of DNA, RNA, and protein have been studied (Holmes and Stewart, 1968; Stewart and Holmes, 1970).

REFERENCES

Abel, J. H., Jr. (1969). Electron microscopic demonstration of adenosine triphosphate phosphohydrolase activity in Herring Gull salt glands. *J. Histochem. Cytochem.* **17**, 570–584.

Acher, R. (1963). The comparative chemistry of neurohypophyseal hormones. *Symp. Zool. Soc. London* **9**, 83–91.

Akester, A. R. (1967). Renal portal shunts in the kidney of the domestic fowl. *J. Anat.* **101**, 569–594.

Akester, A. R., Anderson, R. S., Hill, K. J., and Osbaldiston, G. W. (1967). A radiographic study of urine flow in the domestic fowl. *Brit. Poultry Sci.* **8**, 209–212.

Ash, R. W. (1969). Plasma osmolality and salt gland secretion in the duck. *Quart. J. Exp. Physiol.* **54**, 68–79.

Ash, R. W., Pearce, J. W., and Silver, A. (1969). An investigation into the nerve supply to the salt gland of the duck. *J. Exp. Physiol.* **54**, 281–295.

Ballantyne, B., and Fourman, J. (1966). Cholinesterases and the secretory activity of the duck supraorbital gland. *J. Physiol. (London)* **188**, 32–33.

Ballantyne, B., and Wood, W. G. (1968). ATP-ase and Na^+ transport: Histochemical and biochemical observations on the avian nasal gland. *J. Physiol. (London)* **196**, 125–126.

Bartholomew, G. A., and Cade, T. J. (1963). The water economy of land birds. *Auk* **80**, 504–539.

Bellamy, D., and Phillips, J. G. (1966). Effect of administration of sodium chloride solutions on the concentration of radioactivity in the nasal gland of ducks (*Anas platyrhynchos*) injected with [^{3}H] corticosterone. *J. Endocrinol.* **36**, 97–98.

Berliner, R. W., and Bennett, C. M. (1967). Concentration of urine in the mammalian kidney. *Amer. J. Med.* **42**, 777–789.

Bonting, S. L., Caravaggio, L. L., Canady, M. R., and Hawkins, N. M. (1964). Studies on sodium-potassium-activated adenosinetriphosphatase. XI. Salt gland of the Herring Gull. *Arch. Biochem. Biophys.* **106**, 49–56.

Borut, A., and Schmidt-Nielsen, K. (1963). Respiration of avian salt-secreting gland in tissue slice experiments. *Amer. J. Physiol.* **204**, 573–581.

Braun, E. J., and Dantzler, W. H. (1971). Renal response to acute osmotic load in desert quail (*Lophortyx gambelii*). *Fed. Proc., Fed. Amer. Soc. Exp. Biol.* **30**, 547.

Brown, G. W., Jr. (1970). Nitrogen metabolism of birds. *In* "Comparative Biochemistry of Nitrogen Metabolism" (J. W. Campbell, ed.), Vol. 2, pp. 711–793. Academic Press, New York.

Burock, G., Kühnel, W., and Petry, G. (1969). Über die inaktive Salzdrüse von Enten (*Anas platyrhynchos*), histologische und histochemische Untersuchungen. *Z. Zellforsch. Mikrosk. Anat., Abt. Histochem.* **97**, 608–618.

Cade, T. J., and Greenwald, L. (1966). Nasal salt secretion in falconiform birds. *Condor* **68**, 338–350.

Calder, W. A., Jr., and Bentley, P. J. (1967). Urine concentrations of two carnivorous birds, the White Pelican and the Roadrunner. *Comp. Biochem. Physiol.* **22**, 607–609.

Carpenter, R. E., and Stafford, M. A. (1970). The secretory rates and the chemical stimulus for secretion of the nasal salt glands in the Rallidae. *Condor* **72**, 316–324.

Chan, M. Y., and Holmes, W. N. (1971). Studies on "renin-angiotensin" system in the normal and hypophysectomized pigeon (*Columba livia*). *Gen. Comp. Endocrinol.* **16**, 304–311.

Chance, B., Lee, C., Oshino, R., and Van Rossum, G. D. V. (1964). Properties of mitochondria isolated from Herring Gull salt gland. *Amer. J. Physiol.* **206**, 461–468.

Cooch, F. G. (1964). A preliminary study of the survival value of functional salt gland in prairie Anatidae. *Auk* **81**, 380–393.

Cottle, M. K. W., and Pearce, J. W. (1970). Some observations on the nerve supply to the salt gland of the duck. *Quart. J. Exp. Physiol. Cog. Med. Sci.* **55**, 207–212.

Dainty, T., and House, C. R. (1966). An examination of the evidence for membrane pores in frog skin. *J. Physiol. (London)* **185**, 172–184.

Dantzler, W. H. (1966). Renal response of chickens to infusion of hyperosmotic sodium chloride solution. *Amer. J. Physiol.* **210**, 640–646.

Davis, J. O. (1967). The regulation of aldosterone secretion. *In* "The Adrenal Cortex" (A. B. Eisenstein, ed.), pp. 203–247. Little, Brown. Boston, Massachusetts.

Dawson, W. R., Shoemaker, V. H., Tordoff, H. B., and Borut, A. (1965). Observations on the metabolism of sodium chloride in the Red Crossbill. *Auk.* **82,** 606–623.

deRoos, R. (1961). The corticoids of the avian adrenal gland. *Gen. Comp. Endocrinol.* **1,** 494–512.

deRoos, R. (1963). The physiology of the avian interrenal gland: A review. *Proc. Int. Ornithol. Congr., 13th, 1962 Vol. 2,* pp. 1041–1058.

deRoos, R., and deRoos, C. C. (1963). Angiotensin II: Its effects on corticoid production by chicken adrenals *in vitro. Science* **141,** 1284.

deRoos, R., and deRoos, C. C. (1964). Effects of mammalian corticotropin and chicken adenohypophyseal extracts on steroidogenesis by chicken adrenal tissue *in vitro. Gen. Comp. Endocrinol.* **4,** 602–607.

Diamond, J. M., and Bossert, W. H. (1968). Functional consequences of ultrastructural geometry in "backwards" fluid-transporting epithelia. *J. Cell Biol.* **37,** 694–702.

Dicker, S. E., and Haslam, J. (1966). Water diuresis in the domestic fowl. *J. Physiol. (London)* **183,** 225–235.

Dixon, J. M. (1958). Investigation of urinary water reabsorption in the cloaca and rectum of the hen. *Poultry Sci.* **37,** 410–414.

Donaldson, E. M., and Holmes, W. N. (1965). Corticosteroidogenesis in the fresh water and saline-maintained duck (*Anas platyrhynchos*). *J. Endocrinol.* **32,** 329–336.

Donaldson, E. M., Holmes, W. N., and Stachenko, J. (1965). *In vitro* corticosteroidogenesis by the duck (*Anas platyrhynchos*) adrenal. *Gen. Comp. Endocrinol.* **5,** 542–551.

Douglas, D. S. (1968). Salt and water metabolism of the Adéie Penguin. Antarctic Res. Ser., **12,** 167–190.

Douglas, D. S., and Neely, S. M. (1969). The effect of dehydration on salt gland performance. *Amer. Zool.* **9,** 1095.

Ellis, R. A. (1965). DNA labelling and X-irradiation studies of the phosphatase-positive peripheral cells in the nasal (salt) gland of ducks. *Amer. Zool.* **5,** 648.

Ellis, R. A., Goertemiller, C. C., Jr., DeLellis, R. A., and Kablotsky, Y. H. (1963). The effect of a salt water regimen on the development of the salt glands of domestic ducklings. *Develop. Biol.* **8,** 286–308.

Emery, N., and Kinter, W. B. (1967). Hyperosmolarity of medullary fluids in parakeet kidney. *Fed. Proc., Fed. Amer. Soc. Exp. Biol.* **26,** 376.

Ensor, D. M., and Phillips, J. G. (1970). The effect of salt loading on the pituitary prolactin levels of the domestic duck (*Anas platyrhynchos*) and juvenile Herring or Lesser Black-backed Gulls (*Larus argentatus* or *Larus fuscus*). *J. Endocrinol.* **48,** 167–172.

Ernst, S. A., and Ellis, R. A. (1969). Development of surface specialization in secretory epithelium of avian salt gland in response to osmotic stress. *J. Cell. Biol.* **40,** 305–321.

Ernst, S. A., Goertemiller, C. C., and Ellis, R. A. (1967). Effect of salt regimens on development of (Na^+–K^+) dependent ATPase activity during growth of salt glands in ducklings. *Biochim. Biophys. Acta* **135,** 682–692.

Fänge, R., Schmidt-Nielsen, K., and Robinson, M. (1958a). Control of secretion from the avian salt gland. *Amer. J. Physiol.* **195,** 321–326.

Fänge, R., Schmidt-Nielsen, K., and Osaki, H. (1958b). The salt gland of the Herring Gull. *Biol. Bull.* **115,** 162–171.

Fänge, R., Krog, J., and Reite, O. (1963). Blood flow in the avian salt gland studied by polarographic oxygen electrodes. *Acta Physiol. Scand.* **58,** 40–47.

Fletcher, G. L., Stainer, I. M., and Holmes, W. N. (1967). Sequential changes in the adenosinetriphosphatase activity and the electrolyte excretory capacity of the nasal glands of the duck (*Anas platyrhynchos*) during the period of adaptation to hypertonic saline. *J. Exp. Biol.* **47,** 375–392.

Florey, E. (1966). "General and Comparative Animal Physiology." Saunders, Philadelphia, Pennsylvania.

Folk, R. L. (1969). Spherical urine in birds: Petrography. *Science* **166,** 1516–1519.

Frankel, A. I., Graber, J. W., and Nalbandov, A. V. (1967a). Adrenal function in cockerels. *Endocrinology* **80,** 1013–1019.

Frankel, A. I., Graber, J. W., and Nalbandov, A. V. (1967b). The effect of hypothalamic lesions on adrenal function in intact and adenohypophysectomized cockerels. *Gen. Comp. Endocrinol.* **8,** 387–396.

Frings, H., and Frings, M. (1959). Observations on salt balance and behavior of Laysan and Black-footed Albatrosses in captivity. *Condor* **61,** 304–314.

Frings, H., Anthony, A., and Schein, M. W. (1958). Salt excretion by nasal gland of Laysan and Black-footed Albatrosses. *Science* **128,** 1572.

Gibbs, O. S. (1929). The function of the fowl's ureter. *Amer. J. Physiol.* **87,** 594–601.

Gill, J. B., and Burford, H. J. (1968). Secretion from normal and supersensitive avian salt glands. *J. Exp. Zool.* **168,** 451–454.

Håkansson, C. H., and Malcus, B. (1969). Secretive response of electrically stimulated nasal salt gland in *Larus argentatus* (Herring Gull). *Acta Physiol. Scand.* **76,** 385–392.

Hanwell, A., Linzell, J. L., and Peaker, M. (1970). Salt gland function in the domestic goose. *J. Physiol. (London)* **210,** 97P–99P.

Hanwell, A., Linzell, J. L., and Peaker, M. (1971). Salt gland secretion and blood flow in the goose. *J. Physiol. (London)* **213,** 373–387.

Hart, W. M., and Essex, H. E. (1942). Water metabolism of the chicken with special reference to the role of the cloaca. *Amer. J. Physiol.* **136,** 657–668.

Hester, H. R., Essex, H. E., and Mann, F. C. (1940). Secretion of urine in the chicken. *Amer. J. Physiol.* **128,** 592–602.

Hokin, M. R. (1963). Studies on $Na^+ + K^+$ dependent ouabain-sensitive adenosine triphosphatase in the avian salt gland. *Biochim. Biophys. Acta* **77,** 108–120.

Hokin, M. R. (1967). The Na^+, K^+, and Cl^- content of goose salt gland slices and the effects of acetylcholine and ouabain. *J. Gen. Physiol.* **50,** 2197–2209.

Hollander, P. L. (1969). Salt and water balance and aldosterone production in the Barbary Partridge, *Alectoris barbara*. *Amer. Zool.* **9,** 1080–1081.

Holmes, W. N., and Adams, B. M. (1963). Effects of adrenocortical and neurohypophyseal hormones on the renal excretory pattern in the water-loaded duck. *Endocrinology* **73,** 5–10.

Holmes, W. N., and McBean, R. L. (1963). Studies on the glomerular filtration rate of Rainbow Trout (*Salmo gairdneri*). *J. Exp. Biol.* **40,** 335–341.

Holmes, W. N., and Stewart, D. J. (1968). Changes in the nucleic acid and protein composition of the nasal glands of the duck (*Anas platyrhynchos*) during the period of adaptation to hypertonic saline. *J. Exp. Biol.* **48,** 509–519.

Holmes, W. N., Fletcher, G. L., and Stewart, D. J. (1968). The patterns of renal electrolyte excretion in duck (*Anas platyrhynchos*) maintained on freshwater and on hypertonic saline. *J. Exp. Biol.* **48,** 487–508.

Holmes, W. N., Phillips, J. G., and Wright, A. (1969). The control of extrarenal excretion in the duck (*Anas platyrhynchos*) with special reference to the pituitary-adrenal axis. *Gen. Comp. Endocrinol., Suppl.* **2,** 358–373.

Holmes, W. N., Chan, M. Y., Bradley, J. S., and Stainer, I. M. (1970). The control of some endocrine mechanisms associated with salt regulation in aquatic birds. *Mem. Soc. Endocrinol.* **18,** 87–108.

Hughes, M. R. (1968). Renal and extrarenal sodium excretion in the Common Tern *Sterna hirundo. Physiol. Zool.* **41,** 210–219.

Hughes, M. R. (1970a). Relative kidney size in nonpasserine birds with functional salt glands. *Condor* **72,** 164–168.

Hughes, M. R. (1970b). Cloacal and salt-gland ion excretion in the seagull, *Larus glaucescens,* acclimated to increasing concentrations of sea water. *Comp. Biochem. Physiol.* **32,** 315–325.

Johnson, I. M. (1969). Electrolyte and water balance in the Red-tailed Hawk. *Buteo jamaicensis. Amer. Zool.* **9,** 587.

Johnson, O. W. (1968). Some morphological features of avian kidneys. *Auk* **85,** 216–228.

Johnson, O. W., and Mugaas, J. N. (1970a). Some quantitative and organizational features of the avian renal medulla. *Condor* **72,** 288–292.

Johnson, O. W., and Mugaas, J. N. (1970b). Some histological features of avian kidneys. *Amer. J. Anat.* **127,** 423–436.

Koike, F. I., and McFarland, L. Z. (1965). Urography in the unanesthetized hydropenic chicken. *Amer. J. Vet. Res.* **27,** 1130–1133.

Komnick, H. (1965). Funktionelle Morphologie von Salzdrüsenzellen. Sekretion und Exkretion, funktionelle und morphologische Organisation der Zelle. 2. *Wiss. Konf. Ges. Deut. Naturforsch. Aerzte Schlob Reinhardsbrunn Friedrichroda, 1964* pp. 289–314.

Korr, A. M. (1939). The osmotic function of the chicken kidney. *J. Cell. Comp. Physiol.* **13,** 175–193.

Lange, R., and Staaland, H. (1966). Anatomy and physiology of the salt gland in the Grey Heron, *Ardea cinerea. Nytt. Mag. Zool.* **13,** 5–9.

Lanthier, A., and Sandor, T. (1967). Control of the salt secreting gland of the duck. I. Osmotic regulation. *Can. J. Physiol. Pharmacol.* **45,** 925–936.

LeBrie, S. J., and Sutherland, I. D. W. (1962). Renal function in water snakes. *Amer. J. Physiol.* **203,** 995–1000.

Lonsdale, K., and Sutor, D. J. (1971). Uric acid dihydrate in bird urine. *Science* **172,** 958–959.

McFarland, L. Z. (1959). Captive marine birds possessing a functional lateral nasal gland. *Nature (London)* **184,** 2030.

McFarland, L. Z. (1960). Salt excretion from the nasal glands from various species of the Pelicaniformes. *Anat. Rec.* **138,** 366.

McFarland, L. Z. (1964). Minimal salt load required to induce secretion from nasal salt glands of seagulls. *Nature (London)* **204,** 1202–1203.

McFarland, L. Z. (1965). Influence of external stimuli on the secretory rate of the avian nasal salt gland. *Nature (London)* **205,** 391–392.

McFarland, L. Z., Martin, K. D., and Freeland, R. A. (1965). Activity of selected soluble enzymes in the avian nasal salt gland. *J. Cell. Comp. Physiol.* **65,** 237–241.

McLelland, J., Moorhouse, P. D. S., and Pickering, E. C. (1968). An anatomical and histochemical study of nasal gland of *Gallus gallus domesticus. Acta Anat.* **71,** 122–133.

MacMillen, R. E., and Trost, C. H. (1966). Water economy and salt balance in White-winged and Inca Doves. *Auk* **83,** 441–456.

McNabb, F. M. A. (1969a). A comparative study of water balance in three species of quail. I. Water turnover in the absence of temperature stress. *Comp. Biochem. Physiol.* **28,** 1045–1058.

McNabb, F. M. A. (1969b). A comparative study of water balance in three species of quail. II. Utilization of saline drinking solutions. *Comp. Biochem. Physiol.* **28,** 1059–1074.

McNabb, F. M. A., and Poulson, T. L. (1970). Uric acid excretion in pigeons, *Columba livia. Comp. Biochem. Physiol.* **33,** 933–939.

(E. J. W. Barrington and C. A. Jørgensen, eds.), pp. 47–162. Academic Press, New York.

Maetz, J. (1969). Salt and water metabolism. *In* "Perspectives in Endocrinology"

Marples, B. J. (1932). The structure and development of the nasal glands of birds. *Proc. Zool. Soc. London* **2,** 829–844.

May, D. G., and Carter, M. K. (1967). Unilateral diuresis by infusing arecoline into renal portal system of hens. *Amer. J. Physiol.* **212,** 1351–1354.

Moses, A. M., Miller, M., and Streeten, D. H. P. (1967). Quantitative influence of blood volume expansion on the osmotic threshold for vasopressin release. *J. Clin. Endocrinol. Metab.* **27,** 655–662.

Munsick, R. A., Sawyer, W. H., and van Dyke, H. B. (1960). Avian neurohypophysial hormones: Pharmacological properties and tentative identification. *Endocrinology* **66,** 860–871.

Nechay, B. R., and Lutherer, B. D. C. (1968). Handling of urine by cloaca and ureter in chickens. *Comp. Biochem. Physiol.* **26,** 1099–1105.

Nechay, B. R., Boyarsky, S., and Catacuta, P. (1968). Rapid migration of urine into intestine of chickens. *Comp. Biochem. Physiol.* **26,** 369–370.

O'Dell, B. L., Woods, W. D., Laerdal, O. A., Jeffay, A. M., and Savage, J. E. (1960). Distribution of the major nitrogenous compounds and amino acids in chicken urine. *Poultry Sci.* **39,** 426–432.

Ohmart, R. D., McFarland, L. Z., and Morgan, J. P. (1970a). Urographic evidence that urine enters the rectum and ceca of the Roadrunner (*Geococcyx californianus*) Aves. *Comp. Biochem. Physiol.* **35,** 487–489.

Ohmart, R. D., Chapman, T. E., and McFarland, L. Z. (1970b). Water turnover in Roadrunners under different environmental conditions. *Auk* **87,** 787–793.

Oksche, A., Farner, D. S., Serventy, D. L., Wolff, F., and Nicholls, C. A. (1963). Hypothalamo-hypophyseal neurosecretory system of the Zebra Finch. *Z. Zellforsch. Mikrosk. Anat., Abt. Histochem.* **58,** 846–914.

Paynter, R. A., Jr. (1971). Nasal glands in *Cinclodes nigrofumosus,* a maritime passerine. *Bull. Brit. Ornithol. Club* **91,** 11–12.

Peaker, M. (1971). Intracellular concentrations of sodium, potassium and chloride in the salt gland of the domestic goose and their relation to the secretory mechanism. *J. Physiol. (London)* **213,** 399–410.

Peaker, M., Wright, A., Peaker, S. J., and Phillips, J. G. (1968). Absorption of tritiated water by the cloaca of the domestic duck (*Anas platyrhynchos*). *Physiol. Zool.* **41,** 461–465.

Peaker, M., Phillips, J. G., and Wright, A. (1970). The effect of prolactin on the secretory activity of the nasal salt-gland of the domestic duck *(Anas platyrhynchos). J. Endocrinol.* **47,** 123–127.

Phillips, J. G., and Bellamy, D. (1962). Aspects of the hormonal control of nasal gland secretion in birds. *J. Endocrinol.* **24,** vi.

Phillips, J. G., and Bellamy, D. (1967). The control of nasal gland function, with special reference to the role of adrenocorticoids. *Proc. Int. Congr. Horm. Steroids, 2nd, 1966 Excerpta Med. Int. Congr. Ser.* No. **132,** pp. 877–881.

Phillips, J. G., Holmes, W. N., and Butler, D. G. (1961). The effect of total and subtotal adrenalectomy on the renal and extrarenal response of the domestic duck (*Anas platyrhynchos*) to saline loading. *Endocrinology* **69,** 958–969.

Pitts, R. F. (1968). "Physiology of the Kidney and Body Fluids." Yearbook Publ., Chicago, Illinois.

Porter, P. (1966). Colloidal properties of urates in relation to calculus formation. *Res. Vet. Sci.* **7**, 128–137.

Potts, W. T. W., and Parry, G. (1963). "Osmotic and Ionic Regulation in Animals." Pergamon, Oxford.

Poulson, T. L. (1965). Counter-current multipliers in avian kidneys. *Science* **148,** 389–391.

Poulson, T. L. (1968). Kidney function in marine and terrestrial birds as compared with mammals. "Symposium: Functional Morphology of the Vertebrate Kidney" (unpublished oral presentation to Amer. Soc. Zool.).

Poulson, T. L., and Bartholomew, G. A. (1962). Salt balance in the Savannah Sparrow. *Physiol. Zool.* **35,** 109–119.

Poulson, T. L., and McNabb, F. M. A. (1970). Uric acid: The main nitrogenous excretory product of birds. *Science* **170,** 98.

Roberts, J. S., and Schmidt-Nielsen, B. (1966). Renal ultrastructure and excretion of salt and water by three terrestrial lizards. *Amer. J. Physiol.* **211,** 476–486.

Rounsevell, D. (1970). Salt excretion in the Australian Pipit *Anthus novaeseelandiae* (Aves: Motacillidae). *Aust. J. Zool.* **18,** 373–377.

Sawin, C. T. (1969). "The Hormones." Little, Brown, Boston, Massachusetts.

Schaffenburg, C. A., Haas, E., and Goldblatt, H. (1960). Concentration of renin in kidneys and angiotensinogen in serum of various species. *Amer. J. Physiol.* **199,** 788–792.

Scheiber, A. R., and Dziuk, H. E. (1969). Water ingestion and excretion in turkeys with a rectal fistula. *J. Appl. Physiol.* **26,** 277–281.

Schmidt-Nielsen, B. (1964). Organ systems in adaptation: The excretory system. *In* "Handbook of Physiology" (Amer. Physiol. Soc., J. Field, ed.), Sect. 4, pp. 215–243. Williams & Wilkins, Baltimore, Maryland.

Schmidt-Nielsen, B., and Forster, R. P. (1954). The effect of dehydration and low temperature on renal function in the bullfrog. *J. Cell. Comp. Physiol.* **44,** 233–246.

Schmidt-Nielsen, B., and O'Dell, R. (1961). Structure and concentrating mechanism in the mammalian kidney. *Amer. J. Physiol.* **200,** 1119–1124.

Schmidt-Nielsen, K. (1960). The salt secreting glands of marine birds. *Circulation* **21,** 955–967.

Schmidt-Nielsen, K. (1965). Physiology of salt glands. Sekretion und Exkretion, funktionelle und morphologische Organisation der Zelle. 2. *Wiss. Konf. Ges. Deut. Naturforsch. Aerzte Schloss Reinhardsbrunn Friedrichroda, 1964,* pp. 269–288.

Schmidt-Nielsen, K., and Fänge, R. (1958). The function of the salt gland in the Brown Pelican. *Auk* **75,** 282–289.

Schmidt-Nielsen, K., and Kim, Y. T. (1964). The effect of salt intake on the size and function of the salt gland of ducks. *Auk* **81,** 160–172.

Schmidt-Nielsen, K., and Sladen, W. J. L. (1958). Nasal salt secretion in the Humboldt Penguin. *Nature (London)* **181,** 1217–1218.

Schmidt-Nielsen, K., Jörgensen, C. B., and Osaki, H. (1958). Extrarenal salt excretion in birds. *Amer. J. Physiol.* **193,** 101–107.

Schmidt-Nielsen, K., Borut, A., Lee, P., and Crawford, E. (1963). Nasal salt excretion and the possible function of the cloaca in water conservation. *Science* **142,** 1300–1301.

Schultz, S. G., and Curran, P. F. (1968). Intestinal absorption of sodium chloride and water. *In* "Handbook of Physiology" (Amer. Physiol. Soc., J. Field, ed.), Sect. 6, Vol. III, pp. 1245–1275. Williams & Wilkins, Baltimore, Maryland.

Scothorne, R. J. (1959). On the response of the duck and pigeon to intravenous hypertonic saline solutions. *Quart. J. Exp. Physiol. Cog. Med. Sci.* **44,** 200–207.

Shannon, J. A. (1938). Excretion of exogenous creatinine by the chicken. *J. Cell. Comp. Physiol.* **11,** 123–134.

Shoemaker, V. H. (1967). Renal function in the Mourning Dove. *Amer. Zool.* **7,** 736.

Shoemaker, V. H. (1973). Renal function in the Mourning Dove: filtration, plasma flow and urate excretion. *Comp. Biochem. Physiol.* In press.

Shoemaker, V. H., Licht, P., and Dawson, W. R. (1966). Effects of temperature on kidney function in the lizard *Tiliqua rugosa. Physiol. Zool.* **39,** 244–252.

Siller, W. G., and Hindle, R. M. (1969). The arterial blood supply to the kidney of the fowl. *J. Anat.* **104,** 117–135.

Skadhauge, E. (1964). The effect of unilateral infusion of arginine-vasotocin into the portal circulation of the avian kidney. *Acta Endocrinol (Copenhagen)* **47,** 321–330.

Skadhauge, E. (1967). *In vivo* perfusion studies of the cloacal water and electrolyte resorption in the fowl (*Gallus domesticus*). *Comp. Biochem. Physiol.* **23,** 483–501.

Skadhauge, E. (1968). The cloacal storage of urine in the rooster. *Comp. Biochem. Physiol.* **24,** 7–18.

Skadhauge, E., and Schmidt-Nielsen, B. (1967a). Renal function in domestic fowl. *Amer. J. Physiol.* **212,** 793–798.

Skadhauge, E., and Schmidt-Nielsen, B. (1967b). Renal medullary electrolyte and urea gradient in chickens and turkeys. *Amer. J. Physiol.* **212,** 1313–1318.

Skou, J. C. (1965). Enzymatic basis for active transport of Na^+ and K^+ across cell membrane. *Physiol. Rev.* **45,** 596–617.

Smith, D. P., Fourman, J. M., and Haase, P. (1971a). The effect of an intravenous injection of isotonic saline or sucrose on the secretory activity and butyryl cholinesterase content of the salt glands of *Anas domesticus. Cytobios* **3,** 49–55.

Smith, D. P., Fourman, J. M., and Haase, P. (1971b). The secretory activity and butyryl cholinesterase content of the salt glands of *Anas domesticus* in response to an intravenous injection of hypertonic saline. *Cytobios* **3,** 57–64.

Smith, H. W. (1953). "From Fish to Philosopher." Little, Brown, Boston, Massachusetts.

Smyth, M., and Bartholomew, G. A. (1966a). The water economy of the Black-throated Sparrow and the Rock Wren. *Condor* **68,** 447–458.

Smyth, M., and Bartholomew, G. A. (1966b). Effects of water deprivation and sodium chloride on the blood and urine of the Mourning Dove. *Auk* **83,** 597–602.

Sokabe, H. (1968). Comparative physiology of the renin-angiotensin system. *J. Jap. Med. Ass.* **59,** 502–512.

Spanhoff, L., and Jürss, K, (1967). Untersuchungen zur Genese einiger Enzyme in den Salzdrüsen junger Sturmmöwen. *Acta Biol. Med. Ger.* **19,** 137–144.

Sperber, I. (1948). Investigations on the circulatory system of the avian kidney. *Zool. Bidr. Uppsala* **27,** 429–448.

Sperber, I. (1960). Excretion. *In* "Biology and Comparative Physiology of Birds" (A. J. Marshall, ed,), Vol. 1, pp. 469–492. Academic Press, New York.

Staaland, H. (1967a). Anatomical and physiological adaptations of nasal glands in Charadriiformes birds. *Comp. Biochem. Physiol.* **23,** 933–944.

Staaland, H. (1967b). Temperature sensitivity of avian salt glands. *Comp. Biochem. Physiol.* **23,** 991–993.

Staaland, H. (1968). Excretion of salt in waders, Charadrii, after acute salt loads. *Nytt Mag. Zool.* **16,** 25–28.

Stainer, I. M., and Holmes, W. N. (1969). Some evidence for the presence of a corticotrophin releasing factor (CRF) in the duck (*Anas platyrhynchos*). *Gen. Comp. Endocrinol.* **12,** 350–359.

Stainer, I. M., Ensor, D. M., Phillips, J. G., and Holmes, W. N. (1970). Changes in glycolytic enzyme activity in the duck (*Anas platyrhynchos*) nasal gland during the period of adaptation to salt water. *Comp. Biochem. Physiol.* **37**, 257–263.

Stewart, D. J., and Holmes, W. N. (1970). Relation between ribosomes and functional growth in the avian nasal gland. *Amer. J. Physiol.* **219**, 1819–1824.

Stewart, D. J., Holmes, W. N., and Fletcher, G. (1969). The renal excretion of nitrogenous compounds by the duck (*Anas platyrhynchos*) maintained on freshwater and on hypertonic saline. *J. Exp. Biol.* **50**, 527–539.

Sturkie, P. D., ed. (1965). "Avian Physiology." Cornell Univ. Press, Ithaca, New York.

Sturkie, P. D., and Joiner, W. P. (1959a). Effect of foreign bodies in the cloaca and rectum of the chicken on feed consumption. *Amer. J. Physiol.* **197**, 1337–1338.

Sturkie, P. D., and Joiner, W. P. (1959b). Effect of cloacal cannulation on feed and water consumption in chickens. *Poultry Sci.* **38**, 30–32.

Taylor, A. A., Davis, J. D., Britenbach, R. P., and Hartroft, P. M. (1970). Adrenal steroid secretion and a renal-pressor system in the chicken (*Gallus domesticus*). *Gen. Comp. Endocrinol.* **14**, 321–333.

Technau, G. (1936). Die Nassendrüse der Vögel. *J. Ornithol.* **84**, 511–617.

Teekell, R. A., Richardson, C. E., and Watts, A. B. (1968). Dietary protein effects on urinary nitrogen components of the hen. *Poultry Sci.* **47**, 1260–1266.

Thesleff, S., and Schmidt-Nielsen, K. (1962). An electrophysiological study of the salt gland of the Herring Gull. *Amer. J. Physiol.* **202**, 597–600.

Van Rossum, G. V. (1968). Relation of the oxidoreduction level of electron carrier to ion transport in slices of avian salt gland. *Biochim. Biophys. Acta* **153**, 124–131.

von Lawzewitsch, I., and Sarrat, R. (1970). Das neurosekretorische Zwischenhirn-Hypophysensystem von Vögeln nach langer osmotischer Belastung. *Acta Anat.* **77**, 521–539.

Weyrauch, H. M., and Roland, S. I. (1958). Electrolyte absorption from fowl's cloaca: Resistance to hyperchloremic acidosis. *J. Urol.* **79**, 255–263.

Willoughby, E. J. (1968). Water economy of the Stark's Lark and Grey-backed Finch-Lark from the Namib desert of South West Africa. *Comp. Biochem. Physiol.* **27**, 723–745.

Wood, W. G., and Ballantyne, B. (1968). Sodium ion transport and β-glucuronidase activity in the nasal gland of *Anas domesticus*. *J. Anat.* **103**, 277–287.

Wright, A., Phillips, J. G., and Huang, D. P. (1966). The effect of adenohypophysectomy on the extrarenal and renal excretion of the saline loaded duck. *J. Endocrinol.* **36**, 249–256.

Wright, A., Phillips, J. S., Peaker, M., and Peaker, S. J. (1968). Some aspects of the endocrine control of water and salt electrolytes in the duck. (*Anas platyrhynchos*). *Proc. Asia Oceania Congr. Endocrinol. 3rd, 1967* Part 2, pp. 322–327.

Zaks, N. G., and Sokolova, M. M. (1961). Ontogenetic and specific features of the nasal glands in some marine birds. *Fiziol. Zh. SSSR im I.M. Sechenova* **47**, 120 127 (translation).

AUTHOR INDEX

Numbers in italics refer to the pages on which the complete references are listed.

A

D

E

Y

Z

INDEX TO BIRD NAMES

Q

R

S

SUBJECT INDEX